Chemical Defense in Nature

Structure, Function, and Synthesis of Toxic Secondary Metabolites from Plants and Animals

AF849417

Vladimir V. Kouznetsov
Industrial University of Santander
Bucaramanga, Santander, Colombia

CRC Press
Taylor & Francis Group
Boca Raton London New York

CRC Press is an imprint of the
Taylor & Francis Group, an **informa** business
A SCIENCE PUBLISHERS BOOK

Cover credit: Image created by the author.

Image description: Drawings and structures made in ChemDraw and improved in PowerPoint. The background with flowers was designed by ChatGPT and modified by the author.

First edition published 2025
by CRC Press
2385 NW Executive Center Drive, Suite 320, Boca Raton FL 33431

and by CRC Press
4 Park Square, Milton Park, Abingdon, Oxon, OX14 4RN

© 2025 Vladimir V. Kouznetsov

CRC Press is an imprint of Taylor & Francis Group, LLC

Reasonable efforts have been made to publish reliable data and information, but the author and publisher cannot assume responsibility for the validity of all materials or the consequences of their use. The authors and publishers have attempted to trace the copyright holders of all material reproduced in this publication and apologize to copyright holders if permission to publish in this form has not been obtained. If any copyright material has not been acknowledged please write and let us know so we may rectify in any future reprint.

Except as permitted under U.S. Copyright Law, no part of this book may be reprinted, reproduced, transmitted, or utilized in any form by any electronic, mechanical, or other means, now known or hereafter invented, including photocopying, microfilming, and recording, or in any information storage or retrieval system, without written permission from the publishers.

For permission to photocopy or use material electronically from this work, access www.copyright.com or contact the Copyright Clearance Center, Inc. (CCC), 222 Rosewood Drive, Danvers, MA 01923, 978-750-8400. For works that are not available on CCC please contact mpkbookspermissions@tandf.co.uk

Trademark notice: Product or corporate names may be trademarks or registered trademarks and are used only for identification and explanation without intent to infringe.

Library of Congress Cataloging-in-Publication Data (applied for)

ISBN: 978-1-032-37191-7 (hbk)
ISBN: 978-1-032-37192-4 (pbk)
ISBN: 978-1-003-33577-1 (ebk)

DOI: 10.1201/9781003335771

Typeset in Times New Roman
by Prime Publishing Services

Preface

This book portrays the basis of organic chemistry oriented to biological sciences. Its content corresponds to a one-semester course titled *Chemical Defense in Nature: Chemical Approach* (in Spanish, "*Defensa Química en Naturaleza: enfoque químico*"), which has been dictated for the past several years to the third-year students at the Industrial University of Santander in Bucaramanga, Colombia.

In this book, the biological processes crucial to the survival of living organisms are analyzed from the chemical point of view. By linking the basic elements of organic chemistry (structural analysis and synthesis) with these processes, this book explains how and why animals and plants generate and use toxic organic molecules to develop, live, and successfully reproduce in the natural world. The main idea of this textbook is to demonstrate the great prospects of the union between chemistry (mainly synthetic organic chemistry and analytical chemistry) and biology through the development of phytochemistry, chemical ecology, and related sciences that enrich both natural medicine and modern medicine.

The chemical structures of toxic small molecules isolated from plants and animals, provided in this book, make one understand molecules' immense diversity and unique complexity. A description of their biogenesis and synthesis helps to understand their relationship, which could aid in finding viable, practical applications for these toxic molecules during a chemical transformation or analog preparation.

The content of the book was selected based on the criteria of chemists working in the mixed field of organic chemistry, phytochemistry, and chemical ecology, with particular attention to the structure, function, and synthesis of toxic second metabolites from terrestrial plants, as well as both terrestrial and marine animals and microorganisms. This focus helps understand the complex chemical relationships and clarify the significance of toxic substances. Our goal was to aid students in making sense of the complex integration of life sciences where small organic molecules, with their particular and beautiful structures, play an immense role.

The book's content is organized into seven chapters. All these chapters are analogously organized, first presenting a structural analysis of the discussed toxic molecules and their biological activities, followed by their synthesis and biogenesis. Each chapter has its own numeration of schemes, figures, and tables and ends with conclusions and recommended texts for further reading.

Thus, Chapter 1 reviews the basics of organic chemistry, biochemistry, biology, and ecology, starting with classifying living organisms, which contain, produce, and eliminate specific diverse organic molecules called metabolites. Some terms and notions of the life sciences mentioned above are necessary to understand the complexity of the chemical relationships (allelochemical relationships) of plants and terrestrial animals. They are introduced and explained using the motto that the biological diversity of species generates huge molecular diversity.

Chapter 2 discusses plant-animal interrelations through molecular analysis of vegetable toxins, i.e., oxygenated phenolic and terpenoid metabolites, and their effects on terrestrial animals. This chapter discusses four selected groups of molecules: quinones, furocoumarins, iridoids, and cardenolides.

Chapter 3 presents vegetable nitrogen-containing toxins, such as non-protein amino acids, cyanogenic glucosides, glucosinolates, and alkaloid metabolites. It discusses paying attention to the structure-biological activity relationship, including toxicity, possible biogenesis, and synthetic methods to obtain them.

Chapter 4 presents and discusses three principal groups of small molecules: phytoalexins, phytotoxins, and mycotoxins, which participate actively in plant-microorganism interrelations. This chapter analyzes representative molecules of each group generated during these interrelations will be analyzed, pointing out their biological effects on humans.

Chapter 5 describes toxic heterocyclic molecules metabolized by terrestrial arthropods, invertebrate animals of phylum Arthropoda. Some of the most exciting arthropods that fabricate toxic molecules are the ants, beetles, bees, and spiders, belonging to the respective classes Insecta and Arachnida. In this chapter, our attention is focused on both toxic small nitrogen- and oxygen-containing molecules, which are well-known medically important compounds.

Neurotoxins, molecules coming from the frog and toad skins, and toxic molecules from other mammals, such as passerine birds and serpents belonging to the vertebrate animal group, are presented and analyzed in Chapter 6. In this chapter, interesting chemical intra- and interrelationship of key amphibians, birds, snakes, and some mammals (crested rats, short-tailed shrews, and skunks) are examined through molecular structure analysis. Both invertebrates and vertebrates are part of the animal kingdom Zoa. They obtain energy by feeding other organisms, in part or together, as food, sometimes using their toxic, replete, and poisonous molecules. These toxic molecules serve as chemical weapons in their eternal struggle for survival, and some of these have become drugs and pharmacological agents.

Chapter 7 covers an immense source of toxic secondary metabolites, venomous marine small molecules that have been studied little. Marine organisms are prolific producers of thousands of biologically active secondary metabolites that show a range of biological activities, including cytotoxicity. Marine toxins isolated from the dinoflagellates, sponges, corals, ascidians, mollusks, and fishes are briefly discussed, paying much attention to the structure-activity relationships.

The chemical structures of compounds isolated from plants and animals in this book help us understand the immense diversity and unique complexity of these molecules. Descriptions of their biogenesis and synthesis clarify their relationship, making it easier to see a viable, practical application of these molecules.

The present book tries to answer the following main questions: Why do animals and plants accumulate these toxic substances? Where do these substances come from, and how do they affect their development and growth? What structures do these compounds have? What effects can they have on humans? What is the nature of the interaction between different (or the same) species of living beings? Can they be prepared in the laboratory and used for medicinal purposes?

Acknowledgments

Besides my explicit and implicit teachers, I would like to thank and express my deep gratitude to the doctors of Falla Cardiaca at FCV in Bucaramanga; without them, it would not have been possible to create this book and continue dreaming of other scientific projects.

Vladimir V. Kouznetsov
June 2024, Bucaramanga, Colombia

Contents

CHAPTER 1

Associations Between Chemistry and Biology

1. INTRODUCTION: CLASSIFICATION OF LIVING ORGANISMS

Chemistry and biology, together with physics, explain the fundamental laws of the universe and introduce important concepts that are essential for all other branches of natural science. These sciences study and try to understand the basic processes of the life of living organisms. Living organisms generate and release chemical compounds into the environment, considerably affecting other organisms, including humans, and determining the existence of chemical interactions between these individuals. Thus, all these organisms produce diverse molecules called chemical signals, and all, in mutual benefit, respond to the chemical signals of other organisms. In order to understand how these relationships work, it is necessary to remember some basic things about biology and chemistry. As there is huge basic information in both these natural sciences, it is necessary to use classification systems that help to impose order and a general plan on the principal objects of this book: toxic secondary metabolites produced by living organisms, mainly plants and animals.

From biology, it is known that the simplest unit in living organisms is the cell, sometimes called the building block of life [1]. This smallest part of an animal or plant can function self-sufficiently and is constructed according to a common concept, using similar molecules and metabolic processes. Commonly, two kinds of cells are recognized, eukaryotes and prokaryotes, both of which developed from a common ancestor. Most prokaryotes are unicellular organisms, while eukaryotes are usually multicellular organisms and form four kingdoms: Fungi, Animalia (Zoa), Plantae, and Protista. Prokaryotic organisms are classified into two large domains, Archaea and Bacteria (Figure 1.1).

One principal difference between prokaryotic and eukaryotic cells is that the genetic material of eukaryotes is contained within a nucleus and is more organized in chromosomes. In contrast, prokaryotes have their genetic material less structured. Another difference is its size: eukaryotic cells are usually much more extensive (ranging from 5 to 20 μm up to 150–300 μm in diameter) than prokaryotes (0.1–5.0 μm in diameter). According to the book of Lodish et al. [1], much of the

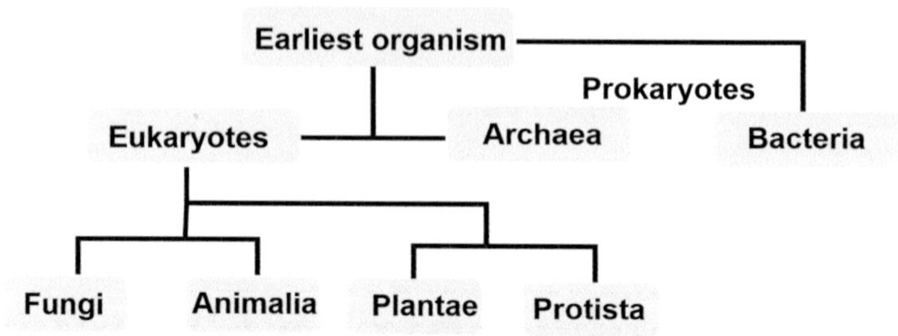

Figure 1.1. The classification levels of living organisms.

cell's contents consist of a watery soup flavored with different molecules and ions, with water occupying 70% of the total wet weight of the cell. On the other hand, the cells can auto-organize in diverse subcellular structures, tissues, and organs to conclude with complex living systems (Figure 1.2).

It is interesting to note that the human body is built from an estimated 100 trillion or 10^{14} cells, and each cell contains around 1,014 atoms organized into different types of molecules. If a group of some organisms has similar features, they are called species. Arranging all living organisms from the largest group (kingdom) to the smallest, the species of organisms results in the hierarchical classification system (Figure 1.3). Similar species form another group, genera (singular: genus), and various comparable genera are grouped into a family, order, class, and phylum, which together form kingdoms. In this system of classification, the different groups are called taxa.

This hierarchical classification system was introduced by the Swedish botanist Carl Linnaeus (1707–1778), who classified 9,000 distinct species in his book *Sistema Naturae*, giving every species two scientific names in Latin, known as the binomial

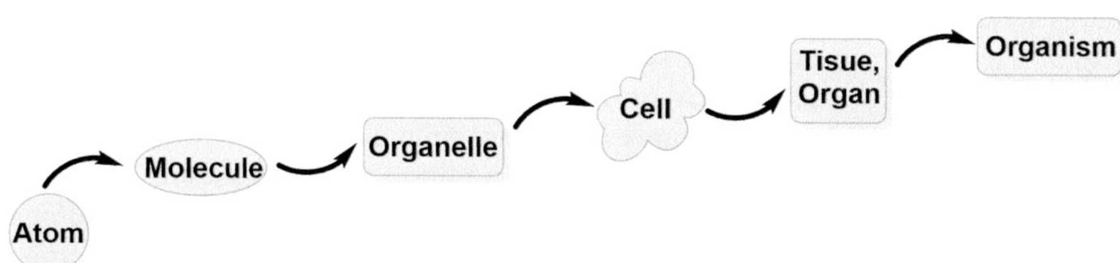

Figure 1.2. The different levels of organization in multicellular organisms.

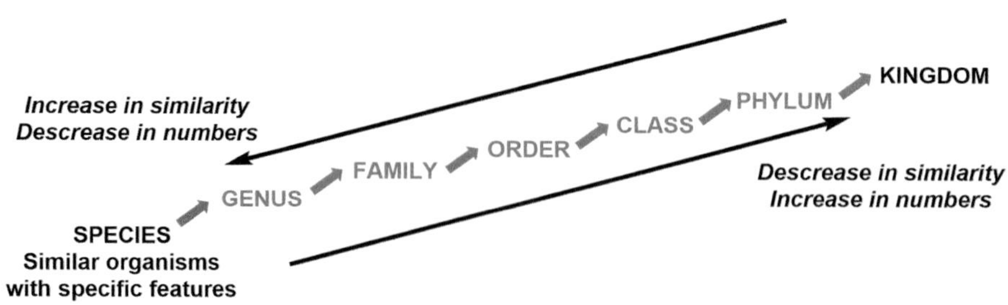

Figure 1.3. The hierarchical system of classification for living things.

system of naming species. Since then, Linnaeus's scientific names, derived from the genus and the species to which organisms belong, have been very helpful and are commonly used in biology. However, it should be noted that one of the first scientists to organize living things was the Greek scientist Aristotle (384–322 B.C.), who developed the first classification system, dividing all known organisms into two groups: plants and animals. He could classify only 400 species. Determining how many species exist on Earth is a tough challenge. Researchers say that the total number of species in any taxonomic group can be predicted. Calculations for all kingdoms of life on Earth predict ~ 8.7 million (± 1.3 million SE) species globally, of which ~ 2.2 million (± 0.18 million SE) are marine. These results suggest that some 86% of the species on Earth and 91% in the ocean still await description [2, 3].

1.1 Biological Diversity and Molecular Diversity: Classification of Secondary Metabolites, Molecular Intra- and Interrelationships

Biological diversity is embodied in the richness of species that are seen in different ways: diversity of species, diversity of genetics, diversity of habitats, and diversity of organic and inorganic molecules (molecular biodiversity). Thus, molecular biodiversity is the richness of molecules found in life. Of the 90 naturally occurring elements, 27 are used by living systems. Among those, a carbon atom is the base of almost all molecules in the cell and is the backbone of life [4]. With four valence electrons, carbon can form four covalent bonds with various atoms. Carbon chains form the skeletons of most organic molecules, varying chains in length and shape. The characteristic properties of organic molecules depend not only on the carbon skeleton but also on the molecular components attached to it. Thus, molecular diversity arises from carbon skeleton, heteroatom, functional group variation, and space orientation. Heteroatoms and functional groups are crucial components of organic molecules most commonly involved in chemical reactions and give each molecule its unique properties. In general, organic compounds range from simple molecules to complex ones. Depending on the shape of carbon chains, all organic compounds can be divided into two immense groups: acyclic (open chain) or cyclic (closed chain) molecules. The first group, also known as aliphatic compounds, includes those with branched or straight chains and can be either saturated or unsaturated. The second group comprises cyclic compounds, which contain carbon atoms connected in a ring (homocyclic) or heterocyclic molecules, which contain atoms other than carbon in their rings (Figure 1.4).

Both natural and synthetic organic molecules could also be classified based on their molecular size (molecular mass or molecular weight): low-molecular-weight (< 900 Daltons, g/mol), medium-molecular-weight (900–10,000 Daltons), and high-molecular-weight compounds (> 10,000 Daltons) [5]. In this context, it is generally considered that the first class of compounds is known as small molecules. In contrast, the other groups of compounds belong to large molecules, such as polymeric and macrocyclic molecules. Now, let us remember that both types of cells (prokaryotic and eukaryotic) consist of all three classes of compounds plus some ions (e.g., sodium, chloride, calcium ions, etc.) and diverse inorganic molecules. The macromolecules

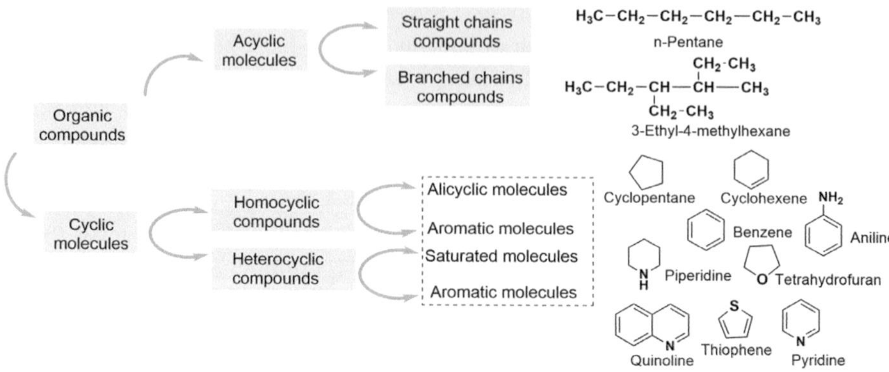

Figure 1.4. Simplified classification of organic compounds.

are classified into four general classes: nucleic acids, proteins, polysaccharides, and lipids, which are crucial to the structure and function of a cell. In contrast, small molecules can regulate biological processes and serve as building blocks for the macromolecules mentioned above.

Every organism has certain features or characteristics that allow it to live successfully in its habitat. First, everyone has to eat, live, and reproduce. The need for survival of each species makes it search for "sophisticated weapons" and different skills and tricks. Within these methods of survival, the chemical methods stand out. The chemical signals produced by organisms are composed of low-molecular-weight organic compounds (secondary metabolites) with a size on the order of 1 nm. These compounds are prepared through the secondary metabolic pathways, which are closely related to the primary pathways that give rise to the primary metabolites (high-molecular-weight macromolecules) with complex molecular architectures.

Low-molecular-weight metabolites are known to us as nutrients, hormones, vitamins, poisons, chemical weapons, spices, perfume aromas, antioxidants, medicines, and biopolymer precursors. During organism growth, these secondary metabolites are also the end products of primary metabolites. Generally, secondary metabolites can be classified based on chemical structure (for example, having rings, containing a sugar), composition (containing nitrogen and oxygen or not), their solubility in various solvents, or the pathway by which they are synthesized (e.g., phenylpropanoid, which produces tannins). Another important classification is related to the presence or absence of nitrogen in their chemical structure. However, the most commonly employed classification of these natural compounds is based on their biosynthetic origin, using structural criteria taken from the classification of organic compounds. Accordingly, there are four main classes: terpenoids (made from mevalonic acid, composed almost entirely of carbon and hydrogen, but sometimes include oxygen), phenolics (made from simple sugars, containing benzene rings, hydrogen, and oxygen), polyketides (which means "many ketone groups"), and nitrogen-containing compounds (extremely diverse, coming from α-amino acids, may also contain sulfur). Each class also forms diverse subgroups of secondary metabolites (Figure 1.5).

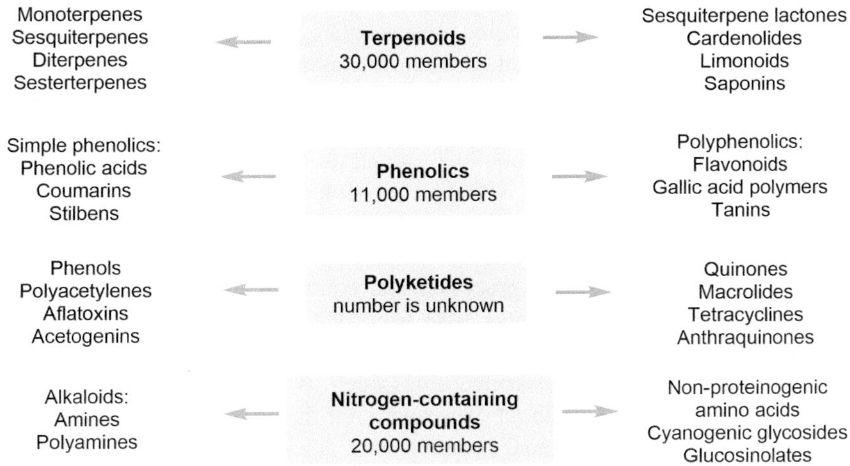

Figure 1.5. Simplified classification of secondary metabolites.

Terpenoids (or isoprenoids) are volatile substances that give plants and flowers their fragrance. They occur widely in the leaves and fruits of higher plants (i.e., conifers, citrus, and eucalyptus). These substances are also abundant in land and marine animals. According to the "isoprene rule" and "biogenetic isoprene rule", the terpenoids are constructed by a repetitive joining of isoprene (2-methyl-1,3-butadiene, C_5H_8) units. A rational classification of the terpenes is based upon the number of isoprene (or isopentane) units incorporated in the basic molecular skeleton: mono-, sesqui-, di-, and sesterterpenes contain the isoprene units linked in a head-to-tail fashion. The triterpenes and carotenoids (tetraterpenes) contain two 15 and 20 units, respectively, linked head-to-head. These terpenes are hydrocarbons, but oxygen-containing compounds such as alcohols, aldehydes, or ketones are also found. The sesquiterpene lactones, cardenolides, limonoids, and saponins also present this large class of secondary metabolites. With over 30,000 members described to date, terpenoids are plants' largest class of small natural products. The complexity of the phenolic compounds ranges from simple molecules such as phenolic acids to highly polymerized compounds such as tannins. Thus, the phenolic compounds (approx. 11,000 members) can be classified into two subgroups: simple phenolic compounds and polyphenolic compounds, which can be observed through the frame shown in Figure 1.5 [6, 7].

Plant phenols can also be classified into two major categories—benzoic acid derivatives, known as phenolic acids (C6-C1 skeleton), and phenylpropanoids (C6-C3 skeleton). The polyketides are also a broad class of structurally diverse (the number of members is unknown), acetate-derived natural products found in plants, microbes, insects, and marine organisms. Finally, the fourth class, nitrogen-containing compounds, contain different groups such as alkaloids, non-protein amino acids, amines, cyanogenic glycosides, and glucosinolates. Among them, alkaloid molecules are among the most influential groups containing mostly basic nitrogen atoms, usually in a heterocyclic ring. More than 12,000 alkaloids are extracted from plants, and over 3,000 have been isolated from marine sources; all of them are derived from amino acids such as ornithine, lysine, phenylalanine, tyrosine, tryptophan, histidine, and aspartic acid. Alkaloids can be classified in terms of their chemical structure, biological activity, the pathway of their biosynthesis, and occurrence into heterocyclic and non-heterocyclic alkaloids.

The term "secondary metabolite" started to appear in textbooks from the 1950s onwards. However, the German physiological chemist Albrecht Kossel (1853–1927), at the turn of the 19th century, unintentionally initiated the division of primary and secondary metabolism. He pointed out that primary metabolites are essential for all life-sustaining processes and common to all organisms, whereas, by contrast, secondary metabolites arise from distinct pathways that are typically limited to only a few organisms. Accordingly, he classified them as less vital and of "secondary" relevance. His concept has been used in scientific literature until now. However, the question about their role is far from satisfactorily answered. All living organisms can synthesize these molecules, albeit not all, and accumulate them in sufficient amounts to stimulate attempts of isolation and structure elucidation, thus knowing their complete roles. Nowadays, the commonly accepted role, "raison d'être" of secondary low-molecular-weight metabolites, provides central metabolism regulation and chemical defense to living organisms [8]. In this context, the term "biological function" is used often. The use of both primary and secondary metabolites by living organisms for their defense against not only their depredators but also the ecological impacts (physicochemical stress from the abiotic and biotic environment) and to procure their food means energy. This is found at all levels of evolutionary development. Thus, it is important to understand the specific intra- and interrelationships that can be described as encrypted messages in a chemical language between living organisms of the same and different species. Studying chemical compounds of biological origin involved in these interactions with living organisms is a central task of interdisciplinary science called chemical ecology. Brower proposed this name in 1969 [9, 10].

What is chemical ecology? Chemical ecology, an interdisciplinary field between chemistry and biology, focuses on chemical communication and other chemical interactions between organisms [11].

What is chemical communication? According to the American biologist and naturalist Edward Wilson (1929–2021), chemical communication is one individual's emission of a stimulus, which induces a reaction in another. These stimuli are known as chemical signals; they are typically low-molecular-weight secondary metabolites and can guide behaviors linked to various vital needs of living organisms. These chemical signals are named semiochemicals (signal is *semeion* in Greek) and are grouped into two subtypes of molecules depending on whether they act as intraspecific or interspecific mediators (Figure 1.6).

Thus, semiochemicals responsible for the communication between individuals of the same species (A-A) are known as pheromones (from the Greek *pherein*, meaning to carry or transfer, and in Arab *horman*, meaning to excite or effect). These were discovered in 1959 by Adolf Butenandt (1903–1995), a German chemist and Nobel Laureate in Chemistry (1939). Pheromones are often classified according

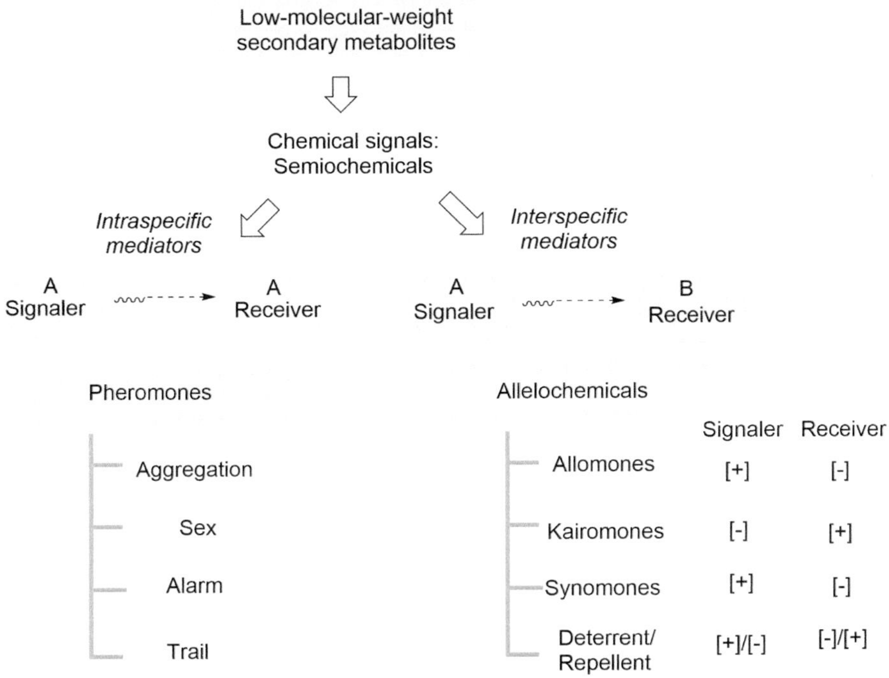

Figure 1.6. Simplified classification of chemical signals and semiochemicals involved in a living organism's specific intra- and interrelationships.

to their function and include sex, aggregation, and alarm pheromones. Other types of semiochemicals are called allelochemicals that act in different species (A-B) and are typically categorized contingent on the benefit(s) [+] and cost(s) [-] to the signaler (or emitter) and the receiver. Allelochemicals include allomones (emitting species benefits), kairomones (receptor species benefits), synomones (both species benefit), and deterrent/repellent, which can be both kinds of receiver and signaler. However, a single chemical signal may act as both pheromone and allelochemical. Moreover, depending on the specific species, some molecules can simultaneously act as repellents and allomones or kairomones. Pheromones and semiochemicals, in general, comprise a wide range of organic molecules that can be volatile or non-volatile. Their efficiency in chemical communication depends on various physical properties, including chemical nature, volatility, solubility, and the lifetime of the molecules in the environment.

The biological properties of these semiochemicals are often related to chemical defense between the same and different species. Therefore, small molecules used in this type of defense can be considered "chemical weapons" that mostly help organisms in the arduous and constant struggle for conservation and survival. However, aggressive behavior is a rare phenomenon in nature (except for human activities). Most living organisms use their chemical weapons as a means of defense and also to procure food. "Eat and not be eaten" is a motto, the most important condition among plants, animals, and other living organisms. To defend and attack, they must know how to fabricate and use these weapons, which means toxic small molecules.

1.2 Toxicity of Natural and Synthetic Substances: LD_{50}, Toxic Molecules, Venoms, and Toxins

The biological effects (beneficial, indifferent, or toxic) of both naturally and synthetically produced chemical substances are dependent on a number of factors, e.g., structure, dose, time, and mode of exposure, among others. Some secondary metabolites of a species have specific negative impacts on other species or can damage them. Being toxic or repellant, they help defend this emitting species.

The toxicity of a chemical substance signifies its capacity to damage a whole living organism or its particular organ system (e.g., heart, liver, or kidneys) as well as its cells (cytotoxicity), generally disrupting a biochemical process (e.g., the blood-forming process) via an enzyme alteration or other targets. So, toxicity is the ability of a substance to damage living tissues when it has reached susceptible places inside or on the body. That is any substance that can induce disease or death when entering the body is considered toxic or poisonous. Toxicity is a property of each compound determined by molecular structure; generally, any organic compound can be harmful to living organisms. But, just as there are degrees of being injurious, there are also degrees of being innocuous. When describing injurious compounds, two terms are usually distinguished: poison and venom [12]. A literature review discloses discrepancies and ambiguity in the definitions of "poison" and "venom". These two terms are frequently conflated with one another, along with the more general term "toxin". Poisonous compounds produced and secreted by living organisms are called toxins; they are natural poisons and toxic metabolites. In a non-technical context, a toxin is any substance (in the context of this book, a metabolite) that, when present in biologically relevant quantities, causes dose-dependent pathophysiological injury to a living organism. This injury reduces the functionality or viability of the organism, potentially causing sickness or injury to a living organism. Venoms are a form of toxins secreted by an animal to cause harm to another. However, it should be noted that little agreement exists on how toxins are classified.

Each synthetic compound or natural chemical substance (metabolite) is characterized by its toxicity, whose quantitative measure is the median lethal dose (LD_{50}). The LD_{50} test was developed in 1927 to measure the acute toxicity of certain compounds in live animals. The lethal dose is the amount of a substance necessary to kill 50% of the experimental animals. It is expressed in the weight of the substance per unit of weight of the animal (milligrams per kilogram; mg/kg). The LD_{50} value of the same toxic substance varies from the mode of its introduction into the animal body (intravenous, intraperitoneal, intramuscular, oral, or parenteral), and obviously, as the basic principle of toxicology states, "the dose makes the poison". The LD_{50} is determined mathematically by studying dose-effect relationships (Figure 1.7A).

Under the same experimental conditions, the LD_{50}s of different poisons may be easily compared. The data for this parameter can be interpreted as follows: a substance with a small LD_{50} (like 5 mg/kg) is highly toxic, while a substance with a

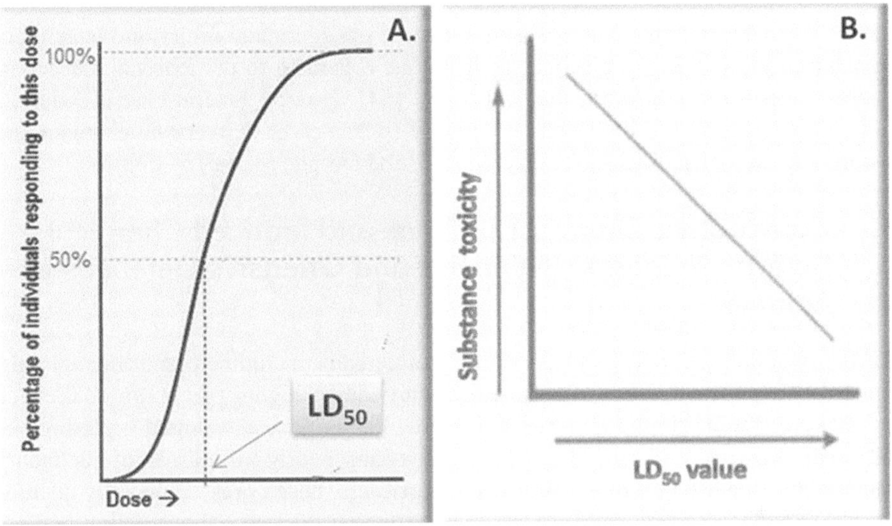

Figure 1.7. Graphical interpretations of LD_{50} and the relationship between substance toxicity and LD_{50} value.

large LD_{50} (1,000 to 5,000 mg/kg) is practically innocuous (non-toxic). There is an inverse relationship between substance toxicity and LD_{50} value (Figure 1.7B).

Systematic studies of toxic metabolites and their effects began only during the Renaissance in the 15th century when Paracelsus (1493–1541) identified the specific chemical components of minerals, plants, and animals responsible for their toxic properties. He also showed that varying the amount of the poison affected the severity of the effects. His historical conclusion, "All substances are poisons: there is none which is not a poison. The right dose differentiates a poison and a remedy", is cited and given in all books on toxicology and pharmacology. Therefore, it is not in vain that Paracelsus is credited as the father of toxicology. His investigations brought a scientific analysis of toxins, and he forged the path of using chemicals and minerals in medicine.

Thus, there are poisonous and venomous organisms [13]. These organisms have generated fascination and loathing since recorded history began. They have also inspired considerable research across a broad range of disciplines. The terms "venomous" and "poisonous" are often used interchangeably; however, this usage is incorrect. Venom and poison are functionally distinct and should not be conflated. There is, in fact, a difference between a venomous organism and a poisonous organism. Both venomous and poisonous animals produce a toxin that can be injurious or even lethal to another organism. The fundamental difference between the two terms involves how that toxin is delivered. Venomous organisms deliver or inject venom into other organisms using a specialized apparatus (usually fangs or a stinger). The venom is produced in a gland attached to this apparatus. Poisonous organisms, on the other hand, do not deliver their toxins directly. The entire body, or significant parts of it, may contain the poisonous substance. These organisms may be harmful when eaten or touched. Recently, a new category of toxic biological secretion was proposed: toxungen. This is defined as a toxic substance (comprised of one or more toxins) that causes dose-dependent physiological injury and is actively transferred via a delivery mechanism from one organism to the external surface of another organism without mechanical injury [14]. Thus, a general classification of toxic organisms should be based on delivery (presence of delivery mechanism and wound), source of acquisition (biosynthesis), and storage (gland) of toxin.

1.3 Constitutive Chemical Defense and Induced Chemical Defense in Plants: Defensive and Offensive Defenses in Animals

Throughout their entire life cycle, all living organisms, including plants and animals, must develop their defense mechanism. Significantly, higher plants are challenged by many different abiotic and biotic stresses. Plants have developed a plethora of different chemical defenses (Figure 1.8), covering nearly all classes of secondary metabolites that represent a significant barrier to herbivores, especially against herbivorous insects; among these are specialists that feed on only a limited number of plant species, or even one single host, and generalists that can feed on numerous species. Co-existing for at least 350 million years, plants and insects first have acquired successful defensive qualities [15]. Toxicity is present in almost all plants, and as it was mentioned above, the toxicity of a substance is always relative, depending on the dose, age, and physical state of living organisms (including humans), the mechanism of absorption, and the mode of excretion (elimination) and other factors. Generally, all plants must be repellent and toxic so as not to "be eaten" by animals. Plants respond to the attack of herbivores by activating defense mechanisms.

In principle, two broad categories of chemical defenses in plants can be distinguished:

(a) toxic metabolites are always present and are stored compounds, conjugated molecules, or precursors of active compounds that can easily activate upon the damage;

(b) toxic molecules or ones with dissuasive effects are induced, and certain aggressors may elicit explicitly.

A further distinction can be made between constitutive and inducible defenses, each of which can be direct or indirect (Figure 1.8) [16]. Direct defenses act by themselves against the aggressor, while indirect defenses act by attracting organisms from an additional trophic level, such as the enemies of attacking herbivores.

However, many defense compounds are toxic to the plant itself, so plants that rely on constitutive chemical defense must be able to synthesize and store these substances without poisoning themselves. As could be seen in plants, secondary compounds such as alkaloids, coumarins, cyanogenic glycosides, and glucosinolates derive from amino acids. Tricarboxylic acid cycle constituents are involved in the formation of polyacetylenes and polyphenols, while glucose, aliphatic acids, and other "primordial molecules" (Figure 1.5) also play a role. In contrast, in animals, such as insects, many defensive secretions are derived from the same amino acids

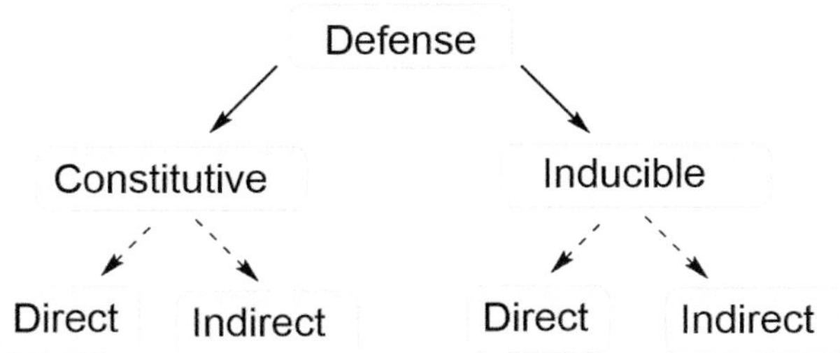

Figure 1.8. Types of plant defenses.

used to construct proteins (among them, quinones in many beetles and cockroaches derive from tyrosine, formic acid in ants from serine, isobutyric acid in swallowtail caterpillars from isoleucine and valine, and alkyl sulfides in ants from methionine). Animals have developed physiological and behavioral strategies to deal with constitutive defenses, and they can either attempt to detoxify the defensive compounds or avoid certain plant types or particular food types of a specific species. Animals can also preferentially consume particular food types to minimize the adverse effects of secondary metabolites on digestion or to avoid consuming highly acidic foodstuffs. Nevertheless, animal secondary metabolites can be considered defensive in function only if they protect their producers from the life-threatening activities of another organism. Distinguishing between offensive and defensive use of these chemicals is problematic.

1.4 Basics of Organic Synthesis and Secondary Metabolite Biosynthesis

Organic chemistry stands out as the principal initiator in biological discovery and the development of new drugs. Among diverse definitions of synthesis (the central part of organic chemistry), the definition of John Cornforth (1917–2013), an Australian-British chemist, Nobel Laureate in Chemistry in 1975, seems to be concise and clear: "The synthesis is the intentional construction of organic molecules through chemical reactions". This construction requires scientific knowledge about the molecules' chemical reactions and physicochemical properties. Therefore, organic chemists determine the structure of organic molecules, study their reactions, and develop procedures to synthesize organic compounds. The ability of man to replicate the molecules of living creatures and create other molecules like them is a remarkable development in human history. As a science, organic chemistry was born at the end of the 18th century when the isolation of organic substances from extracts of natural origin began in 1773. The French chemist Hilaire Rouelle (1718–1779) isolated urea in the urine of humans, cows, and horses, and in 1828, the German chemist Friedrich Wöhler (1800–1882) prepared white, silky crystals of urea, a naturally occurring substance from the living world using potassium cyanate and ammonium sulfate, inorganic substances. This synthesis demonstrated for the first time that organic compounds could be produced from inorganic starting materials. The work of Wöhler launched a new research field in chemistry known as organic synthesis.

As a result, by the end of the 19th century, scientists could synthesize hundreds of organic compounds.

Synthetic organic chemistry is paramount to the technology offering us chemical products. Organic synthesis is essential for biology, medicine, and the chemistry of natural products. Natural products played a crucial role in the emergence and advancement of organic synthesis from its birth to the present day. Thus, from the early days of elemental analysis of natural products, these substances fascinated and challenged organic chemists, first with their structural elucidation and then with their total synthesis.

The jewel of organic synthesis is total synthesis, the endeavor of synthesizing the molecules of living nature in the laboratory [17, 18]. A total synthesis is the complete chemical synthesis of complex natural products from more simple pieces, generally without the help of biological processes. The selection of the desired products to prepare (target molecules) from the myriad natural products for total synthesis by the chemists depends on the novelty of their molecular structure, biological activity, and natural scarcity, among other criteria. Due to the complexity of natural molecules, their efficient construction needs several steps (except in very simple cases), each of which uses chemical reactions that specifically lead to a structure. That is why any total synthesis of complex natural products (target-oriented synthesis, TOS) requires a systematic knowledge of organic reactions that form C-C bonds and those that change one functional group into another. Additionally, a detailed plan of its synthesis is essential, including the use of tactics and strategy and determining where to begin, how to get the desired product, and which concrete reaction to use. Besides total synthesis, there are various kinds of synthesis to obtain a concrete, natural product: (a) semi-synthesis (the synthesis of a given molecule from an advanced natural precursor related to it); (b) formal synthesis (the synthesis of a crucial intermediate that has been already converted into the target molecule); (c) partial synthesis (the synthesis of a portion of the natural product) (Figure 1.9).

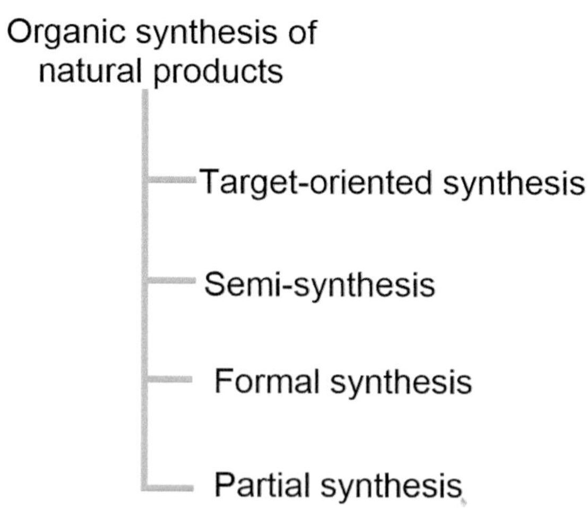

Figure 1.9. Types of organic syntheses of natural products.

Target-oriented syntheses are successfully used in drug discovery efforts involving preselected protein targets. In general, the planned product synthesis must always be "ideal"; that is, it must be rational, effective, safe, economical, etc. In an ideal synthetic route, the desired molecule is prepared from the easily accessible starting material via a simple, safe, ecological operation that proceeds quickly and in an almost quantitative yield.

How can we plan the synthesis of the desired molecule with a complex structure? In 1989, E. J. Corey (Nobel Laureate in Chemistry 1990) explained it, formalizing the terminology for this approach to synthesis and making it an understandable process [19]. This process is called retrosynthesis, in which the structure of the desired molecule (target, TM_0) is dissected to obtain relatively simple starting materials (Figure 1.10). Thus, retrosynthesis that works backward toward the starting material is based on the disconnection approach. Disconnection refers to mentally breaking bonds (strategic bonds) to give successively simpler precursor molecules or commercially available chemicals, but always in a manner in which those bonds can be reformed by known or reasonable chemical reactions. The operation of disconnecting strategic bonds is defined as a transform, which is the exact reverse of a chemical reaction. Therefore, if this process is repeated several times, the structural simplification of TM_0 could lead to a diverse series of molecular fragments that serve as crucial intermediates or precursors (P_1, P_2, P_3, etc.), considered their synthetic targets. So, the synthetic tree (a roadmap of synthetic intermediates) is built (Figure 1.10).

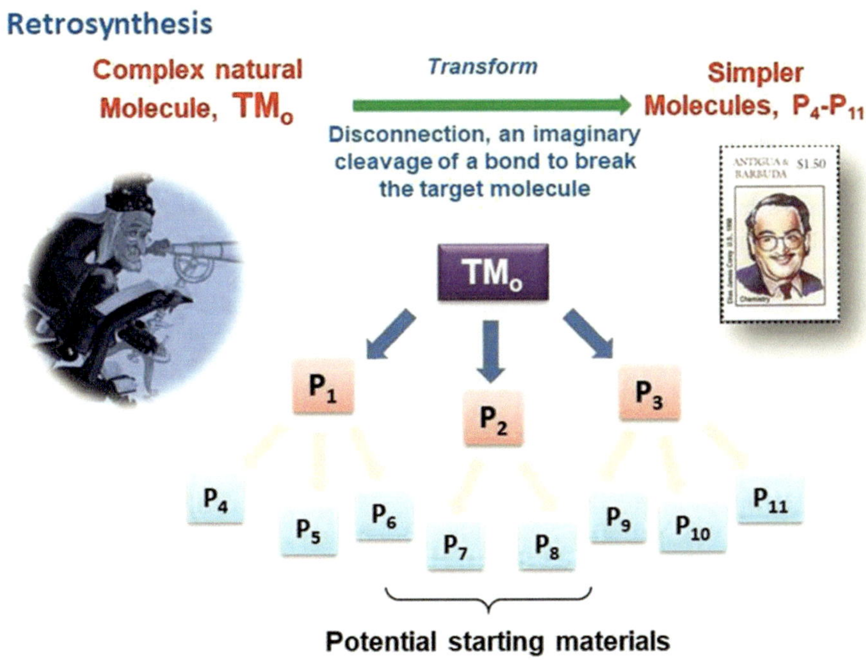

Figure 1.10. Retrosynthetic analysis is based on the disconnection approach developed by Corey in 1989 and the author image in a stamp.

Examining this retrosynthetic pathway, we could note that the choice of starting material is crucial. Thus, how does one correctly select the TM_0? Which of these precursors (P_4–P_{11}) could be the best starting materials? Answers to those questions are the essence of total synthesis, and to respond to them, one must have some experience in synthetic organic chemistry that includes (i) a detailed analysis of the structure of the target, (ii) fundamental knowledge of chemical reactions, (iii) a basic understanding of bonding and reactivity, and (iv) a well-developed chemical intuition. That is to say that the synthesis of complex natural molecules is a special type of intellectual activity since their master methods are based on heuristic solutions. Robert B. Woodward (1917–1979) said, "There is excitement, adventure, and challenge, and there can be great art in organic synthesis". Elias J. Corey (born in 1928) expressed a similar view, stating, "The organic chemist is more than a logician and strategist; he is an explorer strongly to speculate, to imagine, and even to create. These added elements provide the touch of artistry which can be included in cataloging the basic principles of synthesis, but they are very real and extremely important" [20]. Thus, organic synthesis is unusual because of its power to let the practitioner be both an artist and a creator. Phil S. Baran (born in 1977) also explained, "An artist in terms of how one makes that beautiful molecule and its incredible capacity for impact with very simple logic. The charm and appeal of organic chemistry (synthesis) are that there is still so much to be explored and discovered, and if you uncover one of its new secrets, it can have a dramatic impact on all sorts of areas of chemical science that you wouldn't predict" [21].

Thus, the objective of any synthesis is to obtain the desired product in pure form by the most efficient and convenient procedure possible. An essential part of any synthetic strategy is proceeding from the chosen starting material to the product. In general, the synthesis of a final product can be carried out according to consecutive (linear) or convergent strategies.

In a consecutive synthesis, the precursor molecules necessary to achieve the final product are made step by step using a single starting material. Apparently, with the increase in steps, the total efficiency of the preparation of our target product decreases. This problem forces us to perform each step with maximum effort, desiring the best performance and reducing steps. Unfortunately, it is inevitable that "something is lost". In a convergent synthesis, several pieces of a molecule (building blocks) are synthesized individually, and the final product (target) is sequentially constructed using those pieces obtained. These possible approaches are illustrated in Figure 1.11.

The convergent synthesis has fewer steps than the consecutive steps in the overall synthesis (Table 1.1). Besides the efficiency factor, the convergent strategy has another advantage: the great diversity of tactics with which the failure of a final step can be corrected. However, consecutive synthesis is very simple to prepare a TM with a simple structure.

As an example, there are famous classic syntheses of natural products: Willstäter's synthesis of alkaloid tropinone from cycloheptanone consists of 19 steps with a total yield of 0.75%, and Robinson's synthesis of the same alkaloid consists of one step with a yield of 17% (Figure 1.12). Alkaloid tropinone is a bicyclic molecule, and its name, according to the scientific rules, is 8-methyl-8-azabicyclo[3.2.1]octan-3-one.

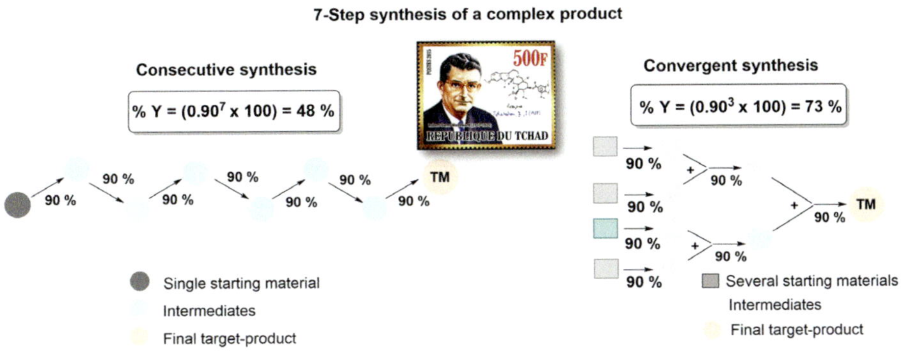

Figure 1.11. Consecutive vs. convergent syntheses. A stamp dedicated to R. B. Woodward.

Table 1.1. Efficiency of consecutive and convergent syntheses.

n	8	16	22	64	80
Y_n, % consecutive	16.8	2.8	0.08	6×10^{-5}	2×10^{-6}
convergent	51.2	41.0	32.8	26.0	24.4

Where n is the number of steps whose performance averages 80 %, the overall yield is calculated as $Y_n = Y^n$ for consecutive synthesis and $Y_n = Y^{\lg n}$ for convergent synthesis.

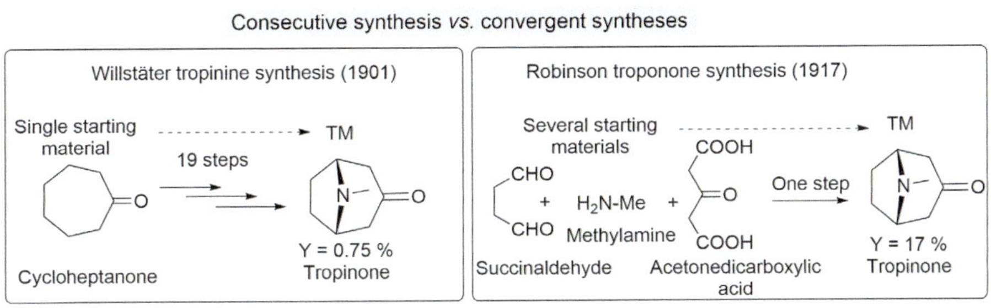

Figure 1.12. Historical examples of consecutive and convergent syntheses of tropinone.

Robinson's synthesis of tropinone is also an example of so-called one-pot synthesis, in which reactants are subjected to successive chemical reactions in just one reactor. This permits avoiding the laborious separation and purification process to obtain the intermediates, saving time and resources while increasing chemical yield.

In modern organic chemistry, scientific efforts pursue efficiency in their processes as the ultimate goal. The meaning of efficiency varies depending on the specific field of application. In developing new synthetic methodologies, efficiency is achieved by generating new reactions or reactions already established with maximum performance and maximum selectivity. What refers to synthesizing a complex natural product, synthetic art consists of knowing how to discern the possibility of building the desired molecules (TM_0) with a new, more effective route. According to Paul Wender (born in 1947), a given compound's "ideal synthesis" should be quickly carried out in a single synthetic operation with a yield of 100%, starting from accessible starting materials and without generating residues. In addition, it should be simple, safe, economical, and environmentally acceptable [22].

Thus, the aim of organic synthetic chemistry today is increasingly focusing on making desired molecules in a practical, scalable manner. The concept of ideality in organic synthesis is not new. In 1975, James Hendrickson (1928–2018), a pioneer in the development of the theory and logic of organic synthesis, defined the ideal synthesis as one which "creates a complex molecule ... in a sequence of only construction reactions involving no intermediary functionalization, and leading directly to the target, not only its skeleton but also its correctly placed functionality." This description of ideal synthesis leads us to the notion of "step economy" or "atom economy", an aspect of ideal synthesis. Atom economy is "the idea of designing a transformation to carry as much of the reactants' mass as possible into the product, maximizing sustainability by minimizing byproducts or waste" (Barry Trost, born in 1941). This can be illustrated, as shown in Figure 1.13. Achieving 100% atom economy means that all atoms in the reactants have been converted into the desired product [23].

Some of the quickest, simplest, and most effective reactions are multicomponent reactions (MCRs), in which three or more reactants react together in one pot to the desired product. These reactions are reactions with a high atom economy. MCRs also involve convergence, easy execution, and operative simplicity and tend to be more environmentally friendly as they use fewer raw materials and produce less waste. In MCR, a product is assembled according to a cascade of elementary chemical reactions and possesses essential structural elements of all reactants [24]. Therefore, a network of reaction equilibriums eventually leads to an irreversible product production step. The challenge is to carry out a multicomponent reaction so that the pre-balanced reaction network leads to the formation of the target product and does not produce byproducts. An example? If we examine again Figure 1.12, we can see that the convergent synthesis of tropinone is based on a three-component reaction.

Suppose one compares the MCR approach with conventional linear-type synthesis. In that case, it is observed that MCRs are far better than linear conventional types concerning time, energy, and yield, respectively. As an example, in this case, we consider the syntheses of cyclooctatetraene (COT) from pseudopelletierine (an alkaloid extracted from pomegranate-tree bark) and from acetylene realized respectively by Willstätter in 1911 and by Reppe in 1948. The difference is very significant: while Willstätter's synthesis consists of 13 consecutive steps, Reppe's synthesis involves the tetramerization of acetylene under high pressure (15 tam, 80–120°C) in the presence of nickel catalysts and is completed in one-pot synthesis (Figure 1.14).

Interestingly, MCRs are involved in the prebiotic synthesis of α-amino acids, sugars, nitrogenous bases (purines and pyrimidines), phosphates, and more complex molecules such as the long, chain-like molecules (like proteins and nucleic acids), which are essential for building living creatures. The proteins are crucial to life and comprise just 20 essential amino acids. For cells to accomplish self-replication, the cooperative action of proteins and nucleic acids is required, and the complex information detailing the specific structure of the proteins inside living things is stored in nucleic acids like RNA and DNA [25]. Therefore, according to the theory reported

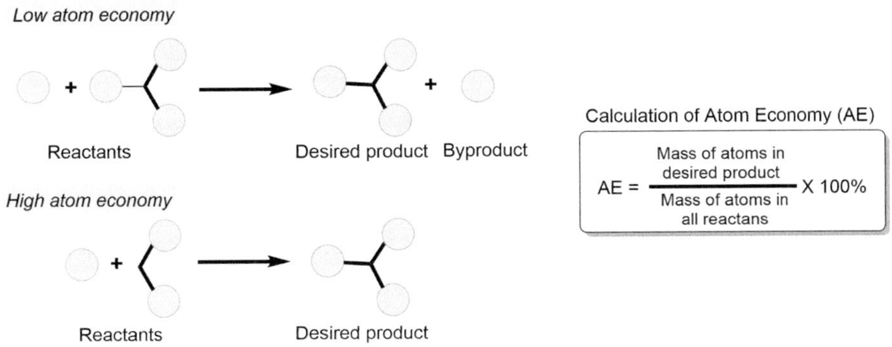

Figure 1.13. An illustration of the concept of atom economy.

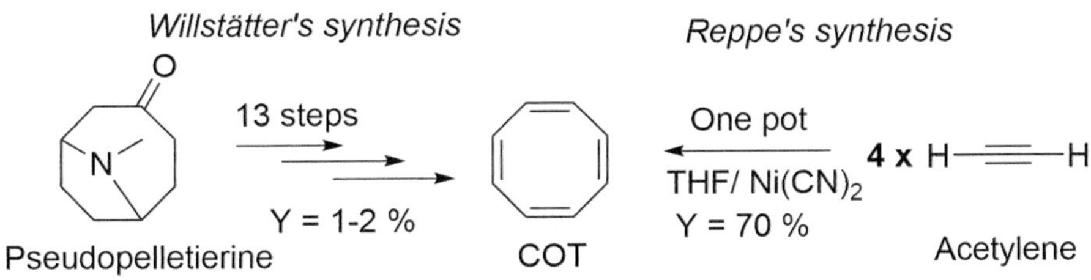

Figure 1.14. A crucial role of step economy in the synthesis: Willstätter's synthesis vs. Reppe's synthesis of cyclooctatetraene.

in 1924 by Russian biologist and biochemist Alexandr I. Oparin (1884–1980), life could be started with simple inorganic molecules such as water vapor, carbon dioxide methane, ammonia, hydrogen sulfide, carbon monoxide, and phosphate that were present in the primitive atmosphere ("reducing" oxygen-less atmosphere) [26].

The Earth and its primitive atmosphere, which lacked free oxygen, were formed approximately 4.6 and 3.5–3.7 billion years ago, respectively. Such chemical life would have been the intermediary between an inorganic rock and the first living biological cell. Thus, the initial condensation reactions of these molecules under "special" conditions reaction, such as electrical activity like lightning or mechanochemical activation by milling with high energy (so-called a "primeval soup"), could have conducted to the formation of hydrogen cyanide (HCN), formaldehyde (CH_2O), cyanoacetylene (HC≡C-CN), urea (NH_2CONH_2) and cyanamide (NH_2CN), among others. All these intermediate molecules are critical reagents for biomacromolecule formation under early Earth conditions.

In 1953, two American chemists, Stanley Miller and Harold Urey, at the University of Chicago demonstrated for the first time that a mixture of gases (CH_4, CO, H_2, and NH_3) in the presence of pools of liquid water H_2O under an electric discharge gave traces of α-amino acids including glycine and α-alanine used in living cells [27]. How could these acids be obtained under these conditions? One-step reactions among this mixture can produce HCN and HCHO that react in the presence of NH_3 to form α-amino nitriles, which can be hydrolyzed by H_2O giving α-amino acids, now known as the Strecker reaction, first reported three-component reaction (1850). Later, in 1961, a Spanish biochemist Joan Oró (1923–2004) showed that the nucleic acid

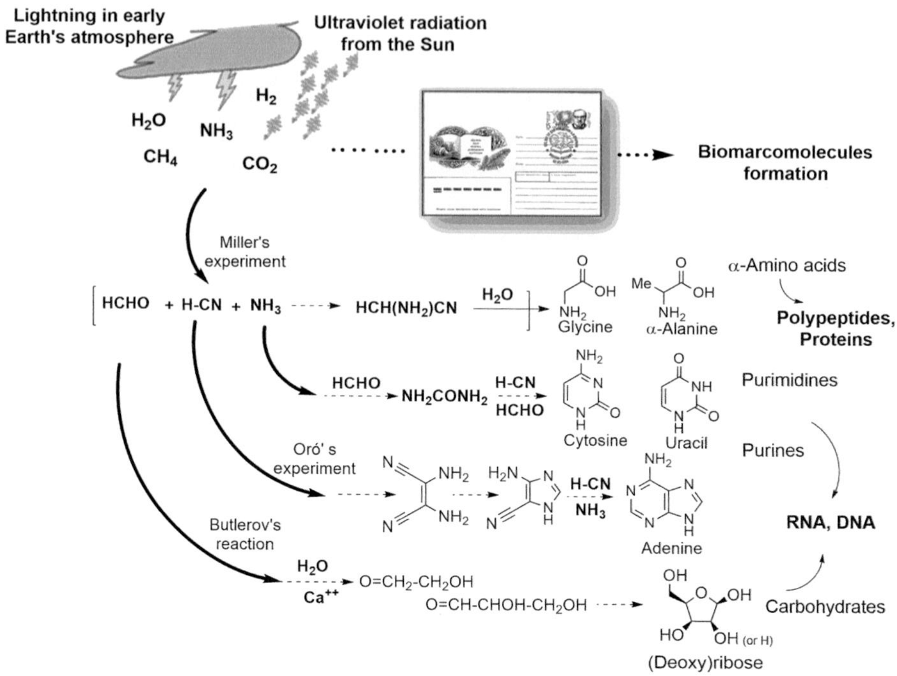

Figure 1.15. Oparin's theory: from simple inorganic molecules to complex biomacromolecules through multicomponent reactions. The cover is dedicated to the memory of Academic A. I. Oparin.

purine base, adenine (a chemical component of DNA and RNA), could be formed by heating aqueous ammonium cyanide (NH_4CN) solutions under conditions that may have been similar to those of primitive Earth. Besides the subsequent polymerization reaction of formaldehyde with itself in the presence of water, the so-called formose reaction, realized by Russian organic chemist Alexandr M. Butlerov (1828–1886), can give rise to ribose sugar via intermediates, such as glycolaldehyde ($O{=}CH_2$-CH_2OH) and glyceraldehyde ($O{=}CH$-$CHOH$-CH_2OH). It is presumed that over a few hundred million years of evolution, these prebiotic molecules evolved into self-replicating molecules by natural selection. In the chemical language, all these descriptions can be presented as depicted in the following scheme (Figure 1.15).

Therefore, the first living things on Earth, single-celled microorganisms or microbes lacking a cell nucleus or cell membrane known as prokaryotes (Figure 1.1), seem to have first appeared on Earth. Cyanobacteria (also known as blue-green algae), one of the earliest types of prokaryotic bacteria, formed into colonies, and fossilized remains of these have been found in Australia dating back to between 3.4 and 2.8 billion years ago.

These early cyanobacteria were the first oxygen-producing, evolving, phototropic organisms. They were responsible for the initial oxygenation of the Earth's atmosphere, producing oxygen and sequestering carbon dioxide in organic molecules during the period from 2.7 to 2.2 billion years ago. Photosynthesizing plants evolved later and continued this process, leading to the buildup of increasing levels of oxygen in the atmosphere and the release of nitrogen into the atmosphere as the oxygen reacted with ammonia. In due course, a layer of ozone (an allotrope of oxygen) formed in the atmosphere, providing better protection for early life forms from ultraviolet radiation. Notably, the rapid explosion of life began to become abundant and was only possible after the presence of oxygen.

The first eukaryotic cells (organisms with one or more complex cells, each containing a nucleus and surrounded by a membrane that holds the cell's genetic material) evolved sometime between 2.5 and 1.7 billion years ago. The nucleus in these cells could hold and protect complex molecules such as RNA and DNA.

Now, it is known that the activity of an organism depends on the total activity of independent cells. There are many activities that a cell has to perform from the time of birth to the death of a single individual. One of the most essential functions of living cell organisms is to grow across their lifespan, reproduce, make copies of themselves, and have a metabolism. The latter function can be defined as a set of biochemical processes catalyzed by multi-enzymatic systems that help an individual or an organism respond to its surroundings. Cell metabolism is mainly involved in the extraction of energy used for many activities like cell growth, cells, tissue repair, etc. Without this process, a cell dies without energy for biological reactions. And, if a cell starts to die, thus, an organism ends.

Metabolic processes proceed through a series of successive reactions; each reaction originates the other, this one the following one. It is believed that there are more than 3,000 types of enzymes and about 3,000 different biochemical reactions. As mentioned, there are two types of metabolism: primary metabolism and secondary metabolism (see Section 1.1).

Thus, the essential metabolic pathways are similar in different living organisms and always reversible reactions. Two groups of these pathways are distinguished: biosynthesis (anabolism, where large and complex molecules are prepared from small precursor molecules) and degradation (catabolism, where large molecules degrade into simple molecules). Like chemical synthesis, these metabolic pathways can be linear, convergent, or divergent. Other very characteristic types of intermediary metabolism are cyclic or spiral forms (Figure 1.16).

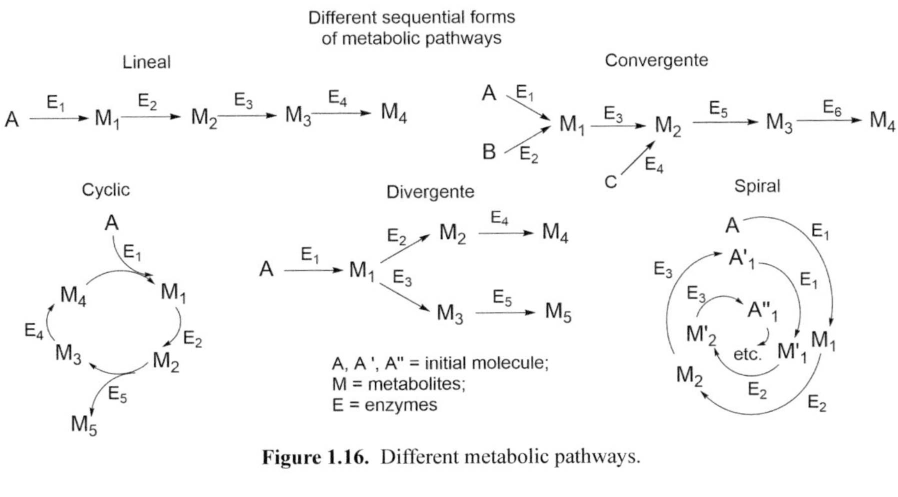

Figure 1.16. Different metabolic pathways.

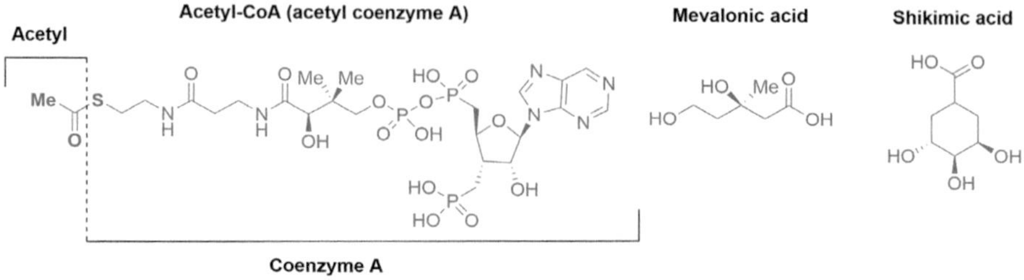

Figure 1.17. Chemical structures of main precursors in the biosynthesis of natural molecules.

Then, natural molecules are biosynthesized by a sequence of reactions that, with very few exceptions, are catalyzed by enzymes. These reactions and their precursors (building materials) can be cataloged in terms of organic chemistry. The three main routes of metabolism use all types of organic reactions (not discussed in detail here). Generally, all small molecules—the products of secondary metabolism—are generated using simple precursors or building blocks: acetyl-CoA (acetyl coenzyme A), mevalonic acid, and shikimic acid (Figure 1.17) [6].

Acetyl coenzyme A is involved in many biochemical reactions in essential biomolecules such as proteins and carbohydrates. The primary function of acetyl-CoA is to deliver the acetyl group to the citric acid cycle (Krebs cycle) to be oxidized to carbon dioxide and water for energy production. The biosynthesis of natural molecules obtained from acetyl-CoA is known as the acetate pathway. Mevalonic acid is a precursor of other biomolecules like terpenes and steroids. According to the scientific rules, its name is (3*R*)-3,5-dihydroxy-3-methylpentanoic acid. This molecule exists in equilibrium with its lactone form, mevalonolactone. The third simple precursor, shikimic acid, is a carboxylic acid. This molecule is a hydroaromatic intermediate in the common aromatic amino acid biosynthesis pathway known as the shikimate pathway or shikimic acid pathway.

Therefore, in the biosynthesis of natural products, these three simple compounds are modified, converted into other compounds, or joined to form macromolecules. However, many natural products, such as alkaloids and mold metabolites, do not seem to have any obvious metabolic or evolutionary function. Some compounds may be formed due to a "metabolic accident" or are byproducts of the synthesis machinery of cellular enzymes. Regardless of their utility to the parent organism, their value to man as drugs, dyes, flavorings, poisons, toxic molecules, and so on is undisputed.

1.5 Conclusion

We observed that the versatility of carbon makes the great diversity of organic molecules possible. Some are involved in the specific inter- and interrelationships of living organisms. Such molecules are called semiochemicals. Their biological properties often relate to chemical defense between the same and different species. In general, each molecule is characterized by its toxicity, whose quantitative measure is the median lethal dose, LD_{50}, and its toxicity, which is always relative and depends on many factors. Therefore, such toxic molecules used in this type of defense can be considered chemical weapons.

Most living organisms use these toxic molecules for conservation and survival. They must know how to produce and use these molecules to defend and attack. Defensive chemicals used by organisms for protection against potential depredators are generally products of secondary metabolism that involve many different biochemical reactions, which are always catalyzed by diverse, specific biomacromolecules called enzymes. Such reactions are like organic reactions that make chemists in synthetic laboratories. Organic synthesis is essential for biology, medicine, and the chemistry of natural products. The jewel of organic synthesis is total synthesis, the endeavor of synthesizing the molecules of living nature in the laboratory.

Synthesis of the desired molecule with a complex structure, including toxic molecules, usually starts with the process called retrosynthesis, in which the structure of the desired molecule is dissected to obtain relatively simple starting materials. The objective of any synthesis is to obtain the desired product in pure form using the most efficient and convenient procedure possible. An important part of any synthetic strategy is how to proceed from the chosen starting material to the product. In general, the synthesis of a natural product can be carried out according to consecutive or convergent strategies. To be known to synthesize the desired molecules, fundamental knowledge of chemical reactions and a basic understanding of bonding and reactivity is needed.

With these introductive basics, we can explore the complex wild world of plant-animal interactions and the relationships between organisms in the kingdoms Animalia and Plantae.

1.6 References

[1] Lodish, H., Berk, A., Kaiser, C.A., Kaiser, C., Krieger, M., Scott, M.P. et al. 2008. Molecular Cell Biology, W.H. Freeman, Basingstoke, pp. 8–10.

[2] Mora, C., Tittensor, D.P., Adl, S., Simpson, A.G. and Worm, B. 2011. How many species are there on Earth and in the ocean? PLoS Biol. 9(8): e1001127.

[3] May, R.M. 2011. Why worry about how many species and their loss? PLoS Biol. 9(8): e1001130.

[4] Campbell, A.K. 2003. Save those molecules! Molecular biodiversity and life. J. Appl. Ecol. 40(2): 193–203.

[5] Hadacek, F. and Bachmann, G. 2015. Low-molecular-weight metabolite systems chemistry. Front Environ Sci. 3: Article 12.

[6] Dewick, P.M. 2009. Medicinal Natural Products: A Biosynthetic Approach, New York: John Wiley & Sons; 550 pp.

[7] Ludwiczuk, A., Skalicka-Wo´zniak, K. and Georgiev, M.I. Terpenoids. 2017. pp. 233–266. *In*: Badal, S. and Delgoda, R. (Eds.). Pharmacognosy. Fundamentals, Applications and Strategies. London: Elsevier Inc.

[8] Hadacek, F., Engelmeier, D., Bachmann, G. and Chobot, V. 2011. Hormesis and a chemical raison d'ĕtre for secondary plant metabolites. Dose-Response 9(1): 79–116.

[9] Bergström, G. 2007. Chemical ecology = chemistry + ecology! Pure Appl. Chem. 79(12): 2305–2323.

[10] Cravatt, B.F. and Gottesfeld, J.M. 2010. Chemical biology meets biological chemistry mini review series. J. Biol. Chem. 285(15): 11031–11032.

[11] Harborne, J.B. 2001. Twenty-five years of chemical ecology. Nat. Prod. Rep. 18(4): 361–379.

[12] Hadley, D. 2019. What Is the Difference Between Venomous and Poisonous? https://www.thoughtco.com/venomous-vs-poisonous-1968412.

[13] Habermehl, G.G. 1981. Venomous Animals and Their Toxins, Berlin: Springer-Verlag, 194 p.
[14] Nelsen, D.R., Nisani, Z., Cooper, A.M., Fox, G.A., Gren, E.C., Corbit, A.G. et al. 2014. Poisons, toxungens, and venoms: redefining and classifying toxic biological secretions and the organisms that employ them. Biol. Rev. 89(2): 450–465.
[15] Mithöfer, A. and Boland, W. 2012. Plant defense against herbivores: chemical aspects. Ann. Rev. Plant Biol. 63: 431–450.
[16] Wittstock, U. and Gershenzon, J. 2002. Constitutive plant toxins and their role in defense against herbivores and pathogens. Curr. Opin. Plant Biol. 5(4): 300–307.
[17] Smith, M.B. 2002. Organic Synthesis, 2nd Edition, New York: McGraw Hill Inc; pp. 818–916.
[18] Nicolaou, K.C. 2014. Organic synthesis: the art and science of replicating the molecules of living nature and creating others like them in the laboratory. Proc. R. Soc. A; 470(2163): 20130690.
[19] Corey, E.J. and Cheng, X.M. 1995. The logic of chemical synthesis, New York: John Wiley & Sons Inc, 436 pp.
[20] Nicolaou, K.C. and Chen, J.S. 2011. Classics in total synthesis III, Weinheim: Wiley-VCH, 746 pp.
[21] Zbikowski, F. 2017. The Charm and Appeal of Organic Chemistry. Nachrichten aus der Chemie. https://www.chemistryviews.org/details/ezine/10581368/The_Charm_and_Appeal_of_Organic_Chemistry.html.
[22] Crow, J.M. 2012. Stepping toward ideality. https://www.chemistryworld.com/features/stepping-toward-ideality/5190.article.
[23] Newhouse, T., Baran, P.S. and Hoffmann, R.W. 2009. The economies of synthesis. Chem. Soc. Rev. 38(11): 3010–3021.
[24] Biggs-Houck, J.E., Younai, A. and Shaw, J.T. 2010. Recent advances in multicomponent reactions for diversity-oriented synthesis. Curr. Opin. Chem. Biol. 14(3): 371–382.
[25] Rana, A.K. and Ankri, S. 2016. Reviving the RNA world: an insight into the appearance of RNA methyltransferases. Front Genetics 7: 99.
[26] Toupance, G., Raulin, F. and Buvet, R. 1975. Formation of prebiochemical compounds in models of the primitive Earth's atmosphere. Origins of Life 6(1): 83–90.
[27] Lazcano, A. and Bada, J.L. 2003. The 1953 Stanley L. Miller experiment: fifty years of prebiotic organic chemistry. Origins of Life and Evolution of the Biosphere 33(3): 235–242.

CHAPTER 2

Vegetable Oxygen-Containing Toxins and their Effects on Animals

Plant-Animal Interactions

2. INTRODUCTION: ABUNDANCE AND IMPORTANCE OF VEGETABLE OXYGEN-CONTAINING TOXINS

Members of the Kingdom Plantae have existed for almost 350 million years. Kingdom Plantae are important, as they are the food source for all other living organisms on Earth, which depend on plants to survive. Among diverse, basic characteristics in common (organisms are multicellular, eukaryotic, autotrophic, and contain chlorophyll; they get their energy from the sun through photosynthesis, and so on), plants develop a self-defense mechanism to protect them from being destroyed by animals, fungi and other plants. It is considered that the vegetable kingdom consists of approximately 295,000–350,000 types of photoautotrophic terrestrial plants.

Plants interact with animals (invertebrates and vertebrates), fungi, microorganisms, and even with other plants using the "chemical language" communicating with the semiochemical substances (Figure 1.6) that these generate during their growth and development. Generally, plants need to protect themselves from various threats. Therefore, some plants use modified parts like thorns, spines, prickles, or trichomes as a defensive medium. This mechanical protection of plants is called physical defense, and it plays a vital role in the survival of mammals and small herbivores. Other plants develop non-digestible cellulose and lignin walls or sticky resin ducts that can help to survive. This tactic allows for reducing the capacity of the herbivores' digestibility. Although these strategies for defending themselves are usually safe, using chemical substances to help plants survive is more effective. This chapter will discuss a small part of the relationships that occur when terrestrial plants and animals use their toxic substances or toxins. It must be noted that what a toxin is in the biological context is problematic because, as mentioned before, what is a poison or toxin for a living organism could be beneficial for another (Sections 1.1–1.3).

In the case of plant-animal relationships (including arthropods), toxins take on the role of repellents, molecules with antifeedant or dissuasive effects. In fact, all plants must be repellent and toxic so as not to "be eaten" by animals. Biological studies have shown that plants respond to the attack of herbivores by activating complex defense mechanisms [1,2], which will not be discussed in this book.

As many plants and animals interact, numerous toxic molecules (i.e., toxic secondary metabolites) are involved in the chemical defense of plants against their predators. Based on their biosynthetic origin, these compounds are usually divided into three major classes: terpenoids, phenolics, and alkaloids, which constitute 55%, 18%, and 27% of the plant's secondary metabolites, respectively. This classification according to molecular structures is more common and helps to better understand the role of each group of molecules in these relationships. This chapter will discuss only the most important plant phenolic and terpenoid metabolites with oxygen atoms.

2.1 Classification of Phenolic and Terpenoid Metabolites

Phenolics and terpenoids have been considered classic defense compounds for protecting plants from herbivores ever since plant secondary metabolites were proposed to have evolved for that reason. In 1959, German entomologist Gottfried S. Fraenkel (1901–1984) first recognized that secondary plant metabolites are not simply the plant waste products but play an essential role in the plant defense against insects and herbivores, describing phenolic compounds as "trigger" substances, which induce or prevent the uptake of nutrients by animal herbivores. These compounds act as protective agents, inhibitors, natural animal toxicants, and pesticides against invading organisms, such as herbivores, nematodes, phytophagous insects, and fungal and bacterial pathogens (interactions of the plants with these latter "enemies" will be discussed in the next chapter). Regarding their chemical structure, phenolic compounds possess at least one aromatic ring with one or more hydroxyl groups (OH), including their functional derivatives.

According to the classification of Harborne and Simmonds (1964) [3], phenolic molecules are usually divided into numerous groups according to the number of carbons contained in their molecule. In plants, they are often present as glucosides. The most common groups are simple phenols, benzoquinones based on the C6 skeleton, phenolic acids (C6-C1 skeleton), cinnamic acids, coumarins, isocoumarins, and phenylpropanoids (all these are C6-C3 skeleton), and more complex derivatives, often containing several aromatic rings linked together or separated by some carbon atoms, like naphthoquinones (C6-C4), stilbenes and anthraquinones (C6-C2-C6), flavonoids (C6-C3-C6) or condensed tannins (C6-C3-C6)n, polyphenols (Table 2.1) [3,4].

The biosynthesis of these compounds occurs via either the shikimate/phenylpropanoid pathway that conduces directly to the phenylpropanoids or the polyketide acetate/malonate pathway, which completes the construction of more complex phenolics. The initial step of this biochemical process, the shikimate pathway, consists of the conversion of the amino acids phenylalanine and tyrosine, known as the phenylpropanoids, in the presence of the enzyme phenylalanine ammonia-lyase (PAL) (Figure 2.1) [5].

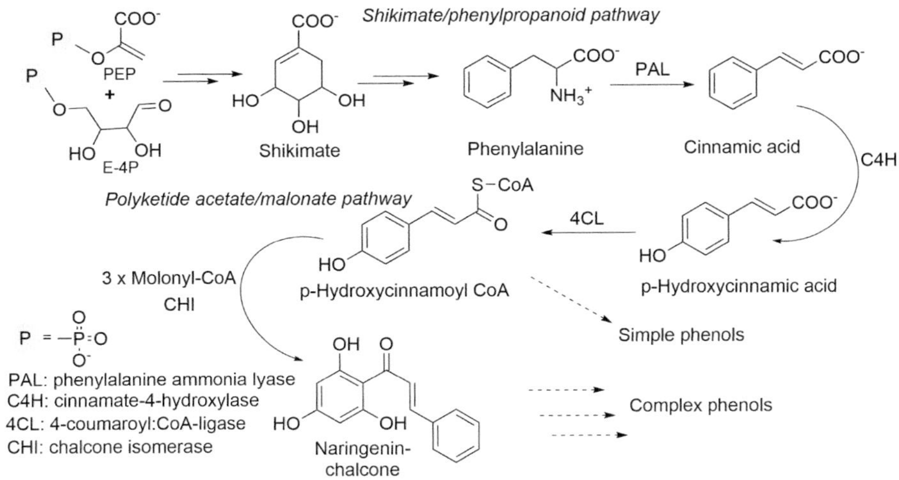

Figure 2.1. Simplified pathway of phenolic compound biosynthesis.

Although it is now generally accepted that plant phenolics play a role in protecting plants from insects, it should be noted that some cotton flavonoids are feeding stimulants for the boll weevil, *Anthonomus grandis*, or oviposition stimulants of a citrus-feeding swallowtail butterfly, *Papilio xuthus* L. Thus, the ecological relationship between plants and insects is a complex one with physical as well as chemical interactions. This relationship is also affected by plant, insect, and some insect-plant factors.

Terpenic molecules are a highly diverse group of carbon-based compounds derived from five-carbon isoprene (C5) units and are ubiquitously distributed [6–8]. They are also known as isoprenoids. Thus, isoprene (2-methylbuta-1,3-diene) is the "building block" of all terpenoids. Here, we must detail that isoprene does not react, but its active forms are isopentenyl pyrophosphate (IPP) and dimethylallyl pyrophosphate (DMAPP). Both are derived from the mevalonate (MVA) pathway or from the 2-C-methyl-D-erythritol 4-phosphate (MEP) pathway (Figure 2.2).

Thus, their initial condensation generates geranyl pyrophosphate (GPP), the precursor of monoterpenoids (two C5 units = C10 units). Condensation of GPP with one IPP gives farnesyl pyrophosphate (FPP), the precursor of sesquiterpenes (three C5 units = 15 units), and the addition of another molecule of IPP generates geranylgeranyl pyrophosphate (GGPP), the precursor of diterpenes (four C5 units = 20 units) and so on [5]. These condensations can occur via head-to-tail, head-to-head, or tail-to-tail joining of isoprene units. The understanding of these condensations has a rich historical context, with the "isoprene rule" first observed in the 1860s by German organic chemist Otto Wallach and later clearly formulated by Croatian-Swiss scientist Leopold Ruzicka. So, the first molecule's tail joins the second molecule's head to form a larger unit. It is interesting to note that a "head-to-tail" fashion term was suggested by Sir Christopher K. Ingold (1893–1970), a famous British chemist, adding to the legacy of the field.

Table 2.1. Classification of plant phenolics and their molecular diversity.

Group	Plant Phenolics Representative Examples and Characteristic Skeletons
C6	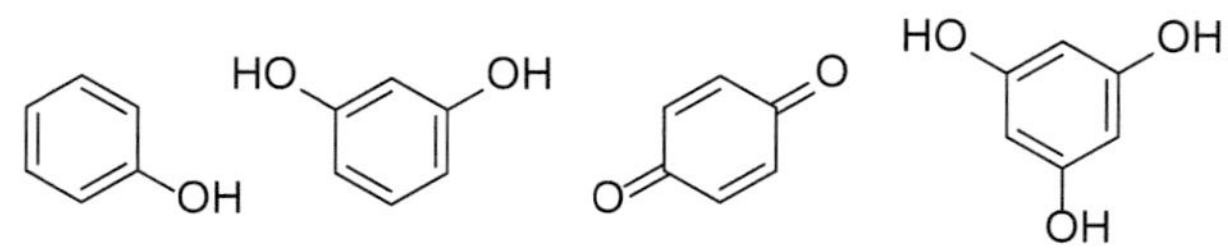 Phenol, Resorcinol, Benzoquinone, Phloroglucinol
C6-C1	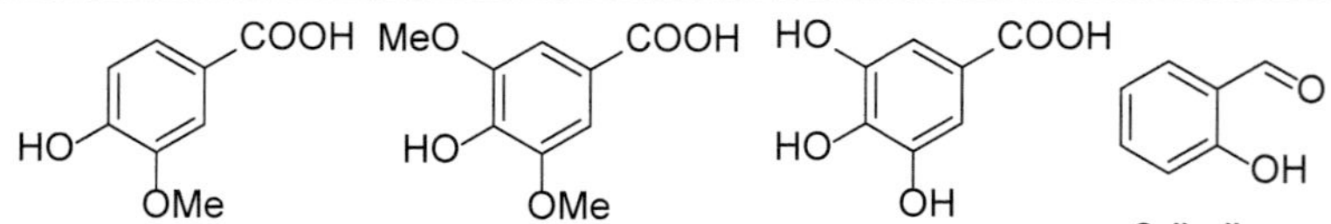 Vanillic acid, Syringic acid, Gallic acid, Salicylic aldehyde
C6-C3	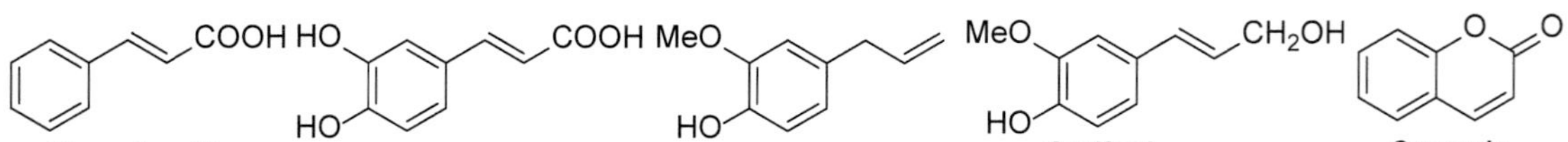 Cinnamic acid, Caffeic acid, Eugenol, Coniferyl alcohol, Coumarin
C6-C4	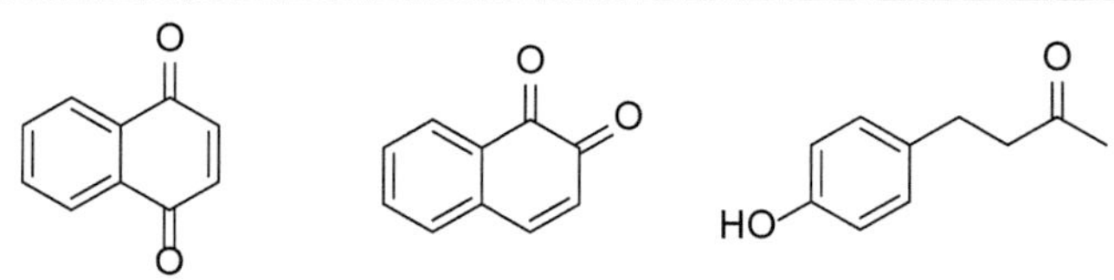 1,4-Naphthoquinone, 1,2-Naphthoquinone, Raspberry ketone

C6-C2-C6	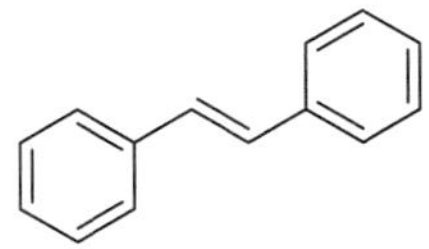 **Stilbene** 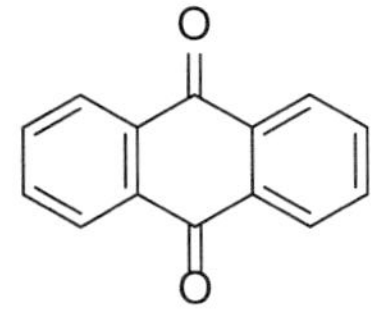 **Anthraquinone** 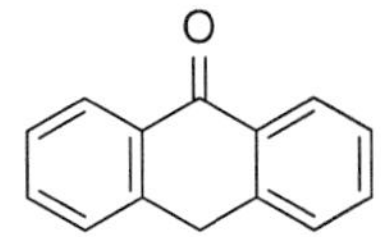 **Anthrone**
C6-C3-C6	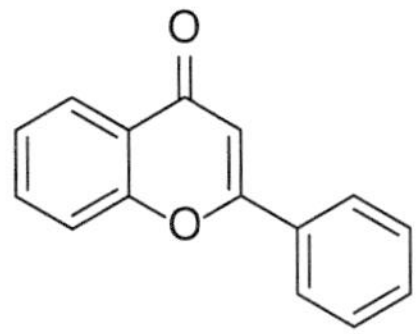 **Flavone** 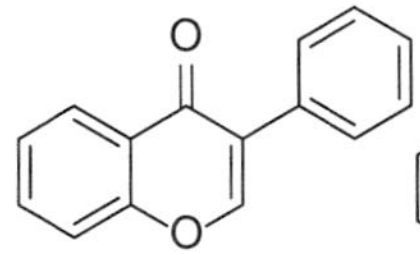 **Isoflavone** 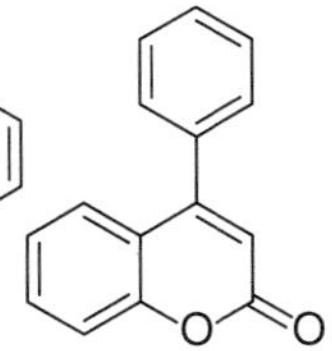 **Neoflavone** 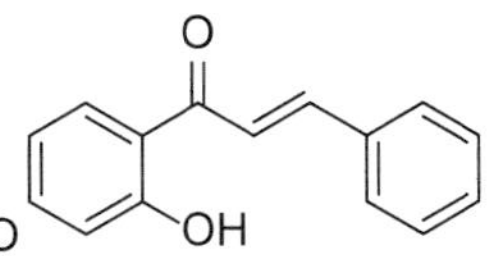 **Chalcone** 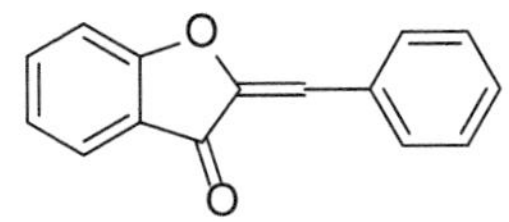 **Aurone**
(C6-C3-C6)n	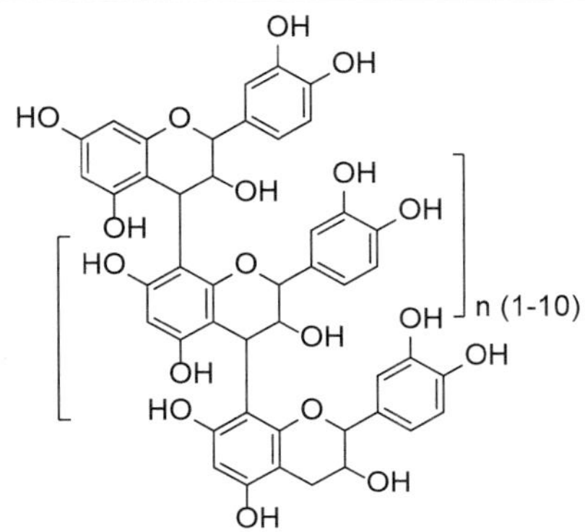

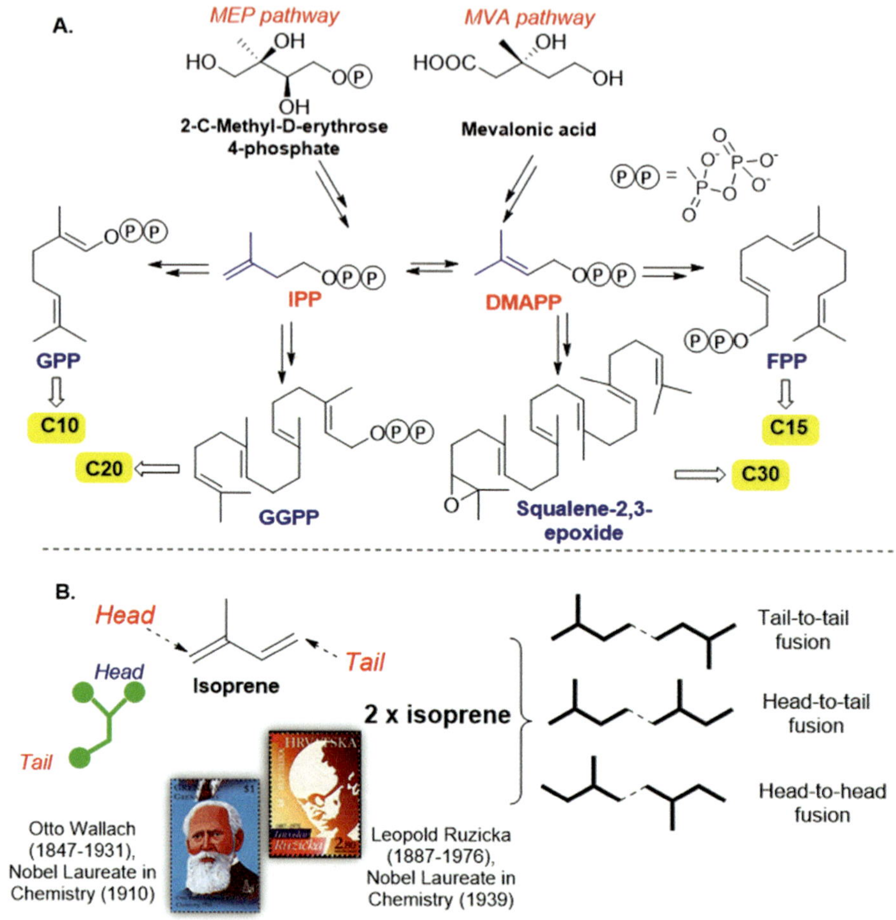

Figure 2.2. A. Simplified pathway of terpenoid synthesis. B. The most common fusions of isoprene units.

It is important to note that the MVA and MEP pathways, while independent, play crucial roles in the synthesis of different terpenoid types. The MEP pathway primarily contributes to the biosynthesis of mono- and diterpenes, while the MVA pathway is responsible for sesquiterpene production.

As shown in Table 2.2, the main terpenoid groups can be further classified into subclasses based on the number of rings in their structure. These include acyclic terpenoids (monoterpenoids) and cyclic terpenoids (bicyclic, tricyclic, and tetracyclic terpenoids). Most terpenoids are colorless, fragrant liquids lighter than water and volatile with steam, with a few exceptions, such as camphor. They are typically soluble in organic solvents and insoluble in water, and most are optically active [6].

Many terpene molecules are oxygenated molecules possessing one or more functional groups like -OH, C=O, or CH=O. Some of them have agreeable properties for use in food, cosmetics, pharmaceutical, and biotechnology industries, forming part of the constituents of essential oils. They are also helpful active ingredients as part of natural agricultural pesticides. Curiously, whereas terpenes and terpenoids

Table 2.2. Classification of plant terpenes and their molecular diversity.

No. of Isoprene Units	No. of Carbon Atoms	General Condensed Formula	Name
2	10	$C_{10}H_{16}$	Monoterpenoids
3	15	$C_{15}H_{24}$	Sesquiterpenoids
4	20	$C_{20}H_{32}$	Diterpenoids
5	25	$C_{25}H_{40}$	Sesterterpenoids
6	30	$C_{30}H_{48}$	Triterpenoids
8	40	$C_{40}H_{64}$	Tetraterpenoids
> 8	> 40	$(C_5H_8)_n$	Polyterpenoids

occur widely, their extraction from natural sources is often problematic. Therefore, they are produced by chemical synthesis, typically from petrochemicals.

Like plant phenolics, terpenoids contribute to direct and indirect defenses, which are crucial in plant-insect, plant-pathogen, and plant-plant interactions. Volatile terpenoids from flowering plants can serve as significant environmental cues to attract or deter pollinators. Mono- and sesquiterpenoids are the major components of essential oils released from plants after herbivore damage, attracting arthropods that prey on or parasitize herbivores to avoid further damage. In addition to volatile terpenoids, several di- and sesquiterpenoids are phytoalexins involved in the direct defense of plants against herbivores and microbial pathogens (Chapter 4).

The most significant characteristic that these molecules have in common is that they are more or less lipophilic. As a compound's physicochemical feature, lipophilicity (or hydrophobicity) depends on various physical and chemical characteristics, e.g., size, molecular surface area, molecular volume, and polarity. As a result of the lipophilic character of these compounds, interactions with hydrophobic parts of the cell play an important role in the mechanism of their toxic action [9].

2.2 Quinones: Cyclic α,β-Unsaturated Ketones

The quinones, products of the oxidation of the phenols, are relatively numerous groups of unsaturated cyclic compounds with two carbonyl groups (C=O) (Table 2.2) [10]. These phenolic metabolites are widely distributed in nature. So far, more than 1,200 quinones have been found in various plant families, such as Ranunculaceae, Aphodelaceae, Fabaceae, Ebenaceae, and Rhamnaceae. Besides mentioned above benzoquinones, naphthoquinones, and anthraquinones, other complex condensed polycyclic aromatic systems like phenanthrenequinones, dianthrones, and naphthodianthrones also belong to this group, often called a quinoid group (Figure 2.3).

Naphthoquinones and anthraquinones are natural pigments found in numerous higher plants, while simple methyl-, hydroxy-, or methoxy-benzoquinone derivatives are produced mainly by insects and fungi. The characteristic of the quinone pigments is their color, which ranges from yellow to orange and red to black.

Since quinones play a role in many physiological and toxicological processes, understanding their intrinsic chemical reactivity is relevant to understanding their

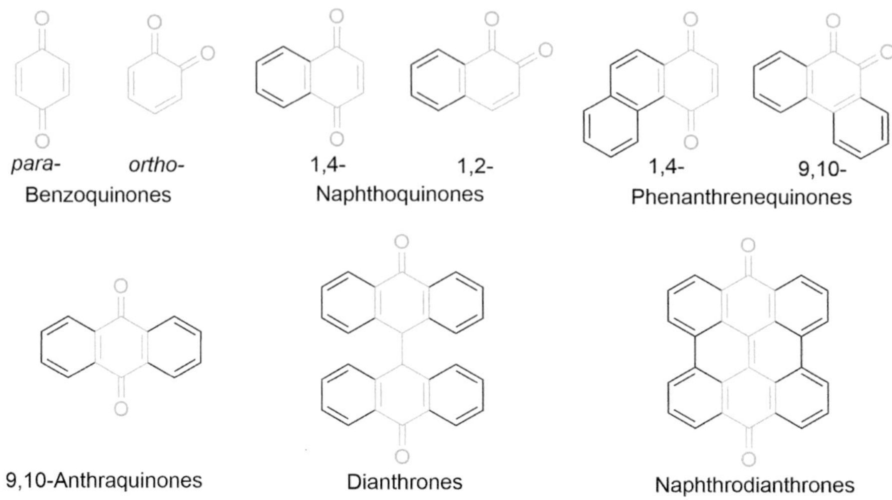

Figure 2.3. Structural classification of quinoid systems.

properties. Quinone molecules have two properties essential for understanding their biological effects:

1. Quinones can undergo reversible oxidation-reduction (redox) reactions;
2. Due to their electrophilic character, many can undergo nucleophilic attack [11].

Thus, an essential property of quinones is their easy conversion to hydroquinones in the presence of mild reducing agents, e.g., Fe, $SnCl_2$, $NaBH_4$, $Na_2S_2O_4$, or others. This process is reversible, meaning that hydroquinones can also be transformed into their respective quinones in the presence of mild oxidation agents such as hydrogen peroxide (H_2O_2), silver oxide (Ag_2O), Fremy's salt ($K_2NO(SO_3)_2$) or others oxidizing agents (Figure 2.4A). In the synthetic laboratory, 1,4-quinones are more easily prepared and are substantially more stable than 1,2-quinones. Milder oxidizing agents are required to prevent further oxidation of 1,2-quinones. According to their standard reduction potentials, some quinones like 2,3-dichloro-5,6-dicyano-1,4-benzoquinone (DDQ) and 2,3,5,6-tetrachloro-1,4-benzoquinone (chloranil) can be used as mild oxidizing agents. DDQ, for example, can oxidize some nonaromatic hydrocarbons to aromatic hydrocarbons or indolines to indoles, heterocyclic molecules.

In biological systems, this type of transformation is common in reductive and oxidative processes to transport one or two electrons from one substance to another, leading to the corresponding semiquinones or hydroquinones, respectively (Figure 2.4B). We recall from the introductory biochemistry course that these processes are catalyzed by a complex enzyme system in which phylloquinone (plants) and ubiquinone (animals) are involved and play pivotal roles in metabolisms. The enzyme ubiquinone-cytochrome *c* reductase catalyzes this complicated process. The enzyme is another gigantic protein complex with a molecular mass of about 500,000 Da. Three different cytochromes and a protein containing an iron-sulfur complex are components of the ubiquinone-cytochrome *c* reductase system. This enzyme belongs to the family of oxidoreductases. Other natural quinone products

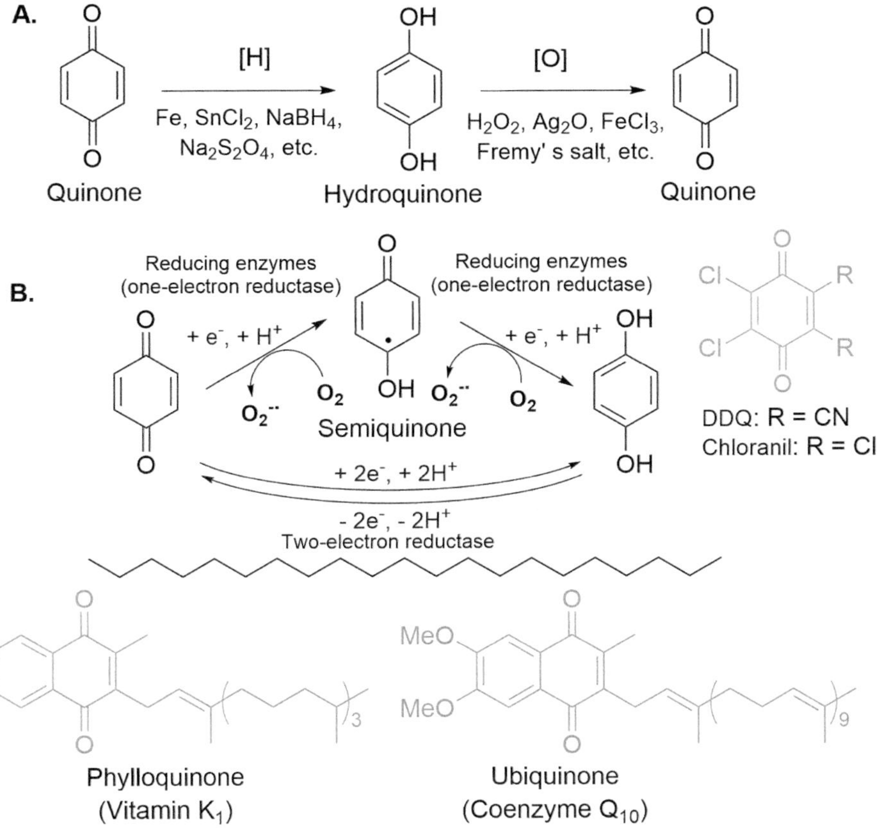

Figure 2.4. A. Oxidation and reduction reactions of quinones and hydroquinones. B. Reversible redox process of quinones.

also possess potent biological activity; for example, adriamycin (doxorubicin) is a front-line chemotherapy treatment.

Figure 2.4 explains that quinones can undergo reduction by cytochrome P450 reductase, leading to captodative-stabilized semiquinone radical anions that can be back-oxidized by molecular oxygen in normal, well-oxygenated cells. Quinones are also suitable substrates for two-electron reductases. Thus, quinones are fast redox-cycling molecules [11].

In addition, they have the potential to bind to substances containing thiol (SH), amine (NH or NH_2), and hydroxyl (OH) groups, which are electron-rich nucleophilic species. The electrophilic character of quinone molecules makes them undergo nucleophilic attack like the classical Michael reaction (Figure 2.5), which may lead to either detoxification or enhanced toxicity. As electrophilic molecules and oxidants tend to adhere to nucleophiles, quinones have different and varied biological properties against the same plants, animals, and microorganisms. Almost all plant quinones are toxic to animals.

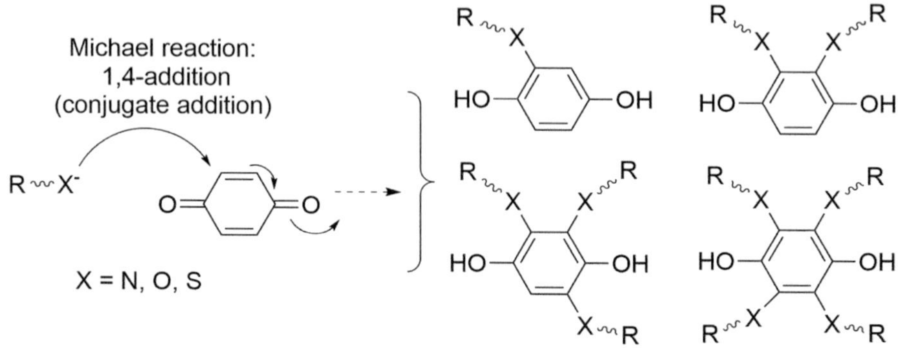

Figure 2.5. Nucleophilic addition reactions of quinone with mono-, di-, tri-, and tetra-substitution formation.

In a biological system, these addition reactions can occur with reactive side-groups of lysine, serine, and cysteine into proteins and the amine group of nucleotides of nucleic acids or with other systems like glutathione (GSH), which is the most abundant non-protein antioxidant present in the cell acting as the first line of cellular defense. Its thiol group is involved in the nucleophilic addition with quinones to give adducts conjugated to the sulfhydryl group of GSH. These processes occur before other functional groups, and such reductive addition represents a major route of quinone elimination. Quinone-GSH conjugation is a detoxification reaction because of the more hydrophilic character of the formed "Michael" adduct than the parent quinone.

The structures of some of the most studied quinone molecules are illustrated in Figure 2.6. In plants, these quinones are usually in the form of glycosides or free form. Primin (2-methoxy-6-*n*-pentyl-1,4-benzoquinone) was found in the leaves of the ornamental plant *Primula obconica* (Fam. Primulaceae), known popularly as primrose [12]. This plant, native to China, can be toxic to cats, dogs, or horses if ingested and may cause vomiting. Allergic reactions, skin irritant effects, and repellent properties are mainly due to the presence of this quinone.

In humans, as a sensitizer, primin is the source of allergic contact dermatitis, which is the so-called primrose dermatitis. Its sensitizing effect and other 1,4-benzoquinones (haptens) have been ascribed to the covalent binding of a receptor protein of epidermal cells to the C-3 and C-5 positions of the quinone ring system, which can be attacked by nucleophiles of the receptor protein, forming a protein hapten-complex (antigen) as shown in Figure 2.5. The 2,5-dihydroxy analogs of primin, embelin (2,5-dihydroxy-3-undecyl-1,4-benzoquinone) and rapanone (2,5-dihydroxy-3-tridecyl-1,4-benzoquinone), which were isolated from respective woody climbing shrubs *Embelia ribes* and *Rapanea pulchra*, exhibit insecticide and anthelmintic properties [13–15].

The astringent leaves and the berries of the species *E. ribes* that contain the embelin are a safe defense against herbivores. This orange 3-substituted *p*-benzoquinone pigment acts as a repellent against the larvae of beetles *Callosobruchus chinensis* L., *Rhyzopertha dominica*, rice weevils *Sitophilus oryzae*, rice flies *Corcyra cephalonica* and flour moths *Ephestia cautella*. All these insects are pests that significantly affect

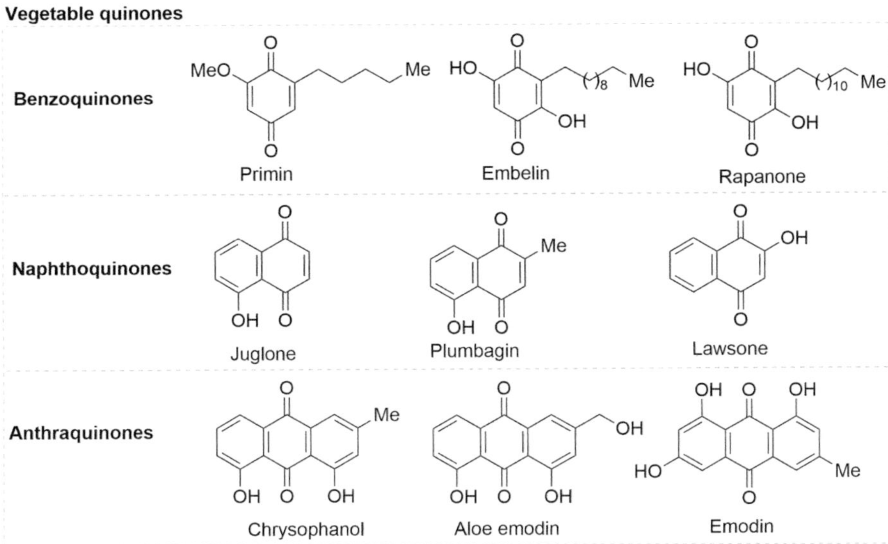

Figure 2.6. Important quinones isolated from higher plants.

the stored products. The extracts of berries of *E. ribes* also possess anti-alimentary activity against the caterpillars *Spodoptera litura*, which inhibits their development. Additionally, embelin has received much attention as a potential male anti-fertility agent, showing fertility-regulating activity in female mammals.

Using a Soxhlet extraction apparatus with light petroleum (bp = 60–80°C), 3.97 g of embelin was obtained in a 0.39% yield from 1 kg of *E. ribes* fruits (called false black piper). In contrast, subsequent extractions with ether, light petroleum, and methanol of the root-bark (1.065 kg) of *R. pulchra* gave 29.6 g (2.8%) of raponone as orange plates [16].

Interestingly, the first representative of simple quinones, *p*-benzoquinone, was described and mentioned by Russian chemist Aleksandr A. Voskresensky (1808–1890) in his doctoral work, “The Discourse on Quinic Acid and the Quinoil Opened in It” (1839), performed in the laboratory of Justus von Liebig (1803–1873). Voskresensky treated quinic acid with manganese dioxide in the presence of sulfuric acid (Figure 2.7).

This simple reaction involves three consecutive chemical processes: dehydration, decarboxylation, and oxidation. MnO_2/H_2SO_4 mixture can also convert aniline through a free-radical condensation mechanism to yield the same product. It is noteworthy that quinic acid is a monobasic polyhydroxy carboxylic acid of composition $C_7H_{12}O_6$, which Hoffmann first isolated in 1790 from the cinchona bark of *Cinchona* L. and used as an anti-malarial medicine. Quinic acid often accumulates in plants in significant quantities. In particular, isolating 13.4% of quinic acid from the young spruce shoots (regarding the weight of dry raw material) was possible.

However, synthetic routes for quinoid cores involve hydroquinone oxidation with the oxidizing agents mentioned above (Figure 2.4A). In light of the interesting pharmacological (antibacterial, anti-protozoal, and antitumor) properties of these simple *p*-benzoquinones and their potential therapeutic applications, easy access to

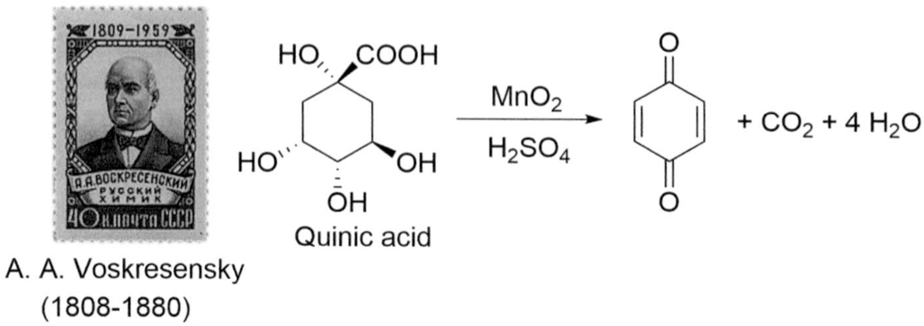

Figure 2.7. Discovery of *para*-benzoquinone by Voskresensky, who named it "quinoil."

them and their derivatives by synthetic chemistry is an important task for organic chemists. The synthesis of primin is briefly discussed below.

Planning a synthesis of primin as a target molecule (TM) can be started with its retrosynthesis. Although the direct introduction of an alkyl side chain of 2-methoxy-benzoquinone is known (a disconnection of C6-C1' bond with conservation of quinone ring, a-type), almost all the syntheses of this molecule are based on *n*-pentyl substituted methoxy-benzene derivatives precursors P_1 which resulted from retrosynthetic analysis using cycle transform (b-type). When needed, *n*-pentyl-methoxybenzene is prepared, and its subsequent oxidation reaction to the *p*-benzoquinone ring completes the construction of the primin skeleton. Thus, further analysis of precursor P_1 structure can give us two possible types of disconnection by breaking strategic bonds: C_6-C_1 (c-type) and $C_{1'}$-C_2 (d-type). The first disconnection involves the use of diverse precursors P_2, while the second implicates the use of precursor P_3. Analyzing benzoquinone chemistry can predict that suitable starting materials for these precursors could be cheap, commercially available reagents such as 4-hydroxy-3-iodo-5-methoxybenzaldehyde (5-iodovanillin) R-1, 1-bromo-3,5-dimethoxy-benzene R-2, 1,3-dimethoxybenzene R-3, 2-methoxyphenol (guaiacol) R-4 (all three reagents for P_2), and 2-hydroxy-3-methoxybenzaldehyde (*o*-vanillin) R-5 (for P_3) (Figure 2.8). Now, we can say that we have the synthetic tree for primin. However, possible alkyl chain precursors will be selected. Diverse pentyl chain precursors like 1-bromopentane ($C_5H_{11}Br$), pentyl-magnesium bromide ($C_5H_{11}MgBr$), hexanoic acid ($C_5H_{11}COOH$), or pentyl boronic acid ($C_5H_{11}B(OH)_2$) and even organoborane prepared *in situ* by hydroboration of 1-pentene with 9-borabicyclo[3.3.1]nonane (9-BBN) have been used in the syntheses of primin. However, primin syntheses with these reagents are long, multi-step procedures and suffer from poor yields and/or expensive starting materials [17–20].

Here, we want to pay your attention to an original and interesting four-step strategy for the synthesis of primin based on combined Mitsunobu reaction-Claisen rearrangement, catalytic transfer hydrogenation, and phenol oxidation starting from guaiacol R-4 and 1-penten-3-ol [21]. The first step of this synthesis is the microwave (MW)-assisted Mitsunobu reaction of guaiacol and 1-penten-3-ol in the presence of triphenylphosphine (PPh_3) and di-isopropyl azodicarboxylate (DIAD) that conduces to the formation of allyl ether, which converts into the Claisen rearrangement product without isolation under such reaction conditions (Figure 2.9).

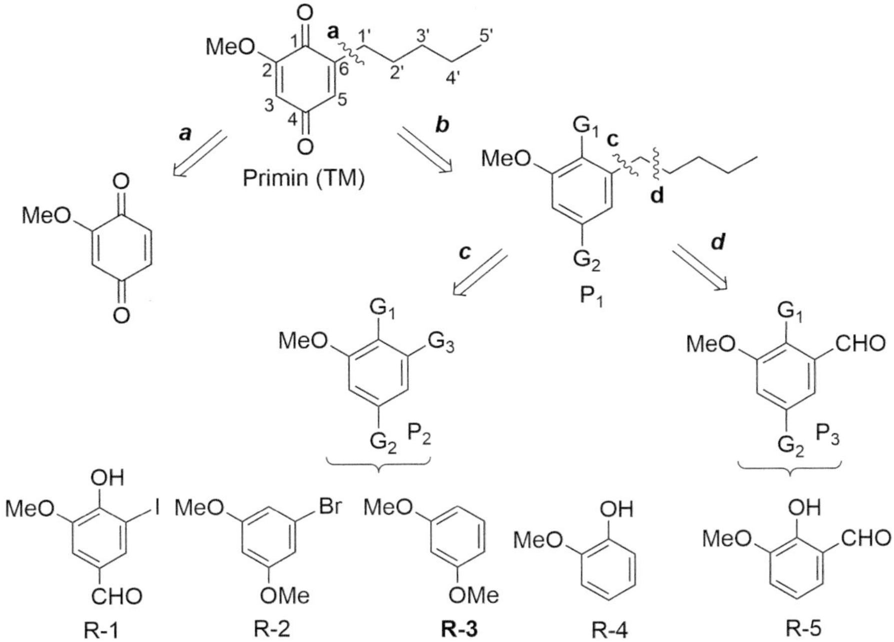

Figure 2.8. Retrosynthesis of primin: suitable disconnection types.

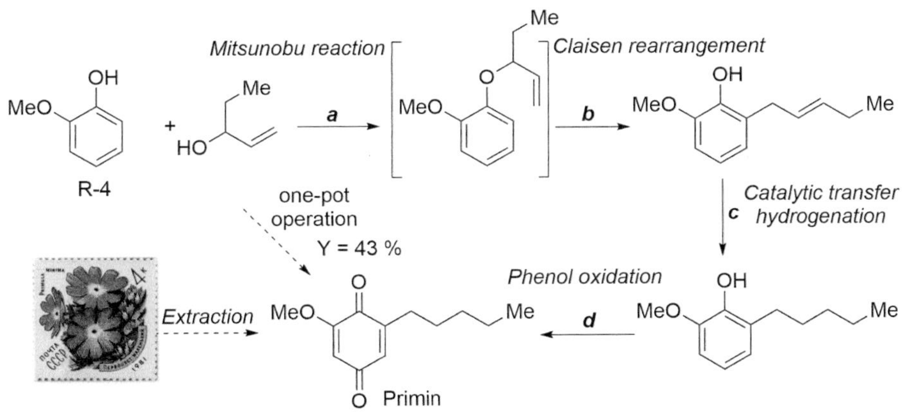

Figure 2.9. Microwave-assisted synthesis of primin: (a, b) PPh_3/DIAD, toluene, 300 W, 220°C, 30 minutes; (c) $HCOONH_4$, Pd/C, 10 W, 10 minutes; (d) Fremy's salt, NaH_2PO_4/aq. acetone, 10 W, 20 minutes.

Then, treatment of this phenolic product with ammonium formate ($HCOONH_4$) and palladium on carbon allows the production of 2-methoxy-6-pentylphenol, which is easily oxidized (Fremy's salt) to give primin with a total reaction time of 1 hour in overall yields of 43%.

Interestingly, the reduced form of primin, 2-methoxy-6-pentyl-1,4-dihydroxybenzene, is also a secondary metabolite known as miconidin, found in various plants, predominantly in the *Miconia* genus. Both primin and miconidin are antifeedant agents against various insects, forming part of a potential antifeedant complex in *Primula* and *Miconia* species. Primin is particularly active against *Pieris brassicae* larvae, while miconidin was found to be effective against *Heliothis armigera* larvae, which are considered a pest in many European and Asian countries [22].

Juglone (5-hydroxy-1,4-naphthoquinone), plumbagin (5-hydroxy-2-methyl-1,4-naphtho-quinone), and lawsone (2-hydroxy-1,4-naphthoquinone) (Figure 2.6), which were isolated for the first time from their respective sources *Juglans nigra*, *Plumbago scandens* L., and *Lawsonia inermis* L., are typical members of numerous vegetable naphthoquinones [23].

Among them, juglone, the more studied natural yellow-brown toxin (LD_{50} = 0.25 mg/100 g bodyweight in mice) from black walnut (*J. nigra*) and English walnut (*J. regia*), presents herbicidal, insecticidal, fungicidal, anticancer and antibiotic activities, among others [24]. Walnut extracts are widely used in folk medicine for treating arthritis, stomach aches, skin disorders, and infectious diseases. Despite its potentially hazardous effects, it has been used as a natural dye and colorant in cosmetics and foods.

Being a powerful herbicide, this naphthoquinone, which occurs naturally in all parts of the black walnut, retards the growth of nearby competing plants. Plants adversely affected by being grown near black walnut trees exhibit symptoms such as foliar yellowing, wilting, and eventual death. All these symptoms are commonly referred to as "walnut wilt". During allelopathic studies done on various vegetables, field crops, fruit trees, ornamental species, and medicinal plants, it was found that diverse plants such as potato, tomato, alfalfa, asparagus, cabbage, magnolia, and blueberries, and various flower species, including alliums, crocus, daffodils, hyacinth, and tulips are all negatively affected by the compound. In contrast, parsnip, lima beans, snap beans, onion, raspberry, and corn are unaffected. Juglone has been shown to inhibit a broad spectrum of microorganisms, including bacteria, algae, and fungi. Interestingly, in the walnut tree, juglone is stored in a non-toxic form, the 5-glucoside of 1,4,5-trihydroxynaphthalene, which is highly labile and easily hydrolyzed to glucose and hydroxynaphthalene derivative.

Juglone-insects' interaction can be quite different. The LD_{50} of juglone is 2.0 mg/g for the caterpillars *Spodoptera frugiperda* and *Lymantria dispar*. Therefore, these insects avoid feeding on Juglandaceae plants. However, some insect species, such as the moths *Actias luna* and *Cydia pomonella*, include these plants in their diet. Why? It is found that some herbivore insects that naturally consume juglone and other pro-oxidant allelochemicals produce special enzymes to prevent the generation of free-radical oxygen during quinone reduction. *Actias luna* and *Cydia pomonella* are capable of detoxification of juglone (and related naphthoquinones) to non-toxic 1,4,5-trihydroxynaphthalene.

Juglone was synthesized and characterized for the first time in 1887. Despite its old history, the biosynthesis of juglone has not been fully established. Still, the shikimate pathway has been accepted as a likely candidate since two essential precursors, *o*-succinyl benzoate (OSB) and 1,4-dihydroxy-2-naphthoate (DHNA), were confirmed during typical experiments with young *J. regia* plant (4-7-month old) using [^{14}C]-radiolabeled reagents (Figure 2.10). This biochemical reaction

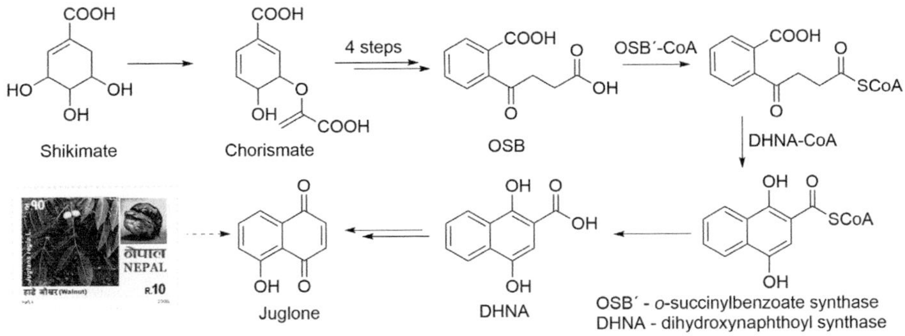

Figure 2.10. OSB pathway for the juglone formation starting from shikimic acid.

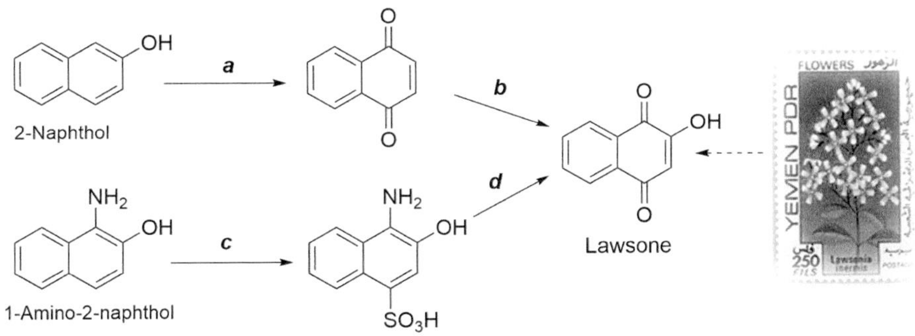

Figure 2.11. Synthesis of lawsone: (a) $FeCl_2$/conc. HCl; (b) 1. conc. H_2SO_4/Ac_2O; 2. KOH/air; (c) $NaHSO_3$, (d) 1. H_2SO_4/MeOH, 2. aq. KOH

sequence is known as the OSB pathway [23, 25]. It is possible that DHNA is also a precursor for lawsone.

Juglone is most commonly synthesized in the laboratory from commercially available 1,5-dihydroxy naphthalene by oxidation using different oxidants. One of the best reagents for this oxidant reaction resulted in being a simple O_2/CuCl system, and the *bis*-(trifluoroacetoxy)iodoperfluorohexane, $C_6F_{13}I(OCOCF_3)_2$ allowed to obtain juglone in 80% and 91% yields, respectively [26, 27]. Its 2-OH analog, lawsone, is easily obtained from natural sources and commercially available in bulk from Yucca Enterprises (India) and Sigma-Aldrich. 2-Naphthol and 1-amino-2-naphthol are usually used as starting materials in the lawsone synthesis (Figure 2.11).

Lawsone, a red-orange pigment, is a remarkable molecule presenting two isomeric forms, better known as 2-hydroxy-1,4-naphthoquinone and 5-hydroxy-1,2-naphthoquinone. Due to this structural feature, lawsone shows three different oxidation states: quinone, semiquinone (the one-electron reduced form of quinone), and catechol (the two-electron reduced form) (Figure 2.12) [28]. Their binding ability with diverse metals such as Fe and Cu is, in part, believed to be responsible for the significant role of quinones in biological systems.

The lawsone-containing species (*Lawsonia inermis*) is popularly known as henna, an extract of dried powdered leaves used as red dye coloring hair and dying nails and skin. What does it reduce the toxicity in the lawsone molecule compared to juglone? Both molecules are based on the 1,4-naphthoquinone skeleton; both

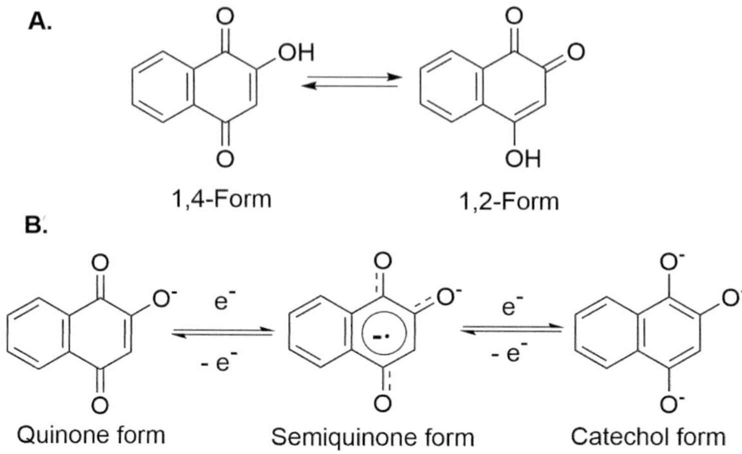

Figure 2.12. A. Isomeric forms of lawsone. B. Redox forms of the lawsone anion.

have a hydroxyl group. There is only one difference between them: the location of the hydroxyl substituent in the ring. The substituent at C-2 has to allow both access to and sufficient electrophilicity of C-3 (so-called Michael addition, Figure 2.5). As a result, lawsone is a comparatively weakly alkylating agent. In contrast, juglone interacts readily with nucleophiles, such as glutathione, causing significant GSH modification. It was also found that the order of toxicity of naphthoquinones against some insects and microorganisms parallels the order of increasing redox potential in these compounds. The alkylation and redox-cycling abilities of these naphthoquinones and other similar appear responsible for most of their cytotoxic effects and other biological properties [29].

In the case of juglone or other 1,4-naphthoquinones and mammalian cells, these quinones could be reduced at the cost of NADH or NADPH, as catalyzed by some alternate enzymes. For example, cytochrome P450 reductase would catalyze its simple reduction to the corresponding naphthoquinoles and semiquinone radical ions. Under aerobic conditions, both molecules formed can auto-oxidize, forming the superoxide radical anion, O_2^{-}. This radical is subsequently disproportioned in O_2 and H_2O_2, promoting the oxidation of lipids, proteins, and DNA. Semiquinone radical ions, active oxygen radical species, and hydrogen peroxide (ROS) are toxic species. They can interfere with the normal functionality of animal cells, eventually altering basic cellular processes such as the biosynthesis of nucleic acid and proteins (Figure 2.13A).

The redox process of quinones is one of the two main mechanisms of the cytotoxic effect of quinones in cellular systems. The high toxicity of juglone compared to other naphthoquinones has also been suggested to correlate with its high redox potential of –93 mV. The second mechanism is related to the alkylating capacity of quinones and radical ions of semiquinone with thiol groups in proteins, particularly with glutathione (GSH) or cysteine residues, where there is the covalent modification of biological molecules at their nucleophilic sites in which quinones act as electrophiles (Figure 2.13B). GSH, a tripeptide synthesized from cysteine with glutamic acid and

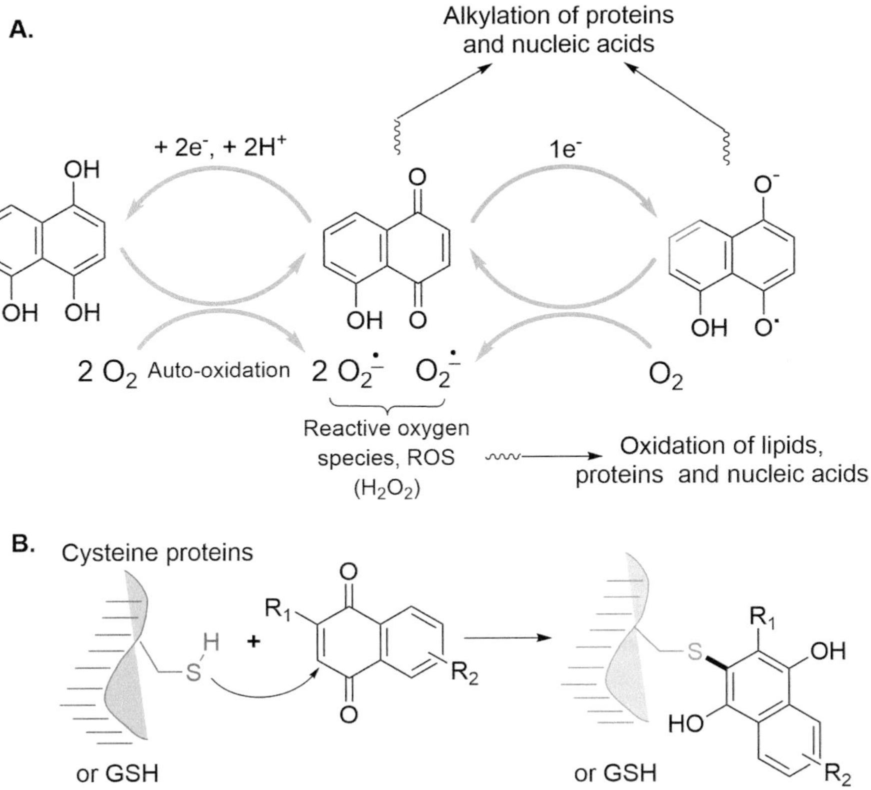

Figure 2.13. A. Alkylation and redox-cycling abilities are the main driving forces for the biological effects of naphtoquinones; B. Alkylation of quinones with cysteine proteins or GSH.

glycine, is a crucial factor in detoxification in herbivores. Its exhaustion caused by these reactions is associated with toxicity.

Anthraquinones, a small group of quinone metabolites structurally related to 9,10-dioxoanthracene (also known as anthracene 9,10-diones), also have numerous biological activities, some of which exhibit a broad spectrum of ecological impacts by mediating biotic or abiotic interactions of plants with their environment. One of the most studied vegetal anthraquinones is the emodin (1,3,8-trihydroxy-6-methylanthraquinone; Figure 2.6) that occurs in varied life forms, including lianas, herbs, shrubs, and trees. Emodin (as well as chrysophanol and aloe-emodin) forms the basis of a range of purgative anthraquinone derivatives and, from ancient times, has been widely used as a laxative compound. Recent studies indicate that emodin exhibits numerous other biological activities, such as anti-inflammatory, antibiotic, and antioxidant activities. However, when discussing the relationships between emodin-containing plants and insects, it is important to highlight its effective feeding deterrent properties. Emodin's potent feeding deterrent property to phytophagous insects was suggested to explain the little insect attack on shrub *Rhamnus alnifolia,* in which emodin concentration is notable. Even in relatively low concentrations of emodin, gypsy moth (*Lymantria dispar*) larvae reduced feeding and prolonged development, and with high concentrations, pronounced mortality occurred in 2–3 days. However, the mode of action of emodin as an insect antifeedant is yet to be studied [30]. It was also found that at certain levels, emodin is toxic and deters vertebrates like house sparrows (*Passer domesticus*), yellow-vented bulbuls (*Pycnonotus xanthopygos*), redwing blackbirds (*Agelaius phoeniceusin*) and starlings (*Sternus vulgaris*). Interestingly, emodin and emodin-containing *Cassia tora* extracts also have a photoprotective function, and their antioxidant activity probably depends on scavenging hydroxyl radicals.

Other members of the quinoid group, hypericins, are fluorescent pigments of the perennial *Hypericum perforatum*, known as Saint John's wort. These pigments based on naphthoadiantrone rings are found in the semisolid state in the black spots scattered on the surface of the leaves, stems, and floral petals of the plant. Hypericin is responsible for the red color of hypericum oil and the blackish coloration of the sprigs of leaves and petals; its content is 0.5–0.7%.

According to biosynthesis studies, it is speculated that these molecules originated from the acetate-malonate metabolic pathway. In this route, a unit of acetyl CoA (Section 1.4) is condensed with seven molecules of malonyl-CoA to form an octaketide chain that subsequently undergoes cyclization and decarboxylation that results in the formation of emodin-anthrone, which is oxidized to the emodin, probably by the emodin-anthrone-oxygenase enzyme. Then, diantrone is produced through a condensation reaction, whose successive oxidations yield protohypericin

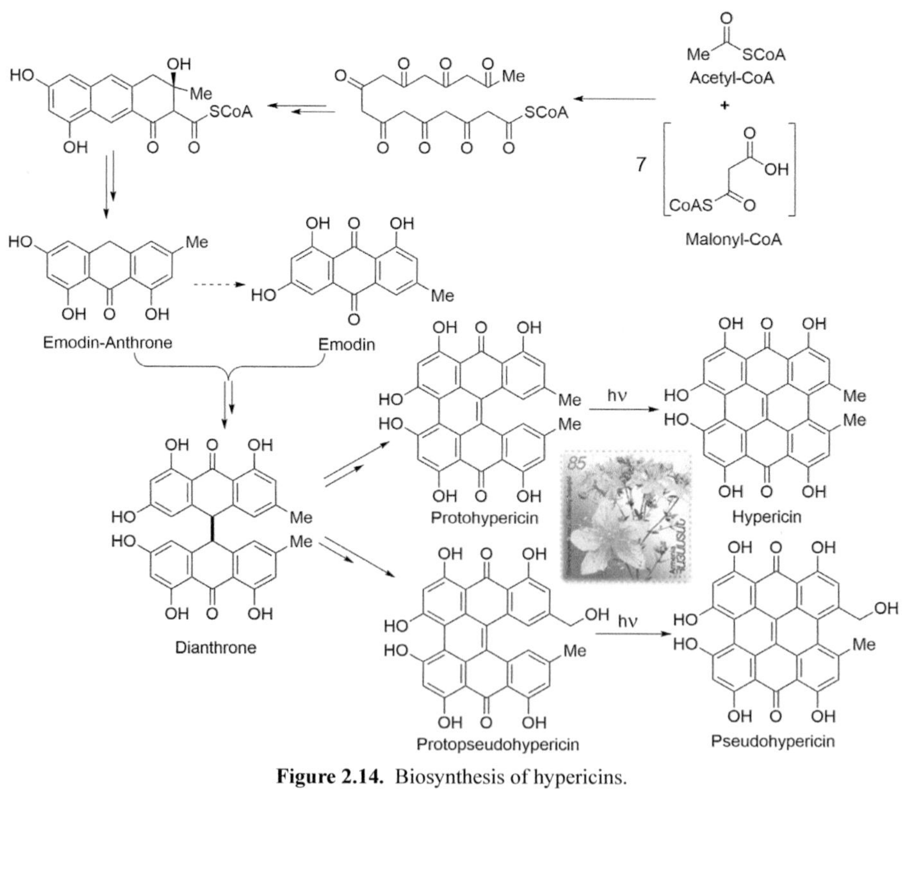

Figure 2.14. Biosynthesis of hypericins.

and protopseudohypericin, which, under conditions of irradiation with visible light, generate hypericin and pseudohypericin (Figure 2.14).

Hypericin is a reactive lipophilic molecule with a molecular weight of 504 Da. It is poorly soluble in non-polar organic solvents, aqueous solutions, and oils. It forms a red solution in most organic solvents, a green solution in basic media, and violet dispersions of fine particles in water. The hypericin molecule has two bands of UV absorption at 591 nm and 510 nm.

Hypericins, being unsaturated polycyclic molecules, are responsible for the side effect of this plant in mammals, mainly grazing animals (equines, sheep, and cattle). This effect is called phototoxic activity. Thus, hypericin and its analogs are phototoxic agents known as photosensitizers, molecules that increase sensitivity to sunlight, which is the main source of ultraviolet (UV) rays. When an animal consumes hypericin-containing plants and is exposed to the sun, it can experience skin rashes and sensitivity to light (photosensitization). Light-colored animals are particularly susceptible. These molecules are phototoxic only when irradiated with visible light (photons) and can generate toxic reactive oxygen species. The photosensitization that results from eating (or absorbing through the skin) a photosensitizing substance like hypericin, which is not modified in the body, is known as primary photosensitization.

Hypericin is believed to also act as an antibiotic and non-selective inhibitor of kinases. Hypericin can also inhibit the action of the enzyme dopamine β-hydroxylase, raising dopamine levels, although it may decrease the concentration of adrenaline and noradrenaline.

Chemically, the absorbed photon excites a suitable molecule like a hypericin molecule (HM) to one energy-rich state, called a singlet excited state. The excited molecule, $^{1}HM^{*}$, undergoes an internal process called intersystem crossing that ultimately results in the chemical alteration of the substrate $^{3}HM^{*}$ (triplet excited state), from which it can return to the ground-state HM_0 with the release of extra energy as photonic radiation (phosphorescence) or transfer it to a suitable acceptor molecule [31]. Here, there are two cases: when the acceptor is another organic molecule, some ROS like hydroperoxy radical $HO\text{-}O^{\cdot}$, hydroxyl radical $HO^{\cdot}$, and hydrogen peroxide could be generated (type I photosensitization reaction). On the other hand, when the energy of the excited triplet state $^{3}HM^{*}$ is transferred to ground-state molecular oxygen ($^{3}O_2$), excited state singlet oxygen ($^{1}O_2$) that is more cytotoxic, is producing (type II photosensitization reaction) (Figure 2.15).

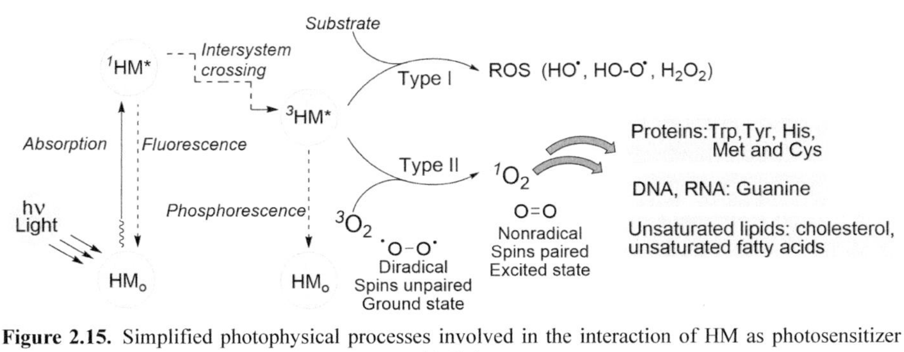

Figure 2.15. Simplified photophysical processes involved in the interaction of HM as photosensitizer with light.

Singlet oxygen then readily reacts with many biological substrates, including specific amino acids in proteins, the guanine bases of DNA and RNA, and various unsaturated lipids, such as cholesterol and unsaturated fatty acids.

The species *H. perforatum*, the source of the hypericins, is less damaged by herbivorous insects. The phototoxic effects of photosensitizers such as hypericin in non-adapted insect species can be dramatic. The dietary administration of hypericin at 50 pg/g concentration to the larvae of *Manduca sexta* and their exposure to photosensitizing wavelengths (i.e., sunlight) can quickly cause black lesions in the insect cuticle. These treatments can lead to acute mortality. Minor doses for sublethal effects can considerably reduce growth feeding and pupal deformity. Remember that *M. sexta* larvae, commonly known as the tobacco hornworm, are a significant agricultural pest, feeding on solanaceous plants. On the other hand, there are specialist insects that feed on this plant. The *Chrysolina quadrigemina* beetle is one of these insects. How do these beetles support the phototoxicity of the plant? They have developed an adaptive and functional behavior against phototoxicity by simply avoiding sunlight after feeding. This behavior involves digging into the ground during the day and feeding only a few hours before sunrise. It is as if they knew that near-UV wavelengths are virtually ineffective in exciting HM. Due to their high selectivity, these beetles are used as a biological control to regulate the growth and development of the Saint John's wort, considered in some countries as weeds.

However, although phototoxicity was reported for several animal species, this side effect is rare in humans [32]. It is noteworthy that hypericin is being studied as a photosensitizer for photodynamic therapy (PDT) to determine its effectiveness as an anti-tumor and antileishmanial agent. In PDT treatment, some natural photosensitizers generate ROS that targets cancer cells. Natural photosensitizers (i.e., color molecules and dyes) are also exploited to search for new dye-sensitized solar cells.

Another group of phenolic molecules, oxygenated tricyclic molecules called furocoumarins, are also phototoxic compounds. Are their action mechanisms different from those of hypericins? We will see when we read the next section.

2.3 Furocoumarins: Phototoxic Heterocycles

Furocoumarins or furanocoumarins are aromatic heterocyclic compounds derived from the condensation of a furan ring with the coumarin ring. Although this fusion can result in several linear and angular forms (in total, 12 forms), the two forms of ring fusion are generally found in nature, namely, 7*H*-furo[3,2-*g*]chromen-7-ones and 2*H*-furo[2,3-*h*]chromen-7-ones. They are regioisomers, which differ by the position of a furan group grafted on a coumarin core molecule at positions C6–C7 for linear molecules or C7-C8 for the angular one. Their simple derivatives without any substitution (parent compounds) are called psoralen and angelicin (Figure 2.16) [33]. The latter compound is a constituent of the roots and leaves of the medicinal plant called Angelica (*Angelica archangelica*). At the same time, the 7*H*-furo[3,2-*g*] chromen-7-one was isolated for the first time from the plant *Psoralea coryfolia,* and that is why it is known as psoralen. It is a metabolite of the ornamental herb *Ruta graveolens* known as rue. Its methoxy analogs, 5-methoxypsoralen (bergapten or 5-MOP) and 8-methoxypsoralen (8-MOP, methoxsalen or xanthotoxin), were

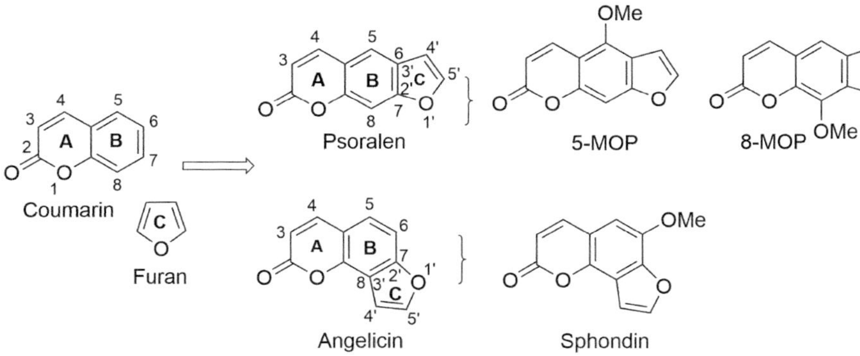

Figure 2.16. Furocoumarins of biological importance.

isolated from the bergamot essential oil of citrus tree bergamot (*Citrus aurantium* var. bergamia or *C. bergamia*) and herb *R. graveolens*, respectively. Methoxy-analog of angelicin, sfondin has 6-methoxy-2*H*-furo[2,3-*h*]chromen-2-one structure. These furocoumarins are also found in diverse plants of families Rutaceae, Moraceae, Fabaceae Apiacea, and Hipericaceas, among others. Probably, the most abundant source of psoralens is *Ficus carica*, a flowering plant native to the Middle East.

These phenolic compounds are known for their photosensitizing effect and phototoxic properties [34, 35], like hypericin. When animals consume furocoumarin-containing plants, these heterocyclic molecules are distributed throughout the body through the blood, in which they are transported in different ways (dissolved in plasma or attached to proteins), reaching the epidermis and dermis of the skin. Exposure to the sun makes the animal more susceptible to sunburn and other damage. In this way, furocoumarins can cause different types of inflammatory skin eruptions in animals and humans. Affected animals often show the following symptoms: depression, salivation, and anorexia, and can usually present with photophobia and seek shade. This phenomenon is observed in mammals and is called phytophotodermatitis. Other plants producing phytophotodermatitis are parsley (*Petroselinum crispum*), chervil (*Anthriscus cerefolium*), dill (*Anethum graveolens*), fennel (*Foeniculum vulgare*), parsnip (*Pastinaca sativa*), celery (*Apium graveolens*), false bishop's weed (*Ammi majus*), among others. The eruption usually begins approximately 24 hours after exposure and peaks at 48–72 hours.

Furocoumarin molecules are consumed mainly in the epidermis and mucous membranes. Once concentrated in the animal's epidermis, these compounds or other photodynamic agents are activated by sunlight (mainly, ultraviolet A irradiation, UVA that can penetrate deeper into the dermis layer of the skin), causing structural changes in the cellular components, which stimulate an inflammatory response in the cutaneous tissue. It is generally accepted that more active, toxic linear furocoumarins (i.e., psoralens) cause cell damage by covalent binding to DNA following UVA irradiation; they exhibit a planar tricyclic structure with two photoreactive sites (3,4-pyrone and 4',5'-furan double bonds). The initial intercalation (i.e., insertion) and interaction with double-stranded DNA are not characterized by covalent bonds ("dark complexes"), but upon absorption of a photon of UVA, a pyrimidine residue

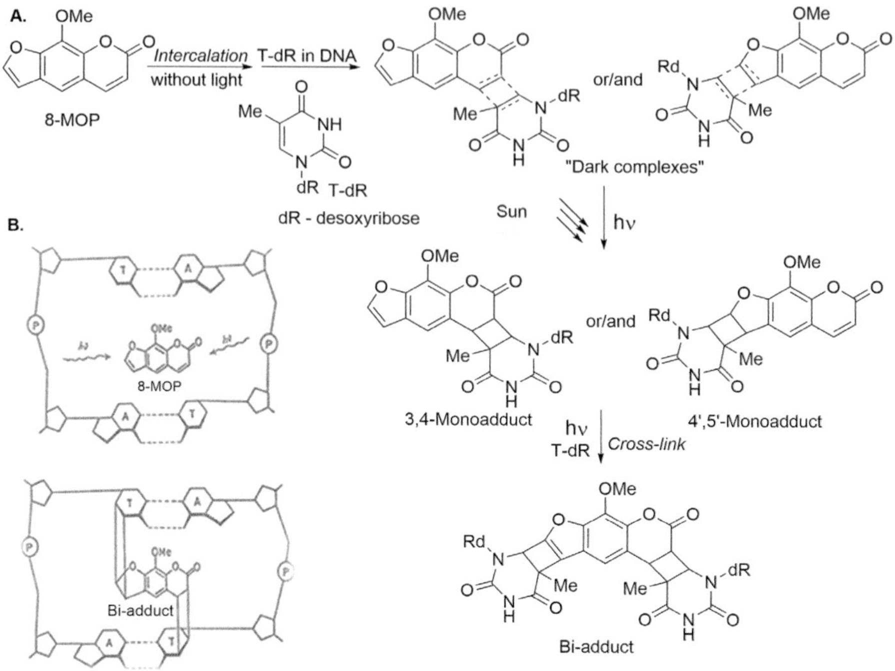

Figure 2.17. A. Intercalation process of furocoumarins (8-MOP) and formation of mono- and bi-adducts. B. Crosslinking process of 8-MOP as an example.

(preferentially a thymine, T) of the DNA covalently binds to the first photoreactive site with a 5,6-double bond or a 4',5'-double bond giving the mono-adducts (Figure 2.17A).

The resulting mono-adducts can form a di-adduct by absorbing a second photon if a new pyrimidine base on the opposite strand of DNA is available for an interstrand cross-link. These adducts result in local distortions to DNA at the site of intercalation. Thus, the planar structure of these furocoumarins helps them to intercalate between nucleic acid base pairs. UV irradiation activates the intercalated complex, forming photo mono-adducts with T-dR bases in cellular DNA. These adducts can react photochemically with another pyrimidine base on the complementary strand of the DNA via the crosslinking process (Figure 2.17B), which is believed to be the primary cause of photoinduced cell killing. This process depends on the structure of the furocoumarin derivatives. Linear-fused molecules give crosslinks efficiently, yielding up to 50% of the overall adducts formed. At the same time, the angular angelicin-like isomers exhibit low toxicity and low DNA-binding activity. The latter compounds only generate mono-adducts with pyrimidine bases [34].

Chemically, we can summarize this complex process in the following sentence: the photoexcited furocoumarins react with biomolecules, especially with pyrimidine bases in cellular DNA, through [2+2] cycloaddition processes to form mono- and bi-adducts.

Like hypericins, activated furocoumarins (mainly psoralens) also interact with molecular oxygen, forming reactive oxygen species. These reactive molecules and free radicals result in cell membrane damage by lipid peroxidation and disruption of mitochondria.

The photodynamic properties of furocoumarins are of great interest in biological sciences. In recent years, research has focused on processes vital to man, such as mutagenicity, carcinogenicity, and tumor inhibition. However, these properties were already used to treat some skin problems in ancient India and Egypt when plant extracts and subsequent exposure to sunlight were applied as a form of treatment for vitiligo, a medical condition in which areas of the skin lose pigment, causing them to appear white. Nowadays, besides vitiligo, several human skin diseases such as psoriasis, cutaneous T-cell lymphoma, and mycosis fungoides are usually treated with a combination of psoralens and UVA radiation with a wavelength of 340–400 nm, generally referred to as PUVA (psoralen plus UVA) therapy. Analogs of 8-methoxy- and 5-methoxy-psoralens are also used as photoactive drugs. However, oral photochemotherapy with 8-methoxy-psoralen (drug name is oxsoralen) and UVA irradiation is not recommended during pregnancy because of possible mutagenic effects.

Furanocoumarins have a mixed biosynthetic origin: the coumarin nucleus is derived from the shikimate chorismate pathway, passing through *p*-hydroxycinnamic acid (*p*-coumaric acid) and ending with umbelliferone (7-hydroxycoumarin). The furan ring comes from the acetate-mevalonate pathway, whose intermediate precursor, dimethylallyl pyrophosphate (DMAPP), binds to umbelliferone to generate the linear furanocoumarins (psoralens) and the angular ones (angelicins) [36] (Figure 2.18).

Biochemical studies using radioactive precursors demonstrated that many steps in the biosynthesis of these molecules are catalyzed by enzymes of the cytochrome P450 family, namely enzyme cytochrome P-450-dependent monooxygenase CYP71AJ and cofactors (NADPH) and molecular oxygen [37].

The role of furocoumarins in plant metabolism is framed within the different defense mechanisms that plants have to survive polyphagous herbivores and pathogenic fungi. This protective action is manifested in the varied toxicity toward the animals, in the antifeedant activity toward insects in the antifungal response, and as phytoalexins. Noteworthy, the distribution of linear and angular furanocoumarins differs between each plant family. It was found that Rutaceae and Moraceae plants produce only linear furanocoumarins, while both linear and angular furanocoumarins have been described in Fabaceae and Apiaceae plants. Curiously, no plant has been identified that produces only angular molecules. It is important because the linear molecules are highly toxic (independent of photoactivation) to a broad spectrum of predators, but the angular molecules are less toxic.

Many Apiaceae are often considered resistant, a difficult food source for several insects due to furocoumarins. Sometimes, they are lethal: the xanthotoxin (8-MOP) in wild parsnip Pastinaca sativa leaves causes 100% mortality in caterpillars *Spodoptera uridania*. However, there are both polyphagous and oligophagous Lepidoptera (*Papilio polyxenes* and *Depressaria pastinacella*, among others) and Diptera (*Phytomyza spondylii*, etc.), which tolerate the toxicity of these plants with furocoumarins. For example, the nocturnal butterfly *D. pastinacella*, which is grayish green color, and its webworm, which is an intense yellow color with black spots, are specialists of the plants of the family Apiacea. His favorite dish is the wild parsnip. They can metabolize approximately 95% of the ingested xanthotoxin. The resistance of these insects to toxic furocoumarin compounds is due to the incredible detoxification capacity through its enzyme system of cytochrome P450 monooxygenases. Other caterpillars-butterflies *P. polyxenes* can rapidly detoxify linear and angular furanocoumarins present in their host plants.

In which compounds are xanthotoxin converted during the detoxification process? How can you know that? In a study, metabolites of xanthotoxin were found in midgut and frass of parsnip webworms. Using combined analysis methods of liquid chromatography and electrospray mass spectrometry (LC-ESI-MS), the two metabolites of xanthotoxin were identified: a compound with molecular weight 266 g/mol in the feces and a compound with molecular weight

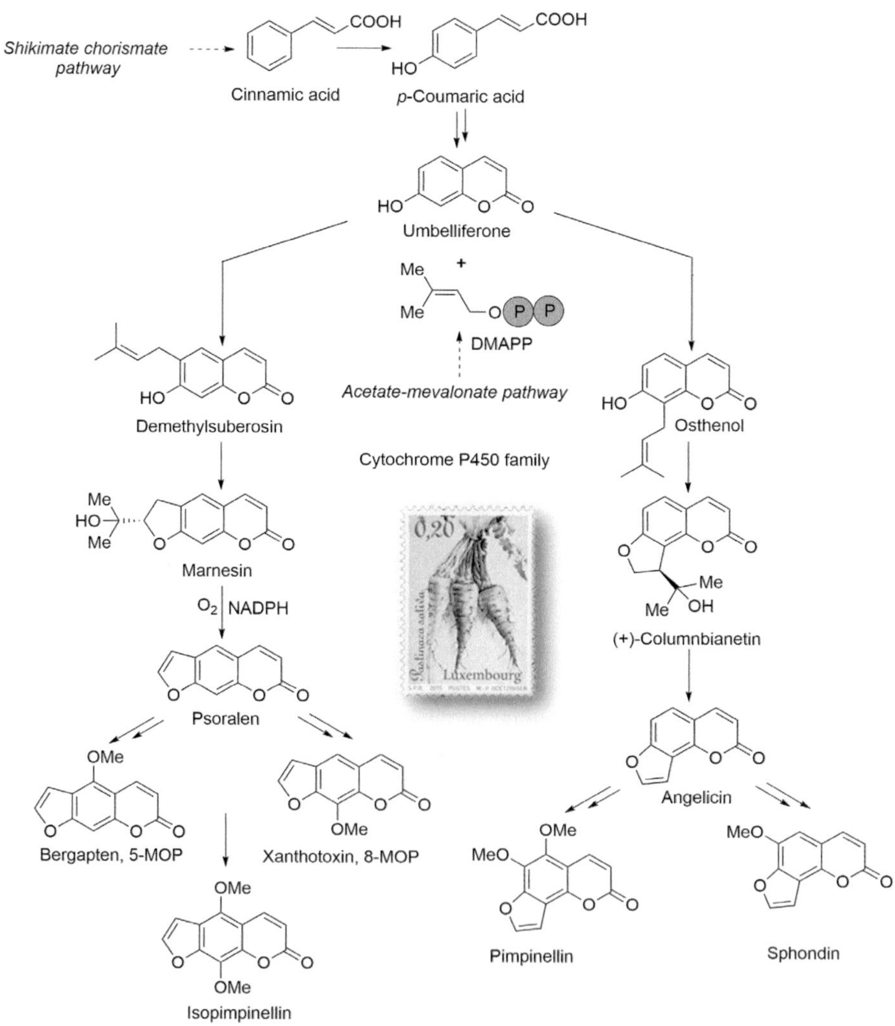

Figure 2.18. Simplified furanocoumarin biosynthesis pathway and an image of parsnip (*Pastinaca sativa*) in a stamp.

252 g/mol in the middle intestine. ^{1}H and ^{13}C NMR confirmed their structures as 6-(7-hydroxy-8-methoxycoumaryl)-hydroxyacetic acid (HCHA) and 6-(7-hydroxy-8-methoxycoumaryl)-hydroxyethanol (HMCH), respectively (Figure 2.19).

These results suggest that the first step of xanthotoxin metabolism involves the formation of an epoxide, assisted by cytochrome P450 in the furan ring, whose subsequent transformation leads to these derivatives.

The development of the synthetic methods for furocoumarins in the chemical laboratory has always been related to the advances in the synthesis of coumarins, whose main methods are the reaction of cyclocondensation between a phenol and ethyl acetoacetate in the presence of an acid catalyst to obtain 4-methyl-coumarin (that was realized in 1883 by German chemist Hans von Pechmann; Figure 2.20A) and the aldol condensation between salicylaldehydes and acetic anhydride in the presence of sodium acetate to obtain coumarin (that was proposed in 1868 by British chemist William H. Perkin; Figure 2.20B).

These two classic synthetic methods of the coumarins have not lost relevance since the reagents are commercially available and cheap products. However, both methods in the initial versions suffer from low overall yields [38, 39]. There is a wide variety of heterogeneous, homogenous, and nanocatalysts for these classic reactions. Besides conventional heating, the use of ultrasound, microwave irradiation, and ball-milling techniques has been introduced in the coumarins syntheses [40, 41].

Interestingly, coumarin is also a natural product found in many edible plants (such as black currant, apricot, strawberry, and cherry) and has been used as an aroma

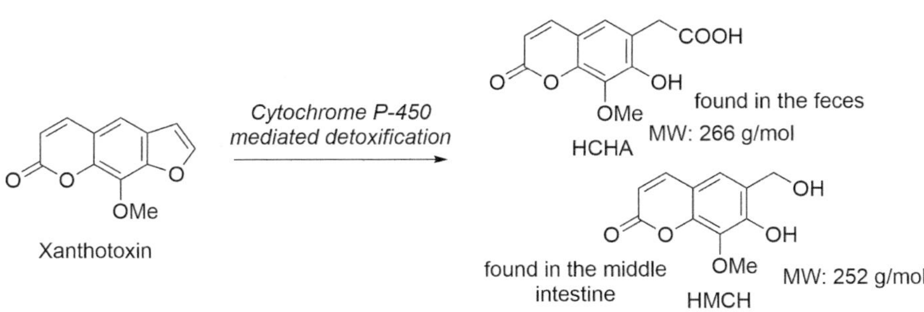

Figure 2.19. Metabolites of xanthotoxin produced by parsnip webworms.

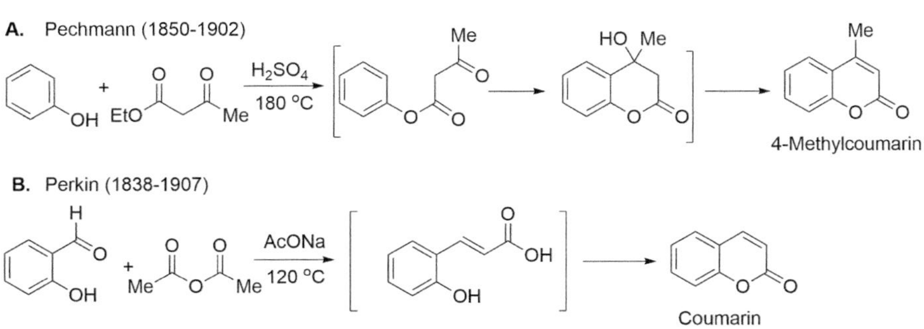

Figure 2.20 Classic methods of synthesis of the coumarins: A. Perkin's synthesis. B. Pechmann's synthesis.

enhancer in pipe tobaccos, alcoholic drinks, and certain artificial vanilla substitutes. But, due to its hepatotoxicity in animal models (LD_{50} = 275 mg/kg), coumarin has been banned as a food additive in several countries since the mid-20th century. Rodents metabolize this molecule commonly to the 3,4-coumarin epoxide, which is a toxic, unstable compound that, after further metabolism, may cause liver cancer in rats and lung tumors in mice. However, humans metabolize coumarin mainly to 7-hydroxycoumarin, a metabolite of lower toxicity. The primary natural source of coumarin is beans of the tree *Dipteryx odorata* (known as tonka beans). Coumarin could be isolated from these beans in 1–3% quantities. Coumarin skeleton forms part of the structure of different phenolics found not only in plants but also in fungi and bacteria. Coumarin derivatives represent a privileged scaffold for medicinal chemistry and pharmacology because of their peculiar physicochemical features that reflect diverse and selective biological properties [42–44].

According to the retrosynthetic analysis for linear furocoumarins (psoralens), the construction of the ABC tricyclic skeleton is carried out from 6-hydroxybenzofuran-5-carbaldehyde (6-HBA), 6-hydroxybenzofuran (6-HBF) or 7-hydroxycoumarin (7-HC), which could be virtually obtained through consecutive disconnections of strategic bonds (a-c types). Subsequent disconnections (e–g types) of these heterocyclic reagents can conduct us to the 2,4-dihydroxybenzaldehyde (2,4-DHB) and 1,3-dihydroxybencene (1,3-DHB, resorcinol), both commercially available reagents (Figure 2.21). The same precursors would be generated by applying the same steps in the retrosynthetic analysis for angular furocoumarins (angelicins).

Analyzing this synthetic tree, we conclude that the b-type disconnection is evident because 7-hydroxycoumarin, a common reagent, is readily available from both 2,4-DHB and 1,3-DHB through the respective Perkin and Pechmann reactions. Indeed, starting with the resorcinol as the initial phenolic substrate and malonic acid as a C3 carbonyl component, the 7-hydroxycoumarins are synthesized by the acid-catalyzed Pechmann reaction. Then, the latter is treated with bromopropanones in the presence

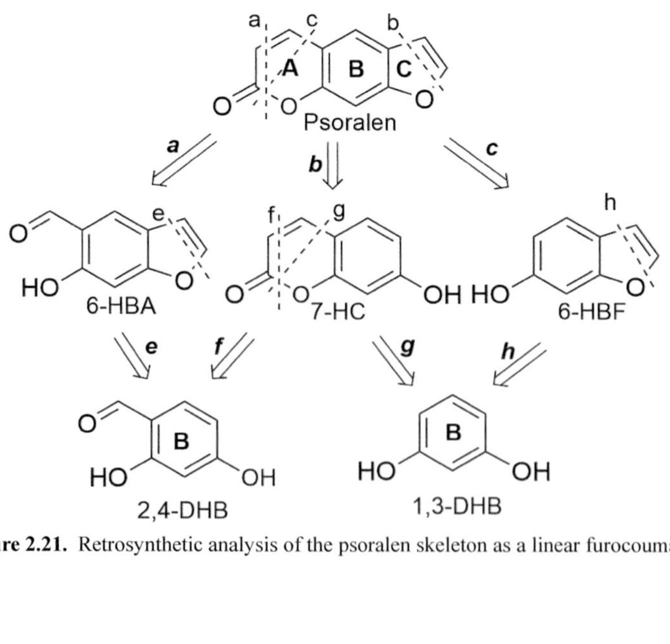

Figure 2.21. Retrosynthetic analysis of the psoralen skeleton as a linear furocoumarin.

of Brønsted or Lewis acids (BA, LA) to be converted into psoralens. However, the last step of this synthesis (intramolecular acylation reaction) is somewhat tricky because O-substituted hydroxy coumarins undergo substitution at the C-8 position rather than at the desired C-6 position, i.e., the formation of angelicins is favored (Figure 2.22).

Several methods for selective synthesis of furocoumarins have been reported during the last decades to overcome these problems. A metal-mediated synthetic approach to furocoumarins using alkynes to generate the furan ring should be mentioned here. Both 6-iodo-7-hydroxycoumarin and 8-iodo-7-hydroxycoumarin were successfully used in the respective psoralen and angelicin synthesis, which involve the use of the same alkyne reagent (ethyl propiolate) and reaction conditions (Figure 2.23). The introduction of the iodine atom allows the regioselectivity of this process.

First reactions of *o*-iodohydroxyquinolines and ethyl propiolate are the vital step in this synthesis of psoralen and angelicin and proceed via the initial alkynylation of respective coumarins using *in situ* formed Cu(I)-acetylides (Cu-C≡C-COOEt), followed by a copper-catalyzed cyclization of the resulting α-hydroxyaryl acetylenes to give carboxy substituted angelicin skeleton that is transformed in angelicin through alkaline hydrolysis of the ester group and subsequent decarboxylation by treatment with copper powder in quinoline.

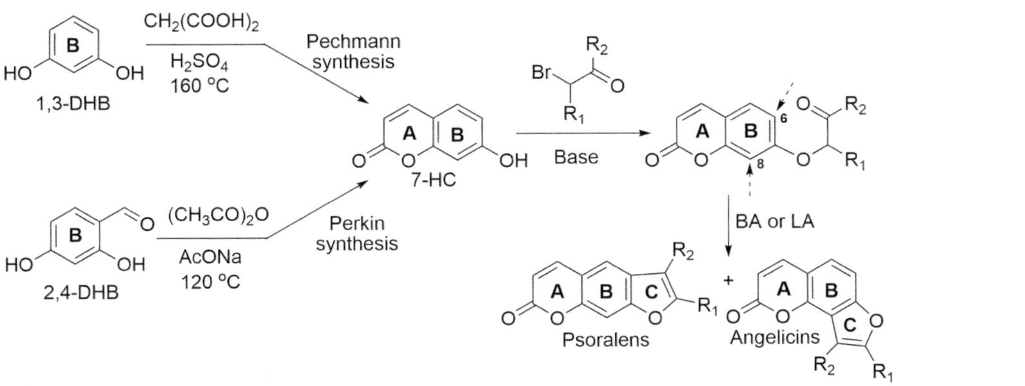

Figure 2.22. Synthesis of furocoumarins using 7-hydroxycoumarin as a crucial precursor.

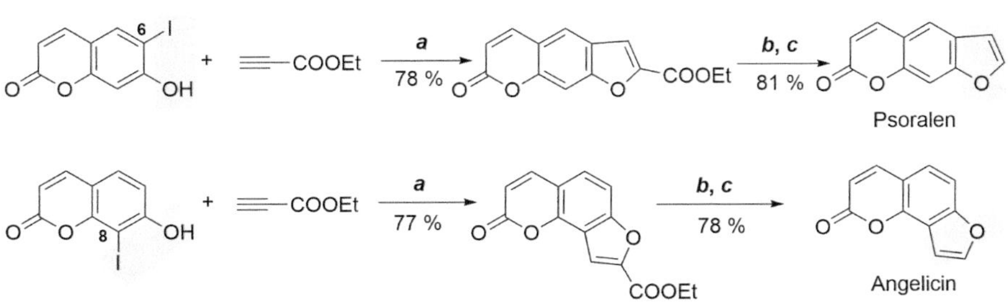

Figure 2.23. Selective synthesis of psoralens and angelicin using *o*-iodohydroxyquinolines and ethyl propiolate: (a) Cu_2O, DMF/100°C, 48 hours; (b) NaOH(aq)/reflux, 2 hours; (c) Cu/Quinoline, DMF/210°C, 1 hour.

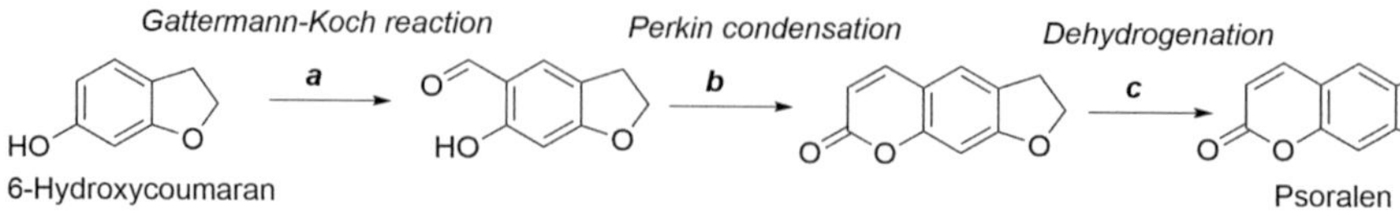

Figure 2.24. Three-step synthesis of psoralens from coumarin: (a) CO + HCl, CuCl/AlCl$_3$; (b) $(CH_3CO)_2O$, AcONa, heat; (c) Pd/C, toluene.

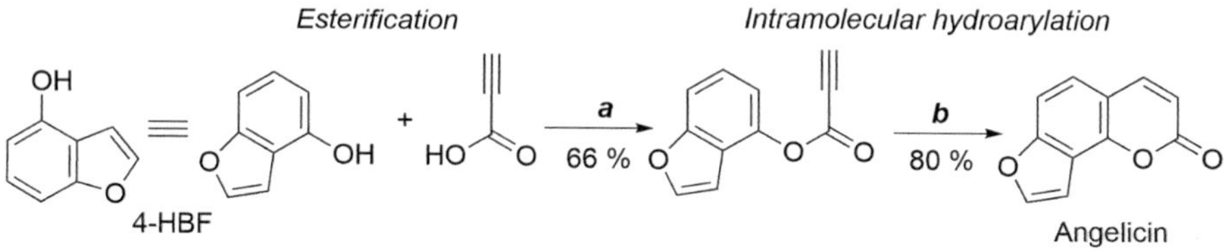

Figure 2.25. Two-step synthesis of angelicin from 4-HBF and propiolic acid: (a) DCC, DMAP (10:1), CH_2Cl_2, rt, 3 hours; (b) 1 mol% $Pd(OAc)_2$, TFA-CH_2Cl_2 (1:1), 30°C, 24 hours.

Linear furocoumarins can also be prepared using 6-hydroxy benzofuran (6-HBF) or its dehydrogenated analog called 6-hydroxy-coumarin (Figure 2.24). This approach involves three common steps: 1. *Ortho*-formylation process (Gattermann-Koch reaction), 2. Perkin condensation using acetic anhydride, and 3. Dehydrogenation of the five-membered ring to produce the furan ring.

The efficient synthesis of angelicin and its derivatives is based on the sequential esterification/intramolecular hydroarylation process that involves 4-hydroxybenzofuran (4-HBF) and propiolic acid as starting materials. The first step, the esterification process that is promoted by a mixture of dicyclohexylcarbodiimide (DCC) and 4-(dimethylamino)pyridine (DMAP), gives 4-benzofuranyl alkynoate, a key intermediate. The second step is its intramolecular hydroarylation in the presence of a palladium catalyst in a mixed solvent of trifluoroacetic acid (TFA) and dichloromethane (CH_2Cl_2), as shown in Figure 2.25.

Thus, furocoumarins, oxygen-containing heterocycles, are toxic molecules for almost all insects and animals, particularly for cattle, sheep, and horses, due to their photodynamic properties. However, due to the same properties, some furocoumarin derivatives serve as photosensitizing drugs prepared in synthetic laboratories, followed by the biogenesis mentioned above and retrosynthetic analysis.

Numerous other phenolic molecules protect plants from animals, which can be found in the specialized literature. Now, we must discuss the most exciting and typical examples of terpenoid molecules. Among them, iridoid glycosides and cardenolides stand out as oxygenated terpenes with significant roles in the interaction between herbivores.

2.4 Iridoids: Cyclopentanopyran Molecules

Iridoid glycosides are oxygen-containing monoterpenes linked with sugars. These molecules were named "iridoid" due to their similarity of structure and biosynthetic origin to iridodial and iridomyrmecin, defensive toxic metabolites with the structure of cyclopentano[*c*]pyran, which were found in the Australian ants of the genus *Iridomyrmex*. It should be noted that iridoid molecules were first isolated in the

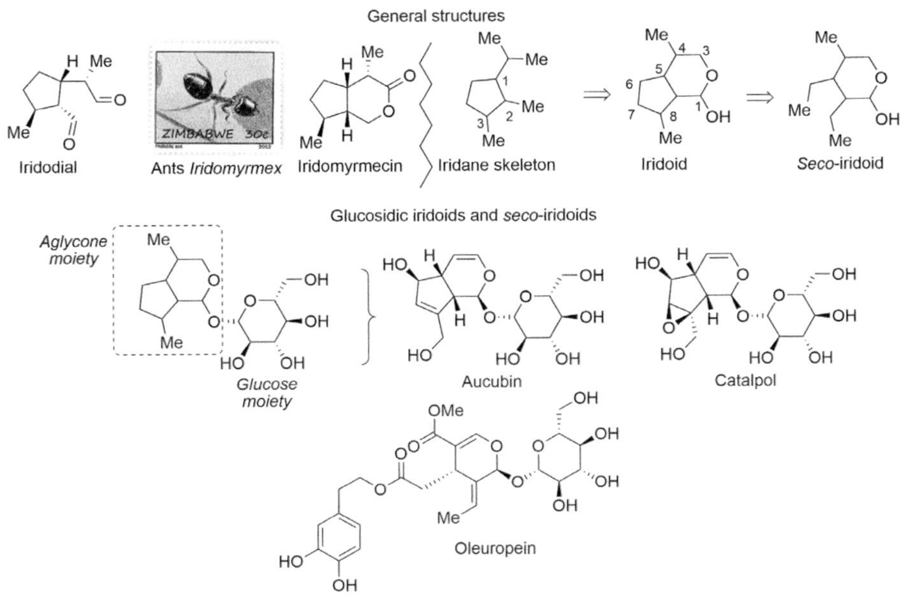

Figure 2.26. Construction of iridoids and structures of some typical iridoids.

mid-1800s from the root of *Rubia tinctorum* without mentioning its chemical nature. From a chemical point of view, iridoids are molecules based on a tetrahydropyran ring, which is fused to a 1-isopropyl-2,3-dimethyl-cyclopentane ring (iridane skeleton). When the cleavage of the bond between the C-7 and C-8 carbons of this skeleton occurs, the *seco*-iridoids, a particular group of "open" iridoids, are produced (Figure 2.26) [6, 45].

There are many molecules with these structures: more than 600 iridoid glycosides were found in 57 families of plants, and the number of plant iridoids is estimated to reach 3,000 molecules. All these are regarded as defense chemicals against herbivores and pathogens since iridoid glycosides usually have a bitter taste and have antifeedant and growth inhibitory activities against insects. They also have antifungal and antibacterial properties. In plants, iridoids such as aucubin and catalpol are mostly found as conjugates. For humans, almost all glycosidic iridoids have a bitter taste. These compounds have beneficial properties on liver and biliary function. They have also shown anti-inflammatory, antimicrobial, antitumor, and antiviral activity. They have been used as an antidote for poisoning produced after consuming poisonous fungi of the genus *Amanita*. Specifically, the iridoid glycoside aucubin, present in the leaves of the *Aucuba japonica*, protects against liver damage caused by this genus of fungi. Thus, for humans, these molecules do not present any danger.

Some plants with glycoside iridoids are widely used and known in kitchens and pharmacies. For example, olive leaves (*Olea europaea* L.) contain mainly secoiridoids. The major *seco*-iridoid is oleuropeoside, also known as oleuropein, which is a bitter compound responsible to a large extent for the antioxidant activity of the leaves of this plant.

The toxic effect of these molecules on invertebrates and microorganisms is probably due to the activity of their aglycone moiety, which is released through enzymatic or non-enzymatic acidic hydrolysis by plant or insect β-glucosidases. The aglycone moieties are reactive intermediates and are usually converted into type-dialdehyde molecules, whose structure is similar to that of glutaraldehyde (pentane-1,5-dial) or formaldehyde. These latter compounds are powerful disinfectants and sterilizing agents due to their strong protein denaturing, protein crosslinking, and lysine-alkylating activities. So, the degradation products of the glycoside iridoids also act as the alkylating agents of sulfhydryl, hydroxy, and amino groups of amino acids, proteins, and nucleic acids, altering the synthesis of DNA, RNA, and proteins, which interrupts the cell's essential functions performance. They can also alkylate amino groups of chitosan, a naturally occurring cationic polysaccharide obtained by deacetylation of chitin, and the latter is a primary component of the exoskeletons of arthropods. This can affect the herbivores that feed the plants with the glycosidic iridoids.

These chemical properties of iridoids can be exploited in biomedical practice. For example, one of the aglycone iridoids, genipin, reacts promptly with chitosan and protein through crosslinking mechanisms, producing blue-colored fluorescent hydrogels (Figure 2.27). These natural blue pigments are used to dye various fabrics, including cotton, silk, and wool. Genipin was obtained from the iridoid glucoside, geniposide, present in the fruit of *Gardenia jasminoides*, a common flower in many parts of Asia. It is a commercially available product. Due to its biodegradability and low cytotoxic properties (LD_{50} = 382 mg/kg in mice), genipin has recently been investigated as a crosslinking material in many biological applications [46].

According to the general metabolism of terpenoids (Figure 2.2), all iridoids come from geraniol, which in initial stages is oxidized, giving 10-hydroxygeraniol (HG) and 10-oxogeraniol (OG). These then cyclize to form iridodial (dialdehyde), which in turn is oxidized to generate several derivatives: iridotrial, 7-deoxyloganetic acid (7-DLA, aglycone form), 7-deoxyloganic acid (7-DLA, glucoside form) and finally, secologanin, a secoiridoid, that can generate indole-terpene alkaloids interacting with L-tryptophan (Figure 2.28).

Returning to the iridoid toxicity toward insects, we find that their effects are not general and must be examined from a co-evolutionary point of view. They can have a clear deterrent effect for non-adapted insects (generalists) but can be phagostimulants for adapted ones (specialists). In the case of generalist insects, feeding with iridoid glycosides reduces the growth rate, increases the duration of larval stages, and decreases survival rates. In contrast, specialist insects can feed on iridoid-producing plants, even with some beneficial effects. Iridoids can act as oviposition and/or feeding stimulants for adults and larvae [47]. Moreover, some specialist insects also sequester plant allelochemicals to protect against their own predators, like birds, ants, stink bugs, predatory wasps, and spiders. The predators preying on iridoid-containing insects could not gain any body weight.

An excellent example of the interactions between a host plant and a specialist insect is the relationship between the buckeye butterfly *Junonia coenia* and the host plant *Plantago lanceolata,* which contains high levels of iridoid glycosides (Figure 2.29A). Females of *J. coenia* were found to use aucubin and catalpol, iridoid

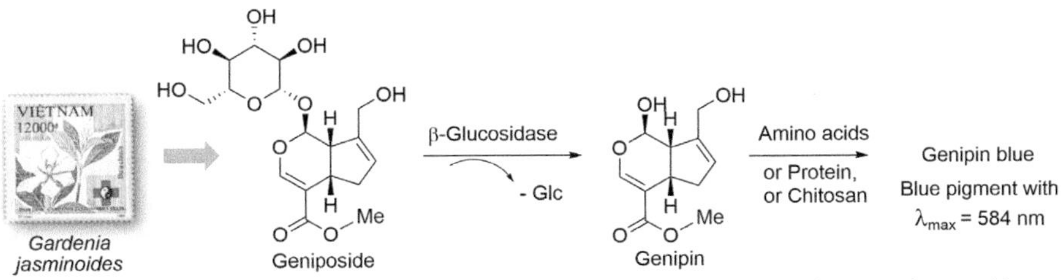

Figure 2.27. Geniposide, the natural precursor of genipin, reacts with amino acids, proteins, or chitosan to give blue-colored fluorescent hydrogels.

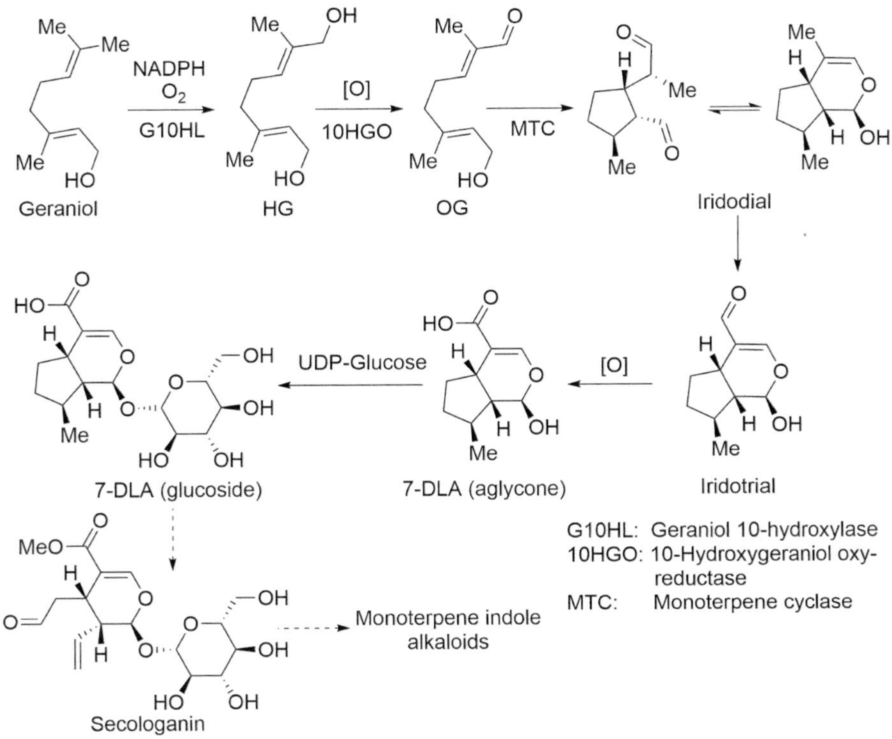

Figure 2.28. General routes of iridoid biosynthesis.

glycosides typical of this plant, as oviposition signals. It was shown that females prefer to lay eggs on substrates containing relatively high levels of the iridoid glycoside catalpol (plant *P. lanceolata*) than on those containing relatively low or no catalpol (plant *P. major*). The *J. coenia* caterpillars may contain up to 20% of the iridoids depending on the larval stage and concentration of these molecules in its host plant, indicating an efficient uptake. However, curiously, they do not retain any iridoids until the adult stage. During metamorphosis, iridoids are either metabolically degraded or eliminated. In contrast, after feeding females, *Euphydryas phaeton* contains both catalpol and aucubin in equal amounts. However, iridoid ratios of the host plant species, *P. lanceolata*, are uneven (aucubin:catalpol 6:1–20:1) (Figure 2.29B).

Another situation is the interactions between the alpine checker-spot butterfly, *E. anicia*, and the same plant *P. lanceolata*. Having a complex mixture of iridoids in this plant, females of *E. anicia* sequester mainly catalpol and some aucubin amounts, even metabolize this latter iridoid glycoside into catalpol [48,49] and could have up to 9% dry weight of these compounds. Larvae of *Ceratomia catalpae* contain only catalpol in high concentrations in their hamolmyph, out of five iridoids available in plant *Catalpa bignonioides*. Thus, catalpol seems to be one of the most bitter-tasting iridoids, playing an important role in chemical defense. It is understandable since it is more toxic to generalist herbivores than aucubin.

Interestingly, a related monoterpene lactone, nepetalactone, present in plant catnip *Nepeta cataria*, is an olfactory attractant to the domestic cat *Felis domestica* without affecting other mammals [50]. When the cats smell the plants *Nepeta* spp., they exhibit behavioral patterns common to the females in heat, regardless of whether the cat is male or female. Why does nepelactone produce this rare effect only in cats? It turns out that the action mode of this molecule is biochemically similar to the sexual attractant molecules (pheromones) of felines. It should be noted that nepelactone and its precursor nepetalactol are too sex pheromones.

Nepetalactone is a principal constituent (70–99%) of the essential oil of the catnip plant and exists as two isomers, *cis-trans* (*Z,E*)- and *trans-cis* (*E,Z*)-nepetalactones (Figure 2.30A). It was found that the latter isomer is the more attractive of the two isomers to cats [51–53]. However, four nepetalactone stereoisomers with differing stereochemistry at the 4α and 7α positions (*cis-cis*, *cis-trans*, *trans-cis*, and *trans-trans*) are observed at different ratios in various Nepeta species, and it seems that the ratio of the stereoisomers appears to play a role in the plant's ability to attract felines and deter insects.

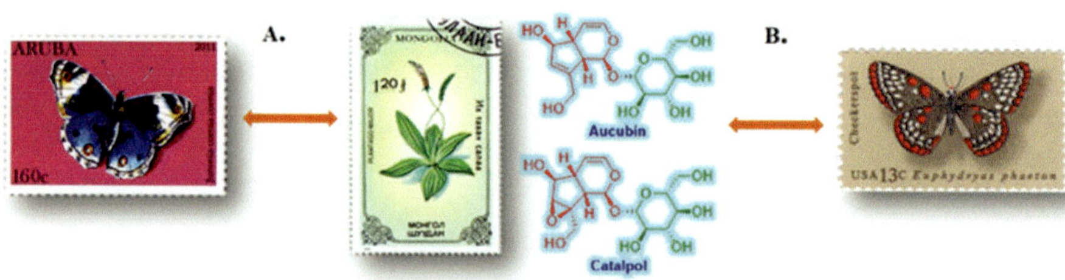

Figure 2.29. Iridoids from *P. lanceolata* as oviposition stimulants for butterfly *J. coenia* (A) and as feeding stimulants for butterfly *E. phaeton* (B) with their images.

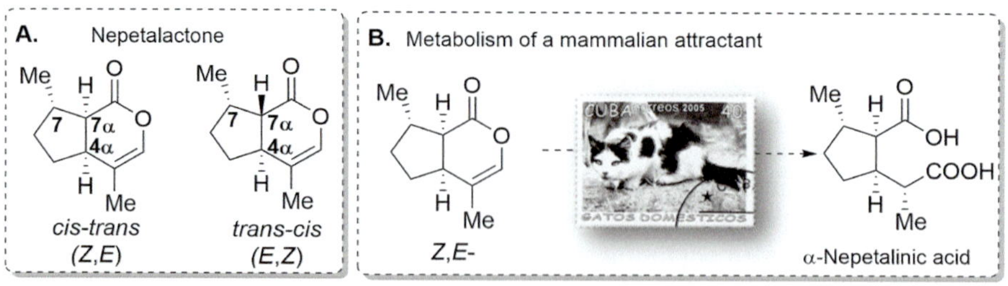

Figure 2.30. A. Stereochemistry of natural nepetalactone; B. α-Nepetalinic acids as a final product of *E,Z*-nepetalactone metabolism.

When ^{14}C-labeled *Z,E*-nepetalactone was orally administered to the domestic cat, α-nepetalinic acid was found as a principal excretion metabolite (50–75%) in the urine (Figure 2.30B). Metabolism of nepetalactone did not produce marked physiological or histological effects.

Interestingly, catnip has folk uses as an insect (mosquito, fly, German cockroaches, etc.) repellent, and some uses have been confirmed scientifically. Nepetalactone vapors were shown to repel insect species in 13 families.

Although diverse biological activities and therapeutic potential of iridoids have been reported, their chemical syntheses are still challenging targets for organic chemists. In 1986, Schreiber and his team [54] reported the five-step synthesis of *cis-trans*-nepetalactone from (*S*)-citronellol. Pickett and co-workers improved Schreiber's protocol ten years later [55]. The method involves two selective oxidation reactions to yield 8-oxocitronellal, intramolecular Diels-Alder type reaction of "enamine and α,β-unsaturated aldehyde" moieties of the molecule formed to give 1-amino-cyclopenta[*c*]pyran derivative, and subsequent hydrolysis and oxidation reactions that allow obtaining *cis-trans*- nepetalactone (Figure 2.31A).

In this method, [4+2] cycloaddition of aldehyde enamines with α,β-unsaturated aldehydes (enals) to afford dihydropyrans is a crucial step, in which *cis*-fused cyclopenta[*c*]pyran skeleton is rapidly and stereoselectively constructed. This intramolecular enamine/enal cycloaddition is recognized as a valuable entry into the iridoid carbon skeleton. Nowadays, both enantiomers of citronellal and citral are successfully used in synthesizing nepetalactones and related iridoids from ants of the genus *Iridomyrmex* and certain Australian butterflies. Notably, the *trans*-fused

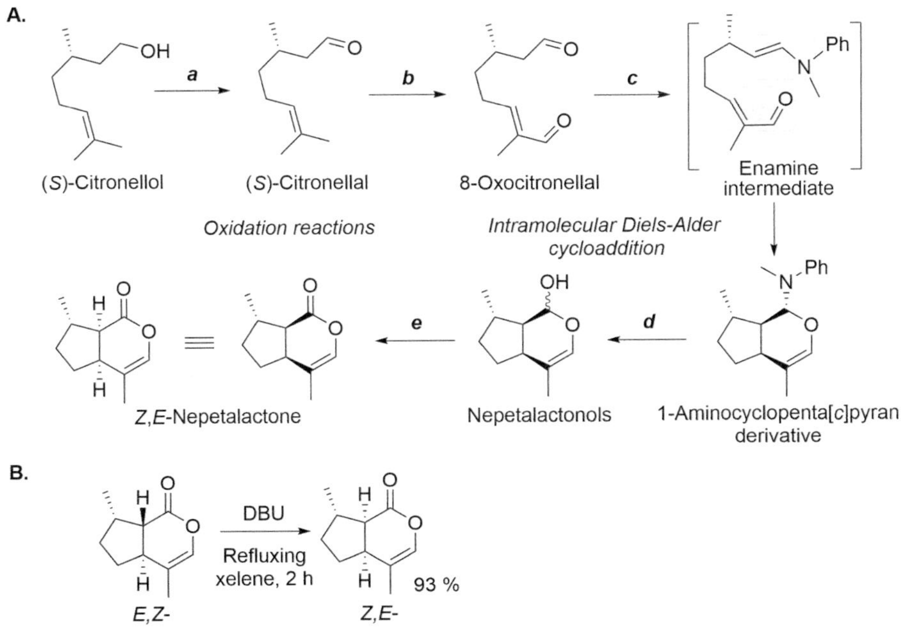

Figure 2.31. A. Linear synthesis of (4a*S*,7*S*,7a*R*)-nepetalactone from (*S*)-citronellol: (a) Pyr·SO_3 complex, DMSO, Et_3N; (b) SeO_2, *tert*-BuOOH; (c) MeNHPh, 4 Å sieves; (d) TsOH, THF:H_2O (1:1); (e) $AgNO_3$, Celite, refluxing benzene. B. Epimerization of nepetalactones.

nepetalactone was converted entirely into the *cis*-fused lactone using the base DBU (1,8-diaza-bicyclo[5.4.0]undec-7-ene) [56] (Figure 2.31B).

Despite iridoids, an important class of terpenoids widespread in nature, being first studied in the middle of the last century, their synthesis and chemistry are still underdeveloped and require new strategies and tactics. Noteworthy, some iridoids possessing antiviral and antibacterial properties were isolated due to pharmacognostic studies of medicinal plants used since ancient times.

2.5 Cardenolides: Toxins or Drugs?

Cardenolides are a particular group of diverse steroids derived from triterpenoids, almost certainly evolved as a defense, primarily in plants belonging to the former family Asclepiadaceae, now subsumed in the Apocynaceae. We recall that steroids, terpenoid metabolites, are fused tetracyclic compounds with three six-membered rings (A, B, and C) and a five-membered ring D and are formally cyclopentano-perhydrophenanthrenes.

In general, this tetracyclic system can present two different forms of ring connection: (a) the A/B/C-ring junctions are all *trans*- and C/D is *cis*- and (b) the A/B and C/D junctions are *cis*-, while the B/C-ring junction provides a rigid *trans*-junction (Figure 2.32A). Thus, there are two configuration forms of cardenolide in plants. Both steroid ring systems are rigid and possess functional groups bonded to ring atoms with well-defined positions. Substituents below the plane of the ring are designated as α; those above the plane of the ring are β. This is because that ring flipping does not occur. In plants, they are steroids linked with sugars (steroidal glycosides). As they are primarily polar compounds, the cardenolides are stored predominantly in vacuoles and may be present in all plant tissues. Thus, the steroid structure of these molecules is called aglycone or genin, and it contains a ring of furan-2-one (lactone, an unsaturated γ-butyrolactone) at the C-17β position of the steroid skeleton, which can be linked to one or more sugars that are known as glycone portion. The union of this skeleton with the sugars is through the hydroxyl group at

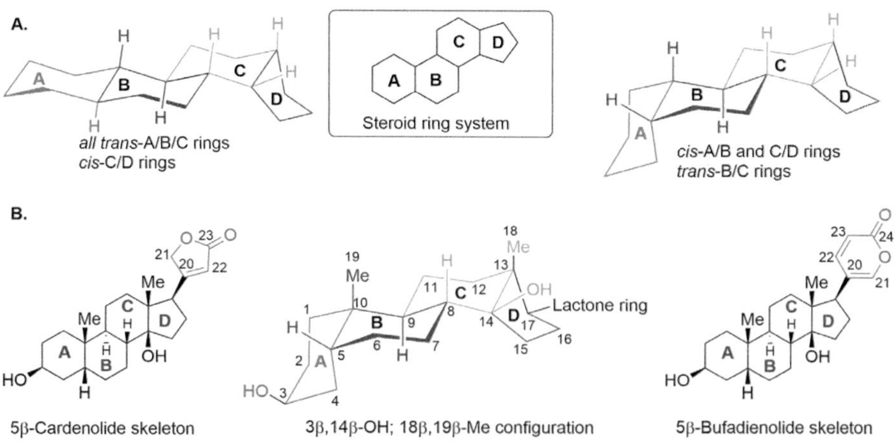

Figure 2.32. Structural characteristics of cardenolides and bufadienolides: A. The general structure of steroids and their configuration. B. Functionality details of common cardenolides and bufadienolides.

the C-3β position. When the ring of γ-butyrolactone is replaced by another larger ring, α-pyrone, the molecules are called bufadienolides (Figure 2.32B); these were discovered for the first time, as skin poisons in toads (Section 6.1.4).

According to IUPAC Steroids Nomenclature, both types of these steroids are classified as 5β-cardenolides and 5β-bufadienolides, which indicate the configuration at position C-5.

Cardenolides (C_{23} steroids) and bufadienolides (C_{24} steroids) form a group of bioactive molecules that affect the normal functioning of a heart and are often called cardiotonic glycosides (CGs) [57]. This class has been in clinical use for more than 1,500 years. The CGs-containing plants have been mainly used in folk medicine as heart tonics, emetics, and diuretics, as well as arrow poisons, abortifacients, and other applications.

CGs increase the force of contraction of the heart via the dose-dependent inhibition of Na^+/K^+-ATPase, an integral plasma membrane protein that results in increased intracellular levels of sodium ions and is used in the treatment of patients with heart failure, even though some are highly toxic to humans and animals. Several plants contain cardiac glycosides, but historically, a famous CG source is *Digitalis* species from the family Plantaginaceae. Among them, foxgloves *D. purpurea* and *D. lanata* are the most important plant sources of digitoxin, digoxin, and gitoxin, precursors of respective aglycones digitoxygenin, digoxygenin, and gitoxygenin (Figure 2.33A).

Interestingly, reparations *of D. purpurea*, a plant native to Europe, as drugs in heart therapy were introduced in licensed medicinal practice by the Scottish physician William Withering in 1785, while *D. lanata* species is the plant harvested in Brazil to produce industrially digoxin. Other important cardenolides are acquired from a decorative shrub, *Nerium oleander* (oleandrin and oleandrigenin), the East African plants *Strophanthus kombé* (k-strophanthoside and strophanthidin) and *Acokanthera schimperi* and *Acokanthera ouabaio* (ouabain and ouabagenin) (Figure 2.33B). These last African plants are toxic and are used by the indigenous people to prepare arrowhead poisons. One of its cardenolide components, ouabain (also known as g-strophanthin), is a highly toxic compound (LD_{50} = 2.2 mg/kg, intravenous administration to mouse) and can cause severe symptoms like vomiting and convulsion. Larger doses may lead to cardiac arrest and death. Ouabain was first isolated from the roots and the bark of *A. ouabaio* in 1888, while ouabagenin, the aglycone of ouabain, was isolated for the first time in 1942.

Cardenolides are remarkable chemical compounds that are at once responsible for the poisoning of livestock and the cure of numerous people with congestive heart failure; they also have been applied to poison arrows in human warfare and mediate antagonistic ecological interactions between plants and herbivores [58]. The latter role of these compounds is the focus of our discussion.

For humans, cardenolides are compounds of bitter taste depending on their polarity and structure. In birds and mammals, the oral ingestion of the first reaction resulting is emesis (vomiting). In mammals, a trigger zone of the chemoreceptors in the bone marrow is responsible for the induction of vomiting in concentrations lower than toxic doses.

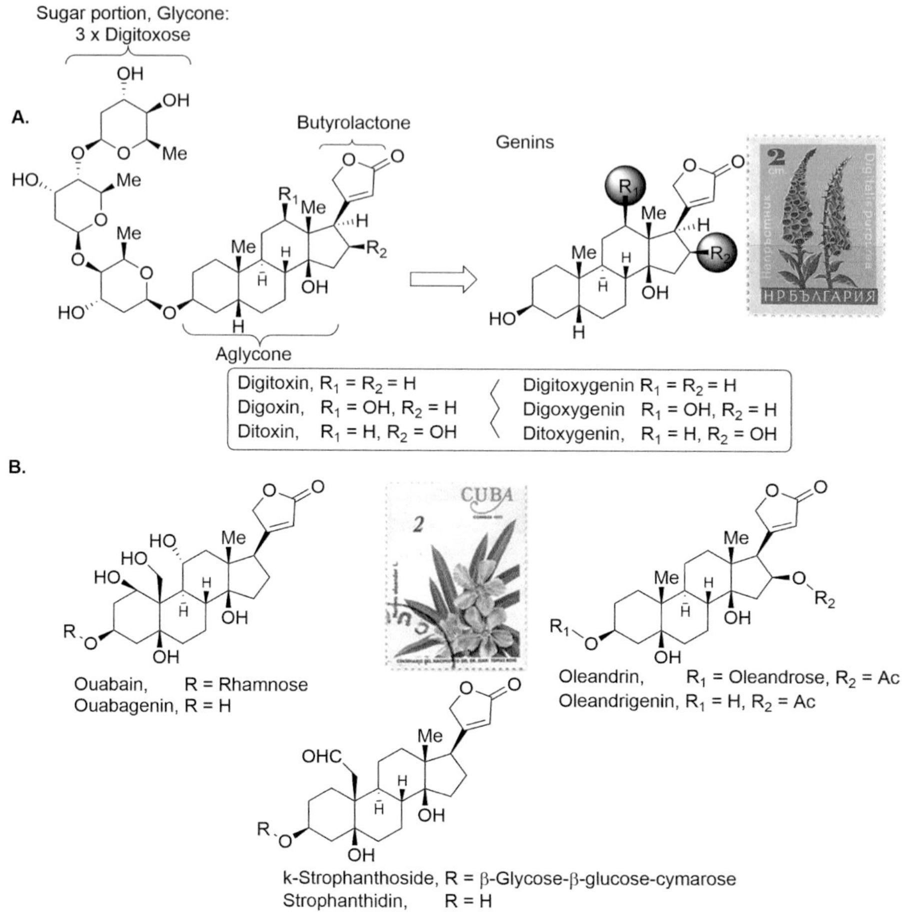

Figure 2.33. Representative examples of cardenolides: A. Several glycosides from *D. purpurea* (their image in a stamp). B. Structural details of other important CGs and *Nerium oleander* species in a stamp.

However, cardenolides can be deadly for some animals, especially insects. Unlike many other toxic plant compounds, cardenolides act in a highly specific manner, as they are specific inhibitors of the sodium-potassium pump, one of the most fundamental systems in all cell animals. When linked, the cardenolides sabotage the transport of vital elements through the cell membrane (Na^+ out of the cell and K^+ into it), damaging the cell [59]. Most particularly, the binding site for cardenolides is located on the α-subunit of the enzyme Na^+/K^+-ATPase. It was found that these plant terpenoid metabolites bind from the extracellular side with the γ-butyrolactone ring profoundly inserted into the transmembrane domain and with the sugar moiety facing the extracellular medium. Notable, the configuration of the A/B and C/D ring junctions is important for toxicity: *cis*-A/B, C/D-*trans*-B/C-form is most toxic, while *trans*-A/B/C-*cis*-C/D-configuration (which are most widely distributed among *Asclepias* species) results to be less toxic. Moreover, cardenolide monosides (i.e., having one sugar) are usually more active than their genins or biosides (i.e., having two sugars), and successive removal of terminal glucose from bi- or triosides can lead to a stepwise increase in toxicity.

Although little data on the toxicity of cardenolides are available, it is known that the LD_{50} of ouabain for the desert locust (*Schistocerca gregaria*) or common cockroaches (*Periplaneta americana*) is below 4.4 or 0.6 μg per animal, respectively, when injected into the body cavity of the animal. In contrast, the caterpillars of the tobacco worm (*Manduca sixta*) show no signs of toxicity effects and tolerate a concentration of high ouabain as 10^{-3} M in the hemolymph. But *Trichoplusia ni* caterpillars suffered from ingesting plant cardenolides solutions (*Asclepias curassavica*).

In general, it is known that at least 18 species of insects belonging to four orders, including beetles, butterflies, and flies, are capable of feeding on plants that contain potent toxic molecules without suffering harmful effects. Thus, these insects must have some adaptation that protects against the toxic effect of vegetable cardenolides. A classic way to protect one's life is the sequestration of cardenolides. Recycling toxic compounds from plants for defensive purposes is a good tactic for survival.

This task is perfectly done by the monarch butterfly *Danaus plexippus*, whose caterpillars sequester the cardenolide glycosides of their host plants (mostly *Asclepias* species, known as milkweeds, named for their latex containing these steroidal molecules) and can transfer them to the adult stage to use these cardenolides as a defense against predators, especially birds. The larvae and adults' lousy taste and toxicity are advertised by conspicuous, warning coloration. When a bird or other predator tastes a monarch, it learns to associate this color pattern with the bad taste and avoids preying on monarchs in the future. This observation was made by Poulton many years ago, in 1909. However, it was not until the 1960s that researchers discovered that cardenolides were the chemicals in milkweed that made monarchs toxic and bitter-tasting [60,61]. Since then, several studies have shown that monarchs can sequester these toxic compounds. Therefore, how do they do that? It seems that in monarch butterflies, resistance is a consequence of a specific mutation called N122H. This mutation reduces the ability of cardenolides to bind to the α-subunit in the sodium-potassium pump [62].

It is interesting to note that the larvae of monarch butterflies feed on tropical milkweed *A. curassavica* (fam. Asclepiadoideae), which contains cardenolides, calotropin, and calactin in a 1:1 ratio and stores these configurational isomers without modifying them. Still, other minor cardenolides, such as uscharidin, are quickly converted into a mixture of calactin and calotropin. It was also found that larvae sequester calotropin up to 1.8 times less than calactin. Both configurational isomers are toxic (LD_{50} = 9–15 mg/kg, intraperitoneal administration to mouse). These remarkable interactions between milkweeds and larvae of monarch butterflies and their predators are graphically present in Figure 2.34.

The cardenolides in Figure 2.34 can be obtained or isolated from plant extracts of *Calotropis gigantea* and *Calotropis procera*, known as Indian and North African milkweeds belonging to the same family Asclepiadoideae. Here, the history is repeated: the colored grasshoppers of North Africa, *Poekilocerus bufonius* (family Pyrgomorphidae), feed on these milkweeds and store them in the tissues of the glands during their early stages of development. Thus, these compounds make them

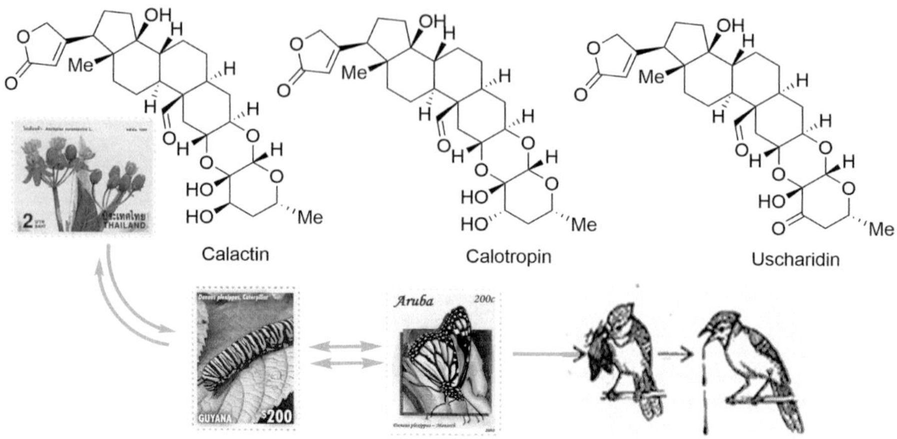

Figure 2.34. Interactions between milkweeds and larvae of monarch butterflies and their main predators.

immune to ingestion by their predators. When a potential predator attacks these grasshoppers, they can defend themselves by expelling a jet from defensive glands. The analysis of this fluid reveals that it contains two principal cardenolides, calactin and calotropin. This demonstrates a sophisticated chemical defense in these animals. However, when a grasshopper is maintained on a diet that includes no milkweeds, the cardenolide content of its protective spray is reduced tenfold.

All leaf beetles of the genera *Chrysochus* (Chrysomelidae) can feed on Apocynum plants (fam. Apocynaceae), but only two Nearctic species *C. cobaltinus* and *C. auratus* contain cardenolides. Both species can sequester cardenolides (e.g., ouabain) from their host plants into their defensive secretions. The secretions of these beetles can contain these compounds in a concentration of 77 to 358 μg/μL. In contrast, sister species, the Asian *C. chinensis* and the European *C. asclepiadeus*, do not contain cardenolides because they do not feed. Why do these beetles of the same genus have different capacities against cardenolides?

The response is in the structural organization of the α-subunit of Na^+/K^+-ATPase of these beetles. The sodium-potassium bomb of *C. cobaltinus* and *C. auratus* possesses the same amino acids at position 111 (valine), 112 (valine), and 122 (histidine) that confer cardenolide resistance as the monarch butterfly has. This allows them to contain these toxins. The cardenolides cannot join this subunit of the sodium pump due to the genetic modification of the beetles. In contrast, the Eurasian species *C. asclepiadeus* and *C. chinensis* show different amino acid positions: 111 (leucine), 112 (alanine), and 122 (asparagine). These exciting findings were possible due to the study interactions between *Chrysochus auratus*, known as the Dogbane beetle, and ouabain-containing plants.

As cardenolides are remarkable molecules with potent biological properties, a considerable interest in constructing these highly oxygenated steroids is always maintained. Thus, numerous works are related to partial (semi-synthesis) and total synthesis [63,64]. The primary starting material for the partial synthesis of steroids, including cardenolides and bufadienolides, are diverse commercially

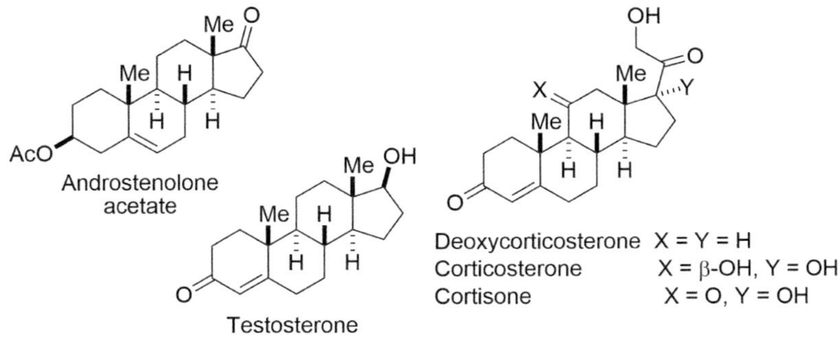

Figure 2.35. Commercially available steroids as starting materials for cardenolide and bufadienolide synthesis.

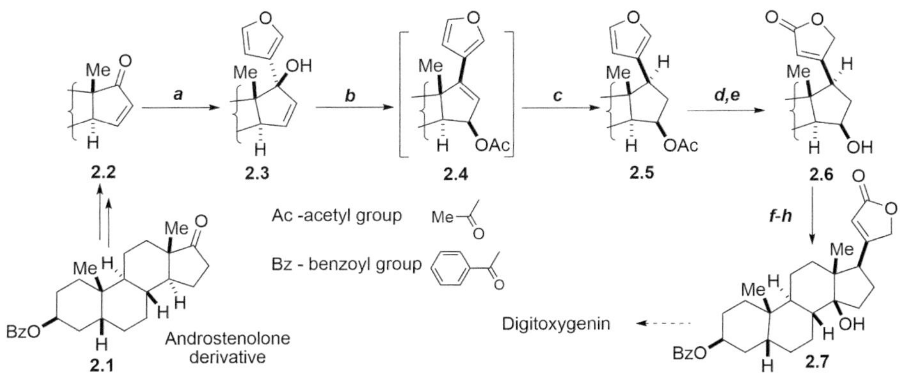

Figure 2.36. The construction of butenolide ring in Wiesner's approach to cardenolide analogs from common 17-oxoandrostane derivatives: (a) 3-Furyl lithium/Et_2O; (b) Ac_2O/Py; (c) H_2/Pd; (d) *m*-CPBA; (e) $NaBH_4$; (f) MsCl/Et_3N; (g) NBS, aq. dioxane; (h) Reduction.

available steroids such as androstenolone acetate, testosterone, deoxycorticosterone, corticosterone, cortisone, among others (Figure 2.35).

The construction of the butenolide ring can be achieved by numerous tactics using these steroid materials. For example, the "furan strategy" for the butenolide ring construction developed by Wiesner and co-workers [65] affords a complete synthetic approach to cardenolide analogs from 17-oxoandrostane derivatives **2.1** (Figure 2.36). The construction starts with the introduction of the C15, C16 double bond and the addition of 3-furyl lithium, which gives an adduct **2.3** (step a). This adduct is then subjected consecutively to acylation, allylic rearrangement, and hydrogenation of the C16, C17 double bond, proving androstane derivative **2.5** (steps b and c).

The furan ring in the saturated derivative **2.5** is then oxidized with a peracid, and the crude oxidation product is reduced with $NaBH_4$ (steps d and e). This allows the construction of the desired butenolide ring of derivative **2.6**. The last three steps "move" in an elegant manner the C15-hydroxyl group to the C-14 position, affording O-benzoyl digoxigenin **2.7** (steps f-h). Realizing the furan-butenolide transformation regioselectively is a crucial step in Wiesner's approach to cardenolide analogs.

Another approach to constructing the butenolide ring involves the Knoevenagel condensation between 17-oxoandrostane acetate derivative **2.8** and ethyl cyanoacetate.

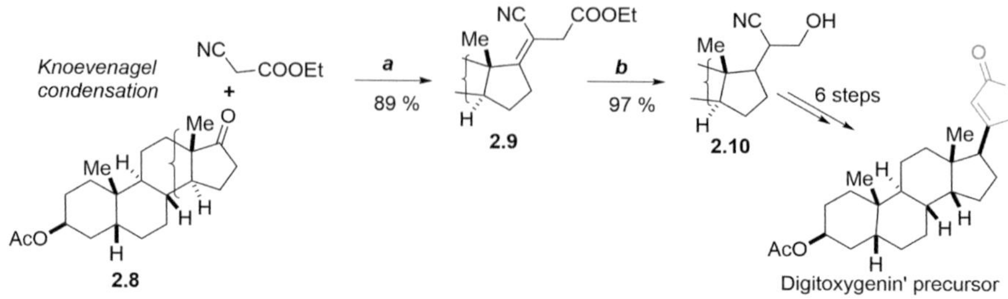

Figure 2.37. The construction of butenolide ring using Knoevenagel condensation: (a) $AcONH_4$, boiling toluene; (b) $NaBH_4$, MeOH.

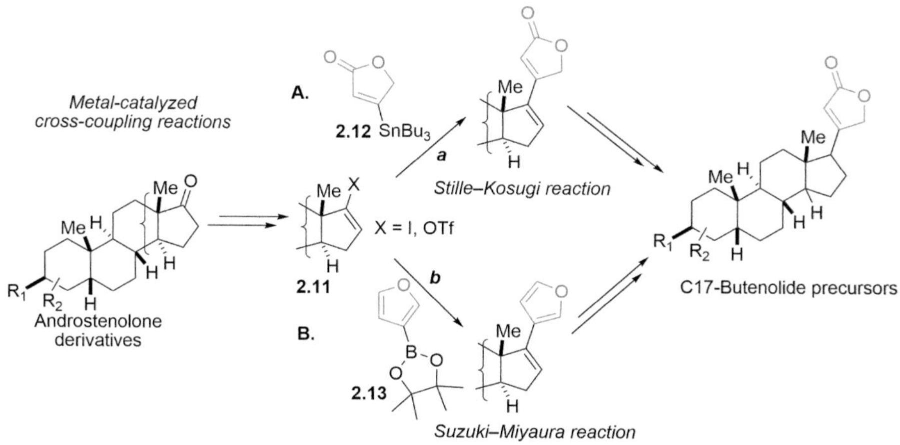

Figure 2.38. General approaches to the construction of the butenolide ring using the Stille-Kosugi (A) and Suzuki-Miyaura (B) cross-coupling reactions: (a) 20 mol% $Pd(PPh_3)_4$/LiCl, CuCl/DMSO/60°C; (b) $Pd(PPh_3)_4/CsCO_3$, dimethoxyethane.

The condensation product, the cyano ester **2.9**, is then reduced with $NaBH_4$ in MeOH to give β-cyano alcohol **2.10**, which can be converted into digitoxygenin's precursor in six steps [66] (Figure 2.37).

These approaches to the construction of the butenolide ring are quite efficient and, thus, were first used in the semi-synthesis of the most important cardenolides, such as digitoxigenin or ouabagenin. Indeed, Figure 2.36 represents key steps in Wiesner's synthesis of digitoxigenin from testosterone [67]. Modern construction of this heterocyclic ring is mainly based on the transition metal-catalyzed cross-coupling reactions like the Heck-Mizoroki coupling, the Stille-Kosugi reaction, and the Suzuki-Miyaura reaction cross-coupling reaction. Such processes are illustrated in a general scheme (Figure 2.38). The Stille-Kosugi reaction involves a cross-coupling of androstane triflate (or vinyl iodide) derivatives **2.11** and 4-tributylstannylfuran-5*H*-2-one (or 3-(tributylstannyl)furane, **2.12**) in the presence of the palladium catalyst $Pd(PPh_3)$, LiCl and CuCl in DMSO (Figure 2.38A), while the Suzuki-Miyaura procedure encompasses the same androstane molecules **2.11** and catalyst, but coupling reagent is different: pinacolborane 3-furyl derivative **2.13** is usually used (Figure 2.38B).

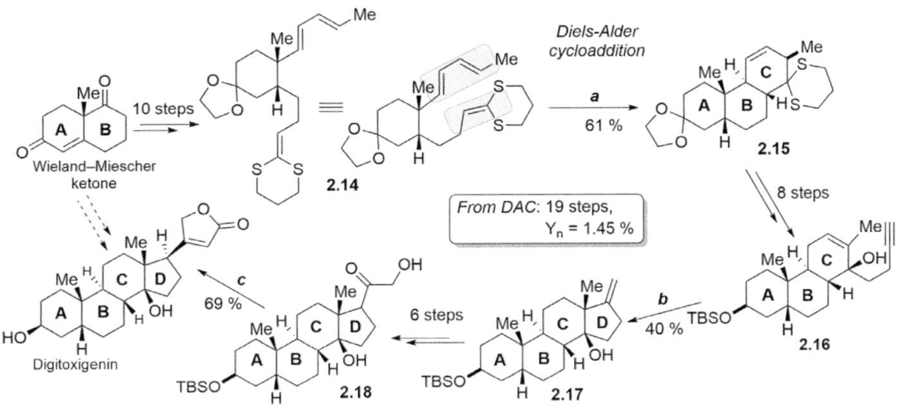

Figure 2.39. Simplified scheme of first total synthesis of digitoxigenin [68]: (a) Toluene, sealed tube, 180°C, 36 hours; (b) Bu_3SnH/AIBN, benzene, then SiO_2; (c) 1. $Ph_2C{=}C{=}O/Et_3N$, benzene, 2. TsOH/MeOH, rt, 24 hours.

Some partial, semi-syntheses, and total syntheses of the above-mentioned cardenolides have been developed since the 1960s. Semi-syntheses of the cardenolides from the abundant plant sterols are considered the most suitable preparative route, while total syntheses are considered less competitive.

As an example of the latter strategy, we will examine briefly the first total synthesis of (+)-digitoxigenin that was realized by Stork and co-workers in 1996. This is a multi-step consecutive synthesis from the Wieland-Miescher ketone (WMK) (Figure 2.39) [68]. This bicyclic diketone is named after two chemists from Ciba Geigy, Karl Miescher and Peter Wieland, who developed its preparation in 1950. Chiral (*S*)-enantiomer WMK has also been accessible since 1971. Prof. Gilbert Stork (1921–2017) selected this enantiomer as the starting material for preparing this biologically important cardenolide. Notably, it is valuable in synthesizing other natural products, e.g., taxol.

The WMK contains the AB-ring structure of the steroid skeleton. In order to obtain ABC- dodecahydrophenanthrene derivative **2.15**, Wieland-Miescher ketone is transformed into substituted cyclohexane **2.14**, which possesses both diene and alkene chains and thus can realize a thermal intramolecular [4+2] Diels-Alder reaction to perform the construction ABC skeleton of **2.15** (step a, Figure 2.39). After eight steps, alkynyl substituted ABC-dodecahydrophenanthrene derivative **2.16** with C14-β-hydroxy group is obtained. Then, another crucial step, alkyne-alkene cyclocondensation under radical conditions, produces an ABCD-androstane derivative **2.17** with an *exo*-double bond in the D ring (step b), which is converted into an androstane skeleton **2.18** with C17β-2-hydroxy-1-ethanonyl fragment) in six steps. Finally, a reaction with phosphoranylideneketene (the Bestmann ylide, $Ph_2C{=}C{=}O$) and deprotection of *tert*-butyl(dimethyl)silyl (TBS) from the C3β-hydroxyl group resulted in digitoxigenin (step c). As we can see, this total synthesis of digitoxigenin is hard work, like all consecutive total syntheses of complex natural products, where the step execution is increased; thus, the total efficiency preparation of our target product decreases.

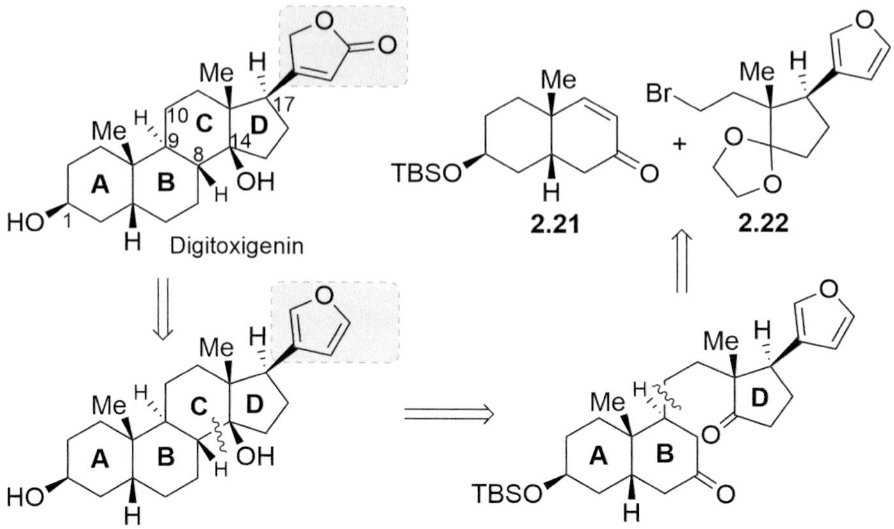

Figure 2.40. Retrosynthetic analysis of digitoxigenin molecule realized by Nakada and Honma.

In 2007, 11 years later, Prof. Masahisa Nakada (born in 1959 in Tokyo, Japan) and his student Masahiro Honma also described an enantioselective total synthesis of (+)-digitoxigenin. However, this total synthesis is accomplished in a convergent manner using two chiral fragments [69]. Their retrosynthetic analysis is outlined in Figure 2.40. First, what they did in their synthesis planning was change the butenolide cycle at C17 for the furan ring. Then, the two consecutive disconnections of the C8-C13 and C9-C10 bonds allowed the identification of the two essential precursors, hexahydronaphthalenone derivative **2.21** and (cyclopentyl)furan (CPF) molecule **2.22**. After this analysis, the appropriate initial materials that offer its preparation must be selected.

Both essential precursors are chiral molecules, which must first be prepared in enantiomerically pure form. The preparation of compound **2.21** begins with Cu-catalyzed cyclization of the prochiral 1-methylcyclohexa-2,5-diene **2.19** with a diazo-butanone chain. After two steps, the chiral hexahydronaphthalenone molecule is built, and its conversion into **2.21** is completed in three steps. The synthesis of CPF starts with the bakers' yeast reduction of the prochiral 2-allyl-2-methylcyclopentane-1,3-dione **2.20** to give substituted oxocyclopentol, which is converted into 3-furylcyclopentane molecule **2.22** after 14 steps in overall yields of 41% (Figure 2.41).

The crucial steps of this convergent synthesis are the conjugate 1,4-addition reaction between α,β-unsaturated ketone **2.21** and bromide **2.22** (Michael addition) that provides furyl naphthalenone derivative **2.23** as the single diastereomer in a 91 % yield (step a, Figure 2.41) and the construction of C-ring to form the fused *trans* B/C and *cis* C/D skeletons of cyclopentaphenanthrenone molecule **2.24** (steps b and c). Its subsequent transformations give a suitable steroidal skeleton **2.25** with OH and furyl groups in β-configuration. Finally, the β-C17 furyl group is converted in the butenolide ring using singlet oxygen under photoirradiation (step d), followed by a

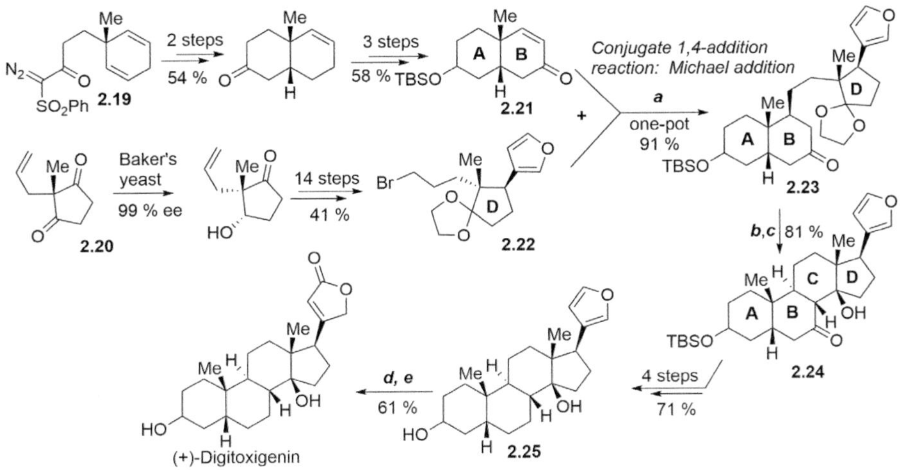

Figure 2.41. Enantioselective total synthesis of (+)-digitoxigenin [69]: (a) to CPF *tert*-BuLi/Et_2O-THF, −78°C then, (2-Th)Cu(CN)Li, −78 to −10°C, again −78°C and then **2.21** is added, warm up to −30°C; (b) TsOH, acetone–H_2O, reflux; (c) Cs_2CO_3, CH_3CN, reflux; (d) hν, O_2, Rose Bengal, CH_2Cl_2, DIPEA, −78°C; (e) $NaBH_4$, CH_2Cl_2, H_2O, rt.

reduction of the resulting hemiacetal with $NaBH_4$ (step e) that produces digitoxigenin in 61% yield. This elegant synthesis is an excellent example of the power of the convergent strategy in total synthesis.

The first total synthesis of ouabain and their aglycone (ouabagenin) realized by the Deslongchamps group in 2008 [70, 71] is also based on a convergent strategy (Figure 2.42). Although it is a convergent synthesis, its preparation is not simple. The key steps were the synthesis of initial *seco*-androstane derivative **2.30** and tetracyclic cyclopenta-phenanthrenone molecule **2.31** with the desired A/B and C/D *cis* rings, together with the B/C *trans*-ring. Both derivatives are products of the tandem sequence of a double-Michael addition (steps a and b), followed by aldol condensation (steps c–e) called polyanionic cyclization. However, the preparation of the starting building blocks **2.28** and **2.29** in this synthesis is very lengthy and laborious: cyclopentyl-hepta-2,4-dienoate derivative **2.28** must be obtained from tetrahydro-indenedione **2.26** in 14 steps in overall yields of 22%, while 6-oxocyclohexenecarbaldehyde **2.29** with protecting group TBDPS from cyclohexanone derivative **2.27** in two steps.

After 14 steps, double protected (G_1 and G_2) derivative **2.31** is now converted into another crucial compound **2.32** that has ouabain' stereochemistry, which was confirmed by the degradation of natural ouabain to obtain the same intermediate and its comparison with the synthetic intermediate. The completion of the synthesis of ouabagenin from this key molecule is realized using appropriate transformations to construct a lactone ring at C17 (butenolide cycle) (Figure 2.42) via the same tactic that made Stork in the synthesis of digitoxigenin (see Figure 2.39, derivative **2.18** is a crucial intermediate in the construction of butenolide ring). Finally, ouabagenin was converted into ouabain, attaching the L-rhamnosyl group to the β-C3 hydroxyl group in six steps. As we can see, this total synthesis of ouabagenin is a nice piece of synthetic art, and many efforts have been made to build this polyhydroxy-substituted molecule from simple materials.

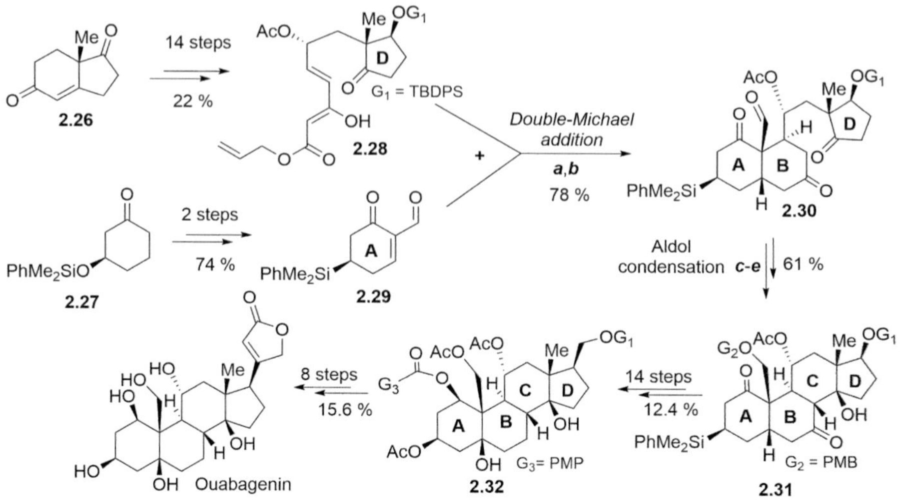

Figure 2.42. First total synthesis of ouabain and ouabagenin [70]: (a) Cs_2CO_3, CH_2Cl_2, 0°C; (b) $[Pd(PPh_3)_4]$, morpholine, THF; (c) $Li(Et_3CO)_3AlH$, THF, −78°C; (d) $PMBOC(NH)CCl_3$, $La(OTf)_3$ (0.1 equivalent), Et_2O, 0°C; (e) KHMDS, THF, 63°C. KHMDS -potassium bis(trimethylsilyl)amide; TBDPS = *tert*-butyldiphenylsilyl; PMB = *p*-methoxybenzyl; PMP = *p*-methoxyphenyl.

Generally, the above-mentioned synthetic approaches for cardenolides are inspired by their biogenesis. The basic structures of these molecules arise biosynthetically by the metabolism of plant steroids like cholesterol. It is worth noting that we can read in many biochemistry textbooks that cholesterol is only rarely found in plants [72]. It is false. While cholesterol is usually a minor constituent of the sterol fraction in plants, it is the principal constituent of some plant surfaces [73]. Cholesterol is a crucial precursor for thousands of bioactive plant metabolites, including discussed cardenolides and the well-known *Solanum* steroidal glycoalkaloids, but its plant pathway has remained enigmatic [74]. It is currently believed that plant cholesterol biosynthetic pathway overlaps with phytosterol metabolism [75]. Thus, we can assume that this steroid acts as a starting material in the biosynthesis of cardioactive cardenolides. Prof. Dewick discussed the general details of this biological process [76]. Some details are briefly presented in Figure 2.43.

Pregnenolone is formed by shortening the cholesterol side chain. Its oxidation gives progesterone, which is transformed into dihydroxy-5β-pregnan-20-one, and then trihydroxy-5β-pregnan-20-one. The lactone ring is created at the last stage, in which the malonyl-CoA is involved in the butenolide ring construction via ester formation and its subsequent aldol addition process (Figure 2.43).

2.6 Conclusion

Oxygen-containing toxins based on oxygenated phenolic and terpenoid molecules are numerous, complex second metabolites living organisms produce. Phenolics and terpenoids are considered classic defense compounds for protecting plants from herbivores. Generally, the biosynthesis of phenolics occurs via either the shikimate/

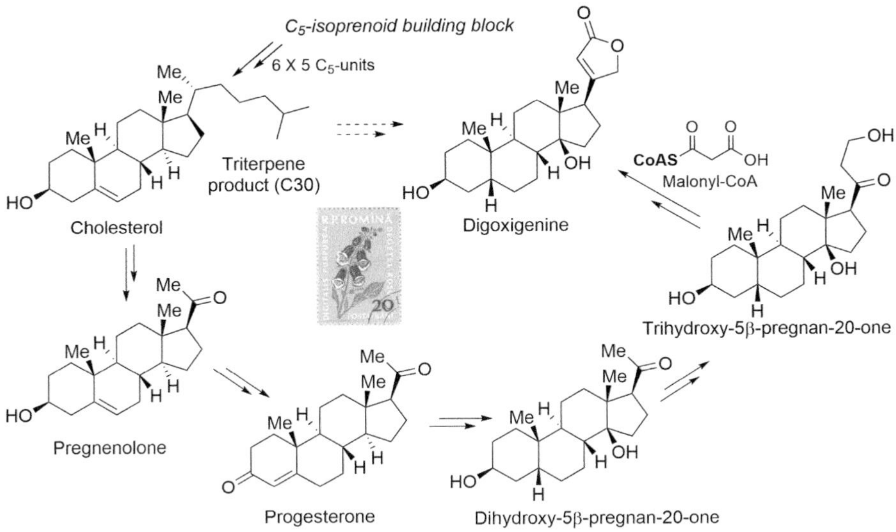

Figure 2.43. General biosynthetic steps from cholesterol to cardioactive cardenolides.

phenylpropanoid pathway, which leads directly to the phenylpropanoids, or the polyketide acetate/malonate pathway, which completes the construction of more complex phenolic molecules. In contrast, natural terpenes are built through the MEP/MVA pathways.

Being cyclic α,β-unsaturated ketones and phenolic molecules, the quinones are widely distributed in nature. Quinone molecules exhibit two important physicochemical properties: they can undergo reversible oxidation-reduction (redox) reactions, and many of them can also suffer from a nucleophilic attack due to their electrophilic character. As a result, quinones have different and varied biological properties against the same plants, animals, and microorganisms. Almost all plant quinones are toxic to animals. Primin, juglone, and lawsone are important phenolic molecules with simple *p*-benzoquinone or naphthoquinone structures, while hypericins are unsaturated polycyclic molecules. The latter compounds are responsible for the side effects of this plant in mammals, mainly grazing animals. This effect is called phototoxic activity. Thus, hypericin and its analogs are phototoxic agents known as photosensitizers.

Another group of phenolic molecules, oxygenated tricyclic molecules, called furocoumarins, are also phototoxic compounds. The role of furocoumarins in plant metabolism is framed within the different defense mechanisms that plants have to survive polyphagous herbivores and pathogenic fungi. This protective action is manifested in the varied toxicity toward the animals, in the antifeedant activity toward insects, and in the antifungal response. However, its photodynamic properties are more exciting and are always of great interest in biological sciences.

Both quinone and furocoumarin compounds are available from synthetic laboratories due to the relatively simple syntheses based on classical organic reactions. In contrast, synthetic methods of preparing iridoid molecules are less developed and thus are more laborious. Iridoid molecules are regarded as defense chemicals against herbivores and pathogens since iridoid glycosides, oxygen-containing monoterpene linked with sugars, usually have a bitter taste and have antifeedant and growth inhibitory activities against insects displaying antifungal and antibacterial properties. They are also known to deter feeding or decrease the growth rate of many generalist insect herbivores. Still, they often act as oviposition cues to specialist butterflies and are feeding stimulants to their larvae. In this context, aucubin and catalpol are the most studied iridoid phenolics.

Cardenolides are other remarkable oxygen-containing terpenic molecules with potent biological properties. Like iridoid molecules, they are terpenes linked with sugars, but they are more complex compounds—oxygenated steroids with the γ-butyrolactone cycle that are based on three six-membered and one five-membered rings. Its tetracyclic system can present two different forms of ring connection: *trans*-A/B/C-*cis*-C/D-configuration and *cis*-A/B,C/D-*trans*-B/C-form. Notable, the latter form is more toxic. Although some cardenolides are highly toxic to humans and animals, they have been used in the treatment of patients with heart failure for a long time, and that is why they are often called cardiotonic glycosides. The most famous cardenolides are digitoxin, oleandrin, ouabain (used as drugs), and calactin or calotropin (involved in exiting interactions between milkweeds and larvae of monarch butterflies as chemical defensive metabolites). As cardenolides are important molecules with remarkable biological properties, enormous interest in constructing these highly oxygenated steroids is always preserved. Therefore, numerous works are related to its partial, semi-synthesis, and total synthesis. Semi-syntheses using abundant plant sterols as starting materials are considered the most suitable preparative route, while total syntheses are considered less competitive.

2.7 References

[1] Basu, S., Varsani, S. and Louis, J. 2018. Altering plant defenses: herbivore-associated molecular patterns and effector arsenal of chewing herbivores. Mol. Plant-Microbe Inter. 31(1): 13–21.

[2] Tanaka, K., Choi, J., Cao, Y. and Stacey, G. 2014. Extracellular ATP acts as a damage-associated molecular pattern (DAMP) signal in plants. Front Plant Sci. 5: Article 446.

[3] Tsimogiannis, D. and Oreopoulou, V. 2019. Classification of phenolic compounds in plants. pp. 263–284. *In*: Watson, R.R. [Ed.]. Polyphenols in Plants, 2nd Ed. London: Academic Press.

[4] Falcone Ferreyra, M.L., Rius, S. and Casati, P. 2012. Flavonoids: biosynthesis, biological functions, and biotechnological applications. Front Plant Sci. 3: Article 222.

[5] Dewick, P.M. 2009. Medicinal Natural Products: A Biosynthetic Approach, New York: John Wiley & Sons; 550 pp.

[6] Ludwiczuk, A., Skalicka-Woźniak, K. and Georgiev, M.I. 2017. Terpenoids. pp. 233–266. *In*: Badal, S., Delgoda, R. [Eds.]. Pharmacognosy. London: Academic Press.

[7] Pichersky, E. and Raguso, R.A. 2018. Why do plants produce so many terpenoid compounds? New Phytologist 220(3): 692–702.

[8] Chen, J., Jiang, Q.D., Chai, Y.P., Zhang, H., Peng, P. and Yang, X.X. 2016. Natural terpenes as penetration enhancers for transdermal drug delivery. Molecules 21(12): 1709.

[9] Sikkema, J., de Bont, J.A. and Poolman, B. 1995. Mechanisms of membrane toxicity of hydrocarbons. Microbiol. Mol. Biol. Rev. 59(2): 201–222.

[10] El-Najjar, N., Gali-Muhtasib, H., Ketola, R.A., Vuorela, P., Urtti, A. and Vuorela, H. 2011. The chemical and biological activities of quinones: overview and implications in analytical detection. Phytochem. Rev. 10(3): 353–370.

[11] López, J., De La Cruz, F., Alcaraz, Y., Delgado, F. and Vázquez, M.A. 2015. Quinoid systems in chemistry and pharmacology. Med. Chem. Res. 24(10): 3599–3620.

[12] Schildknecht, H., Bayer, I. and Schmidt, H. 1967. Struktur des Primelgiftstoffes. Zeitschrift für Naturforschung B 22(1): 36–41.

[13] Kumar, S. and Deshmukh, R. 2017. Embelin as a potential drug molecule: a review. J. Pharmacogn. Nat. Prod. 3(3): 144.

[14] Radhakrishnan, N. and Gnanamani, A. 2014. 2,5-Dihydroxy-3-undecyl-1,4-benzoquinone (Embelin)-A second solid gold of India-A Review. Inter. J. Pharm. Pharmaceutical. Sci. 6(6): 23–30.

[15] Podolak, I. and Strzałka, M. 2008. Qualitative and quantitative LC profile of embelin and rapanone in selected *Lysimachia* species. Chromatographia 67(5-6): 471–475.

[16] Wilkinson, S. 1961. Quinone pigments the isolation of embelin and rapanone from *Rapanea pulchra* Gil and Schellenb. Planta Medica; 9(01): 121–122.

[17] Gunatilaka, A.A.L., Berger, J.M., Evans, R., Miller, J.S., Wisse, J.H. Neddermann, K.M. et al. 2001. Isolation, synthesis, and structure-activity relationships of bioactive benzoquinones from *Miconia lepidota* from the Suriname Rainforest 1. J. Nat. Prod. 64: 2–5.

[18] Watanabe, K., Sugizaki, T. and Tozawa, Y. 2012. A new entry to the synthesis of primin via a B-alkyl Suzuki–Miyaura cross-coupling reaction. Heterocycles 86(2): 985–989.

[19] Nasiri, H.R., Ferner, J., Tükek, C., Krishnathas, R. and Schwalbe, H. 2015. A concise one-step synthesis of primin and iso-primin. Tetrahedron Lett. 56(17): 2231–2233.

[20] Sisa, M., Dvorakova, M. and Vanek, T. 2017. Concise access to primin, miconidin and related natural resorcinols. Tetrahedron 73(35): 5297–5301.

[21] Jacob, A.M. and Moody, C.J. 2005. Microwave-assisted combined Mitsunobu reaction–Claisen rearrangement and microwave-assisted phenol oxidation: rapid synthesis of 2,6-disubstituted-1,4-benzoquinone natural products. Tetrahedron Lett. 46(51): 8823–8825.

[22] Bernays, E., Lupi, A., Bettolo, R.M., Mastrofrancesco, C. and Tagliatesta, P. 1984. Antifeedant nature of the quinone primin and its quinol miconidin from *Miconia* spp. Experientia 40(9): 1010–1011.

[23] Widhalm, J.R. and Rhodes, D. 2016. Biosynthesis and molecular actions of specialized 1,4-naphthoquinone natural products produced by horticultural plants. Horticulture Research 3: 16046.

[24] Strugstad, M.P. and Despotovski, S. 2012. A summary of extraction, synthesis, properties, and potential uses of juglone: A literature review. J. Ecosys. Manag. 13(3): 1–16.

[25] Müller, W.U. and Leistner, E. 1976. 1, 4-Naphthoquinone, an intermediate in juglone (5-hydroxy-1, 4-naphthoquinone) biosynthesis. Phytochem. 15(3): 407–410.

[26] Duchstein, H.J., Müller, K. and Wurm, G. 1988. Zum Reaktionsmechanismus der Oxygenierung von Naphtholen mit dem System $CuCl/O_2$. Arch Pharm. 321(3): 153–157.

[27] Barret, R. and Daudon, M. 1990. An efficient synthesis of juglone. Synth. Comm. 20(18): 2907–2912.

[28] Jordao, A.K., Vargas, M.D., Pinto, A.C., da Silva, F.D.C. and Ferreira, V.F. 2015. Lawsone in organic synthesis. RSC Adv. 5(83): 67909–67943.

[29] Kumagai, Y., Shinkai, Y., Miura, T. and Cho, A.K. 2012. The chemical biology of naphthoquinones and its environmental implications. Annu. Rev. Pharmacol. Toxicol. 52: 221–247.

[30] Izhaki, I. 2002. Emodin–a secondary metabolite with multiple ecological functions in higher plants. New Phytologist. 155(2): 205–217.

[31] Oleinick, N.L. 2010. http://photobiology.info/Oleinick.html.

[32] Larson, R.A. 1986. Insect defenses against phototoxic plant chemicals. J. Chem. Ecol. 12(4): 859–870.

[33] Gambari, R., Lampronti, I., Bianchi, N., Zuccato, C., Viola, G., Vedaldi, D. et al. 2007. Structure and biological activity of furocoumarins. pp. 265–276. *In*: Khan, M.T.H. [Ed.]. Bioactive Heterocycles III. Topics in Heterocyclic Chemistry. Berlin: Springer; vol 9.

[34] Caffieri, S. 2002. Furocoumarin photolysis: chemical and biological aspects. Photochem. Photobiol. Sci. 1(3): 149–157.

[35] Roelandts, R. 2002. The history of phototherapy: something new under the sun? J. Am. Acad. Dermatol. 46(6): 926–930.

[36] Dueholm, B., Krieger, C., Drew, D., Olry, A., Kamo, T., Taboureau, O. et al. 2015. Evolution of substrate recognition sites (SRSs) in cytochromes P450 from Apiaceae exemplified by the CYP71AJ subfamily. BMC Evolutionary Biol. 15(1): 122.

[37] Krieger, C., Roselli, S., Kellner-Thielmann, S., Galati, G., Schneider, B., Grosjean, J. et al. 2018. The CYP71AZ P450 subfamily: A driving factor for the diversification of coumarin biosynthesis in apiaceous plants. Front Plant Sci. 9: 820.
[38] Heravi, M.M. Khaghaninejad, S. and Mostofi, M. 2014. Pechmann reaction in the synthesis of coumarin derivatives. pp. 1–50. *In*: Katritzky, A.R. [Ed.]. Advances in Heterocyclic Chemistry. New York: Academic Press; Vol. 112.
[39] Cadierno, V. 2015. Metal-catalyzed routes for the synthesis of furocoumarins and coumestans. pp. 77–100. *In*: Brahmachari, G. [Ed.]. Green Synthetic Approaches for Biologically Relevant Heterocycles. Amsterdam: Elsevier.
[40] Lončarić, M., Gašo-Sokač, D., Jokić, S. and Molnar, M. 2020. Recent advances in the synthesis of coumarin derivatives from different starting materials. Biomolecules 10(1): 151.
[41] Adimule, V.M., Nandi, S.S., Kerur, S.S., Khadapure, S.A. and Chinnam, S. 2022. Recent advances in the one-pot synthesis of coumarin derivatives from different starting materials using nanoparticles: A review. Top Catal; https://doi.org/10.1007/s11244-022-01571-z.
[42] Stefanachi, A., Leonetti, F., Pisani, L., Catto, M. and Carotti, A. 2018. Coumarin: a natural, privileged and versatile scaffold for bioactive compounds. Molecules 23(2): 250.
[43] Barot, K.P., Jain, S.V., Kremer, L., Singh, S. and Ghate, M.D. 2015. Recent advances and therapeutic journey of coumarins: Current status and perspectives. Med. Chem. Res. 24: 2771–2798.
[44] Garg, S.S., Gupta, J., Sharma, S. and Sahu, D. 2020. An insight into the therapeutic applications of coumarin compounds and their mechanisms of action. Eur. J. Pharm. Sc. 152: 105424.
[45] Liu, Z. and Wen, X. 2013. A systematic review of a naturally occurring iridoid: catalpol. Curr. Bioactive Comp. 9(4): 306–323.
[46] Manickam, B., Sreedharan, R. and Elumalai, M. 2014. 'Genipin' - the natural water soluble cross-linking agent and its importance in the modified drug delivery systems: an overview. Curr. Drug Deliv. 11(1): 139–145.
[47] Ilc, T., Parage, C., Boachon, B., Navrot, N. andWerck-Reichhart, D. 2016. Monoterpenol oxidative metabolism: role in plant adaptation and potential applications. Front Plant Sci. 7: 509.
[48] Opitz, S.E. and Müller, C. 2009. Plant chemistry and insect sequestration. Chemoecol. 19(3): 117–154.
[49] Nieminen, M., Suomi, J., Van Nouhuys, S., Sauri, P. and Riekkola, M.L. 2003. Effect of iridoid glycoside content on oviposition host plant choice and parasitism in a specialist herbivore. J. Chem. Ecol. 29(4): 823–844.
[50] Eisner, T. 1964. Catnip: its raison d'etre. Science 146(3649): 1318–1320.
[51] Waller, G.R., Price, G.H. and Mitchell, E.D. 1969. Feline attractant, cis,trans-nepetalactone: Metabolism in the domestic cat. Science 164(3885): 1281–1282.
[52] Sakurai, K., Ikeda, K. and Mori, K. 1988. Both (4aS,7S,7aR)–(+)-Nepetalactone and its antipode are powerful attractants for cats. Agric Biol. Chem. 52(9): 2369–2371.
[53] Lichman, B.R., Godden, G.T., Hamilton, J.P., Palmer, L., Kamileen, M.O., Zhao, D. et al. 2020. The evolutionary origins of the cat attractant nepetalactone in catnip. Sci. Adv. 6(20): eaba0721.
[54] Schreiber, S.L., Meyers, H.V. and Wiberg, K.B. 1986. Stereochemistry of the intramolecular enamine/enal (enone) cycloaddition reaction and subsequent transformations. J. Am. Chem. Soc. 108(26): 8274–8277.
[55] Dawson, G.W., Pickett, J.A. and Smiley, D.W. 1996. The aphid sex pheromone cyclopentanoids: Synthesis in the elucidation of structure and biosynthetic pathways. Bioorg. Med. Chem. 4(3): 351–361.
[56] Liblikas, I., Santangelo, E.M., Sandell, J., Baeckström, P., Svensson, M., Jacobsson, U. et al. 2005. Simplified isolation procedure and interconversion of the diastereomers of nepetalactone and nepetalactol. J. Nat. Prod. 68(6): 886–890.
[57] El-Seedi, H.R., Khalifa, S.A., Taher, E.A., Farag, M.A., Saeed, A., Gamal, M. et al. 2019. Cardenolides: insights from chemical structure and pharmacological utility. Pharm. Res. 141: 123–175.
[58] Agrawal, A.A., Petschenka, G., Bingham, R.A., Weber, M.G. and Rasmann, S. 2012. Toxic cardenolides: chemical ecology and coevolution of specialized plant–herbivore interactions. New Phytologist 194(1): 28–45.

[59] Jorgensen, P.L., Hakansson, K.O. and Karlish, S.J.D. 2003. Structure and mechanism of Na,K-ATPase: functional sites and their interactions. Ann. Rev. Physiol. 65(1): 817–849.
[60] Parsons, J.A. 1965. A digitalis-like toxin in the monarch butterfly, *Danaus plexippus* L. J. Physiol. 178(2): 290.
[61] Dobler, S., Petschenka, G. and Pankoke, H. 2011. Coping with toxic plant compounds–the insect's perspective on iridoid glycosides and cardenolides. Phytochemistry 72(13): 1593–1604.
[62] Holzinger, F. and Wink, M. 1996. Mediation of cardiac glycoside insensitivity in the monarch butterfly (*Danaus plexippus*): role of an amino acid substitution in the ouabain binding site of Na+, K+-ATPase. J. Chem. Ecol. 22: 1921–1937.
[63] Michalak, M., Michalak, K. and Wicha, J. 2017. The synthesis of cardenolide and bufadienolide aglycones, and related steroids bearing a heterocyclic subunit. Nat. Prod. Rep. 34(4): 361–410.
[64] Pessôa, M.T.C., Barbosa, L.A. and Villar, J.A.F. 2018. Synthesis of cardiac steroids and their role on heart failure and cancer. In Studies in Natural Products Chemistry, Atta-ur-Rahman, Ed, Elsevier 57: 79–113.
[65] Wiesner, K. and Tsai, T.Y. 1986. Some recent progress in the synthetic and medicinal chemistry of cardioactive steroid glycosides. Pure Appl. Chem. 58(5): 799–810.
[66] Kabat, M.M., Kurek, A. and Wicha, J. 1983. Cardiotonic steroids. 5. A synthesis of bufadienolides and cardenolides from 3. beta.-acetoxy-5-androsten-17-one via common intermediates. J. Org. Chem. 48(23): 4248–4251.
[67] Tsai, T.Y.R., Minta, A. and Wiesner, K. 1979. A simple synthesis of cardenolides and their less toxic isomers via Furyl intermediates. Heterocycles 12(11): 1397–1402.
[68] Stork, G., West, F., Lee, H.Y., Isaacs, R.C. and Manabe, S. 1996. The total synthesis of a natural cardenolide:(+)-digitoxigenin. J. Am. Chem. Soc. 118(43): 10660–10661.
[69] Honma, M. and Nakada, M. 2007. Enantioselective total synthesis of (+)-digitoxigenin. Tetrahedron Lett. 48(9): 1541–1544.
[70] Zhang, H., Reddy, M.S., Phoenix, S. and Deslongchamps, P. 2008. Total synthesis of ouabagenin and ouabain. Angew Chem. Int. Ed. 47(7): 1272–1275.
[71] Reddy, M.S., Zhang, H., Phoenix, S. and Deslongchamps, P. 2009. Total synthesis of ouabagenin and ouabain. Chem. Asian J. 4(5): 725–741.
[72] Horton, H.R., Moran, L.A., Ochs, R.S., Rawn, J.D. and Scrimgeour, K.G. 2002. Principles of Biochemistry, 3rd ed.; Prentice Hall: Upper Saddle River, NJ, p. 275.
[73] Behrman, E.J. and Gopalan, V. 2005. Cholesterol and plants. J. Chem. Educ. 82(12): 1791–1793.
[74] Nes, W.D. 2011. Biosynthesis of cholesterol and other sterols. Chem. Rev. 111(10): 6423–6451.
[75] Sonawane, P.D., Pollier, J., Panda, S., Szymanski, J., Massalha, H., Yona, M. et al. 2016. Plant cholesterol biosynthetic pathway overlaps with phytosterol metabolism. Nat. Plants 3(1): 16205.
[76] Dewick, P.M. 2002. Medicinal natural products: a biosynthetic approach. John Wiley & Sons, Chichester, pp. 244–245.

Chapter 3

Vegetable Nitrogen-Containing Toxins and their Effects on Animals

Plant-Animal Interactions

3. INTRODUCTION: NITROGEN-CONTAINING METABOLITES AND THEIR STRUCTURAL DIVERSITY AND A LARGE RANGE OF BIOACTIVITIES

In the previous chapter, we saw that when plants grow, they metabolize different and varied small molecules—secondary oxygenated metabolites—with multiple functions, including defense against herbivorous animals. Therefore, this chapter examines how and why plants generate toxic molecules with an atom and/or more nitrogen atoms, namely, nitrogen-containing metabolites. We have already mentioned that plants have been interacting with animals (especially insects) for several hundred million years, leading to complex chemical defenses against various insect strategies aimed at seeking their food.

These chemical defenses against herbivores are of the utmost importance to plants. As we have already learned, some defenses are constitutive. In contrast, others are induced (see Section 1.3; Figure 1.8). Primary plant metabolism also generates many toxic nitrogen-containing compounds of different classes [1]: amines, alkamides, non-protein α-amino acids, glucosinolates, cyanogenic glycosides, alkaloids, lectins, peptides, and polypeptides that have been implicated in the defense of plants against insect pests. Some of these are discussed in this chapter.

We recall that nitrogen (Latin: *nitrum*; Greek: *Nitron*, meaning "native soda;" genes, meaning "forming") is the main constituent of the Earth's atmosphere and an essential element in all known forms of life. Interestingly, inorganic nitrogen compounds have a long history, with ammonium chloride being known to Herodotus (c. 484–c. 425 BC), the ancient Greek historian. In contrast, organic (i.e., natural) nitrogen molecules became known only in 1773, when the French chemist H. Rouelle isolated urea in urine (see Section 1.4). In living organisms, nitrogen atoms are part of the molecular structures of such vital substances as amino acids, proteins, and nucleic acids. However, we will discuss plant toxic nitrogen-containing metabolites, and our discussion will start with non-protein amino acids.

3.1 Non-Protein Amino Acids: Potent Antimetabolites

Among plant toxins based on nitrogen compounds, non-protein α-amino acids have a simpler molecular structure. The higher plants synthesize hundreds of amino acids, of which only 20 (plus selenocysteine and pyrrolysine) are part of the numerous different proteins. Thus, as necessary allelopathic chemicals, non-protein α-amino acids (NPAAs) do not participate in the normal replication process but play other vital functions in the plant-animal interaction, such as nutrient acquisition, signaling, stress response, etc. However, NPAAs are particularly well-known as toxins [2–4].

However, concerning their toxicity, it cannot be generalized; while some of them are toxic to humans, domestic animals, and some insects, others are known to be ingested by other insects and animals and are excreted unchanged without any action. NPAAs are found in many plant families, including Leguminosae, Liliaceae, Cycadaceae, Rubiaceae, Sapindaceae, Lecythidaceae, and Compositae. They are present, especially in legumes, and are mainly abundant in seeds. More than 250 NPAAs have been reported. The toxic effects of some non-protein amino acids are often related to the mechanism of antimetabolites, which interfere with normal metabolic processes within the cells of living organisms. Thus, in plants, these acids are considered potent antimetabolites. As structural analogs of the 20 protein α-amino acids, NPAAs can exert their toxicity by alternating enzyme systems or incorporating protein synthesis. The most studied and interesting NPAAs selected for our discussion are depicted in Figure 3.1.

Indospicine (L-6-amidino-2-amino-hexanoic acid with (*S*)-configuration), a hepatotoxic arginine (Arg) analog, occurs in leguminous plants of the *Indigofera* genus, e.g., *I. spicata* and *I. hendacphylla* and accumulates in the tissues of grazing animals that consume these plants [5, 6]. Indospicine, being antimetabolite, competes with arginine, inhibiting arginase enzyme and DNA synthesis that ends with liver degeneration. Indospicine hepatotoxicity has been demonstrated in all animal species investigated, with reports of acute and sub-chronic hepatotoxic evidence for mice, rats, guinea pigs, rabbits, dogs, sheep, and cattle. It was also shown that its toxic effects can induce damage to the fetus (teratogenicity). Interestingly, the

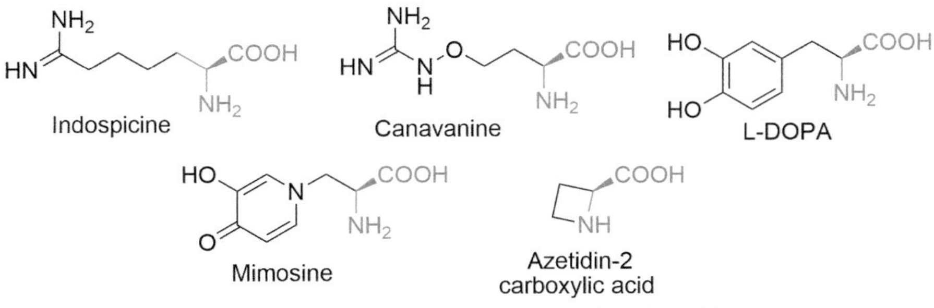

Figure 3.1. Structure of selected non-protein amino acids.

presence of indospicine has been confirmed both in primary consumers (herbivores) and secondary consumers (carnivores, mainly canines). In contrast, the effect of indospicine in humans (if it exists) has not yet been clarified.

Noteworthy is that no known mammalian enzyme can degrade the amidino group of indospicine. However, it is known that amidines such as indospicine could be hydrolyzed firstly to the amide and then subsequently to the corresponding acid (Figure 3.2), with the first step usually being faster than the second. The hydrolysis of pure indospicine to 2-aminopimelamic acid occurs in water at 120°C for 20 hours or with 0.1M Na_2CO_3 at 50°C for 15 hours, while the amide obtained can be hydrolyzed to 2-aminopimelic acid by treatment with 1N HCl at 120°C for 2 hours. Alternatively, acid hydrolysis of indospicine using 6N HCl at 120°C for 20 h directly afforded 2-aminopimelic acid [7].

Chemical microwave-assisted degradation of indospicine under mild alkaline conditions (0.05% sodium bicarbonate, pH 8.8) rapidly (15 minutes) affords 2-aminopimelamic acid as the primary product and 2-aminopimelic acid as a minor product [8]. Interestingly, camels grazing *Indigofera* pasture plants can biodegrade approximately 99% of the indospicine in *I. spicata* within 48 hours of incubation, producing 2-aminopimelamic acid and 2-aminopimelic acid [8] because the camel is unique in being a non-ruminant herbivore animal with compartmental stomach and extensive foregut fermentation processes that facilitate hydrolysis. Due to the biological and pharmacological importance of indospicine, several routes of its synthesis have been proposed. One of them is illustrated in Figure 3.3.

The synthesis begins with the HCl-catalyzed esterification of homoserine lactone **3.1** in methanol to give methyl 2-amino-4-hydroxybutyrate **3.2**, which is transformed into cyanohexanoate **3.3** via two steps. This nitrile is then subjected to a Pinner reaction with anhydrous hydrogen chloride in methanol to obtain the imidate ester **3.4**, whose acid treatment allows the formation of indospicine (Figure 3.3) [9].

L-Canavanine, (*S*)-2-amino-4-(guanidinooxy) butyric acid (Figure 3.1), is one of the most common non-protein amino acids that occur in approximately 350 species of the Fabaceae and is mainly found in legume seeds, particularly in jack bean

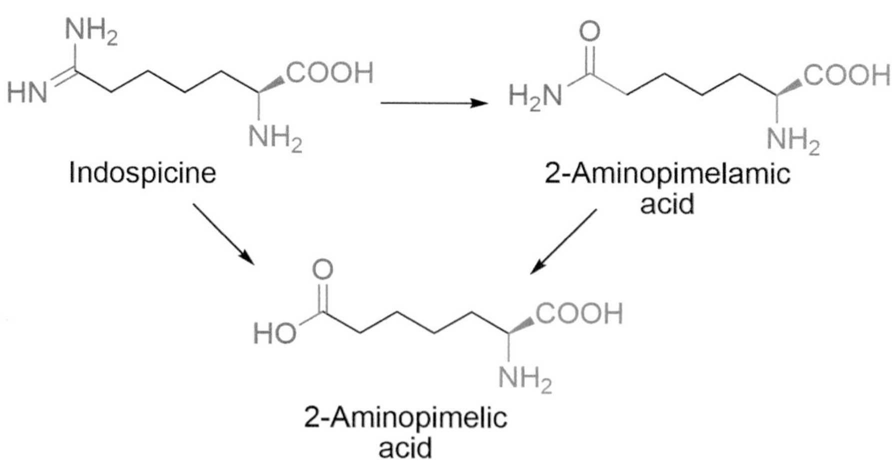

Figure 3.2. Hydrolysis of indoscipine.

Canavalia ensiformis and *Dioclea megacarpa*, where its concentration can reach up to 13% dry weight and constituted more than 95% of free amino acid nitrogen. These beans and their extracts have been shown to inhibit the growth of many neighboring plants. Its roots form symbiotic nodules of nitrogen-fixing bacteria, leading to elimination or reduction in the application of nitrogen fertilizers.

The high content of L-canavanine in these species is the key to their successful survival against numerous herbivores and pests, considering that only 0.02% of canavanine can damage insect larvae. Like indospicine, L-canavanine competes with the protein amino acid L-arginine in protein synthesis. Arginyl-tRNA synthetase readily esterifies L-canavanine to the cognate tRNA-Arg, resulting in the synthesis of abnormal "canavanyl" proteins [10]. Thus, these structural defects of the proteins lead to abnormalities in the development of the insect (or animal) that can cause death. Surprisingly, this natural product was only slightly toxic to rats following a single subcutaneous injection; the LD_{50} was 5.0–5.9 g/kg in adult rats [11]. The L-canavanine was also investigated as a potential antitumor drug because it exhibits physiological potential to disturb Arg metabolism and interferes with the synthesis or accumulation of nitric oxide (NO), a short-lived free radical $NO^{\bullet}$, and is implicated in carcinogenesis.

The good insecticidal and fungicidal activity of canavanine and plants rich in it was confirmed by numerous experimental studies. An interesting example is the action of canavanine on tobacco hornworm, *Manduca sexta*, devastating predators of the Solanaceae plants. As the Fabaceae plants produce canavanine, the tobacco worm has not been able to adapt to these plants; Canavanine significantly reduces larval growth, delaying or/and disrupting larva-pupa metamorphosis. What is a toxin for the *Manduca sexta* larvae is a harmless substance for another species of tobacco budworm, *Heliothis virescens*. This shows its remarkable resistance to canavanine; therefore, it can feed with a diet based on canavanine. In studying this relationship, it was demonstrated for the first time that *H. verescens* uses a constitutive enzyme of the intestine to catalyze the irreversible hydrolysis of canavanine to hydroxyguanidine and L-homoserine, which are eliminated from the body without affecting normal metabolic functions [12] (Figure 3.4).

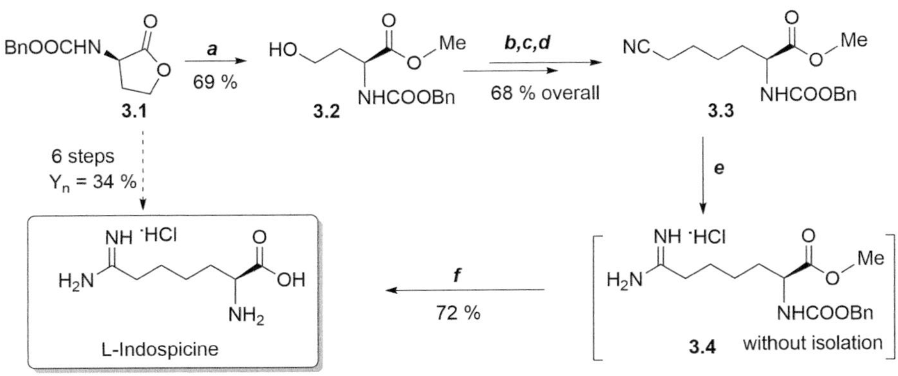

Figure 3.3. Synthesis of indoscipine: (a) MeOH/HCl; (b) MsCl/Et_3N; (c) NaI/DMF; (d) MeCN, Zn/CuI; (e) one-pot: 1. HCl/MeOH, 2. NH_3, 3. HCl; (f) HCl, 20ºC, 8 days.

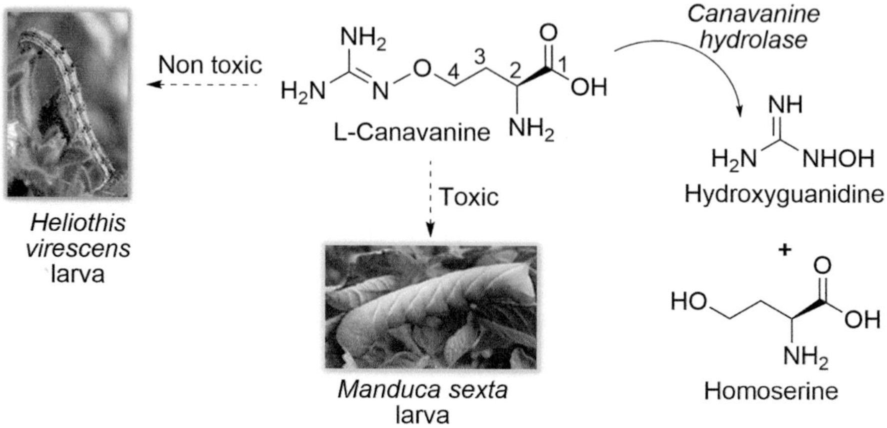

Figure 3.4. Canavanine is toxic for *Manduca sexta* larva, while canavanine is a harmless substance for *Heliothis virescens*, which can hydrolyze it into homoserine and hydroxyguanidine.

This enzyme, called canavanine hydrolase, is a new type of hydrolase that acts on the oxygen-nitrogen bond of the substrate (EC 3.13.1.1). The enzyme has a mass of 285 KDa and comprises two different subunits with a mass of 50 KDa or 47.5 KDa. Canavanine hydrolase has a high degree of specificity for L-canavanine since its activity was negligible in the presence of L-2-amino-5-(guanidinooxy) pentanoate or L-2-amino-3-(guanidinooxy) propionate, the lowest and highest homologs of canavanine, respectively.

However, there are other better-adopted insects than *H. virescens* larvae: the larvae of weevil beetles *Caryedes brasiliensis* demonstrate the most sophisticated bio-adaptation. They use canavanine as the nitrogen source in constructing their own proteins. The *C. brasiliensis* larvae feed on *D. megacarpa,* metabolizing canavanine due to high concentrations of the arginase enzyme in its body. The primary function of this enzyme is to degrade L-arginine in L-ornithine. Still, in high concentrations, it can also transform canavanine into L-canaline (2-amino-4-(aminooxy) butyric acid) and urea (Figure 3.5).

Formed canaline is a toxic molecule, and it is degraded by the enzyme NADPH-Cytochrome P-450-reductase in homoserine and ammonia. The conversion of urea to ammonia is facilitated by the urease enzyme of these insects. Thus, the weevils transform toxic canavanine into ammonia, an essential material in constructing protein amino acids. A fascinating aspect of these biochemical processes is that these weevils have adapted to a harmful substance in their food (canavanine) and use the stored nitrogen in this toxic amino acid as an effective dietary nitrogen source for the biosynthesis of their amino acids [13]. It is interesting to note that alfalfa seeds (*Medicago sativa*), an important crop, contain approximately 0.5% of canavanine, and people with systemic lupus erythematosus disease should avoid the consumption of these seeds since L-canavanine generates a strong antimetabolic activity that aggravates the development of this autoimmune disease, where the immune system attacks healthy cells and tissues [14].

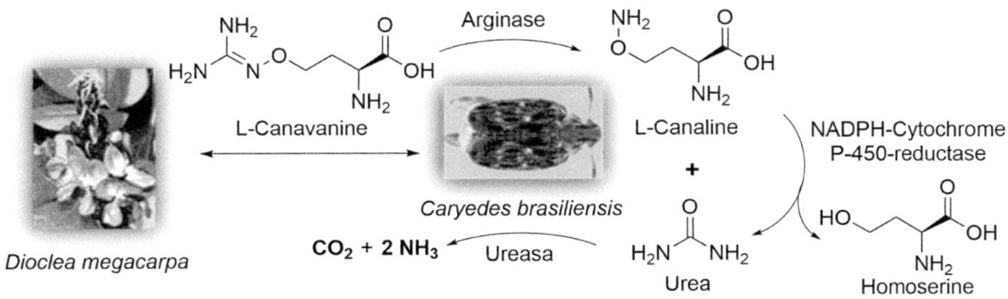

Figure 3.5. Bio-adaptation of *Caryedes brasiliensis* larvae on the *D. megacarpa* species.

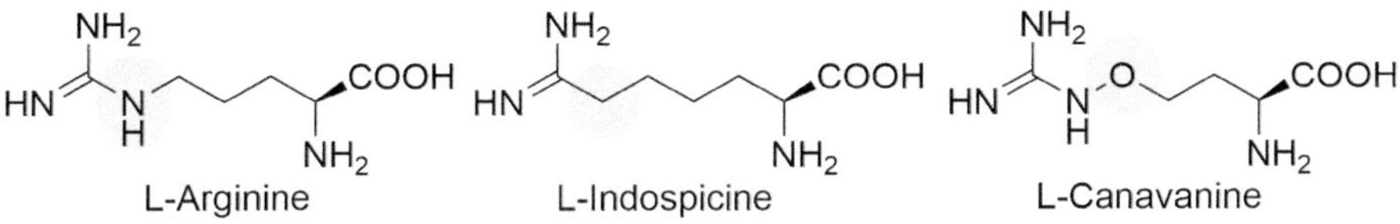

Figure 3.6. Structure of arginine, essential amino acid, and natural analogs.

As we said, the two NPAAs mentioned above are analogs of arginine, an essential amino acid, but they are structurally different: indospicine is a carbon analog, while cavananine is an oxo-analog. This detail is seen well in Figure 3.6.

Among many non-protein amino acids, 3,4-dihydroxyphenylalanine (L-DOPA) (Figure 3.1) stands out for its allelopathic solid activity. It is known that most species of the genus *Mucuna* have a reasonable tolerance to various unfavorable abiotic factors, such as drought, low fertility, and high acidity of the soil. This tolerance is related to the presence of this allelochemical [15]. It was found that these plants can release DOPA from the roots, where their concentration in the vicinity of the roots can reach up to 50 ppm, allowing herbaceous spices, rich in this amino acid, to reduce the growth of neighboring plants.

Therefore, some species of this genus are used as fallow crops to improve soil fertility and combat weeds. Mainly, DOPA is found in large quantities (4–9%) in *Mucuna pruriens* and *Vicia faba*, herbaceous climbing plants, both of the Fabaceae family. Velvet bean (*Mucuna pruriens*) is often used for soil cover and as silage due to the large amount of organic matter it produces with high digestibility. It is estimated that velvet beans can release about 100–450 kg/ha of L-DOPA into the soil.

Due to large amounts of L-DOPA, small mammals or insects do not attack these plants. High concentrations of L-DOPA completely prevent the insect from feeding by acting as a strong repellent agent and can interfere with tyrosinase, an essential enzyme for hardening and darkening the insect's cuticle. An example is the effect of DOPA and velvet bean on the larvae of the *Spodoptera eridania* budworm (Figure 3.7A). Studies revealed that artificially fed larvae containing a mixture of L-DOPA (0.25%) and leaf dust (5%) of *M. pruriens* showed malformations in both larvae and adults, and the mortality was considerably increased. Notably, in a plant, L-DOPA is also a precursor of many alkaloids, catecholamines, and melanin. Its biosynthesis starts with shikimic acid (Figure 2.1) and results from the hydroxylation of L-tyrosine, another important essential amino acid (Figure 3.7B). This reaction occurs with copper-containing enzyme tyrosine hydroxylase in the presence of

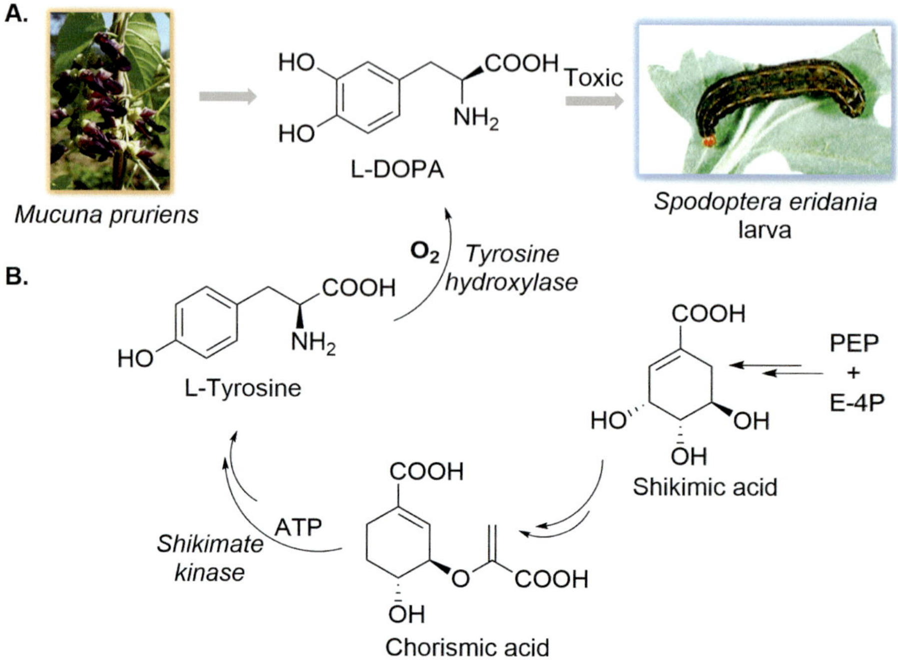

Figure 3.7. Interaction of L-DOPA with insects (A) and its simplified metabolic pathway from shikimic acid (B).

molecular O_2. It can be added that the L-DOPA synthesis pathway in plants is analogous to that in mammals.

DOPA is a precursor of neurologically essential molecules such as the neurotransmitters dopamine, adrenaline (epinephrine), and noradrenaline (norepinephrine). More interesting is that DOPA is not as toxic to mammals and is used as a drug in the treatment of Parkinson's disease. L-DOPA can cross the protective blood-brain barrier, whereas dopamine cannot. That is why this molecule is used to increase dopamine concentrations in treating Parkinson's disease. Therefore, L-DOPA supplements have been used for dopamine replacement therapy, and they remain to be the standard today. In mammals, DOPA may be generated by controlled enzymatic pathways and uncontrolled radical reactions.

L-Mimosine, β-[N-(3-hydroxy-4-pyridone)]-α-aminopropionic acid with (*S*)-configuration (Figure 3.1), is an allelochemical agent with potent growth inhibitory activity against phytopathogenic fungi. It occurs in *Mimosa pudica* and in the leaves and seeds of *Leucaena leucocephala*, a shrub legume native to Central America that is widely used as a pasture legume in the tropics and sub-tropics. *Leucaena* foliage is used as animal feed, and the leaves and seeds are used as human food in Central America, Indonesia, and Thailand. The mimosine content in this plan varies from 2.2% to 10%. This NPAA acts as a tyrosine analog capable of competitive inhibition of tyrosinase and inhibition of thyroxine decarboxylase. Thus, using *Leucaena* for cattle feeding is not without its problems due to the presence of mimosine: it may cause reproductive disorders, teratogenic effects, and organ damage.

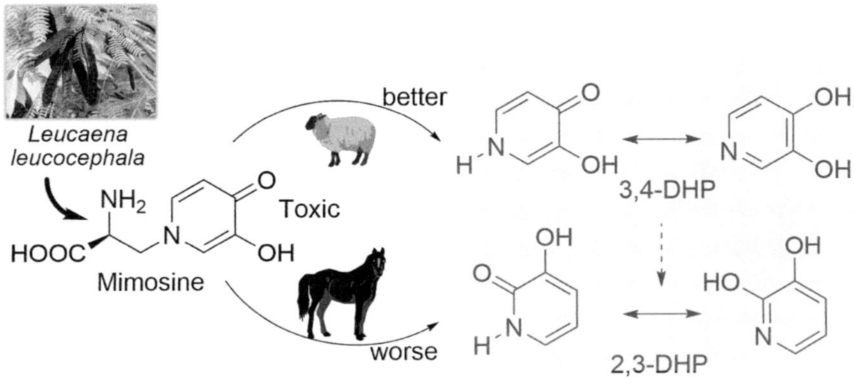

Figure 3.8. Biodegradation of mimosine in animal gut microflora.

Consequently, symptoms of mimosine toxicity include decreased weight gain, cataracts in young animals, infertility, goiter, and, most strikingly, loss of hair. The gut microflora of animals hydrolyzes Mimosine to its immediate secondary metabolite, 3-hydroxy-4-(1*H*)-pyridone (3,4-DHP), which could be further degraded rapidly into its isomer 2,3-dihydroxypyridine (2,3-DHP) (Figure 3.8) [16]. A certain endogenous plant enzyme is present in leaves and seeds that can catalyze this conversion. Mimosine itself is not goitrogenic, but its bacterial metabolite, 3,4-DHP is. Therefore, conversion to these intermediates does not detoxify the toxicity. Generally, ruminants (cattle, sheep, and goats) are better at tolerating these mimosine-containing plants than non-ruminants (horses, pigs, poultry, and rabbits) due to the presence of microflora in the rumen. Sheep find Leucaena forage very palatable. They can tolerate *L. leucocephala* intakes corresponding to only 0.14 g mimosine/kg body weight; however, cattle can tolerate 0.18 g/kg without serious toxic symptoms. Curiously, cattle in Australia and a few other countries do not degrade DHP due to the absence of the required microorganisms.

The last, but not the least NPAA in our discussion is the L-azetidin-2-carboxylic acid (L-Aze) (Figure 3.1), a cyclic analog of proline amino acid, which is first identified from a beautiful flowering plant called lily-of-the-valley (*Convallaria majalis*) [17]. By extracting the fresh leaves of this plant with ethanol in 1956, Fowden, an American chemist, could not only separate, purify, and characterize but synthesize this substance, thus confirming its structure [18]. From 2.2 kg of fresh leaves, 520 mg of L-Aze were obtained, representing 75.7% of the total amount of amino acids. Likewise, the presence of L-Aze was confirmed in the seeds of *C. majalis* (70.1%) and the leaves of *Maianthemum canadense* (58.1%). At the same time, the concentration of proline, its protein analog, was minimal, ranging from 0.1 to 1.8% in the three species mentioned. In the 1970s, studies of its biological effects on insects showed that L-Aze could reduce levels of catecholamines (dopamine and N-β-alanyl dopamine) in hemolymph and cuticles of some insects, indicating this substance has an inhibitory effect on the mechanisms involved in the biosynthesis of dopamine, catabolism or transport.

Today, it is known that L-Aze is a toxic and teratogenic molecule for many species of animals, including man, and is incorporated into a protein instead of

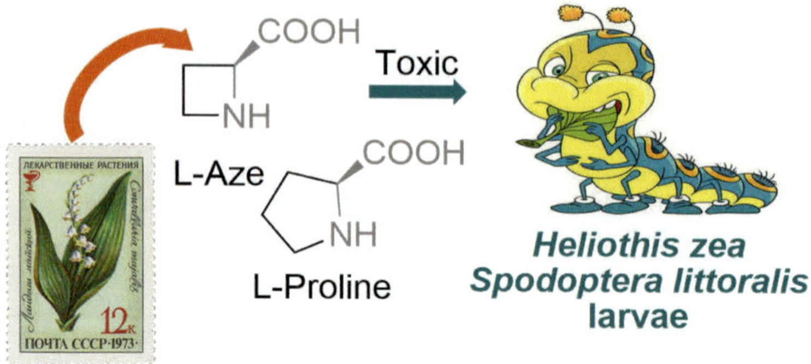

Figure 3.9. *Convallaria majalis,* the first source of L-Aze in a stamp. L-Aze, as an L-proline antimetabolite, is highly toxic for *H. zea* and *S. littoralis* larvae, important pests in agriculture.

proline, altering collagen, keratin, hemoglobin, and protein folding; that is, it acts as an antimetabolite [19]. This unusual amino acid has also been found in many other Liliaceae plants (*Liriope muscari*, *Ruscus hypoglossum,* and *Urginea maritime*) and the roots of sugar and garden beets (*Beta vulgaris*). The plants *L. muscari* and *U. maritime* are highly resistant to corn earworm, *Heliothis zea,* and cotton leafworm *Spodoptera littoralis* larvae, respectively [20–22]. The high concentration of toxic L-Aze is the reason for this resistance. These larvae are one of the worst pests in agriculture, and plants rich in L-Aze may help control these pests (Figure 3.9).

Being a constrained and quite rare amino acid, L-Aze and its derivatives are significant as active pharmaceutical ingredients. Thus, developing efficient synthetic routes for such azetidine derivatives is the main task of organic chemists. Concerning the biosynthesis of this non-proteinogenic amino acid, it was found in 1964 that L-Aze is formed in the plant *C. majalis* from methionine **3.5** through an intramolecular displacement of thiomethyladenosine by the α-amino group of S-adenosylmethionine **3.6**, similar to the role of SAM involved in many biochemical methylation reactions. Administrating ^{14}C-labeled methionine, the radioactive L-Aze is formed, and its decarboxylation gives radioactive carbon dioxide ($^{14}CO_2$) that proves this biosynthetic scheme (Figure 3.10).

Regarding its chemical synthesis, many synthetic routes to racemic (D, L) and L-enantiomer isomers of Aze have been realized using simple acyclic and cyclic precursors [23]. For example, racemic syntheses of Aze from 2-pyrrolidinone **3.7** and γ-butyrolactone **3.10** are given in Figure 3.11 and Figure 3.12, respectively.

Treatment of **3.7** with dimethyl sulfate followed by subsequent bromination with NBS produces **3.8**. Its hydrolysis in diluted HCl solution provides the γ-amino-α-bromobutyric acid **3.9**, which, without further purification, is treated with barium hydroxide followed by a 4 N sulfuric acid solution to adjust pH to 6.0. This allows us to obtain racemic (±)-Aze in a 53% yield (Figure 3.11) [24].

When γ-butyrolactone **3.10** is used as the starting material, the synthesis of Aze begins with the bromination-transesterification sequence, which gives benzyl α,γ-dibromobutyrate **3.11** in good yields (70%). This butylate can be converted in (±)-Aze in a two-step procedure: its reaction with benzhydrylamine (Ph_2CHNH_2) provides the N,O-protected azetidine-2-carboxylic acid **3.12**, which can be hydrogenated

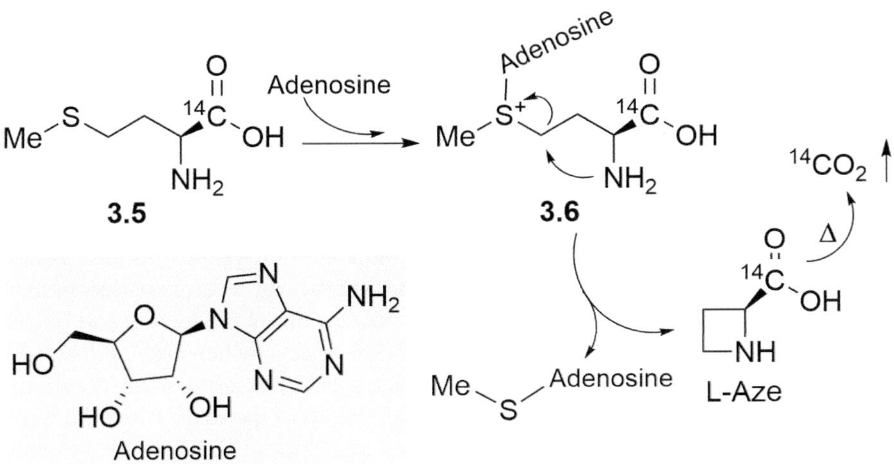

Figure 3.10. Biosynthesis of L-azetidine-2-carboxylic acid in *C. majalis*.

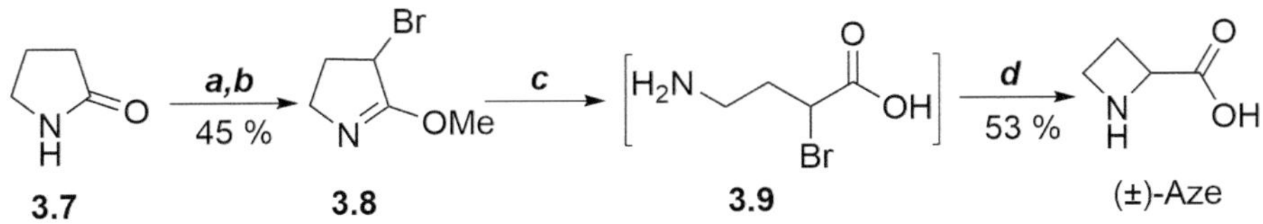

Figure 3.11. Racemic synthesis of Aze from 2-pyrrolidinone: (a) Me_2SO_4; (b) NBS, CCl_4, cat. benzoyl peroxide, reflux, 1 hour; (c) 3 N HCl, reflux, 2 hours; (d) $Ba(OH)_2$, then 4 N H_2SO_4.

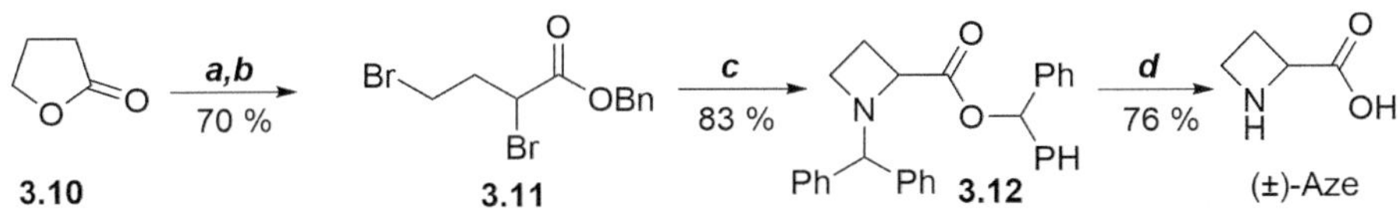

Figure 3.12. Racemic synthesis of Aze from γ-butyrolactone: (a) P, Br_2; (b) BnOH, HCl; (c) Ph_2CHNH_2, MeCN, reflux, 20 hours; (d) H_2, $Pd(OH)_2/C$, MeOH, 20 hours.

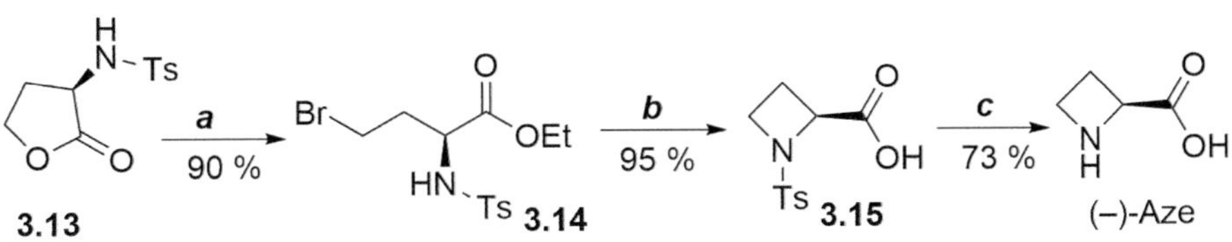

Figure 3.13 Asymmetric synthesis of L-Aze from tosyl-L-homoserine lactone: (a) HBr/EtOH, rt, 12 hours; (b) NaH/DMF-H_2O, 30 minutes.; (c) Na/NH_3 liq.

directly over Pearlman's catalyst (20% palladium hydroxide-on-carbon) to give desired amino acid (Figure 3.12) [25]. The overall yield from γ-butyrolactone was 53.5%, while the 2-pyrrolidinone-based procedure offers only 23.8% of the overall yield having the same quantity of steps.

One of the different approaches to the asymmetric synthesis of L-(–)-Aze consists of using α-amino-γ-bromobutyric acid **3.14**, a reverse precursor (compared with **3.10**), which is easily available from tosyl-L-homoserine lactone **3.13**. Recyclization of **3.14** with sodium hydride provides tosyl-protected acid **3.15**, and then this protecting group is removed by sodium in liquid ammonia to obtain the optically active (–)-Aze (Figure 3.13) [26].

3.2 Cyanogenic Glucosides: Cyanide Producers

Cyanogenic glucosides, glucose-based metabolites capable of producing and releasing toxic hydrogen cyanide, are a group of nitrogenous toxins against herbivores. They are essentially defensive, and numerous molecules are known to be present in more than 2,500 plant species. The process of the production and release of HCN in plants and some animals is called cyanogenesis, which is sometimes also referred to as a cyanide bomb [27]. In plant cells, cyanogenic glycosides, previously biosynthesized, are stored intact in the vacuole; they become toxic when the tissue is affected and the cell membrane breaks, allowing cyanogenic glycosides to meet enzymes in the plant's cytoplasm. This contact first generates cyanohydrins and then hydrogen cyanide (prussic acid, H-C≡N). The above steps provide plants with an effective chemical defense against herbivores and intrusive pathogens that cause tissue damage.

Chemically, the cyanogenesis process begins with three consecutive reactions: N-hydroxylation, decarboxylation, and dehydration of hydrophobic α-amino acids **3.16**, such as L-tyrosine, L-phenylalanine, L-valine, L-isoleucine, and L-leucine. The P450 enzyme system mediates these reactions to form an aldoxime molecule **3.18** through an unstable intermediate hydroxyamyl-carboxyacid derivative **3.17** (Figure 3.14). An α-hydroxynitrile (cyanohydrin, **3.19**) is formed by dehydration and hydroxylation of α-aldoxime carbon. This formed cyanohydrin in plants is unstable, but the plant system can stabilize it for safekeeping and use. This occurs through glycosylation, a reaction catalyzed by a UDP-glucose glucosyltransferase, forming the glucosidic bond Glu-O in **3.20**. The steps mentioned above form the biosynthesis of these nitrogenous toxins.

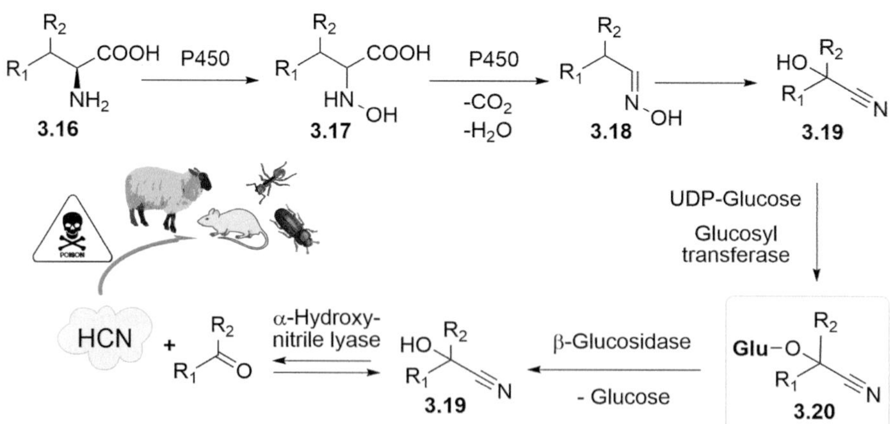

Figure 3.14. Biogenesis of cyanogenic glycosides and their catabolism.

The latter is hydrolyzed by a β-glucosidase, which provides sugar and a cyanohydrin (remember, it is an aglucone or aglycone), which now spontaneously releases hydrogen cyanide and a carbonyl compound when pH is approximately 6.0. Thus, HCN production depends on the biosynthesis of cyanogenic glucosides and the presence (or absence) of degrading enzymes.

Cyanogenic compounds have been found in numerous species of edible plants of first necessity and of more production worldwide, such as corn, barley, rice, wheat, and cassava, among others. Some of these species need treatments that allow the release of HCN before ingestion. HCN molecule is a respiratory inhibitor. Exposure to lethal doses of cyanide leads to cytotoxic hypoxia since cyanide has a strong affinity for iron in the ferric state (Fe^{3+}) and reacts rapidly upon absorption. This interaction with the Fe^{3+} of mitochondrial cytochrome P450 inhibits cellular respiration and ATP production, decreases cellular respiration, and produces lactic acidosis. The toxicity of hydrogen cyanide to humans is dependent on the nature of the exposure. The LC_{50} for gaseous hydrogen cyanide is 100–300 ppm (parts per million). Inhalation of cyanide in this range results in death within 10–60 minutes, with death coming more quickly as the concentration increases. Inhalation of 2,000 ppm HCN causes death within one minute. The LD_{50} for ingestion is 50–200 mg, or 1–3 mg/kg, calculated as hydrogen cyanide, while the LD_{50} of its alkali salts (KCN, NaCN) is 3–8 mg/kg [28].

Due to its well-known and well-documented toxicity, the HCN molecule was assigned only a protective role against herbivores, pathogens, and some natural competitors that were strongly questioned when finding clues that suggested that cyanogenic glucosides could serve as nitrogen storage compounds. The structures of some cyanogenic glucosides are presented in Figure 3.15.

Since they almost always contain a α-D-glucose, we call them glucosides. But other mono- or polysaccharides can also be found. That is to say that a glycoside is usually a monoglycoside (monoglucoside), although there are several diglycosides and triglycosides. Among the most studied cyanogenic monoglycosides are linamarin (present in the leaves and roots of cassava, beans, and flax), lotaustralin (present in clovers), and dhurrin, present in cereal sorghum (*Sorghum vulgare*). These are formed

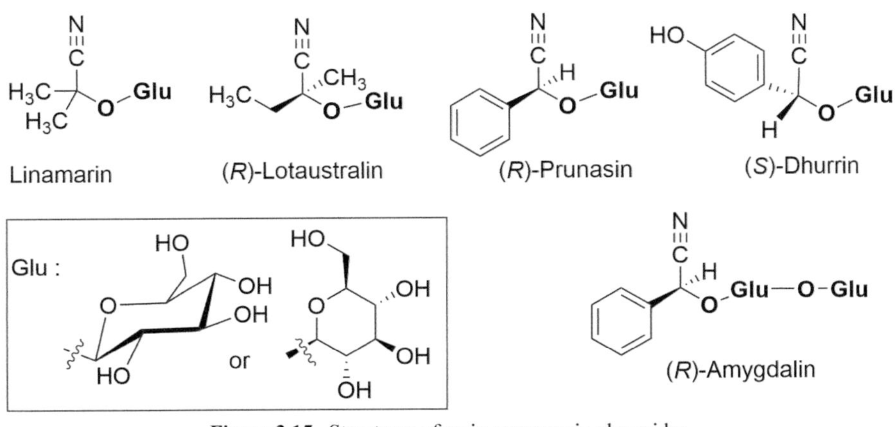

Figure 3.15. Structures of main cyanogenic glucosides.

from L-valine, L-isoleucine, and L-tyrosine, respectively. The relationship between L-amino acids and cyanogenic glucosides has been experimentally demonstrated with the ^{14}C-labeled amino acids.

One of the classic sources of HCN is the almond fruits from the tree *Prunus dulcis* (syn. *Prunus amygdalus*) that contain prunasin (monoglucoside) and amygdalin (diglucoside), which are generated from L-phenylalanine and glucose or gentiobiose, respectively. Fruit can have around 200–400 mg of cyanide per 100 g (dry weight). In ancient times, there have been famous cases of cyanide poisoning. Historians say that the poison preferred by Emperor Nero was cyanide.

In herbivorous insects, the role of cyanogenic compounds can have several purposes: cyanogenesis seems to be an effective food inhibitor for opportunistic or infrequent herbivores but not for specialist herbivores that feed on cyanogenic plants. For example, the mealy bugs *Phenacoccus manihoti* that feed on cassava use linamarin as a phage-stimulant (a substance present in the potential food and induces feeding). On the contrary, generalist insects such as grasshoppers *Schistocerca gregaria* and caterpillars *Spodoptera littoralis* and *Heliconius* avoid feeding on cyanogenic plants.

Butterflies *Zygaena filipendulae* are also generalist insects but can feed on cyanogenic plants, e.g., *Lotus corniculatus*, which is rich in linamarin and lotaustralin. These butterflies, both adults and larvae, not only prevent the damage that HCN could cause them in their tissues due to the presence of these toxins and the enzyme β-glucosidase in immediate action but can also sequester, store these cyanoglucosides and, if necessary, use releasing HCN. Therefore, it is believed that these butterflies are estimated as more modern toxic insects. Their high tolerance to these toxic substances is due to the presence of β-cyanoalanine synthase, which can use cyanide in other metabolic reactions. This strategy allows these butterflies, commonly known as zygena, to overcome the toxicity of cyanogenic plants and to sequester these intact compounds. However, these butterflies are capable of *de novo* synthesizing linamarin and lotaustralin from amino acids such as valine and isoleucine, respectively.

Living organisms have evolved different mechanisms to detoxify HCN and avoid its deleterious effects. One mechanism is through catalysis by β-cyanoalanine synthase, which is a mitochondrial cysteine enzyme. Thus, it catalyzes the reaction of adding cyanide to cysteine (Cys) to generate β-cyano-L-alanine, which then provides L-asparagine (Asp), aspartic acid, and ammonia via CAH-mediated hydrolysis (Figure 3.16A).

It should be noted that β-cyanoalanine accumulation in some plants may deter predators since it is a potent neurotoxin. In the case of zygena, β-cyanoalanine is stored in the defensive droplets in cuticle cavities. Another mechanism for HCN detoxification can be found in many bacteria, plants, and animals, including vertebrates, and is via catalysis by rhodaneses (Figure 3.16B). Rhodaneses transfer sulfur from a thiosulphate ($S_2O_3^{2-}$) to cyanide using cysteine as a sulfur carrier, producing less toxic thiocyanide (SCN^-) than cyanide and sulfite (SO_3^{2-}).

It also happens with some vertebrate animals: HCN adaptation has been studied in farm animals, and there is much evidence that detoxification occurs in sheep and cattle. In humans, cyanogenic glucosides at non-lethal daily doses can invalidate

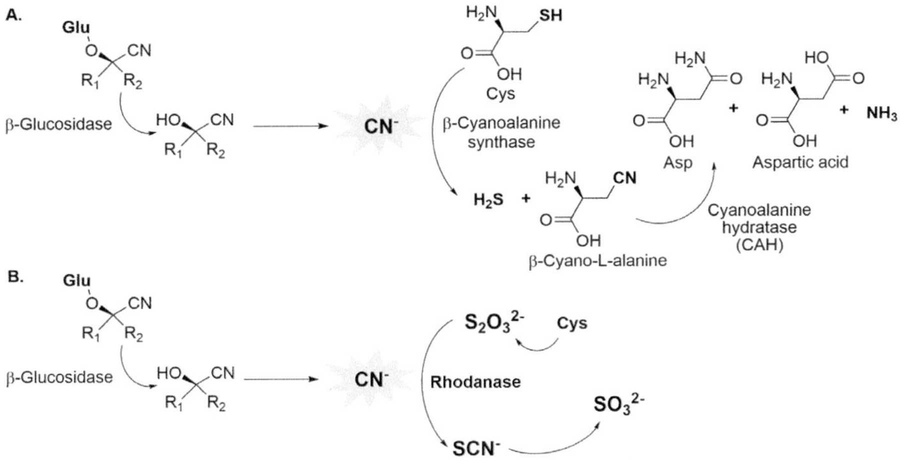

Figure 3.16. Detoxification mechanisms of cyanoglucosides.

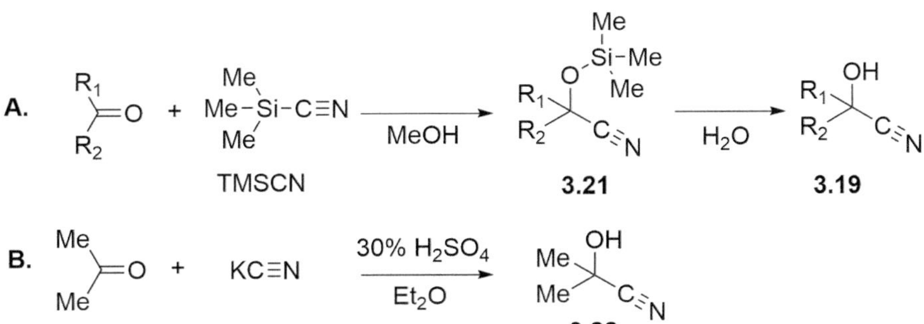

Figure 3.17. Cyanation reactions for obtaining cyanohydrins.

them: linamarin, in concentrations greater than 490 mg/kg in bitter cassava (*Manihot esculenta*), can cause specific health problems for consumers. Similar disorders have also been observed in sheep graze on the highly cyanogenic white clovers *Trifolium repens* in New Zealand.

Aglucone forms of cyanogenic glucosides, cyanohydrins, are readily available in chemical laboratories from aldehyde or ketone and chemical cyanide sources, i.e., sodium cyanide (potassium cyanide) or trimethylsilyl cyanide (TMSCN). The latter reagent adds smoothly across carbon-oxygen double bonds in an aldehyde (ketone), giving an intermediate product called O-silylated cyanohydrin **3.21**, which then can be converted into a cyanohydrin **3.19** (Figure 3.17A). Acetone cyanohydrin **3.22** is prepared from acetone and KCN by slowly adding 30% sulfuric acid to the reaction mixture. The reaction product is extracted with diethyl ether (Figure 3.17B).

The production of HCN for defensive purposes is not limited to the plant kingdom. This mode of protection is also found in animals (mainly arthropods): millipedes *Harpaphe haydeniana*, *Brachydesmus troglobius*, and other species of insects produce HCN in the form of cyanoglycosides to protect against attacks by their predators.

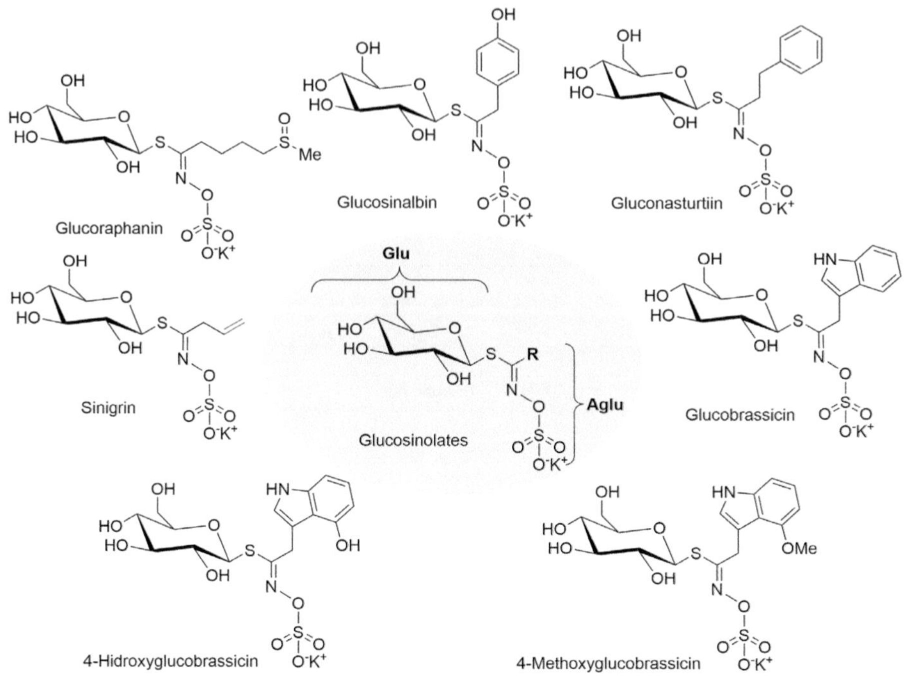

Figure 3.18. Most studied glucosinolates.

3.3 Glucosinolates: Components of the Mustard Oil Bomb

Another class of organic compounds related to cyanoglucosides is glucosinolates, also known as thioglucosides. They are formed by a β-D-thioglucanate group (glucone part, Glu), a sulfonated oxime group (aglucone part, Aglu), and a variable side chain R, which distinguishes the different categories of glucosinolates. They are ionic and occur in the plant as potassium salts [29] (Figure 3.18).

They are found extensively in plants of the family Brassicaceae (formerly Cruciferae), among selected others. Members of this family (especially the genus *Brassica*) are known as cruciferous vegetables, cabbage, broccoli, and horseradish or mustard plants. More than 120 types of glucosinolates (β-thioglucoside-N-hydroxysulfates) have been identified and are classified into four groups: saturated aliphatic, unsaturated aliphatic, and aromatic. The first glucosinolate, isolated and identified in the crucifers (white mustard, *Sinapis alba*), was glucosinalbin (also called sinalbin) (Robiquet and Boutron, 1831).

These compounds are induced after injury and attack by pathogens/insects and under abiotic stresses such as saline stress, UV radiation, etc. Like cyanogenic glucosides, glucosinolates, hydrophilic molecules, are harmless in the vacuoles of the intact plant, but when the plant is being eaten by an animal or digested by a ruminant, a hydrolytic enzyme, myrosinase (thioglucosidase) comes into contact with these compounds, generally hydrolyzing all glucosinolates and generating a wide range of degradation products. These include glucose and an unstable aglucone, which is then rearranged to form sulfate (SO_4^-) and various volatile compounds

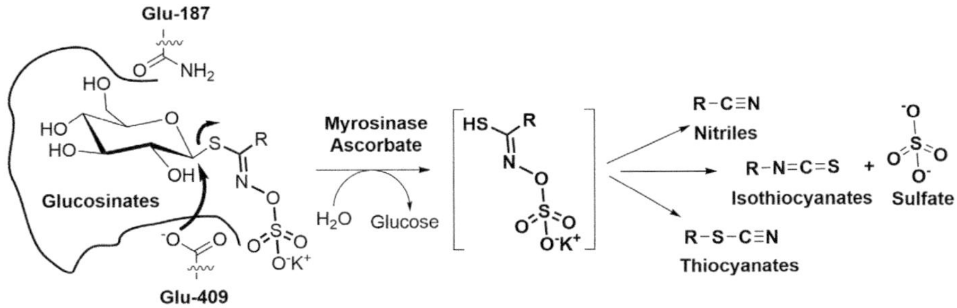

Figure 3.19. General catabolism of glucosinolates and their breakdown products.

(Figure 3.19). Among these compounds, isothiocyanates with functional group –N=C=S are the most studied, responsible for the acute taste of certain vegetables, and are highly irritating to mucous membranes and the bladder.

Myrosinase exists as a dimer with subunits of 60 to 70 kDa each and needs a cofactor (ascorbate) to serve as a base catalyst in the hydrolysis of glucosinolates. It is believed that the enzyme's active sites are the glutamic acid residues Glu-187 and Glu-409. Thus, enzymatic hydrolysis of glucosinolates provides a chemical defense of the crucifers against their predators. Like the binary system "cyanogenic glycoside – β-glucosidase", crucifers have their system, "glucosinolate-myrosinase", called the "mustard oil bomb", whose break down products have potent biocidal activities and can contribute to the defense of the plant against pathogens and generalist herbivores. It is also assumed that they can play roles as allelochemicals in mediating plant-insect interactions.

All glucosinolates are biosynthesized from L-amino acids: substituted alkenyl glucosinolates (R is alkenyl) are formed from methionine via an additional chain extension process; indolyl glucosinolates are formed from tryptophan, and aryl glucosinolates (R is aryl or aralkyl) are split from phenylalanine or tyrosine. For example, glucosinalbin (sinalbin), found in white mustard seeds (*Sinapis alba*), is synthesized from tyrosine, which is oxidized under the P450 enzyme complex to *E-p*-hydroxyphenylacetal oxime **3.23**. The last molecule interacts with another amino acid, cysteine, in the presence of the enzyme C-S lyase, forming a thiohydroximic acid **3.24**. Then come stages of glycosylation (glucosyltransferase enzymes, UDP-

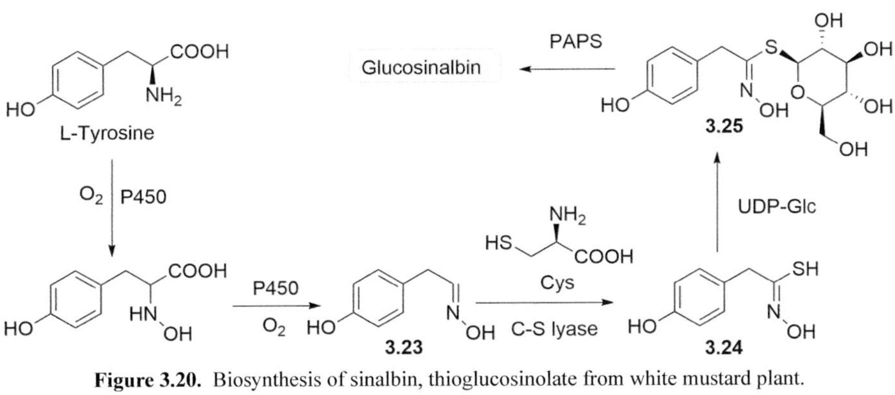

Figure 3.20. Biosynthesis of sinalbin, thioglucosinolate from white mustard plant.

Glc) and sulfonation (sulfotransferase enzymes that utilize 3'-phosphoadenosyl-5'-phosphosulfate (PAPS) as the donor of the sulfonic group) (Figure 3.20).

In the laboratory, substituted aryl glucosinolates are prepared from the corresponding aldehydes and O-acetylated β-D-glucose using a convergent strategy. The reaction of α-chloro-oximes **3.26** and O-acetylated thioglucose **3.27** to afford thiohydroximate products **3.28** is a crucial stage. Next, the sulfation (or sulfurylation) of these products with pyridine–sulfur trioxide complex and subsequent treatment with potassium carbonate were realized and allowed to give the potassium salts **3.29** (step d). Finally, deprotection of acetylated glucosinolates was possible using potassium hydroxide in methanol [30] (Figure 3.21). The completed seven-step synthesis gave 25–40% overall yields from D-glucose. If you compare the chemical synthesis and biosynthesis of these molecules, you can see that the order of the formation of the S-Glu bond is different.

One of the most interesting glucosinolates, sinalbin, sinigrin, and its analogs (progotrin and epiprogotrin) (Figure 3.22) are components of mustard, a condiment made from whole, powdered or crushed dried seeds of mustard plants: *Brassica nigra* (black mustard), *Brassica juncea* (brown mustard) or *Sinapis alba* (white mustard). The seeds are usually mixed with wine, vinegar, water, and various spices to create a paste ranging in color from bright yellow to dark brown [31]. It is also known that these plants have been recognized as functional food for health maintenance and disease prevention, especially cancer [32, 33].

The pungent aroma and flavor of mustard seeds are derived from glucosinolates, which are hydrolyzed by myrosinase to give reactive isothiocyanates or mustard

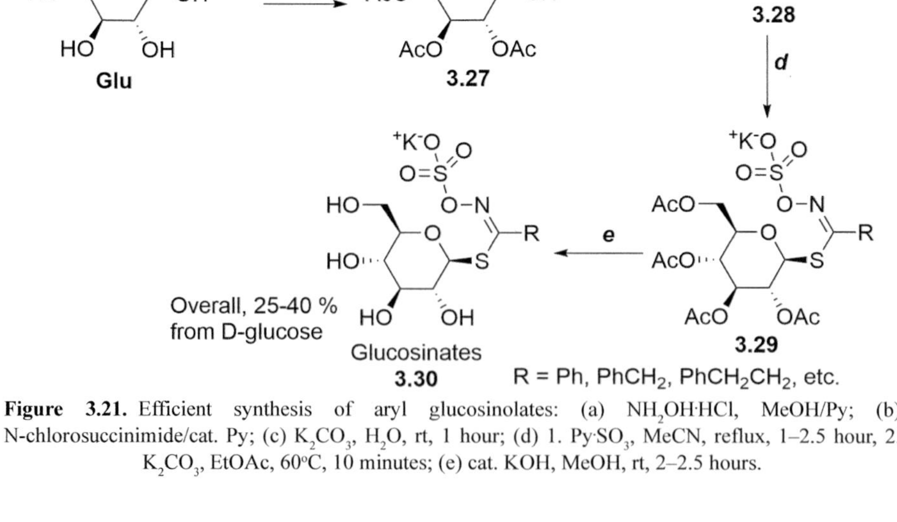

Figure 3.21. Efficient synthesis of aryl glucosinolates: (a) $NH_2OH{\cdot}HCl$, MeOH/Py; (b) N-chlorosuccinimide/cat. Py; (c) K_2CO_3, H_2O, rt, 1 hour; (d) 1. $Py{\cdot}SO_3$, MeCN, reflux, 1–2.5 hour, 2. K_2CO_3, EtOAc, 60°C, 10 minutes; (e) cat. KOH, MeOH, rt, 2–2.5 hours.

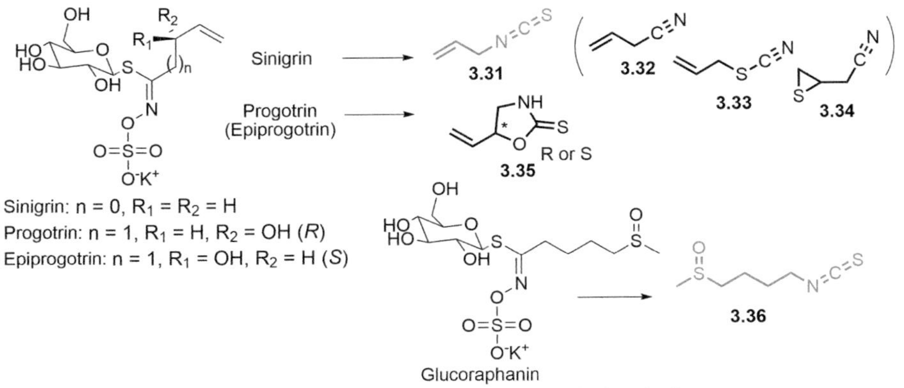

Figure 3.22. Degradation products of mustard glucosinolates.

oils, which impart the mustard-like or garlic-like odors associated with mustard and horseradish. Other metabolites include thiocyanates, nitriles, sulfates, and goitrins (5-vinyl-2-thio-oxazolidines) (Figure 3.22). Interestingly, white mustard seeds contain sinalbin as a main component, which hydrolysis gives *p*-hydroxyphenylmethyl isocyanate that has a softer taste than that produced by the seeds of black mustard with the minor component, sinigrin. However, the ingredients that give black mustard seed its flavor are several degradation products: allyl isothiocyanate (3-isothiocyanato-1-propene) **3.31**, allyl cyanide (but-3-enenitrile) **3.32**, allyl thiocyanate **3.33** and 2-(thioran-2-yl)acetonitrile **3.34**. The first degradation product is a compound responsible for the pungent burning odor and taste [34]. Sinigrin is also found in horseradish root (*Armoracia rusticana*, *A. lapathifolia*), in Brussels sprouts (*Brassica oleracea* var. *italica*), in broccoli (*Brassica oleracea* var. *gemmifera*), and wasabi (green Japanese mustard). Hydrolysis of progotrin (or epiprogrin) affords goitrins **3.35**. Glucoraphanin (4-methylsulfinylbutyl glucosinolate) is a major component of broccoli and other vegetables. Its main degradation product is sulforaphane **3.36**.

The enormous popularity of Brassicaceae plants is due to the fact that regular consumption of cruciferous vegetables, such as broccoli, cabbage Brussels, cabbage, cauliflower, etc., is associated with a reduction in the incidence of cancer. The first evidence that plants containing glucosinolates are beneficial to health was known in the 70s when some scientists discovered that glucosinolates can inhibit the carcinogenesis process in rodent models. Since this time, the scientific interest in the hydrolysis products of glucosinolates as cancer preventive agents has been increasing. Among these, isothiocyanates are more promising and studied, which is understandable since, while glucosinolates are relatively inert molecules, isothiocyanates have high chemical reactivity. They are highly electrophilic molecules that react with N-, O-, and S-nucleophiles. In organic chemistry, such nucleophiles are amines, alcohols, and thiols. Their interactions with the isocyanates give corresponding adducts thioureas, thiocarbamates, and dithiocarbamates (Figure 3.23A).

Therefore, it is not surprising that one of the main cellular targets of isothiocyanates is proteins and peptides containing amino acids with nucleophiles

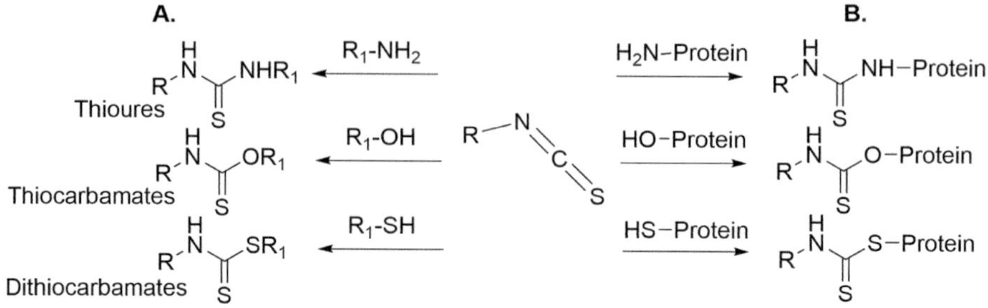

Figure 3.23. Reactions of isothiocyanates and N-, O-, and S-nucleophiles.

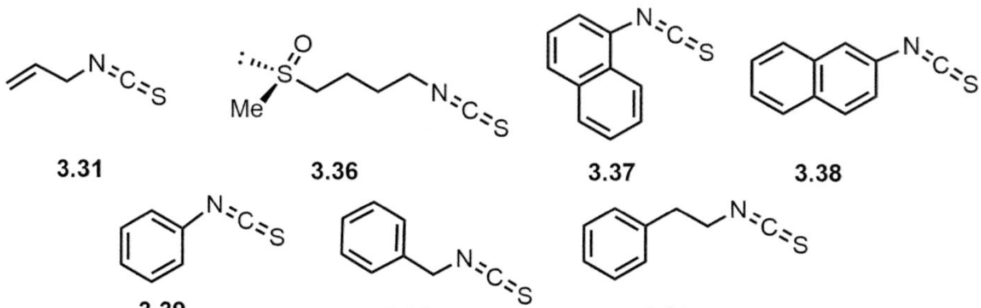

Figure 3.24. Pharmacologically important isothiocyanates: naturally occurring hydrolyzation products and synthetic products.

such as hydroxyl, amino, or thiol groups (e.g., serine, cysteine or glutathione, G-SH) that can readily react to produce adducts (Figure 3.23B).

The reaction with thiol groups is about 1,000 times faster than with amine or hydroxyl groups. While the conjugation of isothiocyanates to small thiol moieties is reversible, isothiocyanates are thought to form irreversible thiourea derivatives with amino groups [35]. Isothiocyanate compounds react best at alkaline pH values where the target amine groups are mainly unprotonated. Many reactions occur in 0.1 M sodium carbonate buffer at pH 9.0. Besides hydrophobicity, the side groups of the isothiocyanate also dictate the overall reactivity by affecting the size, shape, and electrophilicity of the isothiocyanate. After the reaction, the isothiocyanates remain covalently bound to the protein, which changes the tertiary structure of the protein, leading to partial or total loss of enzymatic activity.

For the first time, organic isothiocyanates' ability to block the carcinogenesis process was recognized in 1966, when the biological properties of α-naphthyl isothiocyanate **3.37** were studied. Later, many synthetic isothiocyanates were investigated, looking for possible preventive agents against cancer. Among these, the isothiocyanates of β-naphthyl **3.38**, phenyl **3.39**, and benzyl **3.40** (Figure 3.24) showed tumor growth inhibitory activity in carcinogenesis in animal models. The allyl isothiocyanate **3.31** and phenethyl isothiocyanate **3.41** are also at the center of cytotoxic and genotoxic studies in different animal models. However, one of the most widely studied isothiocyanates is sulforaphane **3.36**. It is found to promote detoxification and prevent and combat cancer-inducing apoptosis [36]. It is an antiviral, antibacterial, and antifungal agent. In plants, sulforaphane deters insect predators and acts as a selective antibiotic. Notably, in biochemistry, phenyl isothiocyanate **3.39** is known as Edman's reagent and is used to sequence amino acids in a peptide [37]. Additionally, phenyl isothiocyanate has been shown to have excellent antibacterial properties.

Apart from their potential anticancer and antiparasitic effects, isothiocyanates are also valuable intermediates for obtaining thioureas. Therefore, the preparation of isothiocyanates has gained prominence, although these types of organic compounds have been known for a long time. There are many synthetic protocols for their preparation [38–43]. Traditional synthetic methods for the preparation of isothiocyanates from amines involve the use of CS_2, thiophosgene ($CSCl_2$), or phosgene ($COCl_2$) reagents, which are all highly toxic (Figure 3.25).

Thus, to mitigate these problems, several reaction conditions were studied to develop a more environmentally benign synthesis of isothiocyanates. Almost all methods of synthesizing isothiocyanates are carried through decomposing dithiocarbamate salts like **3.42** or **3.43**. The combination of RNH_2, CS_2 and Et_3N is quite common and gives salt **3.42**. Green desulfurization agents needed in this transformation can be hydrogen peroxide (H_2O_2), iodine (I_2), and sodium persulfate ($Na_2S_2O_8$). When phosgene is used, carbamate salts **3.44** are formed, and their reaction with Lawesson's reagent [44] gives isothiocyanates (Figure 3.25).

In 2017, Chinese chemists proposed using sodium trifluoromethane sulfonate (F_3CSO_2Na), known as the Langlois reagent [45], in isothiocyanate synthesis. Transformation is carried out in the presence of copper iodide and diethyl phosphonate, which avoids the utilization of toxic sulfur-containing reagents [42] (Figure 3.26).

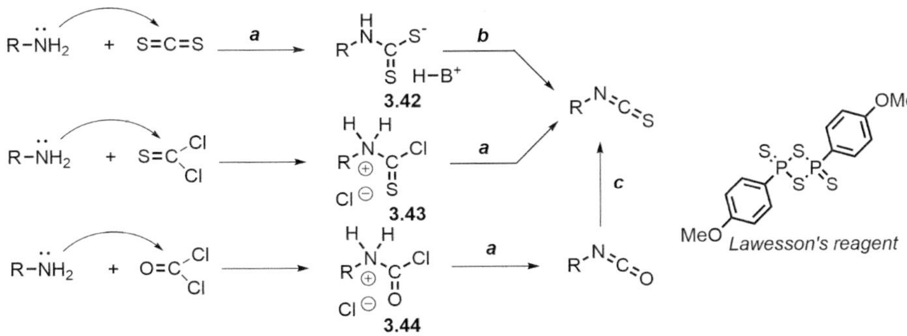

Figure 3.25. Preparation of isothiocyanates using primary amines: (a) base B, e.g., Et_3N/ petroleum ether or ethyl acetate; (b) desulfurization agents, e.g., $CSCl_2$, ClCOOEt, TsCl, MeCOCl, Triphosgene, H_2O_2, I_2, $Na_2S_2O_8$; (c) Lawesson's reagent.

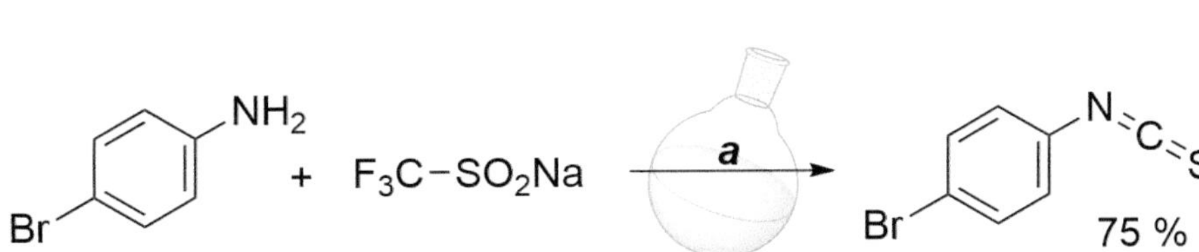

Figure 3.26. Preparation of 1-bromo-4-isothiocyanatobenzene using Langlois reagent: (a) 5 mol% CuI, 2 equivalent $HPO(OEt)_2$, toluene, 110°C, 16 hours.

Under these reaction conditions, thiocarbonyl fluoride may be formed and react with amine derivative-like reactions of amines and thiophosgene. However, a more complete reaction mechanism remains to be examined.

Another approach for isothiocyanates is by the use of the reaction between organic halides (R-X) and sodium thiocyanate (NaSCN) or potassium (ammonium) thiocyanate [46]. Forming thiocyanates can be thermally isomerized into isothiocyanates. This simple transformation was performed in 1875 by G. Gerhlich and O. Billeter [47]. Currently, this isomerization is interpreted as a [3,3]-sigmatropic transposition [48] (Figure 3.27).

Thiocyanate salts are SCN sources often used to introduce a sulfur function on a substrate (R-X), and potassium thiocyanate (KSCN) is probably the most employed for this purpose. Note that allylic thiocyanates are known to easily undergo this rearrangement, leading to the corresponding allylic isothiocyanates. This reaction is also reported as the Billeter-Gerlich rearrangement.

As sulforaphane is an important anticancer agent, we want to briefly describe its synthesis. We recall that sulforaphane was isolated from broccoli and is chiral, possessing the *R* configuration. Still, both *R*-sulforaphane and the synthetic (*R*, *S*) mixture show identical activities in the reported assay, and it is known to be an unstable oil, which creates problems in its preparation and storage. Various methods for preparing sulforaphane and its derivatives have been reported in the literature [49–53], and the commonly used method is based on a protocol developed by Schmid and Prof. Paul Karrer (1889–1971), a Swiss organic chemist who won Novel Price of Chemistry in 1939 [54].

The sulforaphane molecule (4-isothiocyanatobutyl methyl sulfoxide **3.36**), which possesses quite a simple structure, can be prepared through unsophisticated chemical transformations using readily available bifunctional molecules as starting

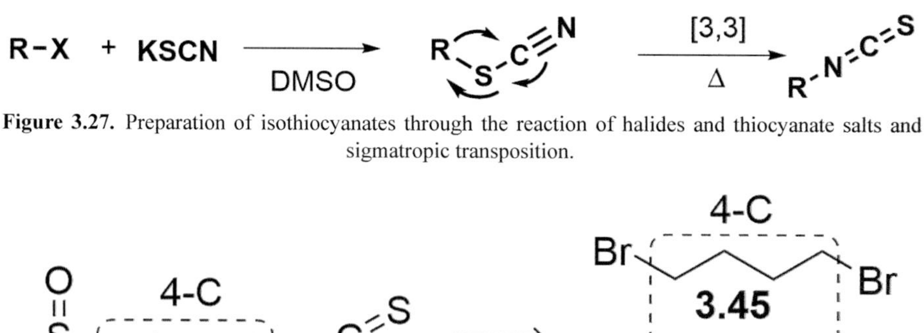

Figure 3.27. Preparation of isothiocyanates through the reaction of halides and thiocyanate salts and sigmatropic transposition.

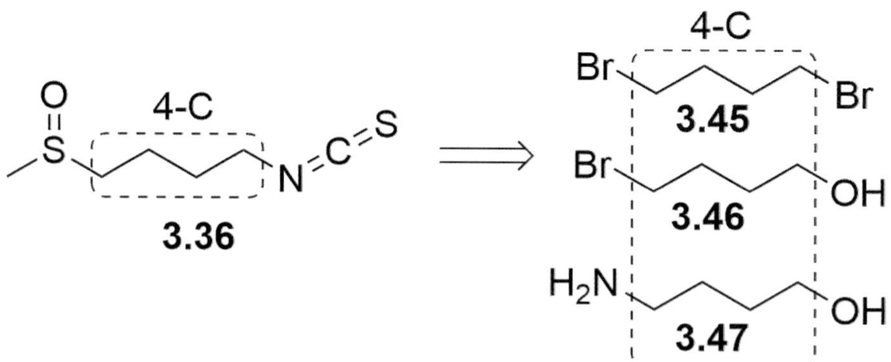

Figure 3.28. Structural analysis of sulforaphane and its close precursors.

materials, such as 1,4-dibromobutane, 4-bromobutan-1-ol, or 4-aminobutan-1-ol. All of these possess a 4-C chain and two functional groups, like sulforaphane (Figure 3.28).

Karrer and Schmid selected the Gabriel reaction for amino group introduction in the carbon skeleton, where phthalimide is usually used. Thus, when a reaction of potassium phthalimide and 1,4-dibromobutane was made, they got mono substitution with the remaining bromide obtaining comp. **3.48** was treated with sodium thiomethoxide (NaSMe) to give sulfide derivative **3.49** (Figure 3.29A). Using its oxidation followed by imide deprotection, 4-methyl sulfinylbutylamine **3.50**, a key intermediate, was easily prepared. The final step of their synthetic scheme involved exposing the amine to a combination of CS_2, iodine, and sodium ethoxide, which allows for the formation of the thioisocyanate function. The yield for this final step was only around 50%, which made the overall yield of the synthesis about 20% [49].

Since this time, several modifications of synthetic strategies for sulforaphane preparation in large amounts have been developed. Among them, the sulforaphane synthesis realized by Meijler et al. [53] calls attention to its simplicity and affordability (Figure 3.30): selected starting material was the 1,4-bromobutanol **3.47**, which was reacted with sodium azide to afford azido alcohol **3.51**. Then, its O-protection with tosylchloride provided the tosyl derivative **3.52**. It was followed by a reaction with NaSMe in dry THF (instead of DMF) with mild heating to afford the unstable azido thioether **3.53**. The latter compound was converted into thioisocyanate erucin **3.54**. Finally, oxidizing the sulfide moiety of erucin with *m*-CPBA (*m*-chloroperbenzoic acid) sulforaphane obtained in good yield.

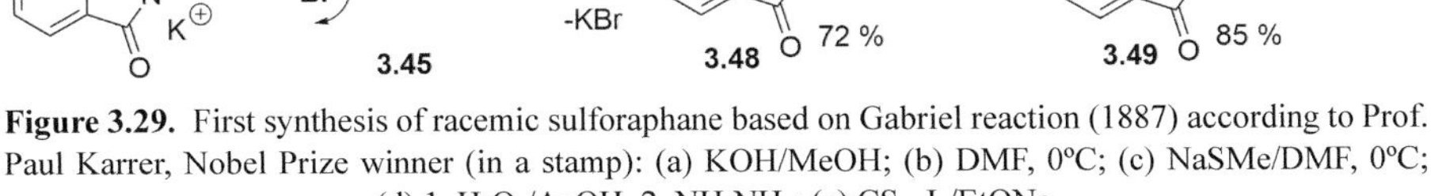

Figure 3.29. First synthesis of racemic sulforaphane based on Gabriel reaction (1887) according to Prof. Paul Karrer, Nobel Prize winner (in a stamp): (a) KOH/MeOH; (b) DMF, 0°C; (c) NaSMe/DMF, 0°C; (d) 1. H_2O_2/AcOH, 2. NH_2NH_2; (e) CS_2, I_2/EtONa.

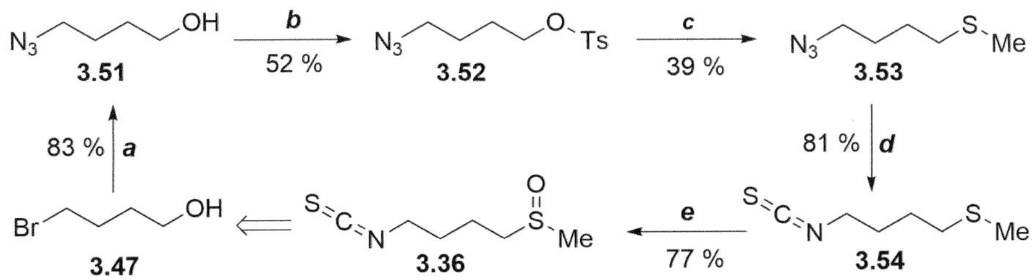

Figure 3.30. Synthesis of (*S*, *R*)-sulforaphane: (*a*) NaN_3, DMF, 70°C, 24 hours; (*b*) TsCl/Et_3N, 4-DMAP, CH_2Cl_2, 0°C; (c) MeSNa, THF, 55°C, 4 hours; (d) PPh_3, diethyl ether, CS_2; (e) *m*-CPBA, CH_2Cl_2.

The above-described synthesis gives a racemic product that can be separated into two antipodes, the S and R forms (stereogenicity at sulfur), with resolution via dibenzoyl D-tartaric acid and L-tartaric acid, respectively.

Now, we can summarize that cabbage, broccoli, radish, mustards, and all these *Brassica* genus plants produce glucosinolates, which, under stress (physical damage, predatory attacks, and myrosinase enzyme), break down into sulfur and nitrogenous molecules that form so-called mustard oil bomb [55, 56]. This mixture is a good chemical defensive weapon against generalist herbivores (mainly insects). More precisely, it is a deterrent to food and a potent poison. Thus, hydrolysis products give a characteristic pungent smell and are toxic to a wide range of organisms, which negatively affects generalist herbivores.

However, some herbivorous specialists have their own biochemical means that allow them to use these plants, full of large amounts of glucosinolates, as food sources. This is the case of the larvae of Pierid butterflies. The interactions between butterflies and caterpillars of this genus and Brassicaceae plants have been studied extensively. Larvae of the cabbage or white cabbage butterfly (*Pieris brassicae*) and their relatives, *P. rapae* and *P. napi*, are pests for the crops of these plants. Glucosinolates, characteristic of cruciferous plants (see Figure 3.18), play a fundamental role as chemical mediators of these interactions. The adult females of *P. brassicae* and *P. rapae*, as well as their larvae, explode glucosinolates as stimuli during the selection of host plants for oviposition and feeding (i.e., glucosinolates are feeding stimulants). It can be said that the caterpillars of these butterflies are addicted to glucosinolates, which are responsible for their fixation in these plants. Why is the mustard oil bomb not effective in this case?

It was observed that *P. rapae* larvae and other larvae tolerate cruciferous plants well as a diet because they use intestinal enzymes to redirect the degradation of glucosinolates to fewer toxic nitriles instead of isothiocyanates [57] (Figure 3.19). It is interesting to add that when the *P. rapae* and *P. brassicae* caterpillars, in turn, are attacked, they may contain nitriles stored in the anterior intestine. The presence of nitriles in high concentration may explain the regurgitation-repellent effect on their natural enemies, like wasps-parasites *Cotesia rubecula* and *C. glomerata*, which lay eggs in the larvae of these butterflies. Thus, a tritrophic system *Cotesia rubecula – Pieris rapae – Brassicaceae* (wasp-caterpillar-plant) is formed, in which glucosinolates and their nitriles play an important role (Figure 3.31).

The sequestration of glucosinolates in specialized herbivores has been little studied. One of these specialists is larvae of the turnip sawfly *Athalia rosae*, which

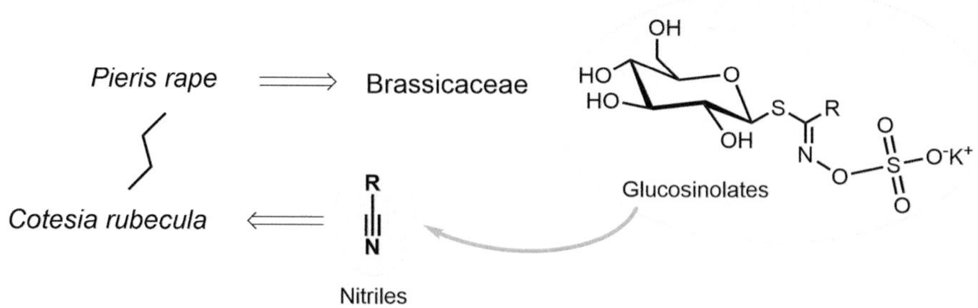

Figure 3.31. Tritrophic system "*Cotesia rubecula - Pieris rapae – Brassicaceae.*"

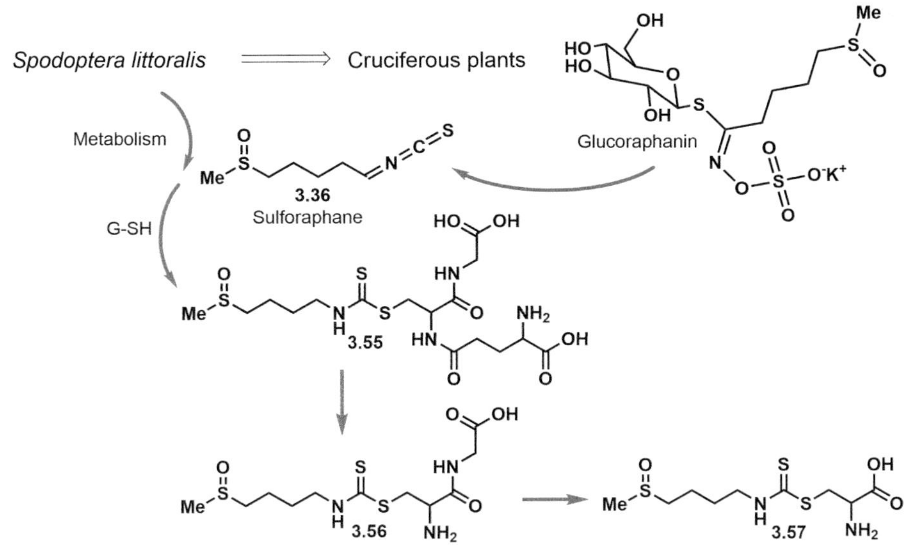

Figure 3.32. Main metabolites of isothiocyanates in larvae *S. littoralis*.

is a pest in cruciferous crops. Some studies indicate that the *A. rosae* larvae can appropriate glucosinolates and effectively concentrate them on their hemolymph. It was experimentally proved that hemolymph drops are easily released after the wound from their predators. This phenomenon is usually related to the sequestration of toxic plant compounds. The hemolymph of these larvae has a high deterrent effect when feeding on ants, spiders, and birds [58].

Many herbivorous insects can excrete toxic unchanged plant compounds with their feces. This physiological process is known as the pharmacokinetic strategy. Concerning the different stages of development, glucosinolates that were kidnapped by *A. rosae* larvae seemingly pass unchanged and in similar concentrations in the stage of adult life.

Generalist insects can also successfully feed on Brassicaceae plants. As an example, we will see the larvae *Spodoptera littoralis* that are generalists and an important pest of cotton, tobacco, and vegetable crops (tomato, pepper, lettuce, broccoli, cauliflower, and artichoke). Adult females lay their eggs in a wide variety of plants. The larvae cause considerable damage since they feed on the leaves, fruits, and buds.

During studies on the relationship between the *Arabidopsis thaliana* plant and the larvae *S. littoral*, it was found that after feeding on *A. thaliana*, feces of larvae *S. littoralis* contained the following conjugates glutathione (G-SH) of sulforaphane: cysteinyl isothiocyanates **3.55**, cysteinylglycine **3.56** and cysteinyl **3.57** conjugates [59] (Figure 3.32). These conjugates are the principal metabolites of isothiocyanates in *S. littoralis*, and their abundance correlates with the amount of isothiocyanates ingested. However, most of the isocyanates these larvae ingest are excreted without metabolizing.

We recall that sulforaphane, like other isothiocyanates, is abundant in the human diet and potently inhibits tumorigenesis induced by various chemical carcinogens in rodents. Therefore, studies on sulforaphane metabolism in humans and laboratory animals are important. Preliminary experiments have shown that sulforaphane is also metabolized mainly by the above-described pathway (Figure 3.32).

3.4 Alkaloids: Multifaceted Molecules

Within the secondary metabolites that plants generate, the alkaloids occupy a big place due to their notable physiological activities in humans and animals. The name "Alkaloid" was first coined in 1819 by German pharmacist Carl F. W. Meissner (1792–1853), referring to natural products of vegetable origin that showed basic properties like those of alkalis (the Arabic name *al-qali*). The definition of alkaloids is still controversial since there are several types of alkaloids. However, it can generally be defined as an alkaloid, a heterocyclic compound with a nitrogen atom from amino acid metabolism that exhibits remarkable biological properties.

The term "alkaloid" covers a vast range of structures, from simple monocyclic compounds to hexa- and heptacyclic molecules. These metabolites can be divided into different classes according to their precursor (e.g., indole alkaloids are alkaloids derived from tryptophan), encompassing more than 20 different classes (e.g., pyrrolidine alkaloids, tropane alkaloids, piperidine alkaloids, pyridine alkaloids, quinolizidine alkaloids, and indole alkaloids, among others). More than 20,000 alkaloids have been registered. These abound, especially in the angiosperm plant species of the Liliacea, Amaryllidaceae, Rubiaceae, Papaveraceae, or Solanaceae families, etc. Alkaloids are the most important class of plant toxins, with their defensive value in plants being indisputable. By acting on several essential systems (central nervous, digestive, respiratory, and circulatory) of animals, alkaloids can protect plants against their multiple and constant attacks. Alkaloids also have an allelopathic value in communicating with plants with similar species, as they inhibit or accelerate the growth of others that could be competitive. They can produce beneficial or lethal effects in humans [60].

Alkaloids usually concentrate on different parts of the plant. For example, nicotine is found in tobacco (*Nicotiana tabacum*) and tea (*Camellia sinensis*) leaves. The seeds of the Strychnos nux–vomica tree (*Strychnos nux–vomica*) and hemlock (*Conium maculatum*) present alkaloids strychnine and coniine, respectively. They are also found in fruits and roots. In plants, alkaloids are generally in the form of salts (citrates, malate, tartrates, oxalates, etc.) of their bases. These salts are soluble in water, while the free bases are insoluble in water but soluble in non-polar organic solvents.

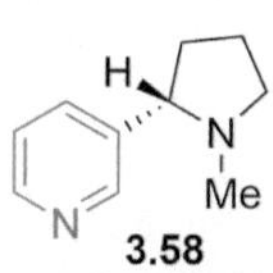

3.58

Nicotine

1561 introduced in France by Jean Nicot
1763 first used as an insecticide
1828 isolated in Germany by W. H. Posselt and K. L. Reinmann
1880 a machine was first patented to mass-produce paper cigarettes
1964 published a study linking smoking with heart disease and lung cancer

Figure 3.33. Structure of nicotine, historical data, and the stamp dedicated to Jean Nicot.

Following well-known examples, morphine of the opium poppy (*Papaver somniferum*), quinine from the cinchona (quina–quina) tree (*Cinchona officinalis*), cocaine from coca (*Erythroxylum coca*) or nicotine could be mentioned. Perhaps one of the most studied toxic alkaloids is nicotine. Chemically, it is a basic chiral molecule made up of a pyridine ring connected with N-methylpyrrolidine, namely, (*S*)-3-(1-methylpyrrolidin-2-yl)pyridine **3.58**. The tobacco plant and nicotine, its principal alkaloid, have a rich history [61], in which fundamental steps are revealed in Figure 3.33.

The average nicotine content of the leaves of the plant *Nicotiana tabacum* (tobacco plant) is 2–6 % dry weight. This plant is indigenous to the Americas and has been used as a medicine and stimulant for at least 2,000 years, and ever since, people have been smoking or chewing its leaves. However, its use as the smoking of pipes and cigars proliferated throughout the 1600s when tobacco was introduced to Europe. Now, it is well-known that the dose of the cigarette (typically containing about 10 milligrams of nicotine) serves to stimulate a human for a few moments. Still, its constant use could produce heart disease and lung cancer.

Nicotine is synthesized in the plant's roots and transported by the sativa to the green leaves. The biogenesis of nicotine is convergent synthesis: simultaneously, two initial precursors, nicotinic acid (niacin **3.59**) and 4-(methylamino)butanal (**3.60**) are formed from L-aspartate and L-ornithine, respectively. Later, they are converted into crucial intermediates, 3,6-dihydropyridine-3-carboxylate **3.61** and 1-methyl-Δ^1-pyrrolidium cation **3.62**, which give the nicotine skeleton through the reductive decarboxylation process with the aid of nicotine synthase [62, 63] (Figure 3.34A).

Due to this intricate nicotine synthesis, the plant obtains a significant advantage in the battle to survive because, as we have mentioned, nicotine is poisonous. However, a detailed answer to the question of how a tobacco plant prepares this toxic molecule is in the regulation of gene expression. When insects attack tobacco plants, producing damage, chemicals from the saliva of the attacking insects (e.g., glutamine) react with fatty acids from the destroyed plasma membranes. The resulting

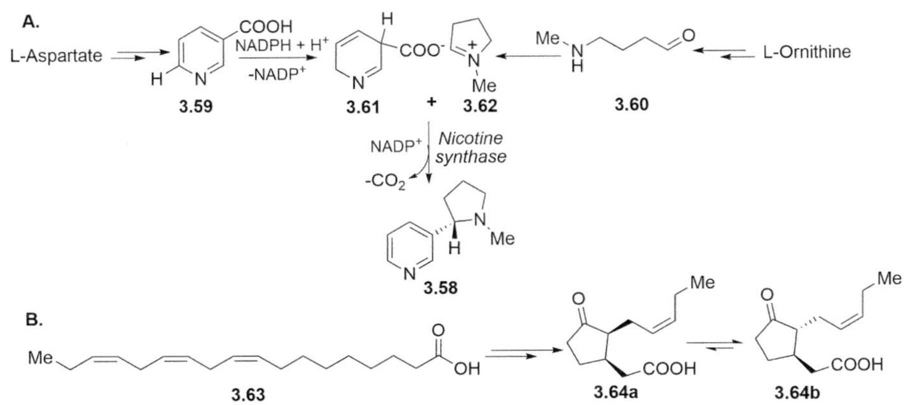

Figure 3.34. A. Simplified biogenesis of nicotine in which a biochemical reducing agent (NADPH) and an oxidizing agent (NADP$^+$) work well as resource-conserving redox cycle. B. Synthesis of the plant hormone jasmonic acid, which is involved in the regulation of gene expression, stimulates the synthesis of the enzymes required for nicotine synthesis.

reaction products trigger the defense mechanism, causing the formation of linolenic acid **3.63** of jasmonic acids: (+)-7-*iso*-jasmonic acid **3.64a** and (–)-7-jasmonic acid **3.64b** (Figure 3.34B). These plant hormone acids are circulated throughout the plant, reaching the roots, where nicotine synthesis is exclusively localized. These hormone acids interfere in the regulation of gene expression, stimulating the synthesis of the enzymes required for nicotine synthesis.

Regarding the chemical synthesis of nicotine alkaloids in a laboratory, there are many protocols for the preparation of racemic nicotine; among them are several strategies that involve the formation of a pyrrolidine ring from 3-substituted pyridines [65, 66]. The first (non-enantioselective) laboratory synthesis of nicotine was reported in 1904. The isomers obtained are generally separated as salts by recrystallization. However, few examples of a synthesis enantioselective of optical (*S*) and (*R*) nicotine isomers have been reported. In 2009, Prof. Helmchen and co-workers reported a short enantioselective synthesis of natural nicotine using Ir-catalyzed ([Ir(COD)Cl]$_2$) allylic amination of methyl (3-(pyridin-3-yl)allyl) carbonate **3.65** with methylamine (step a) and Rh-catalyzed (Rh(acac)CO$_2$) hydroformylation cyclization of chiral allylamine **3.66** (step b) that allowed to obtain (*S*)-nicotine [67] (Figure 3.35). Both reactions require the use of specific metal complex catalysts and chiral phosphorus ligands L_1 and L_2, whose presence guarantees a good enantioselectivity.

Several methods also performed an efficient enantioselective synthesis of the unnatural enantiomer (*R* configuration). We will discuss one of them, a four-step synthesis of (*R*)-nicotine that was realized in 2000 by a French team directed by Lebreton [68]. The first step is the allylation of 3-pyridine carboxaldehyde (**3.67**) with 2.2 equivalents of chiral allylborane reagent (B-allyldiisopinocampheylborane **3.68**) in Et$_2$O at −100°C, with the formation of (S)-allyl alcohol (**3.69**) (Figure 3.36).

In the second step, chiral azide **3.70** is obtained from the allyl alcohol, using the Thompson procedure (involving the use of phospho-azide, (PhO)$_2$P(O)-N$_3$), with a

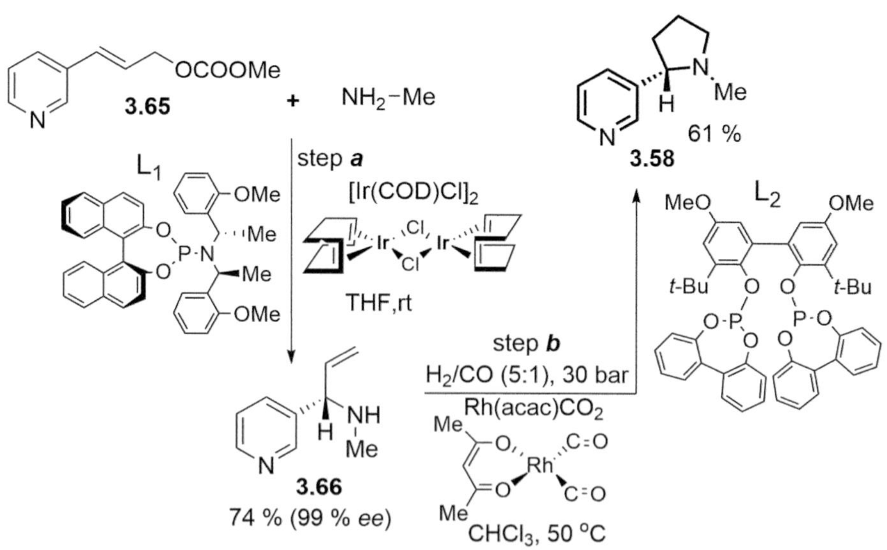

Figure 3.35. Short enantioselective synthesis of the natural (*S*)-(–) nicotine based on allyl amination and hydroformylation reactions.

yield of 90% and without racemization. Then, this azide was purified by chiral HPLC and treated with an excess of freshly prepared dicyclohexylborane ($B(C_6H_{11})_2H$) in THF to afford, after hydrolysis with methanol, (*R*)-nornicotine **3.71** in 85% yield. This step proceeds with the hydroboration of the allyl-azide double bond, followed by forming a boron-nitrogen bond between azide and trialkyl-borane (intermediate A). Finally, the ring closure is produced by migrating the methylene borane group to nitrogen, with the concomitant nitrogen loss. The last step, conversion of (*R*) nornicotine **3.71** to (*R*)-nicotine **3.72**, was carried out using a two-step procedure involving treatment with EtOCOCl and then by an excess of $LiAlH_4$. Extending these straightforward approaches to preparing (*S*) and (*R*) nicotine analogs is useful in drug development and insecticide research.

Nicotine alkaloid is lethal in insects showing LD_{50} values of 24–50 mg/kg for most herbivores; consequently, it was one of the first insecticides used for pest control in agriculture; it is still in use today as an insecticidal fumigant for control of sucking parasites, such as aphids. The levorotatory isomer, i.e., (*S*) nicotine, is ten-hundred times more active than the other (*R*) enantiomer. Their action is to mimic acetylcholine when it binds to its receptor on the postsynaptic membrane of the muscle junction. As acetylcholine interacts with this receptor, it alters the membrane's permeability. Nicotine can compete with this neurotransmitter for the acetylcholine receptors (AChR). Thus, nicotine activity causes the production of new nerve impulses that cause seizures and even death.

It should be recalled that nicotine is toxic to vertebrates: in mice, the LD_{50} was 0.3 mg/kg after intravenous injection and 9.5 mg/kg after intraperitoneal injection; in humans, nicotine is lethal at 1–2 mg/kg. Meanwhile, the larvae of the night butterfly *Manduca sexta*, known as tobacco hornworm, tolerate high doses of nicotine (up to 1 mM in your blood), which is fatal to other herbivores. They are the most nicotine-tolerant species, exhibiting an LD_{50} of 1.5 g nicotine/kg body weight [68]. These insects adapted to nicotine-producing plants (i.e., insect tobacco specialists) have

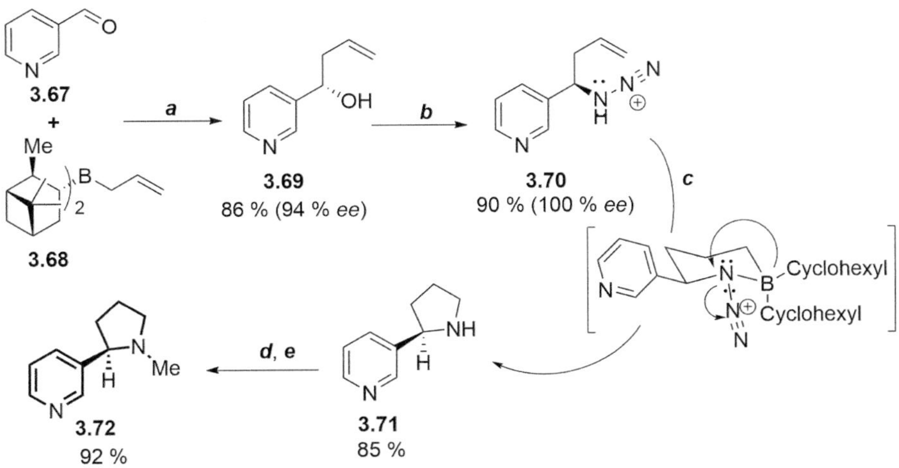

Figure 3.36. Rapid and efficient synthesis of the unnatural (*R*)-(+) nicotine based on an intramolecular hydroboration-cycloalkylation sequence: (a) Et_2O, −100°C; (b) DBU, $(PhO)_2P(O)$-N_3, PhMe, rt; (c) $B(C_6H_{11})_2H$, THF, rt, then MeOH; (d) EtOCOCl (1.2 equivalent), Et_3N, Et_2O, rt; (e) $LiAlH_4$, THF, 0°C.

developed resistance to this alkaloid [69]. How do they tolerate this dose? The larvae developed a blood-brain barrier to protect the cholinergic nervous system from "dietary" nicotine. In larvae, nicotine, once ingested, is transported by a powerful system into the lumen of the Malpighian tubules, where it is excreted without metabolizing.

Interestingly, these nocturnal larvae develop rapidly, reach 12 cm, and only increase food consumption containing nicotine after increasing the P450 enzyme in their digestive tract. Tobacco plants also respond to the larvae attack: it has been found that when larvae attack (eat), the plant tobacco produces more nicotine, from 220% to 400%. Some studies suggest that *M. sexta* defends itself better against its eternal enemies, the parasitic wasps *Cotesia congregate*. It has been shown that the wasps suffer higher mortality when *M. sexta* larvae fed on high-nicotine diets rather than low diets. Thus, nicotine plays a vital role in the tritrophic system *N. tabacum–M. sexta–C. congregate* (Figure 3.37).

The tobacco worms not only attack tobacco plants but can feed on tomatoes (*Solanum lycopersicon*), chili (*Capsicum* spp.), eggplant (*Solanum melongena*), and potato (*Solanum tuberosum*).

In addition to nicotine, the most abundant alkaloid of the *Nicotiana* plant, there are more than ten pyridine alkaloids (Figure 3.38). Like nicotine, some of these have both biological importance and synthetic utility in the search for new pharmacological agents. Alkaloid levels in tobacco are influenced by environmental conditions, interactions with plant pests, and plant genetics.

Anabasine, (*S*)-3-(piperidin-2-yl)pyridine, is an alkaloid of the shrub *Nicotiana glauca*, native of Argentina, Paraguay, and Bolivia. The teratogenic effects in animals of this plant are related to the presence of anabasine. This alkaloid is also isolated from the Asian plant *Anabasis aphylla* (family Chenopodiaceae). Various species of Formicidae ants (*Messor mediorubra*, *M. semirufus*, *M. ebeninus*, etc.) also have this metabolite in their defensive poison. Like nicotine, anabasine is activity-repellent in low concentrations, but in men, anabasine is more toxic than nicotine. Anabasine sulfate has been an important botanical insecticide in the former Soviet Union. Its biosynthesis is like nicotine's, especially the pyridine ring from L-aspartate (Figure 3.34). The piperidine ring of anabasine **3.37** is derived from L-lysine, which gives crucial intermediates cadaverine **3.74** and 2,3,4,5-tetrahydropyridine **3.75**.

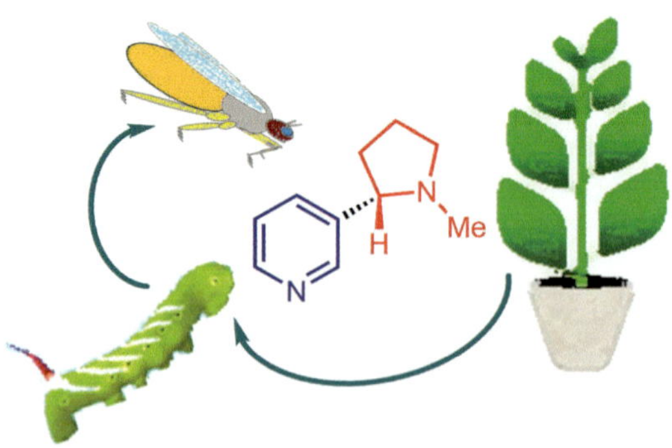

Figure 3.37. Nicotine as a critical player in the tritrophic system *N. tabacum–M. sexta–C. congregate*.

The latter reacts with pyridine intermediate **3.61** to form anabasine skeleton **3.37**. This molecule is created in three steps in the chemical laboratory, from pyridine derivative **3.76** and chiral reagent (*R*)-phenylglycinol **3.77** (Figure 3.39). Other minor components, α-(nor)nicotyrines, myosmin, 2,3′-bipyridyl (isocotein), and cotinine, have poor insecticidal activity and were not studied in detail except cotinine.

Interestingly, nicotinic acid **3.59** and its close analogs, wilfordic (**3.80**), hydroxy wilfordic (**3.81**), or evoninic (**3.82**) acids (Figure 3.40A), form part of the structure of more complex metabolites. These so-called Celastraceae sesquiterpenoids are generally based on a dihydroagarofuran skeleton **3.83** and, mainly on euonyminol (**3.84**) or evoninol (**3.85**) cores (Figure 3.40B).

These pyridine-based sesquiterpenoids can be considered macrocyclic pyridine alkaloids. It is well-known that various species of Celastraceae, which contain these metabolites, have been used in traditional Chinese agriculture for centuries to protect crops against insect damage. First pyridine-based sesquiterpenoids, wilfordine **3.86** and wilforine **3.87**, consisting of the union between euonyminol core and respective (hydroxy)-wilfordic acids, were isolated and characterized in 1953 using extracts of *Tripterygium wilfordii*, a traditional Chinese herb [70, 71]. The sesquiterpene pyridine alkaloid evonine **3.88** was isolated from *Euonymus* species, and its molecular skeleton is based on different agarofuran parts and pyridine diacid components. It consists of a combination of acetylated evoninol and evoninic acid (Figure 3.40C).

Wilfordine and wilforine showed potent toxicity toward the larvae of the European corn borer *Pyrausta nubilalis* at concentrations of 60 ppm, causing 100% mortality after three days. In addition, wilforine displays effective antifeedant activity toward cabbage leaf worm *Pieris rapae* at 1×10^{-5}% dry weight, which attacks plants in the family Cruciferae, causing severe damage. At the same time, wilfordine exhibited potent antifeedant activity against *Spodoptera littoralis*, as we already know that it is a polyphagous cotton leafworm, which also can damage economically important crops [70]. Evonine also displays good insect antifeedant and insecticidal activities [72].

Another toxic molecule related to nicotine is the coniine isolated from Apiaceae family plants (e.g., hemlock *Conium maculatum*). It is a piperidine alkaloid whose

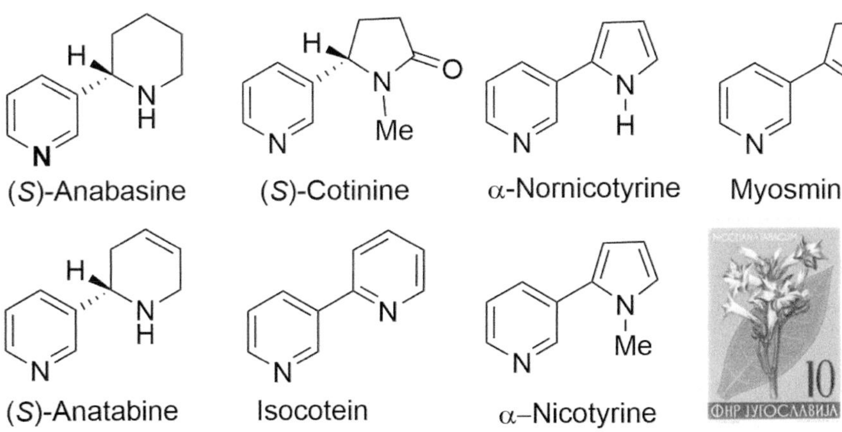

Figure 3.38. Structures of minor nicotine components and *Nicotiana tabacum* plant in a stamp.

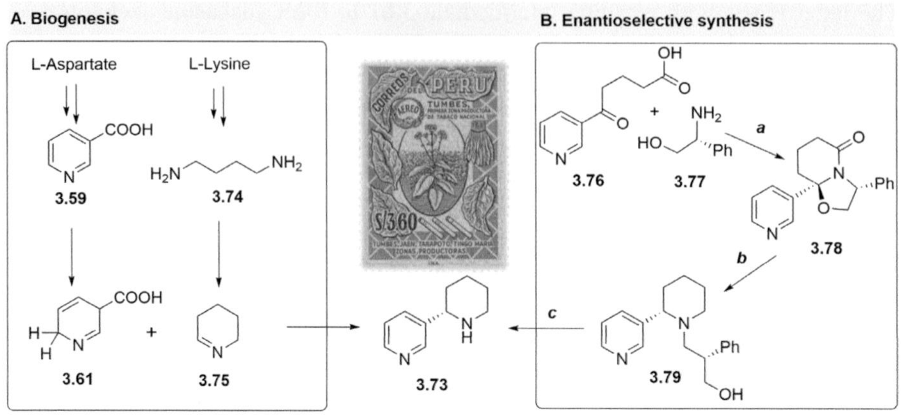

Figure 3.39. A. Biogenesis of anabasine. B. Its enantioselective synthesis: (a) PhMe, reflux, and separation (58%); (b) $LiAlH_4$ and separation (78%); (c) $H_2/Pd(OH)_2/MeOH$ (81%) and tobacco plant in Peruvian stamp.

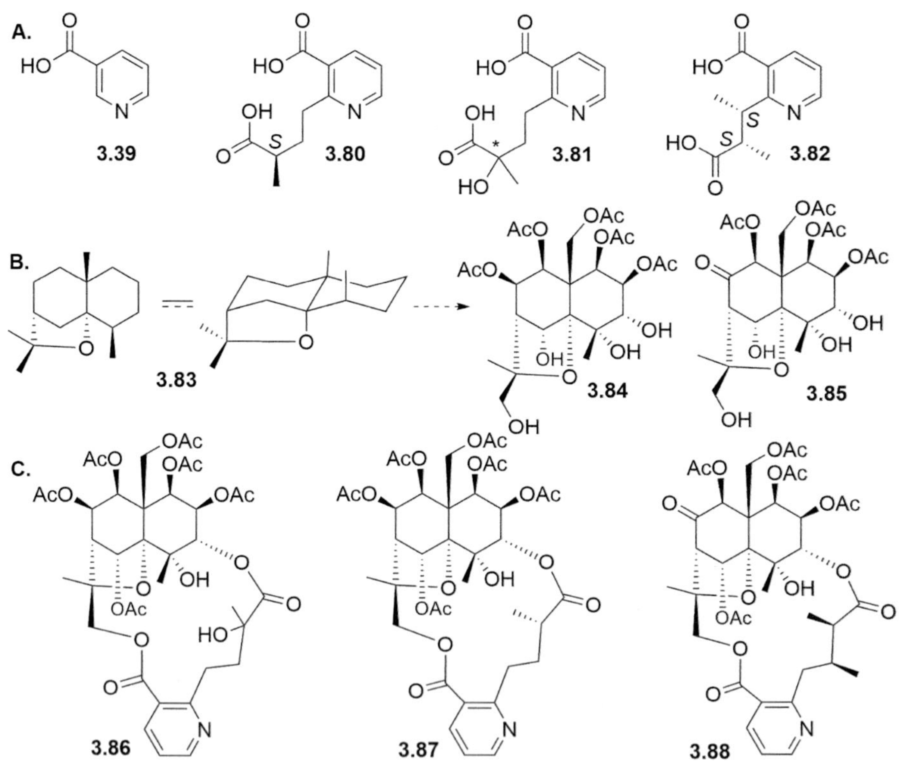

Figure 3.40. A. Pyridine acids are nitrogen-containing components of Celastraceae sesquiterpenoids. B. Euonyminol and evoninol are acetylated members of dihydroagarofuran skeleton. C. Representative macrocyclic pyridine alkaloids.

structure is simple; chemically, it is 2-propyl piperidine [73, 74] (Figure 3.41). Coniine is a nicotinic antagonist, which disrupts the function of the central nervous system and is toxic to humans and all livestock, acting as a neurotoxin. Coniine is more toxic than nicotine but less toxic than anabasine. In plants, it is found as a mixture of the *R*-(−) and *S*-(+) enantiomers, in which the *S*-enantiomer is predominant. Interestingly, the minor enantiomer (*R*-coniine) is biologically more active than the major enantiomer [75]. In mice, the LD_{50} for (−)-coniine is 7.0 mg/kg intravenously, and that for (+)-coniine is 12.1 mg/kg. The oral LD_{50} for coniine is 100 mg/kg. Species *Conium maculatum* is the most studied source of coniine and its analogs. This plant, known as poison hemlock, belongs to a group of widely spread weeds, an invasive species, and can pollute harvested hay as it grows in alfalfa fields, grass pastures, and meadows, generating a risk of poisoning to grazing livestock [74]. Along with the coniine in *C. maculatum*, there are two alkaloids of similar structure, N-methylconiine and γ-coniceine (Figure 3.41), which are also toxic and teratogenic molecules in cattle and equines. The latter compound is poison hemlock's principal alkaloid in leaves and young tissues. Coniine is toxic to humans, and 3 mg produces symptoms. Up to 150–300 mg of coniine can be tolerated, which translates to 6–8 leaves (6 g).

For a long time, the toxicity of the genus *Conium* species, especially *C. maculatum*, was related to the presence of these alkaloids. However, now it is known that another poisonous species, also called water hemlock, *Cicuta virosa*, contains agents responsible for poisoning that are not piperidine alkaloids but polyacetylene molecules, such as cicutoxin [76] (Figure 3.41). According to historical archives, in ancient Greece, criminals sentenced to death should drink the juice or extract of *Conium maculatum* (or *Cicuta virosa*?). The death of the Greek philosopher Socrates (399 BC), produced by this type of poisoning, made these plants universally famous. However, there is no information concerning which hemlock plant the recipe uses: poison hemlock (*C. maculatum*) or water hemlock (*C. virosa*) [77].

The detailed biosynthesis of coniine is not well understood. However, it was postulated that a close biological precursor of coniine is capric acid, a fatty acid, which comes from the condensation of acetyl-CoA and three malonyl-CoA

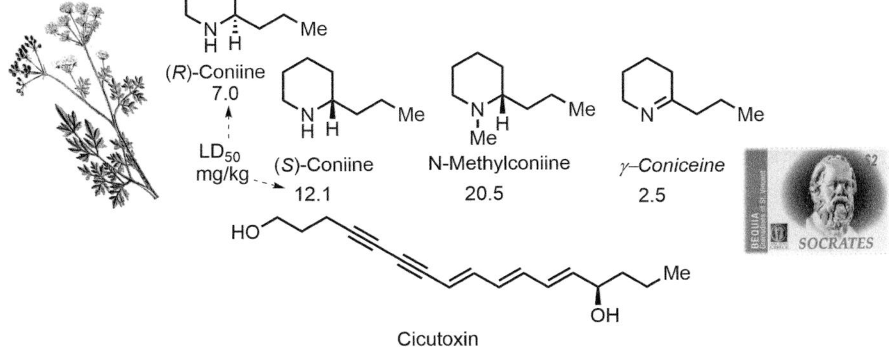

Figure 3.41. Structures of representative piperidine alkaloids from *C. maculatum*, cicutoxin of *C. virosa*, and Greece stamp dedicated to Socrates.

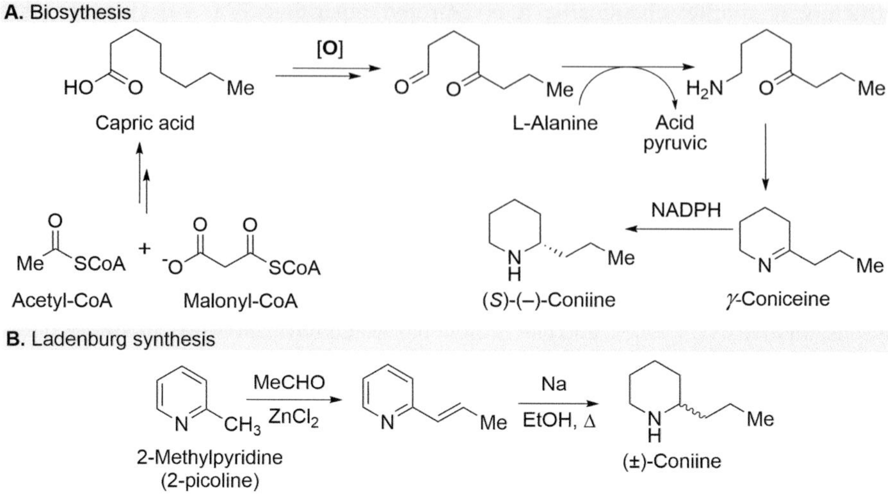

Figure 3.42. A. Biosynthetic pathways of coniine. B. First preparation of coniine: Ladenburg synthesis.

molecules. This suggests that the carbon backbone of coniine is formed via a poly-β-keto acid [78] (Figure 3.42A).

During the oxidation reaction, this acid is converted into 5-oxooctanal, which reacts with L-alanine to form γ-coniceine through an amino-oxo derivative. The nitrogen of L-alanine is subsequently introduced in a transaminase-catalyzed reaction. Coniine is finally derived from γ-coniceine by γ-coniceine reductase catalysis, which is NADPH-dependent. The first synthesis of coniine was carried out by German chemist Albert Ladenburg (1842–1911) in 1886 through the reaction of 2-methylpyridine (2-picoline) and acetaldehyde in the presence of a Lewis base ($ZnCl_2$) to give 2-propenylpyridine, which was reduced to (±)-coniine by reduction with sodium in alcohol (Figure 3.42B). Enantiomeric separation was achieved by fractional crystallization of the diastereoisomers of the salt obtained with (+)-tartaric acid.

Due to the high toxicity of coniine, the presence of both specialist and generalist insects in *C. maculatum* is extremely low compared to other Apiaceae species in their habitats. Interestingly, caterpillars and butterflies of the genus *Papilio* (*P. zelicaon*, etc.) can feed on Apiaceae plants but avoid this toxic plant. However, there are other generalist insects that can survive after invading the *C. maculatum* plant, such as *Trichoplusia ni* moth (family Noctuidae), which can consume diets high in content of this alkaloid. It was postulated that its resistance against the chemical defense of *C. maculatum* could be due to two main mechanisms: (i) effective excretion of ingested alkaloids without metabolizing and (ii) partial detoxification of the alkaloid by the cytochrome P450. It is important to note that *T. ni* larvae are pests in cruciferous plants, known as measuring worms or cabbage gray worms.

Now, let us move on to another type of alkaloid, more complex than coniine and related to Celastraceae sesquiterpenoids. It is about pyrrolizidine alkaloids (PAs), highly toxic heterocyclic molecules, which have the skeleton of the pyrrolizidine (1-azabicyclo[3.3.0]octane) ring in their structure. Naturally, these alkaloids are monoesters and acyclic (open-chain) or cyclic diesters (macrodilactones) of common

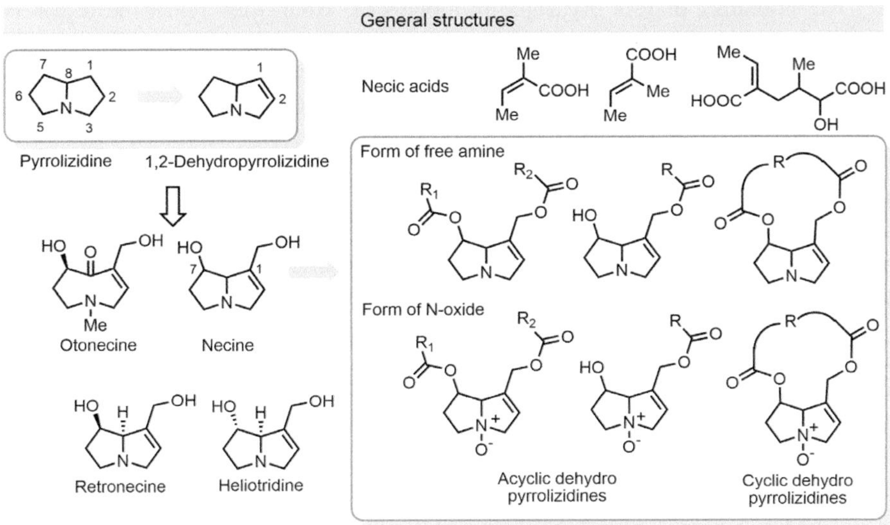

Figure 3.43. General structures of pyrrolizidine alkaloids and the three types of necine.

necine backbone with the 1,7-disubstituted 1,2-dehydropyrrolizidine ring. These alkaloids are found in many plants, sometimes as non-N-oxides toxic or as toxic tertiary amines (Figure 3.43). Pyrrolizidine alkaloids are common constituents of more than 6,000 plants of the three families, Boraginaceae, Compositae (Asteraceae), and Legumionsae (Fabaceae) around the world, and pyrrolizidine alkaloid-containing plants are probably the most common poisonous plants affecting livestock, wildlife, and humans [79].

The hydrolysis of these alkaloids in the laboratory mainly generates two structural components called necine (retronecine and heliotridine) or otonecine, which have rings of 1,2-dehydropyrrolizidine or tetrahydroazocinone, and necic acids derived from branched fatty acids.

In plants, PAs are synthesized in the roots and shoots and transported through the phloem to the stems, leaves, and seeds, where they exert their deterrent effects against herbivores, acting as a constitutive plant defense mechanism. It is interesting that in the roots, the PAs are present almost exclusively as N-oxides, not as free bases, while in the shoots, approximately 35% of the alkaloids are tertiary free bases.

As PAs generally consist of a necine base (three different bases) esterified with a necic acid, biosynthetic pathways of Pas are divergent. Thus, the retronecine base of PAs is derived from α-amino acids (L-arginine or L-ornithine) that form a diamine called putrescine **3.89**, which is converted to homospermidine **3.90** by the enzyme homospermidine synthase (HSS). Homospermidine is then transformed into retronecine (or heliotridine), which reacts with the necic acids, coming of threonine and pyruvate, and gives rise to acyclic or cyclic metabolites. Otonecine-type APs are probably formed from the retronecine through the reactions of C-8 hydroxylation and N-methylation, followed by cleavage of the pyrrolizidine ring [80] (Figure 3.44).

PAs are mainly known for their hepatotoxic and potentially carcinogenic properties [81]. Nevertheless, some PAs show interesting pharmacological properties that are currently under investigation.

The N-oxides of PAs, hydrophilic molecules, when they reach the digestive tracts, often alkaline, of some herbivores, are rapidly reduced to the forms of a toxic lipophilic tertiary amine, which can easily pass through the membranes of animals, are activated by the animal's cytochrome P450, and degrade into iminium ion-like pyrrole fragments such as **3.91**. These are powerful alkylating agents that react with cellular components, for example, DNA and proteins, to give adducts like **3.92**, leading to acute liver damage and, in some animal models, to genotoxicity and carcinogenesis (Figure 3.45). In the case of otonecine-like PAs, these toxic pyrrole fragments can be formed via the oxidative N-methylation reaction, followed by the activation of cytochrome P450.

Here, we would make an interesting question: Why do plants not generate PAs in more toxic free amine form, and why do they generally contain oxidized forms of these alkaloids, which are less bioactive molecules against insect herbivores? It seems pointless when we speak of the self-defense of plants against herbivores. It is believed that an advantage of N-oxides is their higher solubility, which results in more efficient storage and transport. Another probable explanation is the synergistic effects of PA N-oxides with other plant metabolites [81].

Although PAs are found in many botanical families, they are typical structures for three following genera *Senecio*, *Heliotropium*, and *Crotalaria*. Several plants of these genera often grow as weeds in agro-industrial production systems and thus can produce chronic hepatotoxicity and carcinogenic, teratogenic, and genotoxic effects in mammals, mainly horses, cattle, pigs, and poultry. That was explained in the abovementioned illustration (Figure 3.45).

The main sources of the PAs are the following species *Senecio longilobus*, *S. jacobaea* (family Asteraceae), *Crotalaria juncea*, *C. dura*, *C. spectabilis* (family Fabaceae), and *Heliotropium europaeum*, *H. indicum* (family Boraginaceae). The effects of these plants on wild animals can be different [82]. Indeed, the correlation between PA content and feeding is generally negative for generalist insects and various predators (e.g., ants, spiders, lizards, birds, and mammals). However, specialist herbivores, such as moths, butterflies, beetles, and grasshoppers, have evolved adaptations to PAs in their host plant. They have successfully developed biological tactics to feed on poisonous plants. The species that obtain PAs know how to avoid the accumulation of these alkaloids in free form (tertiary amine form), which is toxic. Most of these species store non-toxic alkaloids in the form of N-oxides. With this storage of "material blocks", insects have developed a strategy of using plant defense chemicals to defend against predators. It is important to note that pyrrolizidine alkaloids present in these plant species are also stimulants of strong feed for various adapted insects, i.e., specialists in feeding these plants. These insects' cases of adaptation and utilization of PAs are not so few.

One of the most studied interactions between PA-containing plants and insects is the relationship between the *Senecio jacobaea* plant, known as the herb Santiago, and *Tyria jacobaea*, called the cinnabar moth [82]. The herb Santiago is a weed plant of common wildflowers that can be found throughout Europe. This herb has long been known for its toxic properties, especially for livestock, and it is one of the five plants recognized as harmful weeds. This Arctiidae moth's larvae can convert the pyrrolizidine alkaloids' tertiary amine form to their respective N-oxide form using

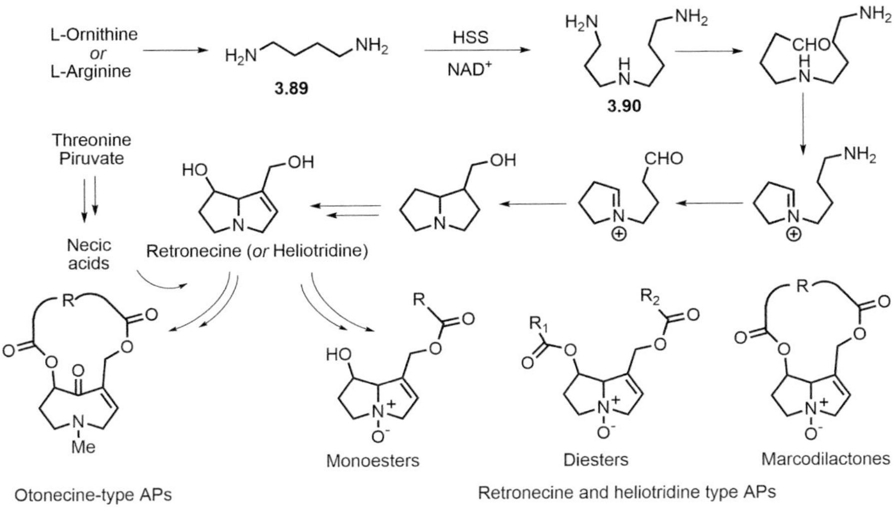

Figure 3.44. General routes of biosynthesis of pyrrolizidine alkaloids.

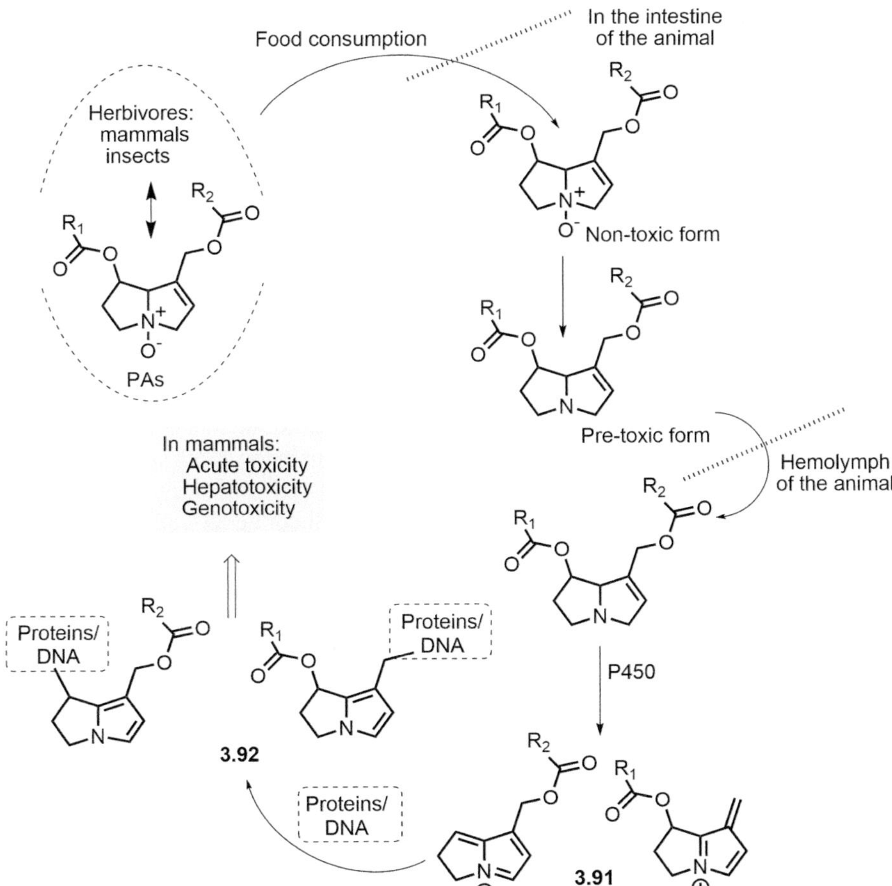

Figure 3.45. Bioactivation of pyrrolidine alkaloids (PAs) via enzymatic transformations.

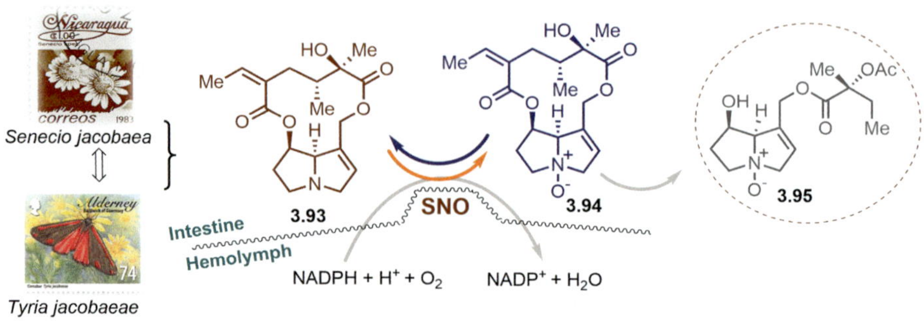

Figure 3.46. Detoxification of Pas from herb Santiago in *T. jacobaeae* moths, both in stamps.

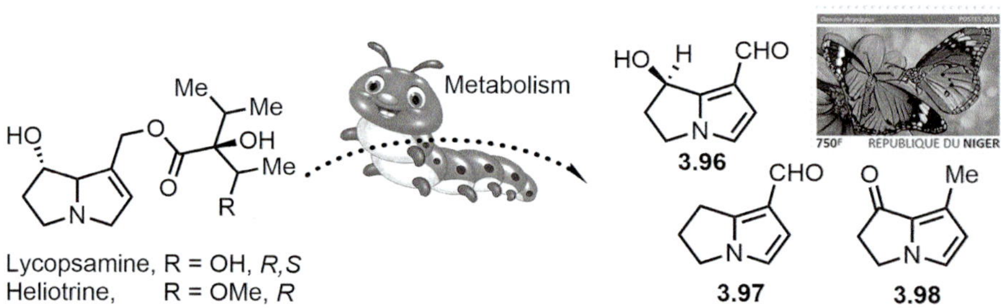

Figure 3.47. PAs as precursors to pyrrolizidine pheromones. *Danaus* butterflies in Nigerian stamp.

an N-oxidation process. When digested PAs are in the insect's intestine in their free forms, they can cross cell membranes and become activated. To prevent bioactivation, larvae of *T. jacobaeae* have the enzyme senecionine N-oxygenase (SNO) in their hemolymph, which converts effectively the amine forms of senecionine **3.93**, in its N-oxide **3.94**, non-toxic molecule. Thus, they can store and use this alkaloid and even transform it into new alkaloids like callimorphine N-oxide **3.95** in their defense against predators [83] (Figure 3.46).

In fact, the sequestered alkaloids are present in all stages of *T. jacobaeae* life (larvae, pupae, and adults). These also serve as stimulants for those moths' oviposition, but not all PAs incite oviposition. The SNO in this moth species is highly specific to PAs. Its activity is highest on senecionine-type PAs, which occur in the host plant *S. jacobaea*; other types of PAs are less well converted [82].

The vast majority of generalist insects avoid feeding on *Senecio* plants. Still, some butterflies of the same family, Arctiidae, are generalists and sometimes include plants with PAs in their diets. Among them are the moths *Estigmene acrea* and *Grammia geneura* of respective *Estigmene* and *Grammia* genera. Their larvae sequester PAs and can transfer them to adulthood, and males can transform them (namely, lycopsamine) into simpler pyrrolizidine molecules, hydroxydanaidal **3.96**, and danaidal **3.97** (Figure 3.47). Both aldehydes are short-distance mating pheromones. These molecules are released as "love dust" just before mating [84].

Indeed, these male pheromones were discovered early during the study of other *Danaus*, *Amauris*, *Creatonotos*, or *Euploea* butterflies. In the case of males of *Danaus* butterflies, which consume plants of the genus *Heliotropium* whose

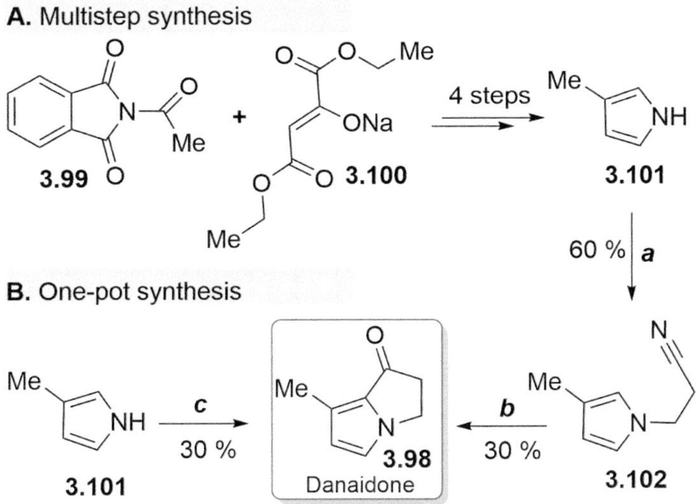

Figure 3.48. A. Multistep synthesis: (a) CH_2=CHCN, 40% Triton B, MeCN, rt, 72 hours, under N_2; (b) 1. Et_2O/HCl(g), 2 hours, 2. AcONa, benzene, 90°C, 1 hour. B. One-pot synthesis: (c) 1. 4 equivalent CH_2=CHCN, 20 mol% DBU, MeCN, rt, 72 hours, 2. HCl(g), Et_2O, $ZnCl_2$, rt, 12 hours, 3. Na_2CO_3/H_2O (pH 9), 80°C, 1 hour.

main alkaloid component is heliotrine, another pyrrolizidine derivative, danaidone (6,7-dihydro-l-methyl-5*H*-pyrrolizine-7-one, **3.98**) is formed during metabolism process. This heterocyclic ketone has also been shown to serve as an aphrodisiac pheromone necessary for successful courtship in *Danaus gilippus berenice* and *D. chrysippus*. These butterflies are known as the plain tiger or African queen, famous in Asia, Australia, and Africa for their showy coloration. Interestingly, the male butterflies pass these alkaloids to their spermatophores (a kind of capsule containing sperm), thus protecting the females and eggs.

Structural identification and the first synthesis of danaidone were realized in 1966 [85]. In this study, Jerrold Meinwald (1927–2018), an American chemist well-known for his work on chemical ecology, isolated this pheromone from 100 male *D. gilippus berenice* butterflies, which were captured in Trinidad and mailed alive in small envelopes to Ithaca, N. Y. where he worked. Interestingly, these tropical butterflies possess elaborated organs called hairpencils, which can be extruded from the end of the abdomen. In these species, the male brushes the hairpencils across the anterior end of the female in flight. During this treatment, danaidone acts, and it appears to induce the female to settle on available herbage. After continued "hairpencilling", copulation occurs.

Thus, the hairs were carefully removed from their extruded hair pencils to get this molecule and extracted directly with methylene chloride or carbon disulfide. After extraction, column purification, and preparative thin layer chromatography, 2.3 mg of danaidone as colorless needles were obtained. Synthetic confirmation of its structure was performed using multistep synthesis involving N-cyanoethylation of 3-methylpyrrole **3.101**, which had to be prepared from N-acetylphthalimide **3.99** and the sodium salt of diethyl oxalacetate **3.100** in four steps [85] (Figure 3.48A). Now,

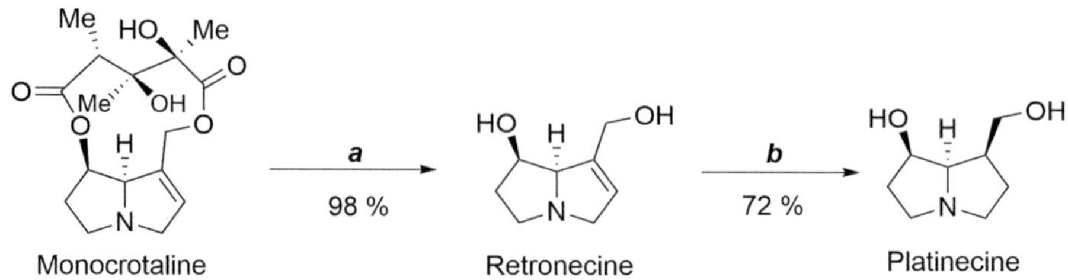

Figure 3.49. Basic hydrolysis of PAs (monocrotaline) as a semi-synthetic method of necine derivatives. Synthesis of platinicine: (a) H_2O, 2 equivalent $Ba(OH)_2$, reflux, 3 hours; (b) H_2, Rh/C, rt, 12 hours.

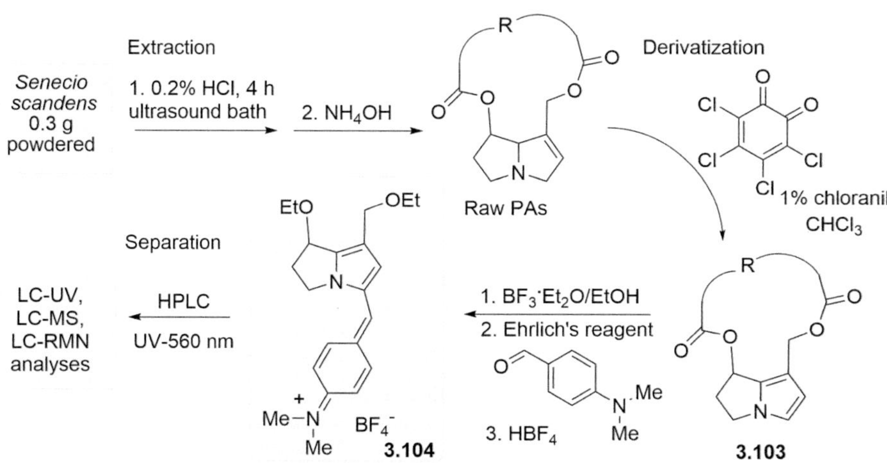

Figure 3.50. Quantitative analysis of PA alkaloids based on the HPLC-UV method.

this key pyrrole is commercially available; hence, one-pot synthesis of danaidone using a DBU-catalyzed cyanoethylation reaction of **3.101** to give N-cianoethyl-3-methylpyrrole **3.102** and its subsequent electrophilic cyclization in the presence of added HCl–$ZnCl_2$ is developed (Figure 3.48B) [86].

Another approach to synthesizing necine bases and their derivatives is by degrading natural PAs, some of which are now available in the market. We recall that these are mono- or diesters and, thus, could be hydrolyzed to afford necine bases. For example, by performing the basic hydrolysis of monocrotaline present in the seeds of *Crotalaria spectabilis* plant, the pure retronecine can be isolated with a 98% yield and used in subsequent transformations. This way, alkaloid platinecine is prepared in 70.6% overall yields from commercially available monocrotaline (Figure 3.49). However, monocrotaline is a toxic molecule. It was noted that it causes pulmonary arterial hypertension in rats.

Due to the high toxicity of PA-containing plants, many identification methods have been developed and quantification of these alkaloids, e.g., NMR, MS, GC-MS, HPLC, HPLC-UV, HPLC-MS, and HPLC-MS/MS. Every method has its advantages and limitations. But, the most commonly used for the analysis of PAs is HPLC with UV detection, despite its limitation. This method cannot detect those alkaloids that do not have chromophores in their structures. However, it is more effective if the PAs

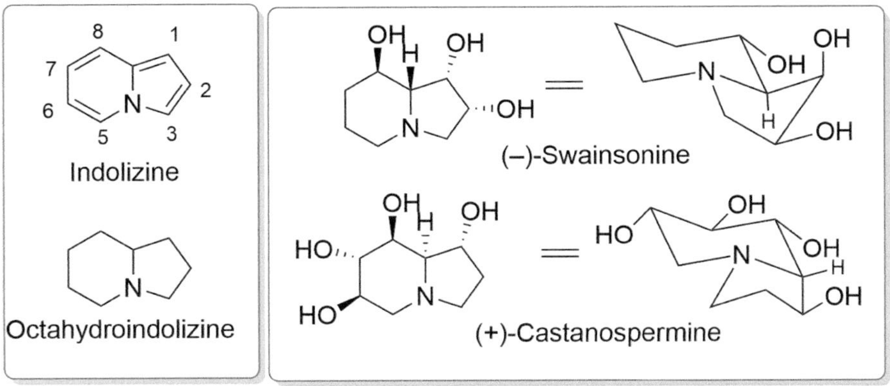

Figure 3.51. Structures of the most studied indolizidine alkaloids, swainsonine, and castanospermine.

are first extracted and derivatized, and then a reagent-chromophore is incorporated into the base (necine or otonecine) of the extracted alkaloid. This process, using Ehrlich's reagent (4-dimethylamino benzaldehyde/H_2O/$HClO_4$/AcOH), is illustrated in Figure 3.50.

When PA extracts, isolated from *Senecio scandens*, are treated with chloranil in chloroform, an intense blue color appears, corresponding to pyrrole derivatives **3.103** (proton NMR and MS confirm pyrrole ring). Then, diester hydrolysis occurs by adding Lewis acid BF3 ·Et^2O and Ehrlich's reagent. The blue solution changes to purple, indicating the formation of the standard marker **3.104**, which can also be analyzed by MS or NMR methods. Finally, purple products are subjected to HPLC analysis at UV-560 nm [87] (Figure 3.50).

Indolizidine alkaloids are heterocyclic molecules structurally related to pyrrolizidine alkaloids. The two alkaloids' most characteristic and studied indolizidines are swainsonine and castanospermine. Their structures are based on the skeleton of indolizine (1,6,7,8-octahydroindolizine) with several hydroxy groups (Figure 3.51). Due to their particular structures, they are also known as "sugar-shaped" alkaloids [88].

Like PAs, indolizidine molecules are involved in many plant-herbivore interactions and livestock poisoning. Both molecules can affect the functions of α- and β-glucosidases enzymes, which regulate the metabolism of carbohydrates in animals. Thus, they are so-called potent inhibitors of these glucosidases. The toxic effects on animals, which feed on indolizidine-containing plants, are due to the ability of these molecules to counteract the breakdown of carbohydrates.

Swainsona plants (*S. canescens*, *S. luteola*, *S. galagifolia*) contain swainsonine, while castanospermine, is found in the only legume plant of the monotypic genus *Castanospermum* (*C. australe*), both belonging to the Fabaceae family and native to Australia. Swaisonine is also found in plants of the genera *Oxytropis* and *Astragalus*, which harbor *Astragalus lentiginosus*, *A. pehuenches*, etc. Swainsonin-containing plants, particularly *A. lentiginosus*, are known as locoweeds. Their ingestion induces severe neurological symptoms including ataxia, trembling, exaggerated fright reactions, general depression, muscle incoordination, nervousness, etc. These

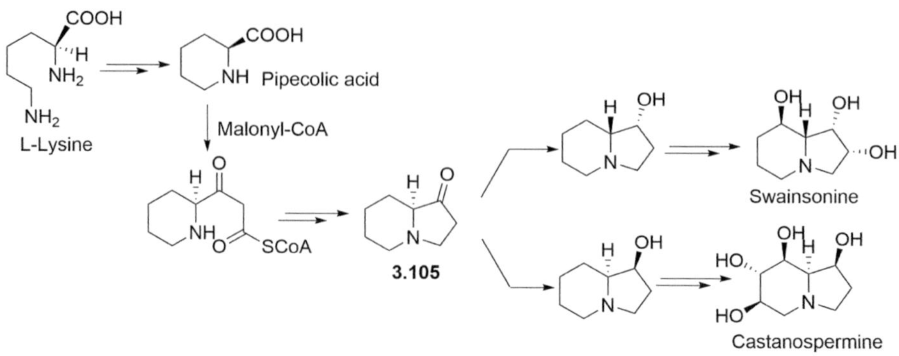

Figure 3.52. Biosynthesis of swainsonine and castanospermine.

symptoms of so-called "locoism" typically occur after 2–3 weeks of continuous grazing on toxic plants. It is believed that toxicity to the central and peripheral nervous system is caused by inhibition of lysosomal α-mannosidase and accumulation of intracellular oligosaccharide.

It is interesting to note that swainsonine was first isolated from *Rhizoctonia leguminicola* fungus in 1973, a mold that infects red clover, alfalfa, and other legumes. Later, in 1979, it was rediscovered in the plant *S. canescens*. Castanospermine was discovered in 1981. In some cases, both amine molecules were isolated as N-oxide forms.

Swainsonine and castanospermine come biosynthetically from L-lysine and pipecolic acid. The latter is condensed with two acetate units or one malonate unit by means of Claisen condensations, in the same way as in polyketides, to give the skeleton of 1-indolizidinone **3.105**, which is converted by means of oxidations in the swainsonine and castanospermine [89] (Figure 3.52).

Both molecules have important antimetastatic, antitumor proliferative, and immunoregulatory activities, among others. Swainsonine has been the first α-mannosidase inhibitor as an anticancer drug (agent GD0039, as HCl salt) and has reached clinical trials (treatment of renal carcinoma) up to phase II [89]. Noteworthy is that being a potent α-mannosidase inhibitor, swainsonine was found to be active against several insects that have this enzyme, e.g., American cockroaches *Periplaneta Americana*, blow flies larva *Calliphora* sp., darkling beetles *Alphitobius diaperinus* and mealworms *Tenebrio molitor*, showing respective inhibition concentrations, IC_{50} 0.066 μM, 0.014 μM, 0.048 μM, and 0.1 μM [90]. Castanospermine has also received much attention due to its promising properties against some viruses and cancer. Since it is a potent inhibitor of α- and β-glucosidases, this molecule can fight dengue viruses and the hepatitis C virus since it inhibits replication and prevents the correct folding of the envelope glycoproteins [91]. On the other hand, *C. australe* seeds contain 0.3% of castanospermine and are toxic to mammals and insects. Studies have shown that castanospermine also acts as a feeding deterrent to insects. It is known that castanospermine is an important feeding deterrent to the Acyrthosiphon pisum aphids, called pea aphids, with an ED_{50} of 1×10^{-4} M in artificial diets. The larval development of cowpea weevil *Callosobruchus maculatus* F. is inhibited when it feeds seeds of *C. australe*. These aphids are pests on stored grains and generate much damage and loss.

Unsurprisingly, these "sugar-shape" alkaloids have been the subject of numerous synthetic studies. However, these studies are not examined here because similar indolizidine alkaloids are also found in different animals, e.g., ants, and various interesting analog molecules, which will be discussed in later chapters.

3.5 Conclusion

A wide array of herbivores attacks plants. In response, plants deploy an arsenal of defensive traits, among which toxins are key "chemical weapons". In this chapter, we examined and discussed four main groups of N-containing toxins: non-protein α-amino acids, cyanogenic glucosides, glucosinolates, and alkaloids. All these are found in different and diverse vegetable sources and can exert their toxicity on herbivores in unusual manners.

Although non-protein α-amino acids do not participate in the normal replication process, they can interfere with normal metabolic processes within the cells of living organisms, including humans. Alternating enzyme systems or incorporating protein synthesis, these small molecules typically act as potent antimetabolites. However, some non-protein α-amino acids are also known to be ingested by other insects and animals and are excreted unchanged without any action.

Cyanogenic glucosides, glucose-based metabolites, are interesting carbohydrate derivatives capable of producing and releasing a well-known toxic molecule, hydrogen cyanide, upon hydrolysis. Since they are inert in plant cells, these compounds are an excellent chemical defense against herbivores acting as latent binary cyanide bombs.

Glucosinolates are characterized by the liberation of mustard oils, a complex mixture of isothiocyanates, thiocyanates, and related nitriles. Glucosinolates and their degradation products are important factors in plant defense against herbivores and pathogens. Thus, the glucosinolate-myrosinase complex is a sophisticated two-component system to ward off opponents.

Alkaloids are known for their numerous pharmacological effects on vertebrates. In our chapter, some of these metabolites were selected according to the author's interests and divided into different classes according to their precursor, i.e., pyridine alkaloids, piperidine alkaloids, pyrrolizidine, and indolizidine alkaloids. In general, the presence of these toxic molecules in plants enhances plant reproductive rates by improving defenses against biotic stresses (herbivore attacks) or by affecting pollinators and seed/fruit disperser visitation. Defensive strategies include predator repellence by toxicity or bitter taste. Nevertheless, the prominent role of alkaloids in plants, like other nitrogenous toxic molecules, is toxicity against predators and pathogens.

It is noteworthy that the same toxic properties observed in the plant defense scenario can often be used in prospection for new drugs. For example, specific toxicity may be used to fight certain tumor cell types or to control specific microorganisms or pests addressed in our text.

3.6 References

[1] Wink, M. 2010. Introduction: biochemistry, physiology and ecological functions of secondary metabolites. *In*: Wink M. [Ed.]. Biochemistry of Plant Secondary Metabolism. Annual Plant Rev. 40: 1–19.

[2] Bell, E.A. 2003. Non-protein amino acids of plants: significance in medicine, nutrition, and agriculture. J. Agric. Food Chem. 51(10): 2854–2865.

[3] Vranova, V., Rejsek, K., Skene, K.R. and Formanek, P. 2011. Non-protein amino acids: plant, soil and ecosystem interactions. Plant Soil 342(1-2): 31–48.

[4] Fletcher, M.T., Al Jassim, R.A.M. and Cawdell-Smith. A.J. 2015. The occurrence and toxicity of indospicine to grazing animals. Agriculture 5: 427–440.

[5] Rodgers, K.J., Samardzic, K. and Main, B.J. 2017. Toxic non-protein amino acids. pp. 263–285. *In*: Gopalakrishnakone, P., Carlini, C. and Ligabue-Braun, R. [Eds.]. Plant Toxins. Toxinology. Springer, Dordrecht.

[6] Tan, E.T.T., Materne, C.M., Silcock, R.G., D'Arcy, B.R., Al Jassim, R. and Fletcher, M.T. 2016. Seasonal and species variation of the hepatotoxin indospicine in Australian *Indigofera* legumes as measured by UPLC-MS/MS. J. Agric. Food Chem. 64: 6613–6621.

[7] Hegarty, M.P. and Pound, A.W. 1970. Indospicine, a hepatotoxic amino acid from *Indigofera spicata*: isolation, structure, and biological studies. Aust. J. Biol. Sci. 23: 831–842.

[8] Tan, E.T., Al Jassim, R., D'Arcy, B.R. and Fletcher, M.T. 2017. *In vitro* biodegradation of hepatotoxic indospicine in *Indigofera spicata* and its degradation derivatives by camel foregut and cattle rumen fluids. J. Agric Food Chem. 65(34): 7528–7534.

[9] Lang, C.S., Wong, S.H., Chow, S., Challinor, V.L., Yong, K.W., Fletcher, M.T. et al. 2016. Synthesis of L-indospicine, [5,5,6-2H 3]-L-indospicine and L-norindospicine. Org. Biomol. Chem. 14(28): 6826–6832.

[10] Bence, A.K. and Crooks, P.A. 2003. The mechanism of L-canavanine cytotoxicity: arginyl tRNA synthetase as a novel target for anticancer drug discovery. J. Enzyme Inhib. Med. Chem. 18(5): 383–394.

[11] Thomas, D.A. and Rosenthal, G.A. 1987. Toxicity and pharmacokinetics of the non-protein amino acid L-canavanine in the rat. Toxicol. Appl. Pharmacol. 91(3): 395–405.

[12] Staszek, P., Weston, L.A., Ciacka, K. Krasuska, U., Gniazdowska, A. 2017. L-Canavanine: How does a simple non-protein amino acid inhibit cellular function in a diverse living system? Phytochem. Rev. 166(6): 1269–1282.

[13] Rosenthal, G.A., Hughes, C.G. and Janzen, D.H. 1982. L-Canavanine, a dietary nitrogen source for the seed predator *Caryedes brasiliensis* (Bruchidae). Science 217(4557): 353–355.

[14] Alcocer-Varela, J., Iglesias, A., Llorente, L. and Alarcón-Segovia, D. 1985. Effects of L-canavanine on T cells may explain the induction of systemic lupus erythematosus by alfalfa. Arthritis & Rheumatism: Official J. Am. Col. Rheumatol. 28(1): 52–57.

[15] Soares, A.R., Marchiosi, R., Siqueira-Soares, R.D.C., Barbosa de Lima, R., Dantas dos Santos, W. and Ferrarese-Filho, O. 2014. The role of L-DOPA in plants. Plant Signal Behav. 9(4): e28275.

[16] Ramli, N., Jamaludin, A.A. and Ilham, Z. 2017. Mimosine toxicity in leucaena biomass: A hurdle impeding maximum use for bioproducts and bioenergy. Int. J. Environ. Sci. Nat. Res. 6(5): 1–5.

[17] Fowden, L. 1955. Azetidine-2-carboxylic acid: a new constituent of plants. Nature 176(4477): 347–348.

[18] Fowden, L. 1956. Azetidine-2-carboxylic acid: a new cyclic imino acid occurring in plants. Biochem. J. 64(2): 323–332.

[19] Bessonov, K., Bamm, V.V. and Harauz, G. 2010. Misincorporation of the proline homologue Aze (azetidine-2-carboxylic acid) into recombinant myelin basic protein. Phytochemistry 71(5-6): 502–507.

[20] Adeyeye, O.A. and Blum, M.S. 1989. Bioactivity of *Liriope muscari* (Liliaceae) extracts to corn earworm (Lepidoptera: Noctuidae). J. Econ. Entomol. 82(3): 942–947.

[21] Okot-Kotber, B.M. and Adeyeye, O.A. 1995. The influence of dietary L-azetidine-2-carboxylic acid on catecholamine levels in the hemolymph and cuticle of the corn earworm, *Helicoverpa zea*. Insect Biochem. Mol. Biolo. 25(8): 959–966.

[22] Hassid, E., Applebaum, S.W. and Birk, Y. 1976. Azetidine-2-carboxylic acid: A naturally occurring inhibitor of *Spodoptera littoralis* (boisd.) (Lepidoptera: Noctuidae). Phytoparasitica 4(3): 173–183.
[23] Couty, F. and Evano, G. 2006. Azetidine-2-carboxylic acid. From lily-of-the-valley to key pharmaceuticals. A jubilee review. Org. Prep. Proced. Int. 38(5): 427–465.
[24] Yamada, Y., Emori, T., Kinoshita, S. and Okada, H. 1973. The Synthesis of (±)-Azetidine-2-carboxylic Acid and 2-pyrrolidinone derivatives. Agric. Biol. Chem. 37(3): 649–652.
[25] Rodebaugh, R.M. and Cromwell, N.H. 1969. A facile new synthesis of dl-azetidine-2-carboxylic acid. J. Heterocyclic. Chem. 6(3): 435–437.
[26] Miyoshi, M., Sugano, H., Fujii, T., Ishihara, T. and Yoneda, N. 1973. A novel synthesis of optically active azetidine-2-carboxylic acid. Chem. Lett. 2(1): 5–6.
[27] Zagrobelny, M., De Castro, É.C.P., Møller, B.L. and Bak, S. 2018. Cyanogenesis in arthropods: From chemical warfare to nuptial gifts. Insects 9(2): 51.
[28] https://www.cyanidecode.org/cyanide-facts/environmental-health-effects.
[29] Blažević, I., Montaut, S., Burčul, F., Olsen, C.E., Burow, M., Rollin, P. et al. 2020. Glucosinolate structural diversity, identification, chemical synthesis and metabolism in plants. Phytochemistry 169: Article 112100.
[30] Lim, Y.W., Ong, M.J.H. and Hewitt, R.J. 2018. Improved synthesis of glucosinolates. Synthesis 50(8): 1640–1650.
[31] Paunović, D., Šolević-Knudsen, T., Krivokapić, M., Zlatković, B. and Antić, M. 2012. Sinalbin degradation products in mild yellow mustard paste. Hemijska industrija 66(1): 29–32.
[32] van Poppel, G., Verhoeven, D.T., Verhagen, H. and Goldbohm, R.A. 1999. Brassica vegetables and cancer prevention. Epidemiology and mechanisms. Adv. Exp. Med. Biol. 472: 159–168.
[33] Kim, H.Y., Yokozawa, T., Cho, E.J., Cheigh, H.S., Choi, J.S. and Chun, H.Y. 2003. *In vitro* and *in vivo* antioxidant effects of mustard leaf (*Brassica juncea*). Phytother Res. 17: 465–471.
[34] Bischoff, K.L. 2016. Glucosinolates. pp. 551–554. *In*: Gupta, R.C. [Ed.]. Nutraceuticals: Efficacy, Safety and Toxicity. Academic Press.
[35] Kaschula, C.H. and Hunter, R. 2016. Synthesis and structure–activity relations in allylsulfide and isothiocyanate compounds from garlic and broccoli against *In Vitro* cancer cell growth. pp. 1–43. *In*: Atta-ur-Rahman [Ed.]. Studies in Natural Products Chemistry. Elsevier, 50.
[36] Mokhtari, R.B., Baluch, N., Homayouni, T.S., Morgatskaya, E., Kumar, S., Kazemi, P. et al. 2018. The role of Sulforaphane in cancer chemoprevention and health benefits: A mini-review. J. Cell Comm. Signaling 12(1): 91–101.
[37] Pehr Victor Edman (1916–1977), a Swedish biochemist who developed a method for sequencing proteins (the Edman degradation).
[38] Luo, B., Wang, J., Li, X., Lu, W., Yang, J., Hu, Y. et al. 2017. New mild and simple approach to isothiocyanates: a class of potent anticancer agents. Molecules 22(6): 773.
[39] Munch, H., Hansen, J.S., Pittelkow, M., Christensen, J.B. and Boas, U. 2008. A new efficient synthesis of isothiocyanates from amines using di-tert-butyl dicarbonate. Tetrahedron Lett. 49(19): 3117–3119.
[40] Mandapati, U., Pinapati, S. and Rudraraju, R. 2017. Copper promoted desulfurization toward the synthesis of isothiocyanates. Tetrahedron Lett. 58(2): 125–128.
[41] Fu, Z., Yuan, W., Chen, N., Yang, Z. and Xu, J. 2018. $Na_2S_2O_8$-mediated efficient synthesis of isothiocyanates from primary amines in water. Green Chem. 20(19): 4484–4491.
[42] Liao, Y.Y., Deng, J.C., Ke, Y.P., Zhong, X.L., Xu, L., Tang, R.Y. et al. 2017. Isothiocyanation of amines using the Langlois reagent. Chem. Comm. 53(45): 6073–6076.
[43] Eschliman, K. and Bossmann, S.H. 2019. Synthesis of Isothiocyanates: An Update. Synthesis 51(08): 1746–1752.
[44] Sven-Olov Lawesson (1926–1985), a Swedish chemist who in 1956 found that 2,4-bis(4-methoxyphenyl)-1,3,2,4-dithiadiphosphetane 2,4-disulfide (Lawesson's reagent) was useful as a mild and convenient thionating agent for ketones, esters, and amides. It is comparable to the reagent P_4S_{10}, which normally needs higher temperatures and a large excess of the thionating agent.
[45] Langlois, B.R. 2017. Once upon a time was the langlois' reagent: A "Sleeping Beauty". pp. 125–140. *In*: Groult, H., Leroux, F.R., Tressaud, A. [Eds.]. Modern Synthesis Processes and Reactivity of Fluorinated Compounds. Elsevier.

[46] Castanheiro, T., Suffert, J., Donnard, M. and Gulea, M. 2016. Recent advances in the chemistry of organic thiocyanates. Chem. Soc. Rev. 45(3): 494–505.

[47] Hansen, H.J. 2000. Otto Billeter oder wie die älteste [3,3]-sigmatrope Umlagerung nach Neuenburg kam: Teil 2. CHIMIA Inter. J. Chem. 54(3): 105–119.

[48] Ilardi, E.A., Stivala, C.E. and Zakarian, A. 2009. [3,3]-Sigmatropic rearrangements: recent applications in the total synthesis of natural products. Chem. Soc. Rev. 38(11): 3133–3148.

[49] Schmid, H. and Karrer, P. 1948. Synthese der racemischen und der optisch aktiven Formen des Sulforaphans. Helv. Chim. Acta 31(6): 1497–1505.

[50] Vermeulen, M., Zwanenburg, B., Chittenden, G.J. and Verhagen, H. 2003. Synthesis of isothiocyanate-derived mercapturic acids. Eur. J. Med. Chem. 38(7-8): 729–737.

[51] Ding, T.J., Zhou, L. and Cao, X.P. 2006. A facile and green synthesis of sulforaphane. Chin. Chem. Lett. 17(9): 1152–1154.

[52] Chen, X., Li, Z., Sun, X., Ma, H., Chen, X., Ren, J. et al. 2011. New method for the synthesis of sulforaphane and related isothiocyanates. Synthesis 2011(24): 3991–3996.

[53] Ganin, H., Rayo, J., Amara, N., Levy, N., Krief, P. and Meijler, M.M. 2013. Sulforaphane and erucin, natural isothiocyanates from broccoli, inhibit bacterial quorum sensing. Med. Chem. Comm. 4(1): 175–179.

[54] Interestingly, Paul Karrer was born in Moscow (Russia) in Swiss family, studied chemistry under Alfred Werner, Nobel laurate in Chemistry and did postdoc study with Paul Ehrlich, another Nobel laurate in Physiology or Medicine.

[55] Bones, A.M. and Rossiter, J.T. 1996. The myrosinase-glucosinolate system, its organization and biochemistry. Physiol. Plant 97: 194–208.

[56] Kissen, R., Rossiter, J.T. and Bones, A.M. 2009. The 'mustard oil bomb': not so easy to assemble?! Localization, expression and distribution of the components of the myrosinase enzyme system. Phytochemistry Rev. 8(1): 69–86.

[57] de Vos, M., Kriksunov, K.L. and Jander, G. 2008. Indole-3-acetonitrile production from indole glucosinolates deters oviposition by *Pieris rapae*. Plant Physiol. 146(3): 916–926.

[58] Müller, C., Agerbirk, N., Olsen, C.E., Boevé, J.L., Schaffner, U. and Brakefield, P.M. 2001. Sequestration of host plant glucosinolates in the defensive hemolymph of the sawfly *Athalia rosae*. J. Chem. Ecol. 27(12): 2505–2516.

[59] Schramm, K., Vassão, D.G., Reichelt, M., Gershenzon, J. and Wittstock, U. 2012. Metabolism of glucosinolate-derived isothiocyanates to glutathione conjugates in generalist lepidopteran herbivores. Insect Biochem. Mol. Biol. 42(3): 174–182.

[60] Matsuura, H.N. and Fett-Neto, A.G. 2017. Plant alkaloids: Main features, toxicity, and mechanisms of action. pp. 1–15. *In*: Gopalakrishnakone, P., Carlini, C.R. and Ligabue-Braun, R. [Eds.] Plant toxins, Toxicology. Springer Science+Business Media Dordrecht.

[61] Streller, S. and Roth, K. 2014. The Chemistry of Tobacco – Part 1. ChemViews Magazine. 07 octubre 2014. https://www.chemistryviews.org/details/ezine/6790291/The_Chemistry_of_Tobacco__Part_1.html.

[62] Streller, S. and Roth, K. 2014. The Chemistry of Tobacco – Part 3. ChemViews Magazine. 02 diciembre 2014.https://www.chemistryviews.org/details/ezine/6790291/The_Chemistry_of_Tobacco__Part_3.html.

[63] Mithöfer, A. and Boland, W. 2012. Plant defense against herbivores: chemical aspects. Ann. Rev. Plant Biol. 63: 431–450.

[64] https://chempics.wordpress.com/2014/04/16/nicotine/.

[65] Wagner, F.F. and Comins, D.L. 2007. Recent advances in the synthesis of nicotine and its derivatives. Tetrahedron 34(63): 8065–8082.

[66] Dübon, P., Farwick, A. and Helmchen, G. 2009. Enantioselective syntheses of 2-substituted pyrrolidines from allylamines by domino hydroformylation-condensation: short syntheses of (*S*)-nicotine and the alkaloid 225C. Synlett 2009(09): 1413–1416.

[67] Girard, S., Robins, R.J., Villiéras, J. and Lebreton, J. 2000. A short and efficient synthesis of unnatural (R)-nicotine. Tetrahedron Lett. 41(48): 9245–9249.

[68] Wink, M. and Theile, V. 2002. Alkaloid tolerance in *Manduca sexta* and phylogenetically related sphingids (Lepidoptera: Sphingidae). Chemoecology 12(1): 29-46.

[69] Steppuhn, A., Gase, K., Krock, B., Halitschke, R. and Baldwin, I.T. 2004. Nicotine's defensive function in nature. PLoS Biol. 2(8): e217.

[70] Spivey, A.C., Weston, M. and Woodhead, S. 2002. Celastraceae sesquiterpenoids: Biological activity and synthesis. Chem. Soc. Rev. 31(1): 43–59.

[71] Aniszewski, T. 2007. Alkaloids - Secrets of Life. Alkaloid Chemistry, Biological Significance, Applications and Ecological Role. Elsevier Science, CHAPTER 2 - Alkaloid Chemistry, pp. 61–139.

[72] Wang, M. and Wu, W. 2002. The insecticide constituents of several Celastraceae plants. Korean J. of Pest Sc. 6(1): 9–15.

[73] https://natoxaq.ku.dk/toxin-of-the-week/coniine/.

[74] Hotti, H. and Rischer, H. 2017. The killer of Socrates: Coniine and related alkaloids in the plant kingdom. Molecules 22(11): 1962.

[75] Lee, S.T., Green, B.T., Welch, K.D., Pfister, J.A. and Panter, K.E. 2008. Stereoselective potencies and relative toxicities of coniine enantiomers. Chem. Res. Toxicol. 21: 2061–2064.

[76] https://natoxaq.ku.dk/toxin-of-the-week/cicutoxin/.

[77] Sullivan, J. 2001. A note on the death of Socrates. Class Q; 51: 608–610.

[78] Hotti, H., Seppänen-Laakso, T., Arvas, M., Teeri, T.H. and Rischer, H. 2015. Polyketide synthases from poison hemlock (*Conium maculatum* L.). FEBS Journal 282(21): 4141–4156.

[79] Fu, P.P., Xia, Q., Lin, G. and Chou, M.W. 2004. Pyrrolizidine alkaloids-genotoxicity, metabolism enzymes, metabolic activation, and mechanisms. Drug Metab. Rev. 36(1): 1–55.

[80] Schramm, S., Köhler, N. and Rozhon, W. 2019. Pyrrolizidine alkaloids: Biosynthesis, biological activities and occurrence in crop plants. Molecules 24(3): 498.

[81] Robertson, J. and Stevens, K. 2017. Pyrrolizidine alkaloids: Occurrence, biology, and chemical synthesis. Nat. Prod. Rep. 34(1): 62–89.

[82] Macel, M. 2011. Attract and deter: a dual role for pyrrolizidine alkaloids in plant–insect interactions. Phytochem. Rev. 10(1): 75–82.

[83] Ehmke, A., Witte, L., Biller, A. and Hartmann, T. 1990. Sequestration, N-oxidation and transformation of plant pyrrolizidine alkaloids by the arctiid moth *Tyria jacobaeae* L. Zeitsch Naturfor C; 45(11-12): 1185–1192.

[84] Honda, K., Matsumoto, J., Sasaki, K., Tsuruta, Y. and Honda, Y. 2018. Uptake of plant-derived specific alkaloids allows males of a butterfly to copulate. Sci. Rep. 8(1): 5516.

[85] Meinwald, J. and Meinwald, Y.C. 1966. Structure and synthesis of the major components in the hairpencil secretion of a male butterfly, *Lycorea ceres ceres* (Cramer). J. Am. Chem. Soc. 88(6): 1305–1310.

[86] Murtagh, J.E., McCooey, S.H. and Connon, S.J. 2005. Novel amine-catalysed hydroalkoxylation reactions of activated alkenes and alkynes. Chem. Comm. (2): 227–229.

[87] Zhang, F., Wang, C.H., Xiong, A.Z., Wang, W., Yang, L., Branford-White, C.J. et al. 2007. Quantitative analysis of total retronecine esters-type pyrrolizidine alkaloids in plant by high performance liquid chromatography. Anal Chim. Acta; 605(1): 94–101.

[88] Fellows, L.E. and Nash, R.J. 1990. Sugar-shaped alkaloids. Science Progress (1933-); 74(2) (294): 245–255.

[89] Ren, Z., Song, R., Wang, S., Quan, H., Yang, L., Sun, L. et al. 2017. The biosynthesis pathway of Swainsonine, a new anticancer drug from three endophytic fungi. J. Microbiol. Biotechnol. 27(11): 1897–1906.

[90] Scofield, A.M., Witham, P., Nash, R.J., Kite, G.C. and Fellows, L.E. 1995. Castanospermine and other polyhydroxy alkaloids as inhibitors of insect glycosidases. Comp. Biochem. Physiol. Part A: Physiol. 112(1): 187–196.

[91] Whitby, K., Pierson, T.C., Geiss, B., Lane, K., Engle, M., Zhou, Y. et al. 2005. Castanospermine, a potent inhibitor of dengue virus infection *in vitro* and *in vivo*. J. Virol. 79(14): 8698–8706.

CHAPTER 4

Principal Players of Plant-Microorganism Interactions

Phytoalexins, Phytotoxins, and Mycotoxins

4. INTRODUCTION: HOW AND WHY DO THE PLANTS AND MICROORGANISMS INTERACT WITH EACH OTHER THROUGH COMPLEX MOLECULES?

One of the most exciting phenomena of interspecific interaction between land plants and microorganisms is the synthesis of defensive substances by plants when infectious diseases appear in them. It is estimated that more than 8,000 species of fungi and 2,000 species of bacteria can act as plant pathogens. Generally, a plant pathogen is a broad term that refers to any organisms, such as fungi, bacteria, protists, and viruses, that grow inside the plant to complete all or part of its life cycle, causing an adverse effect or biotic stress. The process of infection, colonization, and reproduction of the pathogen is called pathogenesis. To protect themselves from pathogen attacks, plants have developed constitutive and inducible defenses (Section 1.3). Constitutive defenses include physical barriers and the production of secondary metabolites and antimicrobial proteins, known as chemical barriers. Thus, with these barriers, most higher plants resist microbial attack or coexist in a complex relationship with parasites (mainly fungi, bacteria, viruses, etc.). The biochemical bases of this resistance are of special interest to biochemists, chemists, biologists, and agriculturalists, as they are key to the development of plant biochemistry and modern agriculture because plant pathogens are responsible for important crop yield reductions. It has been estimated that between 13% and 16% of crop production worldwide is lost directly (yield, quality, etc.) every year due to pathogens.

Marshall Ward (1854–1906) proposed the first theory of resistance, suggesting that some compounds must be in plants acting as inhibitors (toxins) to counteract parasitic invasion. Another important mechanism of plant defense against pathogens is tolerance [1, 2]. While plant resistance is defined as the host's ability to limit pathogen multiplication, tolerance is the ability of a plant to reduce the effect of infection on its fitness, regardless of the level of pathogen multiplication [3]. Both concepts first appeared in the literature over a century ago (*ca.*, 1904–1905). There is a lot of biochemical information about this process in scientific literature. In our book, we cannot delve into these issues. We briefly note that plant-pathogen interactions are described according to the respective specific molecular signals called *Damage-Associated Molecular Patterns* (DAMPs) and *Microbe-* or *Pathogen-Associated Molecular Patterns* (MAMPs or PAMPs) [4]. We recall again that DAMPs are plant-derived substances and breakdown products, which indicate tissue damage, whereas PAMPs are molecules from pathogen activity during an invasion. These resistance mechanisms are complex and mainly involve two aspects: pre-infectious and post-infectious factors in the affected plant. In brief, pre-infectious factors are the plant's passive or constitutive defense before the attack of pathogens (pre-invasion). Such molecules are now called phytoanticipins, which are constitutive plant secondary metabolites with defensive roles against microbes [5]. Let us recall how some plant metabolites, such as cyanogenic glycosides, function. It was previously shown that these compounds, stored in the plant, can release HCN molecules when the plant is attacked by its predators (Section 3.2). In the case of pathogen progression (pathogenesis), plant biochemical machinery can generate another type of toxic compound only after different types of phytopathogenic microorganisms have stimulated them. In other words, such molecules are immediate molecular responses of plants to the pathogen invasion, i.e., post-invasion products. Molecules of this type are called phytoalexins or metabolites of biological, chemical, or physical stress [6]. Such molecules will be firstly discussed in this chapter.

However, it is important to clarify that the mentioned processes are not as simple as we described. When pathogens attack plants, this stress can induce biochemical and physiological changes in plants. These changes are not only expressed in the production of phytoalexins, which are the results of this process, but in the generation of pathogenesis-related (PR) proteins that subsequently prevent various pathogen invasions. Thus, plants enhance defense responses by inducing the activity of a broad spectrum of defense enzymes, which are PR proteins, namely peroxidase, β-1,3-glucanase, chitinase, polyphenol oxidase, and phenylalanine ammonia-lyase (PAL) among others. We are now ready to begin discussing the first group of plant protective molecules: phytoalexins.

4.1 Phytoalexins: "Invisible" Protective Molecules of the Plants

For the first time, the concept of phytoalexins was mentioned by Müller and Börger in 1940. Their theory was based on the experiments on resistance factors of potatoes *Solanum tuberosum* versus the parasitic fungus *Phythophthora infestans*, which produces a disease known as late blight. They initially infested potato tuber tissue with an incompatible race of *P. infestans*. Repeating the infection process to this tissue, they observed that the tissue induced resistance to a compatible race of the same parasitic fungus [7]. Thus, it was correctly hypothesized that the tuber tissue produced substances that inhibited the pathogen and protected the tissue against later infection by other compatible pathogen races in response to the incompatible interaction. These substances were called phytoalexins. The phytoalexin term is derived from the Greek *Phyto*, meaning plant, and *Alexein*, meaning protective action. It is well-known that phytoalexins are numerous heterogeneous compounds considered molecular markers of disease resistance [8].

Most of the known phytoalexins are toxic to phytopathogenic fungi. Some are also harmful to bacteria, nematodes, and other pathogens. Phytoalexins have several peculiar characteristics, namely (i) they are not detected before infection; (ii) they are produced quickly, a few hours after the microbial attack; (iii) they are synthesized in sufficient concentration to counteract the invasion; and (iv) their formation is restricted to a local area around the infection site.

Since 1940, research on phytoalexins has been developed extensively, not only for the study of their roles in defense against pathogens and pests but also concerning the effects that promote health. More than 300 phytoalexins are known; half are characterized and isolated from Fabaceae and Solanaceae family plants. The most intensive studies recently focused on the phytoalexins of plant families of significant economic importance: corn, rice, oats, cotton, and grape. Considering those mentioned above, we can address the study on several groups of phytoalexins.

4.1.1 Resveratrol and other Stilbenoids from Vitaceae Plants

In grapes and other species in the Vitaceae family, the main component is resveratrol (3,5,4'-trihydroxy-*trans*-stilbene) **4.1**, which appears in response to fungal infection; for example, from the pathogenic fungus *Botrytis cinerea* or the fungus *Plasmopara viticola*. Both fungi cause serious diseases, rot gray (botrytis), and mildew, which affect the entire plant, leaves, buds, and clusters. Along with resveratrol are the pterostilbene **4.2**, dimethoxy analog, and dimeric metabolites, *trans*-ε-viniferin **4.3**, *trans*-δ-viniferin **4.4**, and *trans*-dehydrodimer **4.5** (Figure 4.1).

Chemically, both resveratrol and pterostilbene are simple derivatives of stilbene known as monomeric stilbenoids, whereas products **4.3**–**4.5** are dimeric stilbenoids based on dihydrobenzo[*b*]furan skeleton, which is obtained during oxidative dimerization of resveratrol with the aid of plant peroxidase (a process called plant oxidative stress). As is observed from Figure 4.1, stilbenes are phenolic compounds

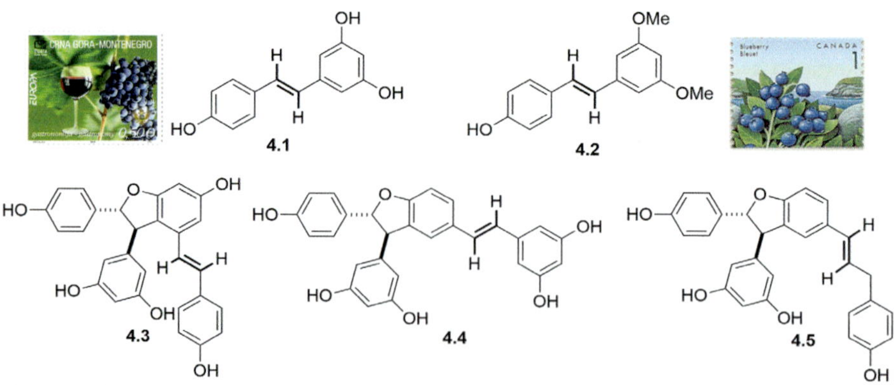

Figure 4.1. Structures of representative grape phytoalexins, phenolic compounds. Images of grapes in diverse stamps.

having two aromatic rings with hydroxy groups and are linked by a double-bonded ethylene bridge (in *trans*-configuration). In nature, i.e., higher plants, it is usual to find stilbenes exist as monomeric **4.1** and **4.2**, dimeric such as **4.3–4.5**, trimeric, oligomeric, and polymeric forms, or as glycosides which these stilbenes are bound to simple sugars.

Main monomeric stilbenes of grapes, resveratrol, and pterostilbene exhibit many similarities in pharmacological properties, including antioxidant, neuroprotective, anticancer, cardioprotective, analgesic, anti-atherosclerosis, anti-aging, anti-diabetic, anti-inflammatory, and anti-obesity activities. However, as a more lipophilic molecule, pterostilbene is a metabolite product with stronger pharmacological properties than resveratrol. That has been attributed to its two methoxy groups, which enhance membrane permeability, bioavailability, and thus biological potency of pterostilbene [9]. However, studies have shown that resveratrol and pterostilbene can cross the blood-brain barrier and influence brain activity, protecting against cerebral ischemic injury.

Both stilbenes commonly occur in the *trans*-form rather than the *cis*-form. Interestingly, resveratrol is present in red grapes and red wines, its primary dietary sources, whereas pterostilbene occurs in blueberries and grapes. Curiously, the concentration of *trans*-resveratrol in red wine is six times higher than in white wine, which contains high levels of *cis*-resveratrol. When resveratrol is exposed to ultraviolet and visible light, *trans*-to-*cis* isomerization occurs.

Another interesting fact is that resveratrol and pterostilbene were first isolated from the roots of poisonous *Veratrum grandiflorum* herbs and the bark extracts of *Pterocarpus santalinus* tree, respectively, and characterized in 1940, the same year when the phytoalexin theory was announced. Later, both phytoalexins were identified in *Vitis vinifera* (grapevine). The third source of resveratrol is the roots of the *Polygonum cuspidatum* plant, an important traditional medicine in China and Japan. Its root extracts treat diseases related to the cardiovascular system (remedy *ko-jo-kon*). Surprisingly, their antifungal activity against fungus *B. cinerea* is moderate (resveratrol, ED_{50} = 200 μg/mL; pterostilbene ED_{50} = 52–60 μg/mL).

Among the many pharmacological activities of resveratrol, its anticancer properties are the most well-known. Since its first report in 1997, there have been many investigations. Resveratrol is a promising agent for cancer prevention and treatment, especially prostate cancer. Therefore, red wine is rich in *trans*-resveratrol, and its moderate consumption has the health benefits of lower rates of this cancer type. The pharmacological potency of both resveratrol and protostilbene in the treatment of neurodegenerative disorders, such as Parkinson's, Alzheimer's, and Huntington's diseases, as well as amyotrophic lateral sclerosis, has also been recognized.

The biosynthesis of resveratrol and its derivatives is initiated through the shikimate pathway involving aromatic amino acids like L-phenylalanine **4.6a** or L-tyrosine **4.6b** [10]. These are firstly converted to the phenylpropanoid acids: cinnamic acid **4.7a** or *p*-coumaric acid **4.7b**, respectively, via non-oxidative deamination by phenylalanine ammonia-lyase (PAL) and tyrosine ammonia-lyase (TAL). Cinnamic acid **4.7a** can be transformed into *p*-coumaric acid **4.7b** by coumarate 4-hydroxylase (C4H). Then, *p*-coumaric acid **4.7b** converted to *p*-coumaroyl-CoA **4.8** by 4-coumaroyl-CoA ligase (4-CL). Finally, three molecules of malonyl-CoA are condensed with *p*-coumaroyl-CoA to produce resveratrol **4.1** (Figure 4.2).

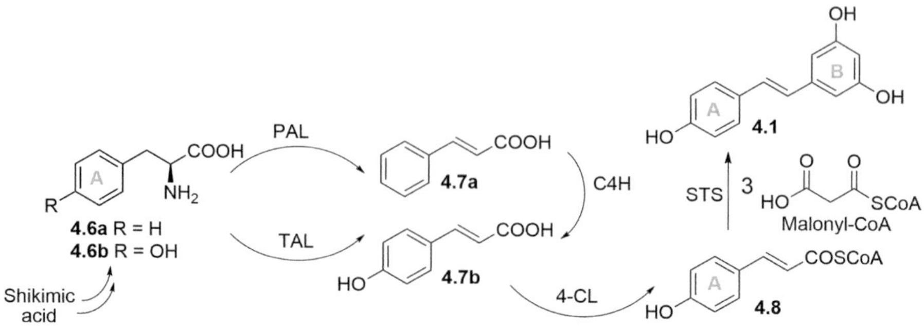

Figure 4.2. Biosynthesis of resveratrol through the shikimate pathway.

Resveratrol can be a platform for achieving several derivatives through modifications by tailoring enzymes, such as hydroxylases, *O*-methyltransferases, and glucosyltransferases, whereby diverse resveratrol analogs are generated. In the case of biosynthesis of pterostilbene, there is currently no scientific evidence that the pterostilbene is derived directly from resveratrol. However, it is believed that this conversion can take place by *O*-methyltransferase.

With the discovery of these physiological properties of resveratrol and pterostilbene, a tremendous interest arose in synthesizing their numerous analogs. Note that these molecules are unsymmetrical *trans*-stilbenes. The first syntheses of resveratrol and its analogs are based on the Wittig reaction or its modifications with forming the ethylene bridge between two aromatic rings of this molecule [11]. Let us examine the reaction of 3,5-dimethoxy benzyl bromide **4.9** and triethyl phosphite $P(OEt)_3$ that gives diethyl (3,5-dimethoxy benzyl) phosphonate **4.10** (Figure 4.3). This phosphonate reacts with *p*-anisaldehyde **4.11** to form the crucial intermediate 3,4,5-trimethoxy-*trans*-stilbene **4.12**.

This two-step conversion of **4.10** to **4.12** is known as the Wittig-Horner reaction, which allows better *trans*-selectivity control than the Wittig reaction. Finally, methoxy groups in **4.12** can be removed by freshly prepared aluminum triiodide ($AlBr_3$). Then, the resveratrol-aimed product can be separated by thin-layer chromatography [12].

However, many possibilities exist for forming a double C=C bond using diverse reagents and reaction conditions. One of the most attractive double bond constructions is using the Heck reaction [13]. For example, the initial reaction of 4-acetoxy-iodobenzene **4.13** and 3,5-diacetoxystyrene **4.14** in the presence of palladium catalyst furnishes triacetate-protected intermediate **4.15**, which is easily converted into resveratrol (Figure 4.4A). This is a two-step synthesis, and the first

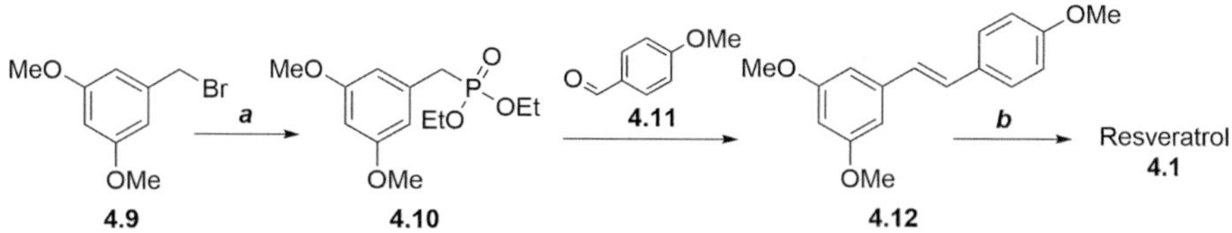

Figure 4.3. Synthesis of resveratrol through Wittig-Horner reaction: (a) $P(OEt)_3$, 130°C, 5 hours; (b) MeONa, DMF, 0°C to rt, overnight; (c) $AlBr_3$.

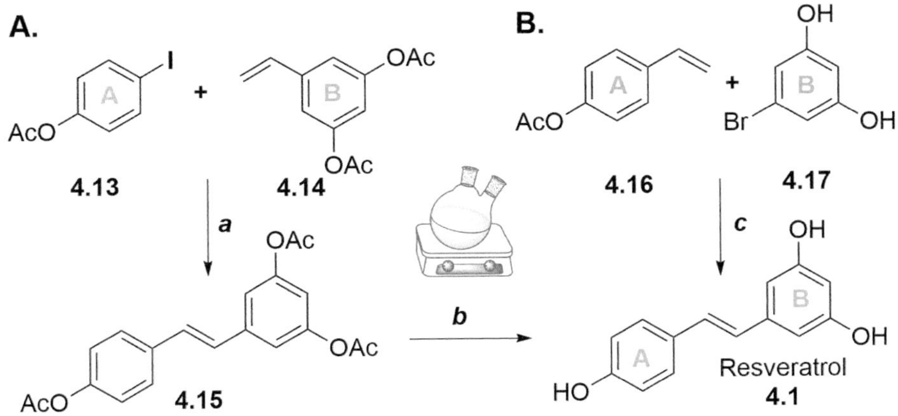

Figure 4.4. Heck reaction application in the synthesis of resveratrol. A. Two-step synthesis and B. Simple, short synthesis: (a) $Pd(OAc)_2$, PPh_3/Et_3N, MeCN (70%); (b) MeONa, THF-MeOH (91%); (c) $Pd(OAc)_2$, Et_3N, 100°C.

step is the Heck arylation of styrenes (comp. **4.14** with a ring B of resveratrol), in which the arylation reagent is **4.13** with a ring A of the final product.

Another variant of the same Heck arylation of styrenes allows obtaining desired resveratrol merely (Figure 4.4B). However, the construction of the stilbene system is different, i.e., in this case, arylation agent **4.17** possesses a ring B while styrene **4.16** has ring A of resveratrol skeleton. This is beautiful; we can choose reagents according to their availability. Moreover, the arylation reagent has no *O*-protected groups, and the AcO-protected group in the styrene is removed during the reaction process. An interesting modification of this Pd-catalyzed reaction is *one-pot* two sequential Heck-type reactions of three commercially available reagents, 4-methoxyiodobenzene **4.18**, vinyltrimethylsilane **4.19** (in access), and 3,5-dimethoxyiodobenzene **4.20**, which are consequentially reacted in the same flask to form trimethoxy resveratrol derivative **4.12** (compare Figure 4.3). Thus, this elegant one-flask reaction provides (*E*)-3,5,4-trimethoxystilbene **4.12** in a 72% yield. Then, the demethylation method involving the boron trichloride/tetra-*n*-butylammonium iodide reagent allows for cleanly obtaining resveratrol **4.1** in 85% yield [14] (Figure 4.5). This synthetic methodology for constructing unsymmetrical *trans*-stilbene derivatives based on two sequential Heck-type reactions is an efficient and convenient approach to natural stilbene construction.

Here, we briefly described the properties of resveratrol as an example of a stilbenoid group that is generated as defensive molecules in some Vitaceae plants during their fungal infection. Another big plan family, Fabaceae plants, elaborates on numerous phytoalexins. The following section provides structurally different and more complex molecules than resveratrol derivatives but closely related to them, i.e., polyphenolic molecules.

4.1.2 Pterocarpans and other Isoflavonoids from Fabaceae Plants

Most of the phytoalexins produced by Fabaceae plants belong to a large isoflavonoid subclass, which includes isoflavones, isoflavanones, pterocarpans, pterocarpenes,

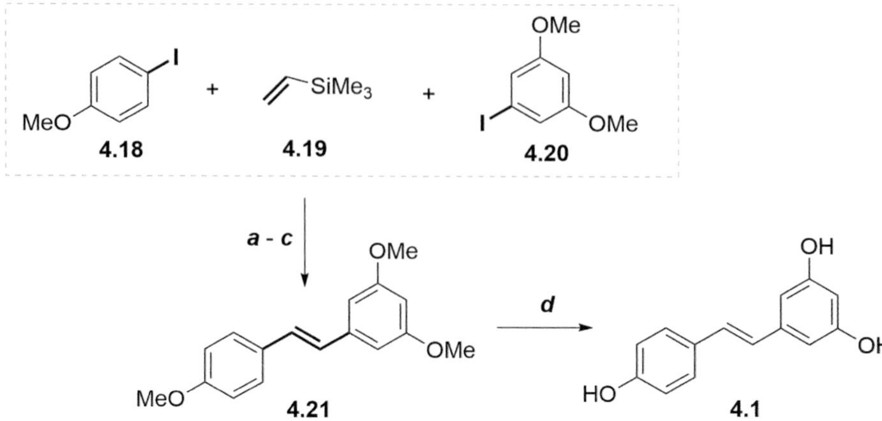

Figure 4.5. One-pot two sequential Heck-type reactions to obtain resveratrol: (a) cat. $Pd(dba)_2$, KF, n-Bu_4NCl, toluene, rt; (b) removal of CH_2=CH-$SiMe_3$ excess; (c) K_2CO_3, DMF, 70°C; (d) BCl_3, n-BuNI/CH_2Cl_2.

and coumestans, among others. This chemical diversity arises from botanical diversity; this family contains approximately 730 genera of species. Economically essential crops, such as beans, lentils, chickpeas, and peas, are part of this family and generate their phytoalexins with complex and diverse molecular architecture.

Despite huge chemical diversity, this chapter focuses on the most studied and interesting phytoalexins from Fabaceae plants, pterocarpans. Pterocarpans and isoflavones play a vital role in plant-microbe interaction. Structurally, these phytoalexins are oxygenated heterocyclic molecules. The name "pterocarpan" was proposed by Harper in 1965 because the first two compounds of this type, pterocarpan and homopterocarpan, were isolated in 1874 from species of the genus *Pterocarpus*. The pterocarpan skeleton (6a,11a-dihydro-6*H*-benzofuro[3,2-*c*][1]benzopyran) contains two fused oxygenated rings: benzopyran AB (or chromane) and 2,3-dihydrobenzo[*b*]furan CD (Figure 4.6). These molecules have two chiral centers (i.e., four stereoisomers exist), but only two stereoisomers with the *cis* fusion of the heterocyclic rings, (*R*, *R*) and (*S*, *S*) forms, are stable. In nature, dextro- and levorotatory molecules usually occur in addition to racemic forms [15].

The structures of the most important phytoalexins of the Fabaceae family are illustrated in Figure 4.7. The first three pterocarpans (maackiain, medicarpin, and pisatin) were identified by studying the extracts of the species *Pisum sativum*. Pisatin was the first chemically characterized phytoalexin (1960, Cruickshank and Perrin). This molecule belongs to the group of 6a-hydroxypterocarpans. Its isolation from *P. sativum* pods was possible when this species was inoculated with conidia of the brown rot fungus *Monolinia fructicola*.

The pisatin molecule is slightly soluble in water and has high solubility in organic solvents. In neutral or slightly alkaline solutions, it is stable and loses water in the presence of acid to convert to anhydropisatin. Medicarpin is the biogenic precursor to pisatin. It was also isolated from the small *Medicago truncatula* plant of the same Fabaceae family. Due to its prolific seed production and small diploid genome, *M. truncatula* is the model organism in plant biology studies. The structure of

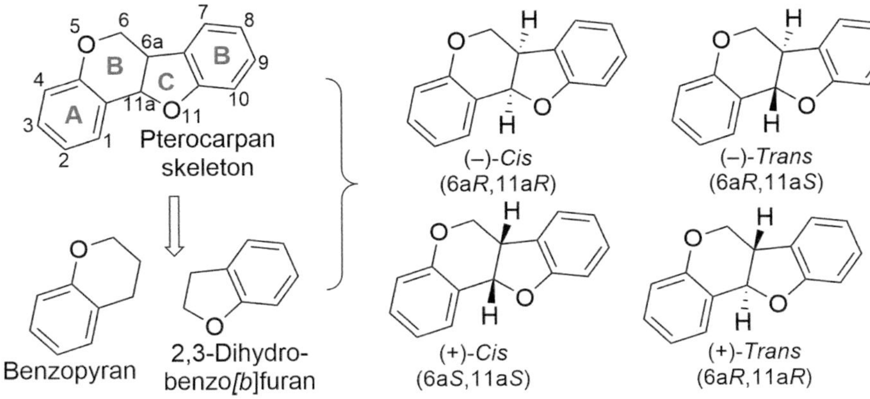

Figure 4.6. Heterocyclic pterocarpan system.

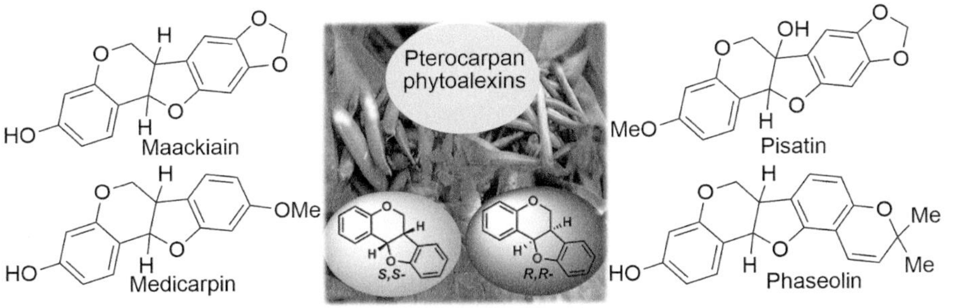

Figure 4.7. Selected pterocarpan phytoalexins from Fabaceae plans.

phaseolin was determined in 1964 when it was isolated from the species *Phaseolus vulgaris* L.

In addition to antifungal activity, pterocarpans exhibit various biological activities (antiviral, antibacterial, antioxidant, anti-proliferative activity in various human cancer cell lines, etc.) [16, 17]. Interestingly, some pterocarpans isolated from the dichloromethane extract of the "Burma padauk," an Asian tree (*Pterocarpus macrocarpus*), also show good insect antifeedant against both common cutworms, *Spodoptera litura,* and subterranean termites, *Reticulitermes speratus* [18].

The biogenesis of pterocarpans begins with the transformation of phenylalanine **4.6** followed by the condensation of a coumaroyl-CoA molecule **4.8** (Figure 4.2) with the three malonyl-CoA molecules in the presence of chalcone synthetase, which makes it possible to form first a chalcone **4.22** and then a flavanone **4.23** and then daidzein **4.24a** or formononetin **4.24b**, which undergo consecutive enzymatic transformations (hydroxylation, reduction, and intramolecular cyclization) to give dihydroxypterocarpan **4.25a** and medicarpin **4.25b** (Figure 4.8). Thus, the pterocarpan structure is formed at the last stages of the biosynthesis of flavonoids, as a part of the isoflavonoid branch of the pathway.

Due to the biological significance of these heterocycles, numerous synthetic methodologies are reported in the literature for their synthesis [15, 19]. One of the most straightforward synthetic pathways for pterocarpan systems like homopterocarpan

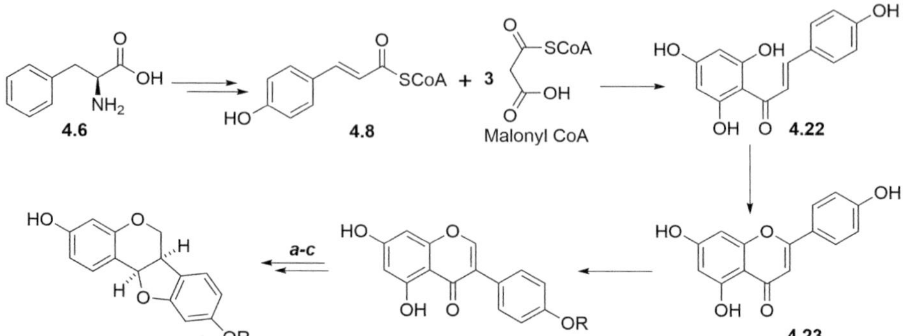

Figure 4.8. The simplified biological pathway to pterocarpan phytoalexins: (a) hydroxylation in the presence of isoflavone 2′-hydroxylase; (b) reduction in the presence of isoflavonereductase; (c) intramolecular cyclization in the presence of pterocarpan synthase.

4.28 is reducing 2'-hydroxyisoflavanone **4.27** using sodium borohydride ($NaBH_4$) in THF and ethanol. Isoflavones, which are obligatory intermediates in the formation of the pterocarpan four-ring system, are prepared in two steps from the corresponding chalcone **4.26** (Figure 4.9). From this figure, it can be observed that this synthesis is accomplished in a linear manner, similar to a biological pathway. However, the final product is racemic.

Another common key step in generating the pterocarpan skeletons has involved mild oxidation of the 2′-hydroxyisoflavans **4.29**, which possess the same ABD ring system of 2'-hydroxyisoflavanone **4.27**. This oxidative method is used in the synthesis of racemic (±)-medicarpin **4.31** and (±)-maackiain **4.32** and involves the simple treatment of benzene solution of **4.29** with 2,3-dichloro-5,6-dicyanobenzoquinone (DDQ) (Figure 4.10). This conversion probably proceeded through a quinone methide intermediate **4.30**, which underwent a nucleophilic addition by the 2′-OH group.

Other approaches allow the preparation of these natural pterocarpans and their synthetic analogs. Among these are Heck-type coupling reaction-oxyarylation between 3-chromene derivatives **4.33** and 2-chloromercuriophenols **4.34** in the presence of stoichiometric amounts of Li_2PdCl_4 that allows obtaining different pterocarpans **4.35** in good yields (Figure 4.11). Horino and co-workers proposed this route to the pterocarpan skeleton in 1976 [20].

Promoter Li_2PdCl_4 is usually formed *in situ* from $PdCl_2$ and LiCl. The required phenolic moiety **4.34** is prepared by treatment of a suitable phenol derivative with mercury (II) acetate in methanol, followed by treatment with lithium chloride. The preparation of the 2*H*-chromene ring depends on the kind of pterocarpans synthesized in each case, but many of them use resorcinol as the starting material. The same pterocarpans can be obtained by reaction of 2*H*-chromenes **4.33** and 2-iodophenols **4.36** using base (K_2CO_3 or Et_3N), DMF as solvent, $Pd(OAc)_2$ as catalyst, and *n*-Bu_4NCl as chloride source. This approach to the pterocarpan system was developed by Larock and co-workers in 1991. Later, they patented the improved method [21]. Both synthetic methods allow for constructing a pterocarpan system in

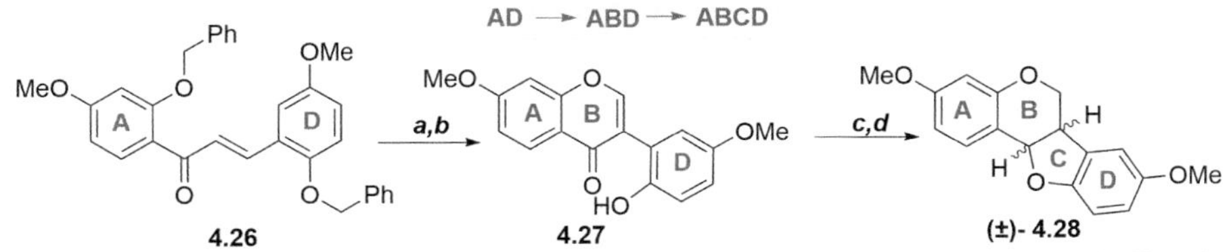

Figure 4.9. Synthesis of (±)-homopterocarpin using reduction of 2'-hydroxyisoflavanone: (a) $CH(OMe_3)_3$, CF_3COOH/PIFA; (b) $BF_3{\cdot}OEt_2$, CS_2/CH_2Cl_2; (c) H_2, Pd-C; (d) $NaBH_4$, THF-EtOH, then HCl 10%.

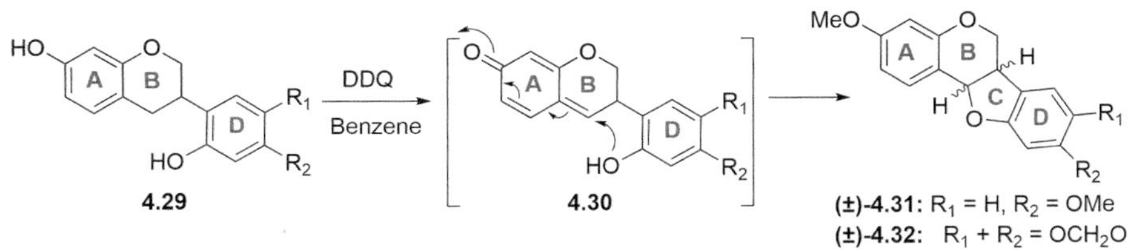

Figure 4.10. Synthesis of (±)-medicarpin and (±)-maackiain via oxidation of the 2′-hydroxyisoflavans.

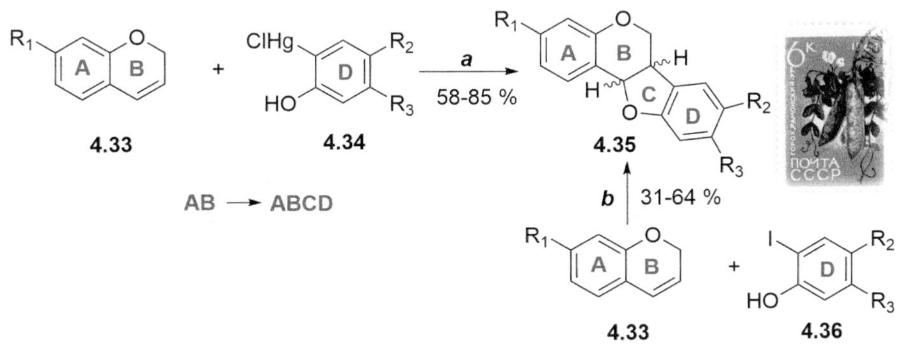

Figure 4.11. Synthesis of the pterocarpans by Heck-type coupling reaction: (a) Li_2PdCl_4 ($PdCl_2$ and 2xLiCl), MeCN, rt; (b) $Pd(OAc)_2$ (5% mol), n-Bu_4NCl, base, DMF, 100°C) and pea plant, *Pisum sativum* L. in an old stamp of USSR.

a one-pot manner. Still, both suffer certain drawbacks, e.g., main reagents are toxic (2-chloromercuriphenols) or are not always readily available (2-iodophenols), excess of catalyst (Li_2PdCl_4), and both would allow the induction of enantioselectivity.

Similar 2-iodophenols **4.36** were used in the Mitsunobu coupling reaction with 3-hydroxymethylbenzofurans **4.37** followed by an intramolecular Heck reaction of the intermediates **4.38** to give the pterocarpene skeleton **4.39** (Figure 4.12). (+)-Medicarpin has been synthesized asymmetrically for the first time in a linear scalable process with an overall yield of 11% [22].

4.1.3 Sulfur-Containing Indole Heterocycles from Cruciferae Plants

Indole heterocycles play a crucial role in the self-defense of Cruciferae family plants, which are also excellent sources of glucosinolates (Section 3.3). Chemical and biological studies of these plants revealed more than 40 sulfur derivatives of

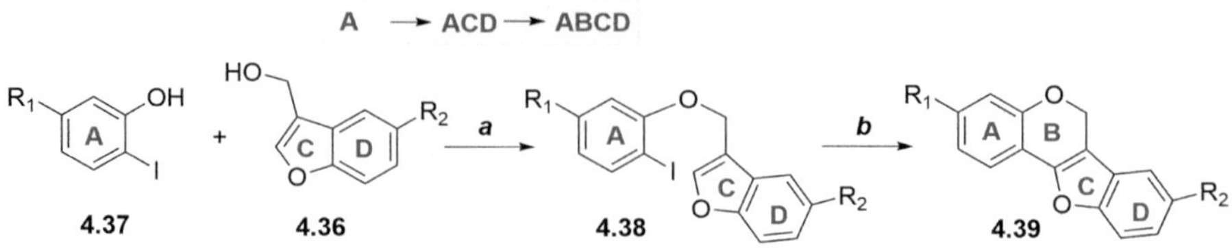

Figure 4.12. Synthesis of the pterocarpenes based on Mitsunobu and Heck-type coupling reactions: (a) DIAD, Ph_3P/THF; (b) $Pd(OAc)_2$, KOAc/DMF. DIAD = diethyl azodicarboxylate.

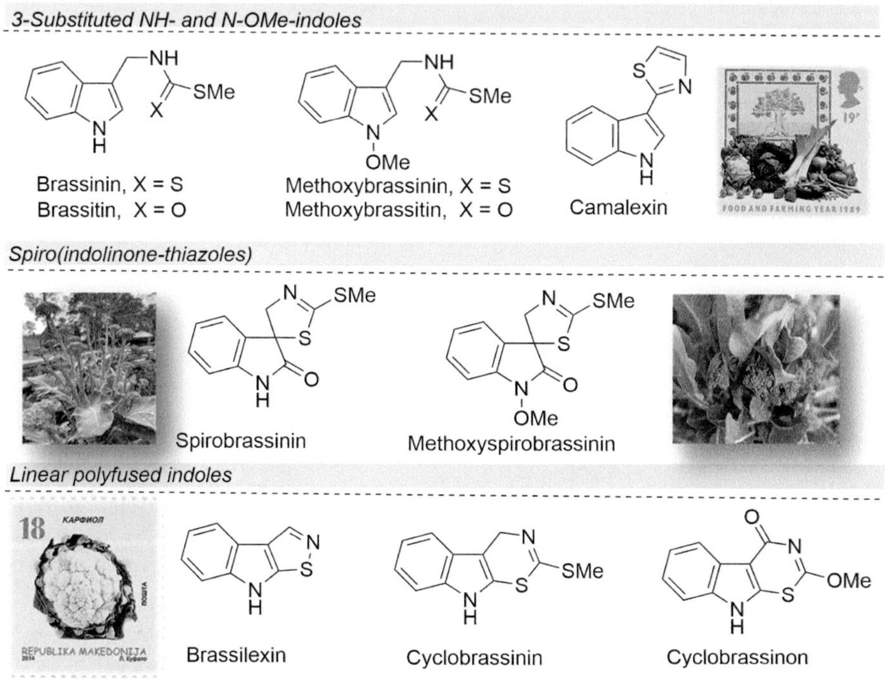

Figure 4.13. Typical representatives of cruciferous phytoalexins and their structures.

indole that act as phytoalexins, which, to repeat, are antimicrobial low-molecular-weight secondary metabolites produced by plants after their exposure to biological, chemical, or physical stress [23, 24]. Phytoalexins from crucifers (brassinin, cyclobrassinin, and 1-methoxybrassinin) were first reported in 1986 by Takasugi and co-workers isolating from Chinese cabbage which had been infected by the *Pseudomonas cichorii* bacterium [25]. Since then, although a relatively larger number of cruciferous species remain to be investigated, close to 30 cruciferous phytoalexins have been isolated and their structures elucidated (Figure 4.13).

Most phytoalexins have been isolated using multiple chromatographic columns with bio-guided thin-layer control to detect bioactive components. In most cases, the antimicrobial activity of these compounds was established at the time of isolation, and no further biological evaluations were made. They were found to inhibit the spore germination of fungal pathogens, such as fungi *Phoma lingam*, *P. wasabiae*, *Bipolaris leersiae*, *Rhizoctonia solani*, etc. In addition to antimicrobial and antifungal activities, cruciferous phytoalexins exhibited remarkable anticancer, chemopreventive, and antiproliferative activity [26, 27] (Section 3.3). They displayed significant antiproliferative effects on intracellular amastigotes of *Trypanosoma cruzi* and could be prospective candidates for antichagasic drug design and development [28].

As seen in Figure 4.13, there are three types of indole derivatives: NH-(or N-methoxy)-3-substituted indoles with a dithiocarbamate group (NH-CS-SR) (brassinin, its analogs, and camalexin), spiro-indolinones (spirobrassinin and its N-methoxy analog presented here as racemic mixtures), which possess two molecular rings with only one typical carbon atom (spiro-atom), and linear polyfused indoles (brassilexin, cyclobrassin, etc.) that contain two or more fused rings sharing a common bond [29]. What is remarkable in the two latter of indole phytoalexins is the presence of the sulfur atom and SMe group. Generally, the structure of phytoalexins depends on the type of plant that synthesizes them and partially on the elicitor inducing their synthesis.

Cruciferous crops are cultivated worldwide and constitute an extremely valuable group of plants. Enormous quantities of vegetable crucifers, such as broccoli, kale, cauliflower, Brussels sprouts, mustard greens, and various cabbages, are consumed annually. Cruciferous crops benefit the human diet partly due to their anticarcinogenic properties, which could reduce the risk of several types of cancer. However, the mechanism of the chemopreventive effects of the same plant and its indole constituents is unknown [27].

Biosynthetic studies of these indole derivatives revealed that their biogenesis begins with the conversion of (*S*)-tryptophan to indolyl-3-acetohydroxamic acid and then opens into several metabolic pathways, including indolyl glucosinolates, brassinins, and camalexins involving more than ten enzymes of cytochrome P450 and NADPH complexes [30–32]. Noteworthy, these phytoalexins are a group of elicited plant metabolites with vital ecological roles related to defense, usually not detected in healthy plants. Therefore, it is not easy to investigate this topic. On the other hand, molecular understanding of plant defenses and their metabolic pathways helps to design practical synthesis of these molecules and their analogs, valuable entities in drug research. Notably, brassinin is a moderate inhibitor of indoleamine 2,3-dioxygenase, a promising cancer immunosuppression target.

The synthesis of brassinin consists of three stages, starting from indole-3-carboxaldehyde **4.40**: the formation of the respective oxime **4.41**, its reduction (Devarda's alloy – Cu/Al/Zn) to tryptamine **4.42** and the conversion of the aminomethyl fragment into a dithiocarbamate group [carbon disulfide (CS_2), methyl iodide (CH_3-I) in the presence of pyridine (Py) and triethylamine (Et_3N)] to give phytoalexin brassinin [29]. Tryptamine can also be synthesized from indole **4.43** via the formation of gramine **4.44**, an alkaloid found in sprouting barley and a useful synthetic intermediate, which can be synthesized using the modified Mannich reaction with $ZnCl_2$ catalyst [33, 34] (Figure 4.14). The starting material, indole-3-carboxaldehyde, can be prepared from indole through the Vilsmeier-Haack formylation reaction ($POCl_3$/DMF, toluene). Moreover, the preparation of **4.44** was efficiently carried out by the reaction of **4.40** with sodium borohydride ($NaBH_4$) in a 50% aqueous dimethylamine-methanol (1:1, v/v) mixture [34].

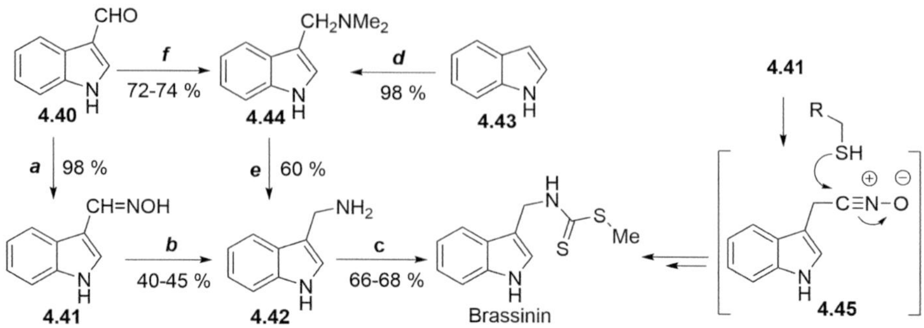

Figure 4.14. Synthetic routes for obtaining brassinin: (a) $NH_2OH{\cdot}HCl$, Na_2CO_3/H_2O-EtOH, 50–80°C; (b) reduction conditions, Devarda's alloy, NaOH, MeOH; $NaBH_4$, MeOH or $NiCl_2{\cdot}6H_2O$, $NaBH_4$, MeOH, rt; (c) CS_2, CH_3I, pyridine/Et_3N, 0–3°C; (d) CH_2O, Me_2NH, $ZnCl_2$, EtOH, rt, 90 minutes; (e) CH_3-I, NH_4OH, *tert*-BuOH, 15°C, 2 hours; (f) $NaBH_4$/50% aq. Me_2NH-MeOH, 20–60°C, 2.5 hours.

Interesting to note that the proposed biosynthetic pathway of brassinin could be started with the oxidation reaction of indolyl-3-acetaldoxime **4.41** that affords the corresponding nitrile oxide **4.45**, an intermediate, which may suffer from an alkylthiol RCH_2SH (e.g., cysteine) to yield brassinin via sequential processes [32] (Figure 4.14).

Like brassinins, camalexin and 6-methoxycamalexin are also 3-substituted indole phytoalexins but possess a heterocyclic ring, 3-thiazolyl substituent instead of the dithiocarbamate group. They were isolated from the leaves of *Camelina sativa* (known as false flax) infected by the fungus *Alternaria brassicae* [35]. Despite the apparent simplicity of this plant metabolite, their synthesis has evidenced several drawbacks concerning yield, scale-up potential, or availability of necessary reagents. Their first synthesis is based on the Grignard reaction and C-3 indole alkylation reaction using indole **4.43** and 6-methoxyindole **4.46** as available starting materials (Figure 4.15). Thus, the Grignard reaction of indole derivatives with methylmagnesium iodide produced intermediate N-indolylmagnesium iodides **4.47**, which were then treated with 2-bromothiazole **4.48** in refluxing benzene to obtain respective 3-(thiazol-2-yl) indole phytoalexins in reproducible yields of 75–76% [36].

The proposed preparation of camalexin and 6-methoxycamalexin allowed us to evaluate their antifungal activities, which were examined using the bioautographic assay with the phytopathogenic fungus *Cladosporium* sp. as the test organism. Thiabendazole fungicide **4.49**, based on the 2-(thiazol-4-yl)benzimidazole skeleton, was used as a structurally similar positive reference standard. The study revealed the following order of antifungal activity at an application level of 1 μg: thiabendazole >> camalexin >> 6-methoxycamalexin. Additionally, camalexin displayed antimicrobial activity against some bacterial and fungal plant pathogens, such as *Pseudomonas syringae*, *Alternaria brassicicola*, *Alternaria brassicaeae*, and *Leptosphaeria maculans*, disrupting their cell membranes. This showed the potential of camalexin as a lead compound in biological research, being a nontoxic molecule.

Camalexin also exhibited potent growth inhibition of the human breast cancer cell line SKBr3, showing high IC_{50} values of 2.7 μM, higher potency than the traditional anticancer agents cisplatin (IC_{50} = 7.4 μM) and melphalan (IC_{50} = 13.0 μM) [37]. Moreover, it induced apoptosis in human T-leukemia Jurkat cells at the micromolar

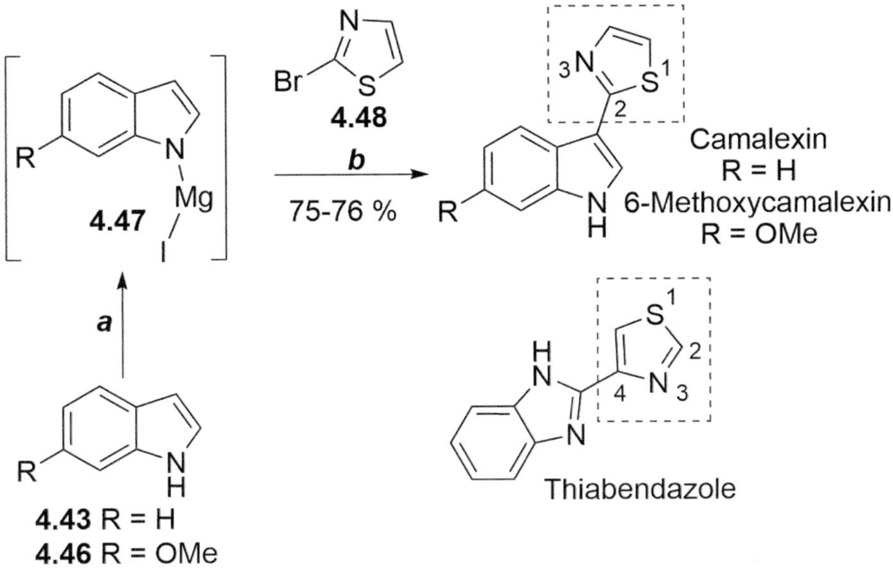

Figure 4.15. First one-pot synthesis of camalexin and 6-methoxycamalexin: (a) CH_3-MgI, Et_2O, mild reflux, 67 hours; (b) benzene, reflux, 20 hours.

concentration (IC_{50} = 46.2 μM) [38]. The above-commented facts encouraged organic and medicinal chemists to study synthetic routes of camalexin and its analogs to investigate in detail the mode of action of these molecules on cells of parasites, including cancer. Thus, several interesting camalexin analogs from the pharmacological point of view are synthesized and tested. Synthetic analogs **4.50–4.53** are attractive and characteristic examples (Figure 4.16). The first couple of 3-(thiazol-4-yl)indoles **4.50–4.51**, easily obtained using the reaction of 3-bromoacetylindole with thioformamide or thioacetamide in ethanol at reflux, showed less anticancer activity (SKBr3 cancer cell line, IC_{50} = 18.0 and 4.9 μM, respectively) than camalexin, while benzo-analogs **4.52–4.53**, substituted at the respective C-2 and C-3 positions of the indole ring, exhibited more anticancer activity (IC_{50} = 2.6 μM for 2-(1-methyl-1*H*-indol-2-yl)benzo[*d*]thiazole **4.52**) [37].

Moreover, benzocamalexin (2-(1*H*-indol-3-yl)benzo[*d*]thiazole) **4.53**, prepared by cyclocondensation reaction of 1-(*tert*-butoxycarbonyl)indole-3-carbaldehyde with 2-aminobenzenethiol [39], significantly decreased survival of several cancer cell lines (HeLa, Jurkat, A-549, MCF, MDA, CEM) with IC_{50} ranging from 23.3 to 30.0 μM and triggered downregulation of microtubule-associated proteins and Bcl-2 family proteins, leading to apoptosis, i.e., cancer cell death [40]. This is an example of how the fusion of benzene to thiazole ring of camalexin structure (simple modification) significantly enhanced its cytotoxicity. Unfortunately, these synthetic analogs' antibacterial and antifungal activity was not evaluated.

A special subgroup of indole phytoalexins comprises spiroindolin-2-one phytoalexins with spiro-attached thiazoline ring in the C-3 position of the indolin-2-one skeleton. A case in point is the spiro[indoline-3,5'-thiazol]-2-one system, a key carcass of the spirobrassinin and 1-methoxyspirobrassinin structure. By possessing a chiral carbon atom (spiro-atom), both phytoalexins exist in two enantiomeric forms.

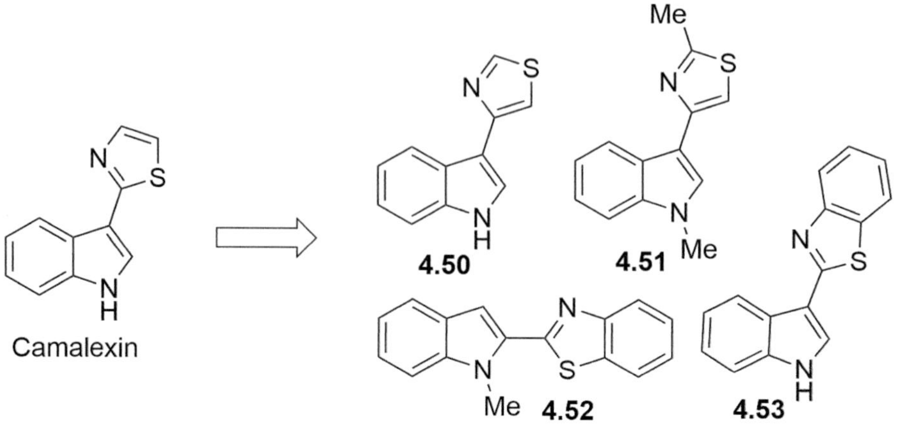

Figure 4.16. Camalexin, a natural plant metabolite, and its synthetic structural analogs.

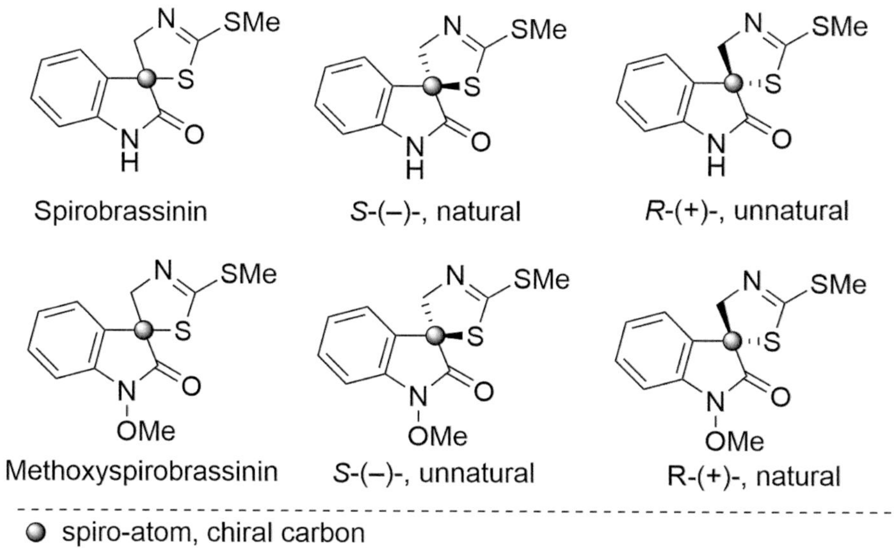

Figure 4.17. Spirobrassinins and their enantiomeric forms.

Whereas natural spirobrassinin has *S*-(-)-form, and natural 1-methoxyspirobrassinin has an *R*-(+)-configuration (Figure 4.17). (*S*)-(-)-Spirobrassinin was first isolated as a constituent of Japanese radish (*Raphanus sativus*) inoculated with the bacterium *Pseudomonas cichorii* [41, 42].

The synthesis of racemic (±)-spirobrassinin (non-asymmetric synthesis) was realized using 3-(aminomethyl)-3-hydroxy-oxindole **4.56b** as a suitable starting material from commercial isatin **4.54** (1*H*-indole-2,3-dione). When treated with CS_2 and Et_3N in pyridine, racemic dioxibrassinin **4.57b**, a minor phytoalexin-related metabolite found in cabbage, was formed (Figure 4.18A). Its subsequent cyclization with thionyl chloride ($SOCl_2$) in dry pyridine afforded racemic (±)-spirobrassinin in 66% yield [42]. The total synthesis of natural (S)-(−)-spirobrassinin realized by

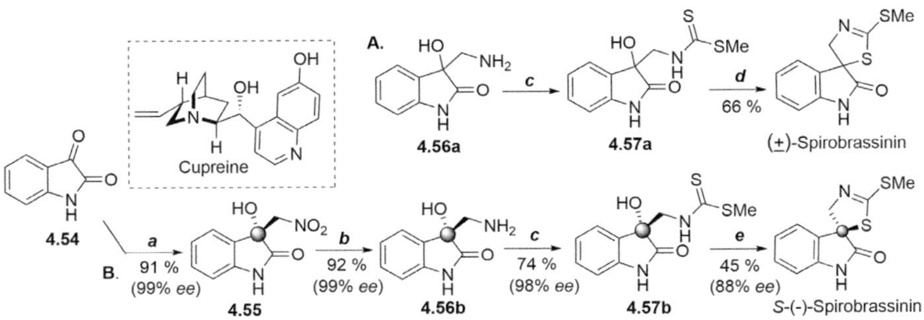

Figure 4.18. Synthesis of racemic (±)-spirobrassinin and (*S*)-(−)-spirobrassinin: (a) CH_3-NO_2/10 mol% cupreine, $PhCO_2H$, DMA, −15°C, 3 hours; (b) H_2, Pd/C, MeOH; (c) 1, CS_2/Et_3N, Py, 2, MeI, rt, 4 hours; (d) SO_2Cl, dry Py; (e) CH_3-SO_2Cl, dry Py.

Wang and co-workers was also based on isatin derivatives [43]. They selected the catalytic asymmetric Henry reaction of isatin **4.54** with nitromethane (CH_3-NO_2) as a key reaction, which was performed in polar dimethylacetamide (DMA), and the presence of cupreine, an alkaloid found in cinchona bark, as an organocatalyst. This simple aldol reaction created a chiral quaternary center (in our case, spiro carbon) to obtain a highly enantioenriched adduct **4.55a** with the needed configuration. The nitro group in **4.55a** was then reduced by Pd-catalyzed hydrogenation to give amine **4.56a** (but in enantiomeric (*R*)-(+)-form), which was treated using the similar reaction conditions of synthesis of racemic (±)-spirobrassinin to afford natural (*R*)-(+)-dioxibrassinin **4.57a** and, then (S)-(−)-spirobrassinin molecule after cyclization with methanesulfonyl chloride (CH_3-SO_2Cl) (Figure 4.18B). Thus, natural product (*S*)-(−)-spirobrassinin was prepared in four steps without requiring protecting groups in 27.9% overall yield with excellent enantioselectivity.

Curiously, an examination of the antifungal activity of racemic (±)-spirobrassinin and (*S*)-(−)-spirobrassinin against the fungus *Bipolaris leersiae* revealed only moderate activity at a concentration of 1 mmol/L, compared to brassinin. However, these spiro-compounds and 2′-amino analogs showed interesting antiproliferative effects inhibiting leukemic cells Jurkat with IC_{50} from 27 to 82 μM [44, 45].

Among different phytoalexins, several penta- or hexacyclic heteroaromatic indole molecules such as brassilexin, sinalexin, cyclobrassinin, and cyclobrassinon stand out due to their special interest from the structural and biological point of view. Brassilexin and sinalexin (N-methoxy substituted brassilexin) phytoalexins are based on the planar heteroaromatic structure of isothiazolo[5,4-*b*]indole, while cyclobrassinin and cyclobrassinon have a 1,3-thiazino[6,5-*b*]indol-4-one skeleton (Figure 4.14).

Analyzing the antifungal activities of brassilexin and sinalexin against four of the major pathogens of crucifers, i.e., Alternaria brassicae, *Phoma lingam* (*Leptosphaeria maculans*), *Rhizoctonia solani*, and *Sclerotinia sclerotiorum*, it was found that brassilexin caused complete inhibition of fungal mycelium growth in all four species at a concentration of 5 M for the duration of the experiment (up to 3 weeks). In contrast, sinalexin showed less growth inhibition (60%) to *P. lingam* at this concentration. Now, synthetic brassilexin and sinalexin are available due to their efficient synthesis. One of them involves a one-pot protocol based on the Vilsmeier-

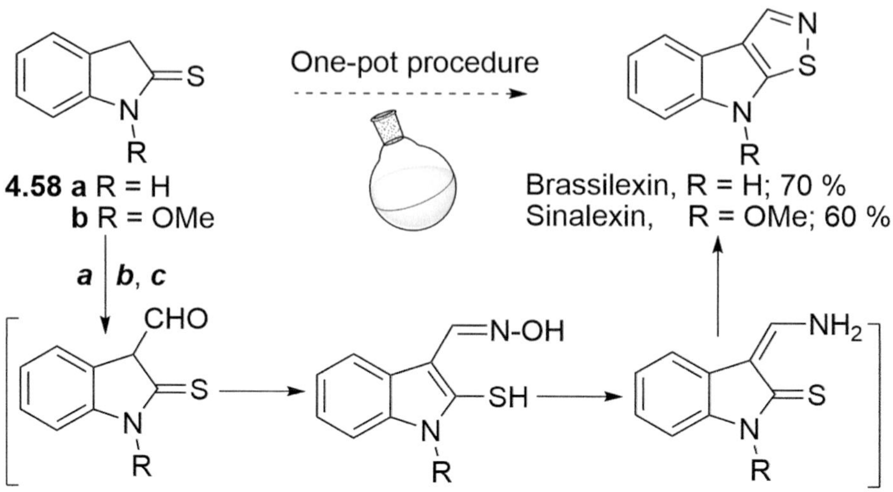

Figure 4.19. One-pot synthesis of sinalexin and brassilexin: (a) $POCl_3$/DMF, 40°C, 1 hour; (b) NH_3/NH_4OH; (c) I_2/Py.

Haack formylation reaction of indoline-2-thiones **4.58** with $POCl_3$ in DMF (1 hour at 40°C), followed by ammonia workup and then oxidizing by iodine in pyridine [46] (Figure 4.19).

The Vilsmeier-Haack formylation procedure was also utilized to prepare cyclobrassinon from 2-chloro-1*H*-indole in seven steps [47, 48]. In contrast, cyclobrassinin was prepared by cyclizing brassinin with pyridinium bromide perbromide in dichloromethane at room temperature [49]. It is noteworthy that cyclobrassinin was found active against seven cancer cell lines, such as Jurkat (acute T-lymphoblastic leukemia), MCF-7, and MDA-MB-231 (mammary gland adenocarcinoma), HeLa (cervical adenocarcinoma), CCRF-CEM (acute T-lymphoblastic leukemia), and A-549 (non-small cell lung cancer), exhibiting IC_{50} values of 26.2–72 μM depending on the cell line, with leukemic cells being the most sensitive (IC_{50} = 26.2–29.5 μM) [50].

As we can see, research in developing and synthesizing new derivatives of indole phytoalexins, which exhibit more favorable antiproliferative and chemopreventive properties than natural substances, is a very important and promising area. Due to their simple structure and antiproliferative activity, indole phytoalexins are potentially effective models in developing new antitumor drugs.

4.1.4 Avenanthramides from Oat Grains

One of the most important crops of the Poaceae family is oats, *Avena sativa*, which is the ideal food supplement to improve hormonal levels, increase energy, prevent heart disease, etc. In addition to human food, the multifunctional uses of oats include forage, fodder, straw for bedding, and hay. Oat is an annual 1.5-m-high grass widely cultivated in cool and moist regions of Northern Europe and North America. When the species *A. sativa* is infected with pathogenic fungi, for example, *Puccinia coronata* f. sp. *avenae*, oat leaves produce phenolic phytoalexins called avenanthramides

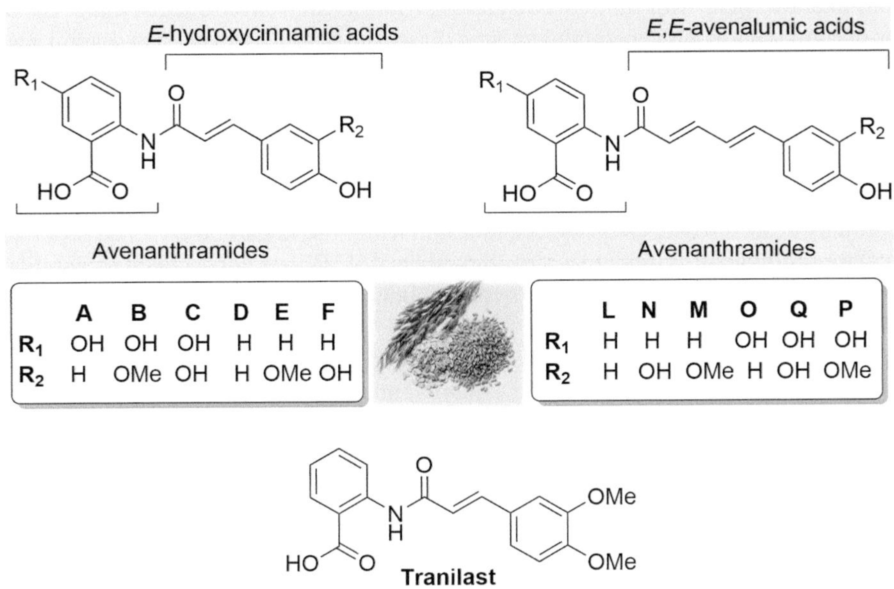

Figure 4.20. Structures of the most significant and abundant AVNs as phytoalexins with potent antioxidant properties and tranilast drug.

(AVNs). As their production in oat leaves is directly induced by fungal pressure [51], these effective antifungal properties imply their primary biochemical role in plants is fungal defense. This was confirmed; avenanthramides inhibited the spore germination of the crown rust fungus by approximately 50% at a concentration of 200–300 μg/mL. These phytoalexins can be accumulated by treating oat leaves with various elicitors, including chitin derivatives, toxins, and heavy metal ions.

Chemically, they are amides, N-cinnamoyl anthranilates, formed between *trans*-hydroxycinnamic acids (*p*-coumaric, ferulic, and caffeic acids) and the amino group of anthranilic acid or 5-hydroxy- and 5-hydroxy-4-methoxyanthranilic acids and the lesser-known avenalumic acid [(2*E*,4*E*)-5-(4-hydroxyphenyl)-penta-2,4-dienoic acid] derivatives. Therefore, they are alkaloidal substances. It is noteworthy that AVNs are normal constituents of healthy oat grains. More than 40 compounds were identified by Collins and Mullin using the high-performance liquid chromatographic technique [52, 53]. The structures of the most studied avenanthramides are illustrated in Figure 4.20.

Oats are the only cereal that contains AVNs, which are metabolites with powerful antioxidant properties [54, 55]. It was found that the total content of avenanthramides in husked oats ranges between 0.003 and 0.008% (on a dry basis), depending on oat cultivars, and the oat bran and oat flakes whole grain contained 13 and 27 mg of avenanthramides per kg, respectively. It was also established that the concentration of avenanthramides A–C in the oat bran was 4.1–4.4 mg/kg (on fresh weight), while their concentration in oat flake whole grain was more (8.6–9.0 mg/kg from the fresh weight). Thus, they are difficult to isolate in sufficient quantities. It should be remembered that antioxidants are related to age-linked deterioration, heart disease, and cancer. In this context, it is worth noting that tranilast (rizaben), a drug based on N-(3,4-dimethoxy-cinnamoyl)-anthranilic acid skeleton, is used as an anti-allergic agent with limited side effects. Still, its anticancer effects, either alone or in combination with chemotherapeutic drugs, have been evidenced in several pre-clinical studies [56].

However, we must return to oat avenanthramides, which could increase the antioxidant capacity in healthy older adults. In addition, AVNs showed anti-inflammatory (through inhibition of lipoxygenases activity), antiproliferative, anti-genotoxic, and anti-itching activities that may provide additional protection against coronary heart disease, colon cancer, and skin irritation. In particular, it was reported that the antioxidant activity of avenanthramides A–C (R_1 = OH) declined in a similar order as that for the corresponding hydroxycinnamic acids, as caffeic acid > ferulic acid > *p*-coumaric acid, i.e., avenanthramides C > B > A. Notably, the antioxidant capacity of avenanthramides C is higher than α-tocopherol, one of the eight isoforms of vitamin E, is the most potent fat-soluble antioxidant known in nature [55, 56]. Avenanthramides D–F (R_1 = H), formed from anthranilic acid, are weak antioxidants. Thus, their potential as antioxidants depends on the number and arrangement of hydroxyl groups and the nature of the substituents in the ring structures. Exerting antioxidant and antiproliferative effects that help prevent or treat cancer, AVNs are promising chemopreventive and anticancer phytochemicals [57, 58].

Due to their poor content in the plant, they are difficult to purify individually. With these encouraging biological properties, efficient preparation on a large scale is an important task. Direct amide formation by condensation of anthranilic acids (N-component) and hydroxycinnamic acids (COOH-component) is possible. Still, the acyl chloride derivative of the suitably protected hydroxyaromatic acid and the appropriate free aminobenzoic acids are needed, and then the protecting groups must be removed [53]. Moreover, it is an unsustainable process since it implies using nonrenewable solvents and the generation of toxic by-products. Thus, this chemical approach is merely a poor copy of what nature creates. Encoding enzyme-catalyzed biosynthesis of AVNs is based on the use of two precursors, L-phenylalanine (L-Phe) and anthranilic acid, which are derived from chorismic acid, that is, they come from the shikimate pathway. Subsequent transformations include the condensation of *p*-coumaroyl-CoA with hydroxyanthranilic acid in the presence of hydroxycinnamoyl-CoA:hydroxyanthranilate N-hydroxycinnamoyl transferase (HHT) enzyme to generate AVN-A. The C3'H-catalyzed transformation of *p*-coumaroyl-CoA into caffeoyl-CoA, which could then be condensed with 5-hydroxyanthranilic acid (5-HO-AA) to form AVN-C by HHT. Finally, AVN-C might be methylated to AVN-B by a CCoAOMT enzyme (caffeoyl-CoA O-methyltransferase). In order to complete the biosynthesis mechanism, it should be added that the amino acid L-Phe is initially catalyzed by phenylalanine ammonia-lyase (PAL) and cinnamic acid 4-hydroxylase (C4H) enzymes to afford *p*-coumaric acid, which could be activated into its CoA thioesters by 4-coumarate-CoA ligase (4CL) [59–61] (Figure 4.21).

The chemical synthesis of these molecules does not involve the key condensation between cinnamic acids and the amino group of anthranilic acids and uses acid-catalyzed aldol reaction between benzaldehyde derivatives **4.60** and 2-methylbenzoxazin-4-ones **4.59**; both reagents can be easily prepared from the respective 4-hydroxybenzaldehydes and anthranilic acids refluxing for 4 hours [55, 62]. The aldol condensation affords the protected (*E*)-2-styryl-benzoxazinones

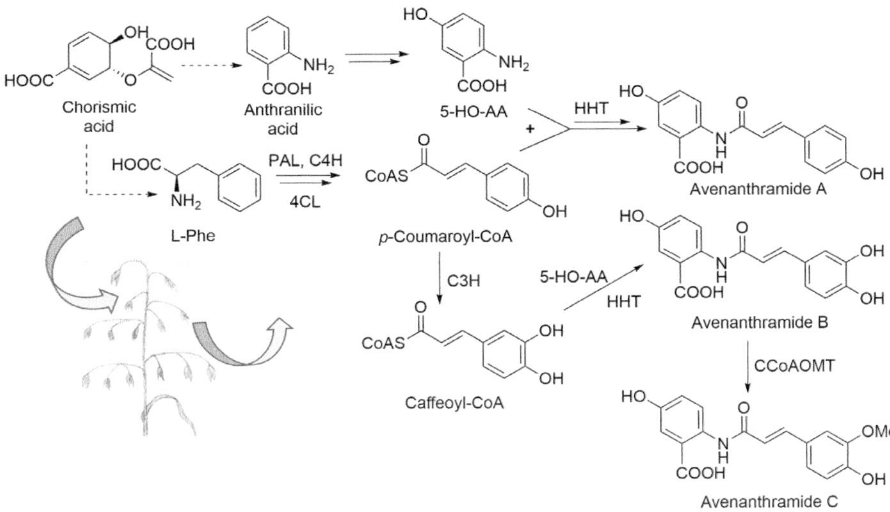

Figure 4.21. The proposed biosynthetic pathway of three major avenanthramides in oat [60].

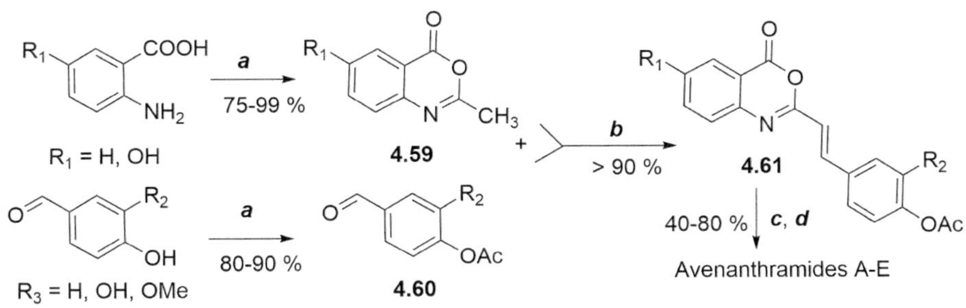

Figure 4.22. Chemical synthesis of avenanthramides and related compounds from 2-methylbenzoxazin-4-ones: (a) Ac_2O, reflux, 4–5 hours; (b) toluene, *p*-TsOH, reflux, 48 hours; (c) Na_3PO_4/H_2O; (d) HCl.

4.61, so-called avenalumins [63], and a final treatment with phosphate-promoted hydrolysis results in the avenanthramides A–C [55] (Figure 4.22). All compounds were crystalline, with colors varying from pale yellow to brown.

In conclusion, a combination of chemical and biological synthesis can extend the repertoire of needed phytochemicals by applying a genetic engineering strategy as an alternative to conventional chemical synthesis. For instance, AVNs and their synthetic analogs (e.g., dihydroavenanthramides) can be produced through genetically engineered microorganisms, e.g., a unicellular fungus *Sacharomyces cerevisiae*, known as "Baker's Yeast" and a bacterium *Escherichia coli* [58].

The representative example is the microbial production of the drug tranilast and several of its analogs using modified yeast strain to co-express a 4-coumarate/CoA (4CL5) ligase from and a hydroxycinnamoyl/benzoyl-CoA/anthranilate N-hydroxycinnamoyl/benzoyltransferase (HCBT) from the respective plants *Arabidopsis thaliana* and *Dianthus caryophyllus*. Recall that tranilast corresponds to the 4′-methoxylated form of N-(3′-methoxy-4′-hydroxycinnamoyl)-anthranilic acid (derived from ferulic acid), which is produced by our recombinant yeast strain when grown in the presence of 3,4-dimethoxycinnamic acid **4.62** and anthranilic

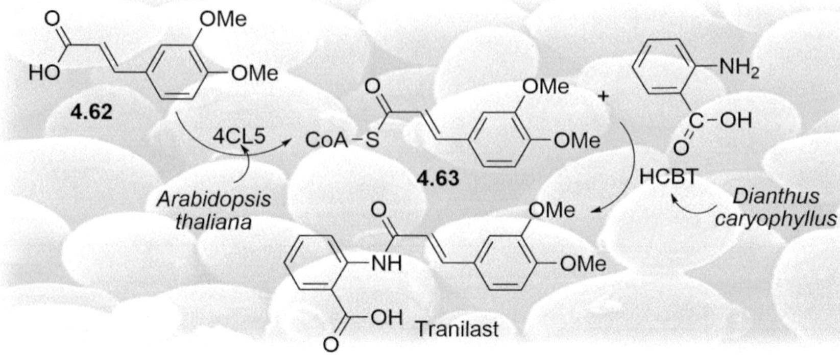

Figure 4.23. Biological production of the drug tranilast in yeast *Saccharomyces cerevisiae* on the background of *S. cerevisiae* cells.

acid. Thus, for its production, 4CL5 substrate, DMHA, and anthranilate were supplied independently as precursors to the culture medium of recombinant yeast HCBT grown on the respective medium without amino acids supplemented with 6% glucose (overnight, 30°C room), and then the cultures were shaken at 200 rpm in a 30°C room for 15 hours for the production of tranilast via the formation of 3,4-dimethoxycinnamoyl-CoA (DMH-CoA) **4.63**, an indispensable intermediate [64] (Figure 4.23).

The tranilast production presented in this example using the 4CL5-HCBT yeast strain as a catalytic system is a promising chemical synthesis alternative. It could be further optimized in various ways.

The biological production of natural AVNs is also possible using metabolic engineering in *Escherichia coli* as a valuable resource. For example, Lee and colleagues synthesized nine AVNs, including AVNs A-C in *E. coli* from hydroxycinnamic acid-CoA (HCA-CoA) and anthranilates using two genes: Os4CL, which converts hydroxycinnamic acids into corresponding HCA-CoA, and HCBT, which synthesizes AVNs from HCA-CoA and anthranilates [65], similar to the process in yeast *S. cerevisiae*.

4.2 Phytotoxins vs. Mycotoxins: Fungal Metabolites that Inhibit Plant Growth and/or Metabolites that Cause Disease and Death in Humans and Animals

Plant-microorganism co-evolutionary relationships are complex. So far, we have seen that the formation of phytoalexins is the response of plants to the invasion of endophytes (fungi, bacteria, etc.). However, another side of the coin is that the response of these pathogenic agents to the plants is the development of their toxic substances, which represent the expression of the virulence of the pathogens. These compounds are generally called phytotoxins. They are represented mainly by low-molecular-weight secondary metabolites capable of disrupting the vital activity of plant cells or causing their death at concentrations below 10 mM.

Moreover, phytotoxins that affect plants are usually further categorized by the target range, from the host's range to a wider radius outside the host plant; these are termed host-selective toxins (HSTs) and non-host-selective toxins (NHSTs). HSTs are toxic only to host plants, while NHSTs can affect many plants. Regarding toxic fungal secondary metabolites, besides the term phytotoxins (more correctly, fungal phytotoxins), mycotoxins are generally used, especially in modern times. The latter substances are defined as low-molecular-weight fungal metabolites toxic to vertebrates and exert severe adverse effects on the health of farm animals and humans who eat contaminated agricultural products. Thus, many fungal metabolites are both mycotoxic (toxic to animals) and phytotoxic (toxic to plants). However, the border division between these last groups is illusory. They were initially categorized as distinct toxic chemicals, but this could be considered an inappropriate classification for toxins from plant pathogenic fungi.

On the other hand, numerous phytotoxic studies found that each genus of fungi generates different toxic substances, although they have similar metabolic characteristics [66]. Thus, we are going to examine some of the most important genera of fungi (*Alternaria*, *Aspergillus*, *Phoma*, *Claviceps*, *Fusarium,* and *Penicillium*), trying to describe the structures of heterocyclic secondary metabolites, classify them, and their chemical, and biological, ecological, and pharmaceutical roles in the development of life sciences.

4.2.1 Alternariol, its Derivatives, and other Toxins from the Genus Alternaria

Secondary metabolites are crucial in fungal development and actively shape interactions with other organisms. Metabolites of the genus *Alternaria* are no exception. This genus has more than 300 species, with *A. alternata* being the most studied species. *A. alternata* alone infects more than 100 plant species. Several *Alternaria* species can produce secondary metabolites, such as phytotoxins and mycotoxins. Fungal disease induced by *Alternaria* in agricultural production has caused significant concern worldwide. For example, *A. alternata* and *A. tenuissima* can lead to black-point disease of wheat, and *A. solani* can infect tomatoes and contribute to tomato early blight (a devastating fungal disease).

The fungal metabolism of this genus produces many compounds, approximately 70 phytotoxins/mycotoxins. Among these compounds may be mentioned dibenzo-α-pyrone derivatives (also named 6*H*-benzo[*c*]chromen-6-ones and 6*H*-dibenzo[*bd*] pyran-6-ones). Alternariol **4.64**, its mono 9-methyl ether **4.65**, and alternuene **4.66**, isolated and reported in 1953 [67], are the three most studied dibenzo-α-pyrone derivatives (Figure 4.24). They are non-host-selective toxins (NHSTs) [68].

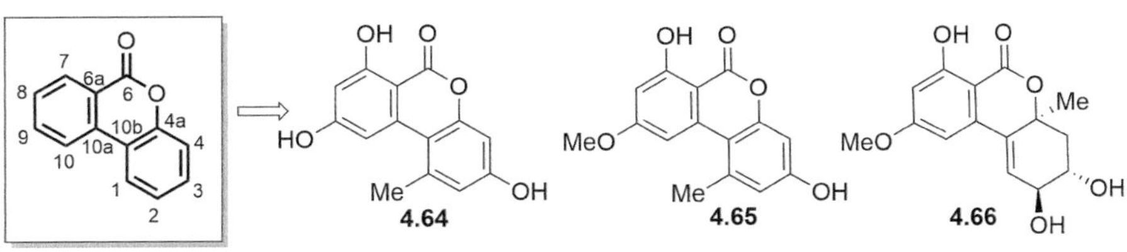

Figure 4.24. Structures of the most studied dibenzo-α-pyrone derivatives.

The first two substances cause weakly acute toxic effects exhibiting LD_{50} higher than 400 mg/kg body weight (mice), while alternuene showed considerable toxic action with LD_{50} = 50 mg/kg. Regarding their cytotoxic activity against Henrietta Lacks's cervical cancer cell line HeLa cells, less toxic alternariol and alternariol-9-methyl ether were the most cytotoxic molecules (LD_{50} = 6 and 8–14 μg/mL, respectively), and alternuene displayed an LD_{50} value of 28 μg/mL [69]. Alternariol was also cytotoxic to human colon carcinoma cell lines (HCT116 cells), inducing cell death, i.e., apoptosis, in these cells [70], as well as Caco-2 cells, causing both apoptosis and necrosis of these cells [71]. It was also cytotoxic to L5178Y mouse lymphoma cells with an EC_{50} value of 1.7 μg/mL [72]. Most importantly, it could produce chronic effects affecting DNA integrity, leading to genotoxicity and teratogenicity [66].

Black molds of the genus *Alternaria* occur ubiquitously and can grow on various substrates. These toxic metabolites, so-called "Alternaria toxins", are relevant as potentially harmful food contaminants [73]. These toxins have been encountered in cornflakes, cereals, fruits such as tomatoes, apples, and other crops.

How is the principal component of *A. alternata*, alternariol, produced in nature and the chemical laboratory? It is hard to imagine that in nature, alternariol and its 9-methyl ether, being complex dibenzo-α-pyrone molecules, are built on the simple precursor, acetyl-CoA, which reacts through repeating Claisen-type condensations with six molecules of malonyl-CoA to form the linear enzyme-bound sixketide chain **4.66** [74] (Figure 4.25).

This process is catalyzed by the concerted action of several enzymes (ketosynthase, acyltransferase, and acyl carrier protein). The aromatization and lactonization process of **4.66** promoted by aldol condensations yields alternariol **4.64**. The subsequent enzyme-catalyzed methylation reaction affords alternariol-9-methyl ether **4.65**. This acetate-malonate metabolic pathway is similar to hypericin biosynthesis (Section 2.2).

In the laboratory, efficient preparation of alternariol relies on convergent synthetic tactics identifying aromatic precursors containing polyfunctionalized A and C rings that are obtained using the disconnection approach (retrosynthetic analysis) that involves the two strategic simple bond O-C(O) and C_{Ar}-C_{Ar} breaks separately [75–77] (Figure 4.26A). To complete the convergent synthesis of alternariol, orcinol (1,3-dihydroxy-5-methylbenzene) **4.67** and 1-bromo-3,5-dimethoxybromobenzene **4.68** were chosen as starting precursors.

These building blocks were properly transformed in an orcinol-derived boronic acid **4.69** and brominated resorcylic aldehyde **4.70**, which reacted via the palladium-catalyzed Suzuki coupling reaction to yield biaryl derivative **4.71** in 70% yield. Thus, alternariol **4.65** was synthesized via a two-step sequence. First, biaryl aldehyde **4.70** was oxidized using sodium chlorite ($NaClO_2$) in the presence of sulfuric acid in DMSO to give the corresponding acid **4.72**, which was used for the next step without further purification. Then, its intramolecular lactonization and demethylation reactions, which were conducted by adding BBr_3 in dichloromethane, allowed affording the ABC tricyclic system of alternariol **4.65** as a white solid (mp = 268–269°C) in 90% yield (over two steps from **4.71**) [76] (Figure 4.26B).

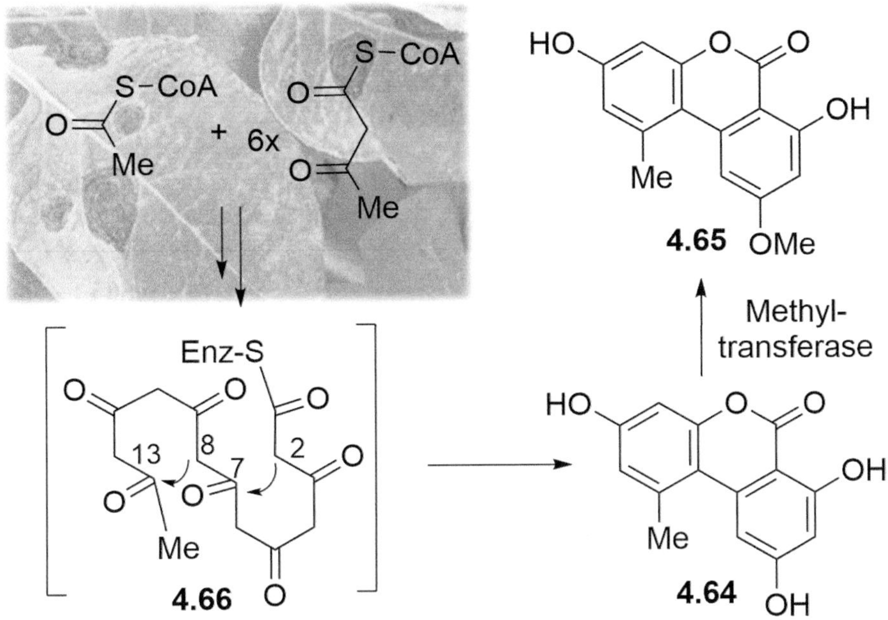

Figure 4.25. Proposed simplified biosynthetic pathway of alternariol and its methyl ether. Cyclization occurs between C-2, C-7, and C-8, C-13, and lactonization, resulting in alternariol. An image of tomato early blight, which is caused by *A. solani*, the principal pathogen threat.

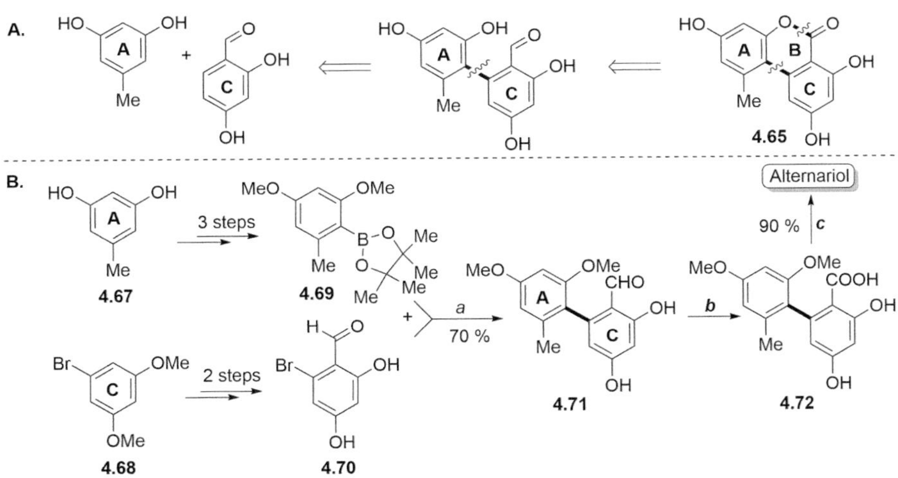

Figure 4.26. A. Retrosynthetic analysis of alternariol. B. Its convergent synthesis: (a) $Pd(PPh_3)_4/K_2CO_3$, DMF, 100°C, 12 hours; (b) $H_2SO_4/NaClO_2$, DMSO, rt, 4 hours; (c) BBr_3, CH_2Cl_2; 0°C to rt, 12 hours.

This convergent synthesis allows sufficient quantities of this mycotoxin to be prepared for further bioassays.

In addition to dibenzo-α-pyrone toxins, other components of a different chemical nature are found in the metabolism of *Alternaria* species; for example, so-called tetramic acid [79], L-tenuazonic acid (TeA) **4.73** (Figure 4.27), originally

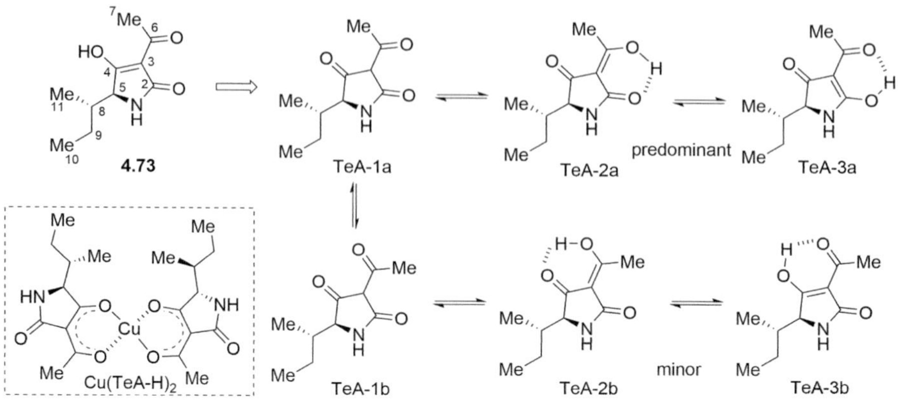

Figure 4.27. Proposed scheme for the tautomerism of tenuazonic acid and structure of tenuazonic acid complex with Cu(II), Cu(TeA-H)$_2$, which is prepared by treating TeA with Cu(OAc)$_2$ in MeOH-H$_2$O mixture.

isolated from the culture filtrate of *A. tenuis* but it was also found in other species, such as *A. alternata*, *A. longipes*, and *A. tenuissima*. Its general structure was first described in 1957 as a 3-acetyltetramic acid, i.e. (5*S*,8*S*)-3-acetyl-5-sec-butyl-4-hydroxy-3-pyrrolin-2-one (TeA-1). TeA is soluble in organic solvents, including petroleum ether, and is sparingly soluble in water. This compound is a monobasic acid with pK_a 3.5 and shows optical activity $[\alpha]^{20}_D$ = –121°C (c = 2.0, $CHCl_3$). However, its varied structures are usually used in the literature because this small and simple functionalized molecule presents different tautomers. This tetramic acid exists typically predominantly in the 2,4-diketo form, i.e., (5*S*,8*S*)-3-acetyl-5-sec-butylpyrrolidine-2,4-dione structures (TeA-1a and TeA-1b) equilibrating fast through the external and internal keto-enol-tautomerization process to afford the respective predominant TeA-2a/TeA-3a pair and minor TeA-2b/TeA-3b forms in gas phase and solution [79] (Figure 4.27). Thus, the first pair of tautomers TeA-2a/TeA-3a, are thermodynamically more favorable than those of TeA-2b/TeA-3b. Tautomers TeA-2a/TeA-2b are called internal tautomers, while TeA-3a/TeA-3b pair is termed external. Pyrrolidine-2,4-dione TeA-1a/TeA-1b forms are usually not observed. It is noteworthy that bearing a β-diketone moiety, TeA is a suitable metal-chelating ligand. Metal chelation provides this mycotoxin with different physicochemical properties, like stability and increased lipophilicity. It is particularly significant in the physiological context, i.e., their transport in biological tissues and across membranes [79, 80].

These latter forms are called (5*S*,8*S*,*Z*)-5-sec-butyl-3-(1-hydroxyethylidene) pyrrolidine-2,4-dione and (5*S*,8*S*,*E*)-5-sec-butyl-3-(1-hydroxyethylidene) pyrrolidine-2,4-dione, respectively (Figure 4.27). Consequently, the TeA-2a/TeA-3a forms coordinate the metals (calcium, magnesium, copper, iron, and nickel ions) through their 3-acyl and 2-carbonyl oxygen atoms. At the same time, the TeA-2b/TeA-3b tautomers possess the metal ion sandwiched between the enolic oxygen of the 3-acyl group and the 4-oxygen atom. Interestingly, TeA is usually stored as a copper salt, a stable, chloroform-soluble, green needle.

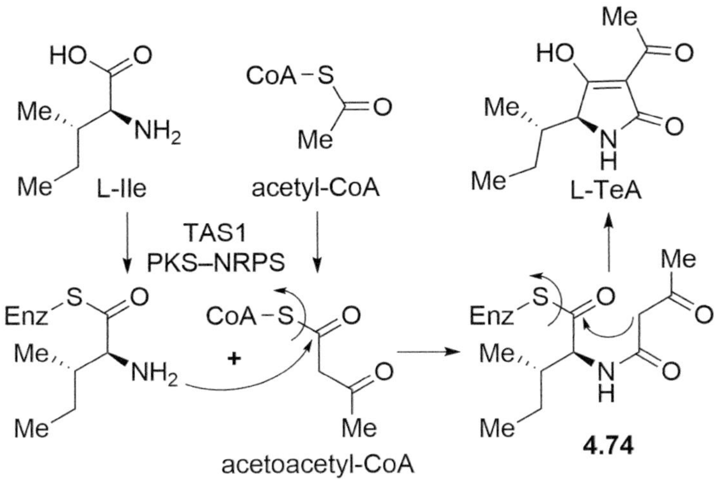

Figure 4.28. Simplified scheme of biosynthesis of TeA by TAS1.

TeA has long been reported to be toxic in animals, exhibiting antibacterial, antiviral, phytotoxicity, and entomotoxic effects. For example, it acts as an inhibitor of *Paenibacillus* larvae, a Gram-positive bacterium that infects honey bees (honeybees' disease). TeA displayed a MIC of 32 μg/mL, comparable with oxytetracycline, an antibiotic currently used to prevent this disease [68]. Among the *Alternaria* toxins, it is regarded as the potent toxic compound exhibiting LD_{50} values of 81 mg/kg body weight (mice, applied orally). In contrast, it exhibited moderate cytotoxic effects on HT-29 cells at concentrations as low as 100 μM.

On the other hand, TeA showed inhibitory activity against 4-hydroxyphenylpyruvate dioxygenase (HPPD) with an IC_{50} value of 18 μM. This enzyme is a key enzyme in tyrosine catabolism, which is important for the biosynthesis of plants. Thus, TeA inhibits photosynthetic activity by inhibiting photosystem II, which prevents plant growth and could be used for weed control in corn, barley, oats, rice, sorghum, sugarcane, etc. [81].

The biogenesis of L-TeA involves using an activated L-isoleucine (L-Ile) and acetoacetyl-CoA to synthesize N-acetoacetyl-L-isoleucine **4.74**. Further cyclization of this intermediate then produces TeA with the aid of gene TAS1, which consists of complex hybrid enzymes such as polyketide synthase (PKS) and non-ribosomal peptide synthetases (NRPS) [82] (Figure 4.28).

L-isoleucine benzyl ester hydrochloride **4.75** was used as a starting chiral source in its chemical synthesis. It was subjected to domino acylation-cyclization upon treatment with a twofold excess of (triphenylphosphoranylidene) ketene **4.76** immobilized on polystyrene (PS) in THF at 60°C to give the benzyl pyrrolidinone derivative **4.77** with the configuration of initial amino acid. Its hydrogenation over Pd/C allowed obtaining pyrrolidine-2,4-dione **4.78**, which was converted into L-(−)-tenuazonic acid in good yield via the Lewis acid-mediated direct C-acylation reaction ($BF_3 \cdot OEt_2$/acetyl chloride) and subsequent basic hydrolysis with 1% NaOH aq. [83] (Figure 4.29A). This straightforward three-step synthesis of optically pure

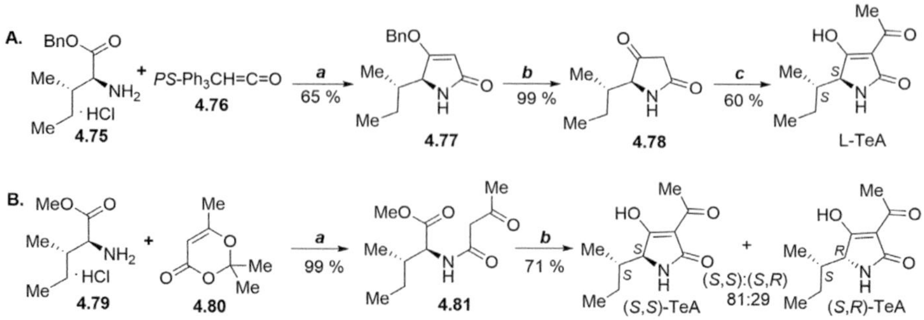

Figure 4.29. A. Three-step synthesis of natural L-TeA: (a) THF, 60°C, 14 h; (b) H_2 (1 atm), 5%-Pd/C, MeOH, rt, 2 hours; (c) 1. $BF_3{\cdot}OEt_2$, AcCl, 80°C, 8 hours, 2. ice-cold 1% aq. NaOH, then cold 1M HCl. B. Two-step preparation of the synthetic mixture TeA epimers: (a) *m*-xylene/Et_3N, 138°C, 5 hours; (b) MeONa/MeOH, 65°C, 4 hours.

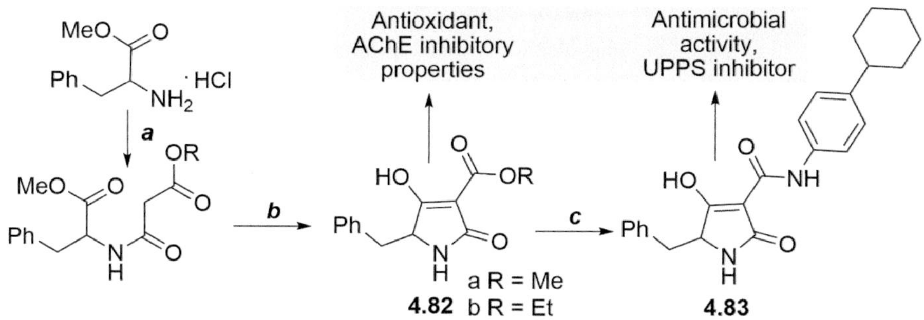

Figure 4.30. Synthesis of selected TeA derivatives with promising biological activity: (a) methyl (ethyl) malonyl chloride, CH_2Cl_2, Et_3N, rt, 2 hours; (b) MeONa/MeOH, reflux, 2 hours; (c) MW, EtOH, 100–120°C, 5–8 minutes.

L-TeA is very efficient (38.6% overall yield). However, it employs a toxic and hardly available phosphorus reagent that decreases its synthetic value.

In another synthesis based on a biomimetic approach, well-established synthetic methods for TeA preparation were used. It involves a simple two-step procedure: (1) N-acetoacetylation reaction of the α-amino group of an L-isoleucine methyl ester hydrochloride **4.79** with diketene acetone adduct **4.80** (2,2,6-trimethyl-4*H*-1,3-dioxin-4-one) in boiling *m*-xylene in the presence of triethylamine to afford a key N-acetoacetyl amino derivative **4.81** like biosynthetic **4.74**. (2) So-called Lacey-Dieckmann cyclization of a latter derivative under basic treatment with sodium methoxide in hot methanol to give a mixture of the 5*S*- and 5*R*- epimers (81:19) of TeA [84] (Figure 4.29B). These synthetic procedures do not lose importance since TeA was identified as a potential multitarget ligand with antimicrobial, anti-cholinesterase (AChE), anti-amyloidogenic, and antioxidant activities [85, 86]. The selected TeA derivatives were prepared analogously in three steps from commercially available racemic α-amino acid, phenylalanine methyl ester hydrochloride using ethyl (or methyl) malonyl chloride to provide the intermediates like **4.81**, which were subjected to Lacey-Dieckmann cyclization conditions to yield 5-benzyl-3-methoxy(ethoxy)carbonylpyrrolin-2-ones **4.82a–b**. Formation of 3-carboxamide

4.83 was best achieved via amidation condensation of **4.82a** under microwave (MW) irradiation conditions (Figure 4.30).

Synthesized derivative **4.82b** showed antioxidant activity (EC_{50} = 6.1 μmol/μmol for the 1,1-diphenyl-2-picryl-hydrazyl-DPPH assay) and AChE inhibitory activity (IC_{50} = 57 μM) [85]. At the same time, compound **4.83** exhibited antibacterial activity against Gram-positive bacteria *Enterococcus faecalis*, *Staphylococcus aureus*, and *Streptococcus pneumonia* with MIC values of 8.0, 0.5, and 1.0 μg/mL, respectively, inhibiting undecaprenyl pyrophosphate synthase (UPPS), an enzyme essential for bacterial viability [86].

4.2.2 *Cytochalasins, Complex, Diverse Perhydro-isoindolinone Molecules, and Ophiobolins based on the Octadipentacarbotriclycic Skeleton With a 1,4-Keto Aldehyde Function from the Genera Phoma and Bipolaris (Helminthosporium)*

Cytochalasins and ophiobolins are complex metabolite molecules produced by several fungi, especially those belonging to the genera *Phoma* and *Bipolaris* (*Helminthosporium*). Understanding the taxonomic classification of these fungi is difficult since their characteristics are constantly examined, and some species are listed under different names. However, in this section, our task is to study their metabolites structurally and have knowledge of the relationships between structure and activity.

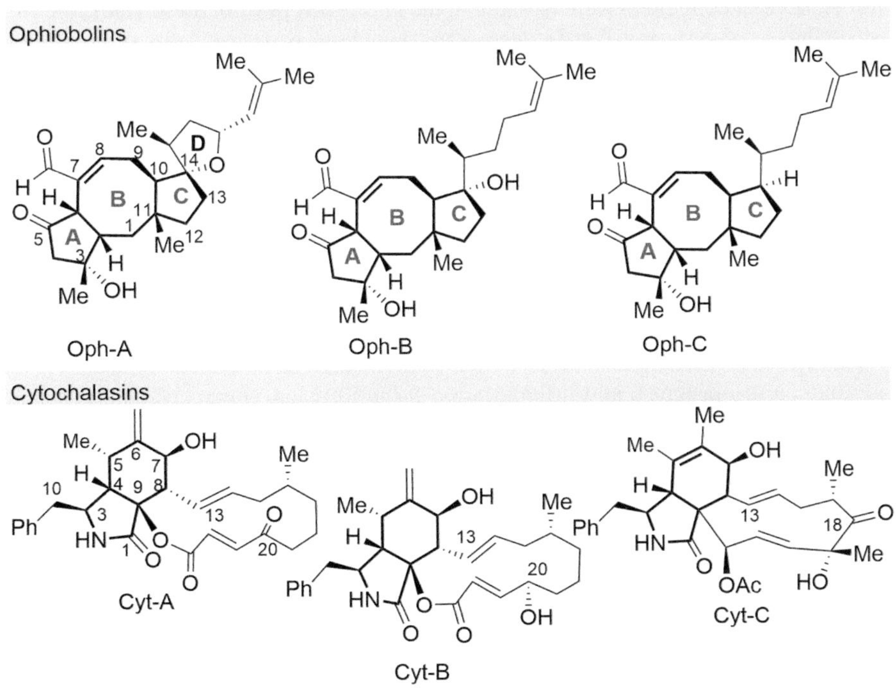

Figure 4.31. Structures of the representative ophiobolins and cytochalasins. Their main skeleton rings are in bold.

Ophiobolins are part of these phytotoxic sesterterpenes isolated as secondary phytotoxic metabolites produced by fungal pathogens of several crops, such as rice, maize, and sorghum. More than 50 ophiobolins were characterized. They all consist of a C25 skeleton with a dicyclopenta[*ad*]cyclooctane system with methyl and hydroxyl groups or transformed methyl groups at positions C-3, C-7, and C-11. The conformation of the respective *cis*- and *trans*-junction of this system's A/B and B/C rings appears typical. However, some molecules with the *trans-trans* conformation exist (Ophiobolin A contains an additional ring D in the spiro-junction). In contrast, cytochalasins with phenylalanine (L-Phe) residue are derivatives of perhydro-isoindolone (bicyclic lactam), linked with macro-lactonic or macrocyclic unsaturated systems. There are more than 120 cytochalasins with phenylalanine (Figure 4.31). The name comes from the Greek *kytos* (cell) and *chalasis* (relaxation), pointing to the best-known property of cytochalasins, which will be discussed later. On the other hand, the term ophiobolins was adopted when this compound was isolated from the pathogenic plant *Ophiobolus miyabeanus,* which was reclassified as *Bipolaris oryzae* [87].

Although their structures are different, both groups of compounds have the same function: to protect the microorganisms, attack and kill other invasive species, prevent maximum development and growth of these fungi, and generate competition. They are interesting compounds from a pharmacological point of view.

The first member of the ophiobolins group, Oph-A (also called cochliobolin), was characterized by the X-ray diffraction method in 1966 as a phytotoxin produced by the rice pathogen *Bipolaris oryzae* (syn. *Helminthosporium oryzae or Drechsler oryzae or Cochliobolus miyabeanus*) and *Bipolaris maydis* (*H. maydis*). After Oph-A, several congeners were isolated in the late 1960s [88]. In these years, the first cytochalasins, Cyt-A and Cyt-B, were also described isolating from fungi *Phoma* strain S 298 and *Helminthosporium dematioideum*, respectively [89]. Note that brown spots caused by the fungus *Bipolaris oryzae* with Oph-A as a main metabolite are a common and devastating disease of rice in many rice-growing regions of the world. These ophiobolins are also toxic to animals. The LD_{50} doses of Oph-A for mice are 238 mg/kg when administered subcutaneously, 21 mg/kg intraperitoneally, 12 mg/kg intravenously, and 73 mg/kg orally. The LD_{50} of Oph-B in mice is 4.4 mg/kg when injected intraperitoneally.

Being a sesquiterpenoid metabolite, Oph-A was investigated initially as a phytotoxic molecule that affects crops. Later, Oph-A and analogs were proven to possess many other biological properties, such as antimicrobial, nematocidal, cytotoxic, anti-influenza, and inflammation-promoting activities. Despite numerous studies on these biological functions, our knowledge regarding ophiobolin biosynthesis and their structure-activity relationship is limited. However, it was known that based on a core structure of 5-8-5 tricyclic carbon skeleton with C25 carbon frameworks, the biosynthetic origin of ophiobolins is derived from head-to-tail linkages of five isoprene units (five C5 units = 25 units), where the cyclization of GFPP (geranyl farnesyl pyrophosphate) with the aid of terpene cyclase (TC) followed by rearrangement is a crucial step. Recall that dimethylallyl diphosphate (DMAPP) and isopentenyl diphosphate (IPP) are building blocks (see Section 2.1), which suffer the prenyltransferases (PTs) catalyzed-chain elongation transformation [88] (Figure 4.32A).

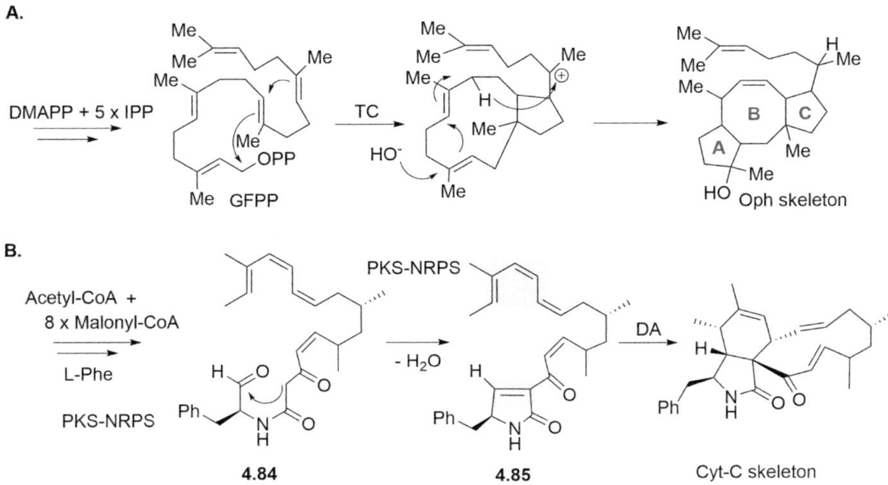

Figure 4.32. Proposed biosynthetic assembly of ophiobolin and cytochalasin backbones.

Further biosynthetic modifications of these compounds are catalyzed by various enzymes, including cytochrome P450 monooxygenases, oxidoreductases, hydrolases, and different group transferases. Not all details of the bio-origin of cytochalasins are known either, despite extensive studies started in 1970. It was proposed that in the early stages of cytochalasin biosynthesis (e.g., Cyt-C), a polyketide chain with an aldehyde group **4.84** was produced by condensation of acetyl-CoA and eight malonyl-CoA units followed by the incorporation of L-Phe under the action of the PKS-NRPS hybrid enzyme. Then, this β-keto amide derivative is thought to react in a Knoevenagel condensation to produce pyrrolinone unit **4.85**, a key intermediate in the middle stage of cytochalasin biosynthesis, which is defined as the conversion from the linear PKS-NRPS intermediate to the isoindolone fused macrocyclic structure. Thus, the intramolecular [4+2] Diels-Alder (DA) cycloaddition reaction of **4.85** is possible (Figure 4.32B). The DA reaction is believed to be catalyzed by specific enzymes [89].

In general, it could be said that Ophs are mixed bioproducts formed from the mevalonate and methyl erythritol phosphate pathways. At the same time, cytochalasins result from a mixed biosynthetic pathway involving a polyketide and L-Phe formation.The first total asymmetric synthesis of the 14-membered macrolactone cytochalasin B by Stork and Nakamura is based on the crucial step of the DA reaction in cytochalasin biogenesis [90]. The starting materials of its construction were triene **4.86** and N-acetylhydroxypyrrolone **4.87**. Thus, first, the esterification reaction between **4.86** and **4.87** was performed to afford a diene ester based on 3-pyrrolinone **4.88**, which possesses dienophile and diene moieties. The second, its subsequent intramolecular Diels-Alder reaction, was realized under harsh conditions (heating in mesitylene at 180–190°C) to provide the macrocyclic DA cycloadduct **4.89** in only modest yields (30%, endo/exo 4:1). The latter was converted into cytochalasin B (also called phomin) (Figure 4.33).

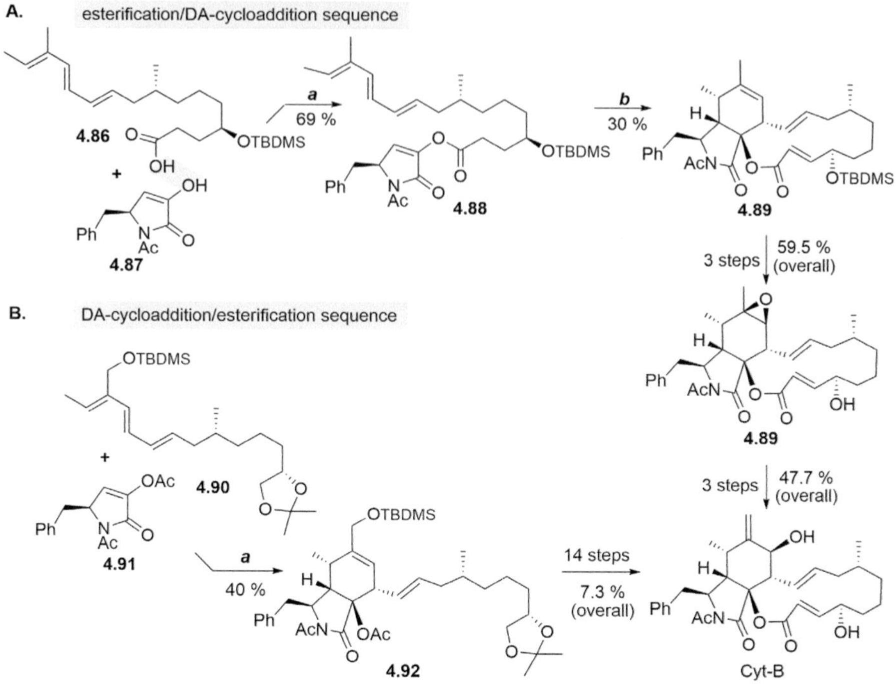

Figure 4.33. A. Total synthesis of Cyt-B based on an esterification/DA-cycloaddition process sequence (simplified scheme): (a) 2-chloropyridine methiodide, 20 mol% DMAP, NEt_3, CH_2Cl_2, rt; (b) mesitylene, 180–190ºC, 6 d. B. Total synthesis of Cyt-B involving a DA-cycloaddition/esterification sequence (simplified scheme): (a) xylene, 170ºC, 4 d. TBDMS = *tert*-butyldimethylsilyl.

Examining the synthetic scheme of Figure 4.33A, it is not difficult to see that Stork-Nakamura's synthesis of Cyt-B contains an esterification/DA-cycloaddition process sequence. The same Stork team developed an alternative approach using a DA-cycloaddition/esterification sequence [91]. In this case, prolonged heating of a mixture of 3-acetoxy-3-pyrrolin-2-one **4.90** and triene **4.91** in xylene allowed obtain isoindolone cycloadduct **4.92** with moderate yields (40%, endo/exo 4:1). Laborious elaboration of the latter completed the formal total synthesis of Cyt-B in lineal 14 steps (Figure 4.33B). Cyt-B, originally named phomin, is the most cytochalasin studied and has been used as a tool for the study of cell division and cell motility. It and other cytochalasins inhibit cell division and are known to induce an apoptotic response, activities that are thought to arise from the ability of these molecules to inhibit the polymerization of actin, which is the most abundant protein in most eukaryotic cells and is also key to the process of infection by some pathogenic microorganisms. Moreover, Cyt-B produces G2/M cell cycle arrest and apoptosis, inclusive, which is applied to test the genotoxicity of substances *in vitro*. In 2022, Cyt A–C were identified as leads for neglected tropical diseases, i.e., Chagas disease and leishmaniasis. They exhibited IC_{50} values ranging from 2 to 20 μM against intracellular *Trypanosoma cruzi* and *Leishmania infantum* amastigotes, values comparable to those of the standard drugs benznidazole and miltefosine for Chagas disease and leishmaniasis, respectively. Furthermore, these fungal alkaloids presented moderate cytotoxicity against standard cell lines (HFF-1, THP-1, and HepG2) and an excellent overall profile for oral bioavailability assessed *in vitro* [92]. Thus, it could be used as antiparasitic and/or anticancer drugs.Oph-1 was the most widely studied molecule in the Ophs family. Besides its potent phytotoxic effects on almost all of the tested plants (corn, rice, sorghum, soybean, tobacco, etc.), even at the lowest concentration, Oph-1 exhibited antibacterial actions (*Staphylococcus aureus*, *Mycobacterium intracellulare*), antifungal properties (*Aspergillus flavus*, *Trichophyton mentagrophytes*, etc.), and selective cytotoxic activities on multiple types of mammalian cells (26 human cancer cells) showing high IC_{50} values *in vitro* (e.g., chronic lymphocytic leukemia (CLL), IC_{50} = 1 nM; pancreatic cancer P335, IC_{50} = 63 nM; glioblastoma Hs683, IC_{50} = 620 nM). Oph-1 is also active in mouse glioblastoma and melanoma models [93]. This broad spectrum of biological activity makes this sesterterpenoid-type compound a promising drug candidate. Although many researchers have studied their chemical synthesis, there are no suitable synthetic procedures for its large-scale preparation. The total synthesis of (+)-Oph-A reported by Nakada and co-workers in 2011 is the first and remains the only synthesis of this sesterterpenoid compound. Even if this is a convergent synthesis, it is based on a complex multistep reaction [94, 95]. Preparation of a molecule as complex as Oph-A was not possible without retrosynthetic and structural analysis. Recall that this phytotoxin contains a unique 5-8-5-5 tetracyclic ring system (ABCD) with a specific 1,4-keto aldehyde function, including a spiro-tetrahydrofuran ring, which comprises a total of eight stereogenic centers. Note that there are the *cis*-fused AB and the *trans*-fused BC ring junctions, including a spirocyclic CD ring system. Nakada retrosynthetic analysis of Oph-A is shown in Figure 4.34, where the key intermediate **4.94** with an ACD skeleton has been generated after the disconnection of a problematic eight-membered carbocyclic ring B (i.e., it is difficult to construct)

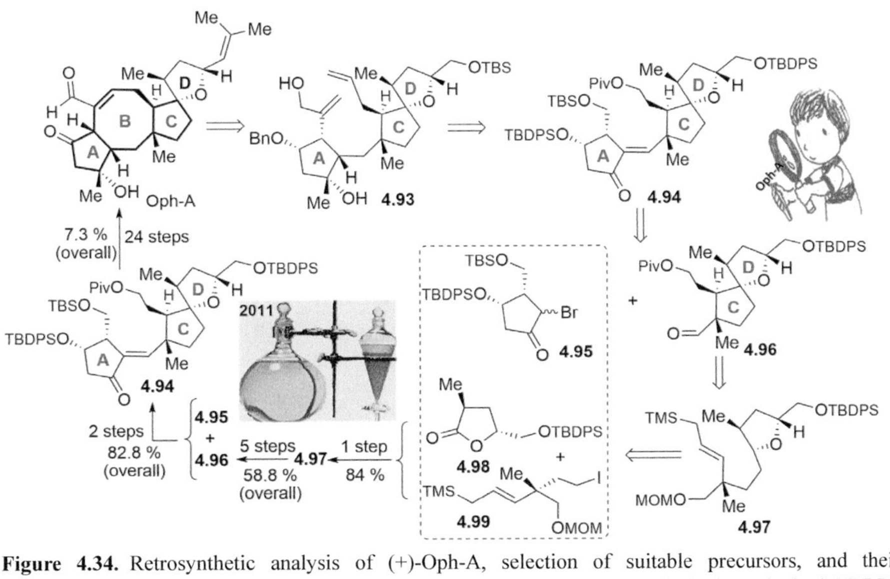

Figure 4.34. Retrosynthetic analysis of (+)-Oph-A, selection of suitable precursors, and their transformations in the simplified scheme. Bn = benzyl, MOM = methoxymethyl, Piv = pivaloyl, TBDPS = *tert*-butyldiphenylsilyl, TBS = *tert*-butyldimethylsilyl.

and the making a skeleton **4.93**. Subsequent disconnection of **4.94** allowed us to find the first precursor, **4.95**, for ring A and a sporadic CD system **4.96**, whose disconnection led to a polyfunctionalized tetrahydrofuran derivative, **4.97**. Two other valuable precursors, **4.98** and **4.99**, were generated from the latter. However, neither of these precursors is commercially available and must be prepared first, as well as **4.95**.

Therefore, starting with the appropriately selected and prepared chiral precursors **4.95**, **4.98**, and **4.99**, Japanese chemists get a key tricyclic ACD molecule **4.96** in eight steps using a convergent strategy. By having this compound in their hands, the construction of the ophiobolin eight-membered carbocyclic ring B through ring-closing metathesis of **4.93** was completed by performing 24 steps with a 7.3% overall yield from **9.94** (Figure 4.34). After examining the retrosynthetic scheme of Oph-A, we believe it would be worth mentioning the valuable contribution of Prof. Yoshito Kishi (1937–2023), who directed numerous successful syntheses of toxins, including total synthesis of (+)-Oph-C, which was completed in 1989 [96].

The efforts and expenses of these total syntheses and each synthesis of natural products with complex architecture are colossal and important for developing organic chemistry; however, this victory is partial from a medicinal point of view. It is difficult to analyze some modifications of a molecule prepared in this way. Still, studying the relationship between activity and structure is necessary, looking for the part of its structure responsible for the activity (pharmacophore unit). Simplified intermediates generated during the retrosynthetic analysis of Oph-A are more accessible for basic biological assays than the same natural product. We can now closely align the Corey retrosynthetic analysis to the looked biological activity, identifying simplified pieces of the natural product with comparable potency or potentially new functions during a total synthesis investigation.This strategy is called "pharmacophore-directed retrosynthesis" (PDR), where the key first step is identifying possible pharmacophore units [97]. Applied to our examined phytotoxin, Oph-A, PDR approach led to bicyclic derivatives **4.100**–**4.101** and the simplest form of the pharmacophore, monocyclic 1,4-keto aldehyde **4.102** to confirm the importance of approach (Figure 4.35) [98].

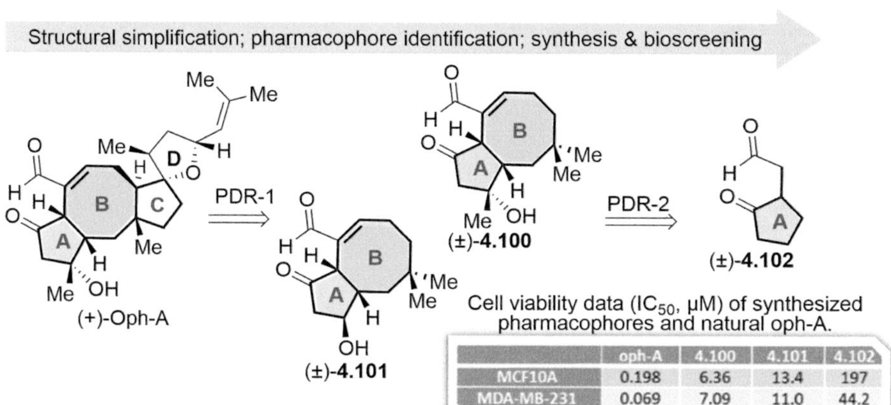

	oph-A	4.100	4.101	4.102
MCF10A	0.198	6.36	13.4	197
MDA-MB-231	0.069	7.09	11.0	44.2

Figure 4.35. Pharmacophore-directed retrosynthesis of Oph-A and cell viability data of generated pharmacophore units.

Note that this PDR approach was impossible without some preliminary biological data of Oph analogs that suggests the importance of retaining the integrity of the AB bicycle with potential 1,4-ketoaldehyde and hydroxyl functions to maintain the desired biological effects. That is why these hypothetical pharmacophores of Oph-A were proposed in this Figure. Thus, based on previous synthetic knowledge about these carbocyclic systems, simplified derivatives **4.100–4.102** were prepared through a lineal strategy and tested in two cell lines, a non-cancerous cell line (MCF10A) and a breast cancer cell line (MDA-MB-231) to evaluate the cytotoxicity of the synthesized derivatives as well as their selectivity.

Cell viability assays toward MDA-MB-231 breast cancer cells confirm the proposed hypothesis that significant cytotoxicity observed for the ophiobolins originates from structural features of the A/B-ring system. These model PDR studies are likely suitable for constructing tri-, bi-, or monocyclic derivatives of Oph-A and closely related compounds that are not easily accessible from the natural product. However, to apply PDR generating suitable pharmacophores of the studied toxic molecule, we must have several data, e.g., existing structure-activity relationship data of natural product congeners or biosynthetic precursors and the bioactivity of structurally related natural derivatives.

Now, we can advance in examining fungal indole-based mycotoxins, produced by the genus *Claviceps*, which grow on many herbaceous plants of the Poaceae family (400 species of plants, including cereals such as sorghum, barley, millet, oats, etc.), especially on rye (*Lolium temulentum*), a primary ingredient in making bread. The fungus-plant relationship is usually classified as parasitic because the fungus prevents the entry of essential nutrients and directly causes the loss of seeds.

4.2.3 Indole Metabolites Produced by Fungi of the Genus Claviceps

In past centuries, poisonings of humans or animals during ingestion of infected grass or grains (especially rye) by the species *Claviceps purpurea* were quite frequent. This parasitic fungus secretes toxic ergot alkaloids, recognized as important natural toxins in human history. They are related to "ergotism", an illness (Saint Anthony's fire), presenting two main characteristics: gangrene (in chronic ergotism), with loss of infected tissues, and convulsions (in acute ergotism), with paranoia and hallucinations [99]. Such epidemics, which caused an incalculable number of deaths, no longer occur in humans; nevertheless, outbreaks of ergotism in livestock do still take place. However, ergot poisoning in animals remains a socio-economic problem. One of the most common symptoms of poisoning in cattle is lameness, which can progress to gangrene. Chronic toxicity is also manifested in decreased weight gain and lower reproductive efficiency. Acute ergot toxicity can cause vomiting, diarrhea, cardiovascular collapse, seizures, and abortions.

When used deliberately as medicines, the same alkaloidal compounds have alleviated the suffering caused by migraine headaches, treated symptoms of Parkinson's disease, etc. This is one of the many beneficial and harmful aspects of one natural molecule.

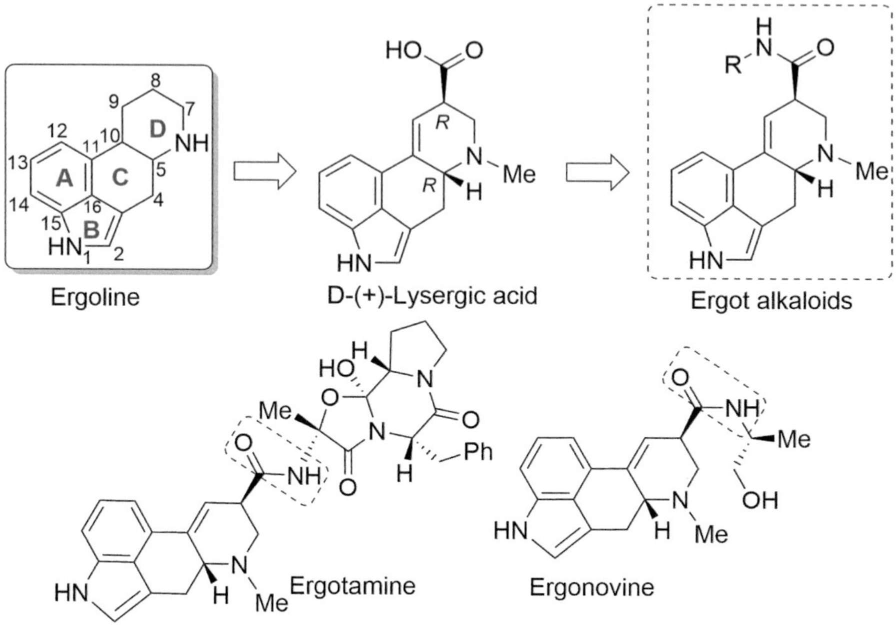

Figure 4.36. Major ergot alkaloids as secondary amines of lysergic acid based on tetracyclic ergoline system.

The chemical and biological interest in these alkaloids has been constant for over 250 years (note that the first connection between ergotism and infected rye was discovered in 1853). Their structural elucidation indicated that the base element is the heterocyclic compound—indole[4,3-*fg*]quinoline, indole AB, linked to a perhydro-quinoline ring CD. However, it is now popularly known as ergoline; this trivial name comes from the French word "Ergot," used for "spur," named after the similarity between sclerotia (sclerotium) of fungus. Only in the 1930s was it found that ergot alkaloids are characterized by a tetracyclic ergoline ring ABCD, secondary amide- or peptide-like amide derivatives of D-(+)-lysergic acid. Thus, ergot alkaloids are classified within a large family of indole alkaloids; moreover, they are 3,4-substituted indole derivatives. The main components of this type are ergotamine and ergonovine (ergometrine) (Figure 4.36). The first is a water-insoluble tripeptide derived from L-alanine, L-phenylalanine, and L-proline, being an example of the ergopeptine subgroup, and the second is a simple e-nonpeptidic amide, with L-(+)-2-aminopropanol, and belongs to the subgroup of ergoamides, which are water-soluble amino alcohol derivatives. It is interesting to note that ergot sclerotia contain about 0.15–0.5% alkaloids, which usually contain two centers of chirality of R-configuration at C-5 and C-8 positions.

The most famous component of the ergot alkaloids, ergotamine, was isolated in 1925, but the complete elucidation of its structure was not achieved until 1951. Ergonovine was discovered in four different laboratories almost simultaneously, with four diverse names (ergometrine, ergotocine, ergosterine, and ergobasine) being assigned to the alkaloid, whose structure was elucidated in 1935.

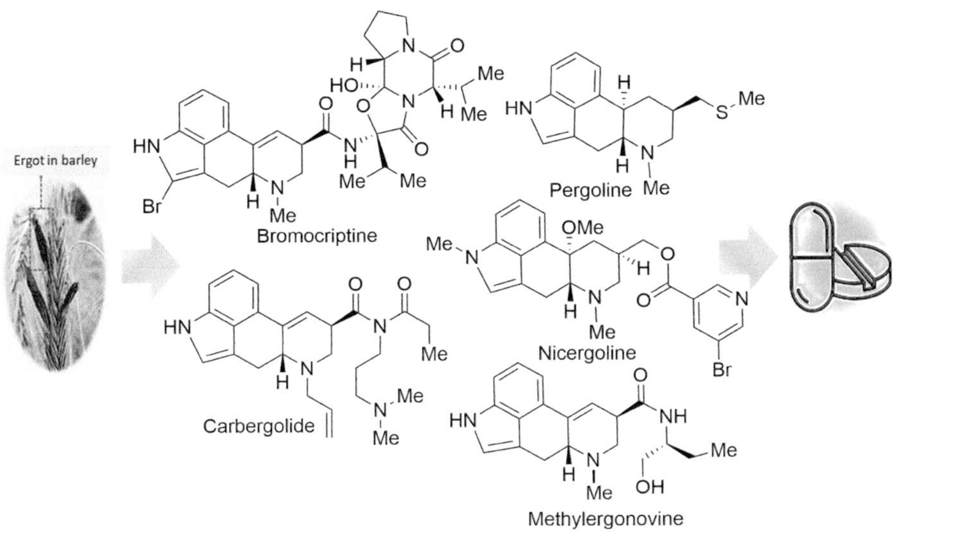

Figure 4.37. Presentative semi-synthetic ergot alkaloids as drugs.

Both ergotamine and ergonovine are naturally obtained ergot alkaloids of medical importance. At this time, ergotamine, in the form of maleate or tartrate with caffeine, is manufactured as the drug Cafergot, which can treat migraine and decrease tachycardia. Caffeine facilitates the absorption of ergotamine. Note that ergotamine is considered a toxic compound, although its toxicity is not high. The expected median lethal dose (LD_{50}) is between 60 and 100 mg/kg for small mammals (rats, mice, and rabbits) and intravenous application. On the other hand, ergonovine is one of the ergot-based drugs used in obstetrics as uterine stimulants to control postpartum hemorrhage [100].

Semi-synthetic drugs of ergot alkaloids are also used successfully in medical practice (Figure 4.37). Bromocriptine, a brominated ergocryptine alkaloid, was introduced in 1974 to treat Parkinson's disease. Since then, the simpler semi-synthetic derivatives, pergoline and carbergolide, have been used as alternative agents in the fight against this disease. Currently, bromocriptine is a medication that treats type 2 diabetes by decreasing your blood sugar. Nicergoline, a methyl ester derivative with a bromonicotinic acid residue, is an antihypertensive agent in treating senile dementia.

Methylergonovine, a synthetic ergonovine analog, was introduced for medical use (the treatment of postpartum and post-abortion hemorrhages) in 1946. The drug is currently prepared via a reaction of (+)-lysergic acid with L-(+)-aminobutanol, using different coupling reagents [101].

As can be seen, these compounds have quite complicated structures. For medicinal purposes, ergot is cultivated artificially (biotechnology methods). No synthetic method can compete efficiently with the "Laboratory of Nature" regarding stereo- and enantioselective synthesis. Annual world production of ergot alkaloids has been estimated at 5,000–8,000 kg of all peptidyl ergot alkaloids and 10,000–15,000 kg of lysergic acid, mainly used to manufacture semi-synthetic derivatives [101].

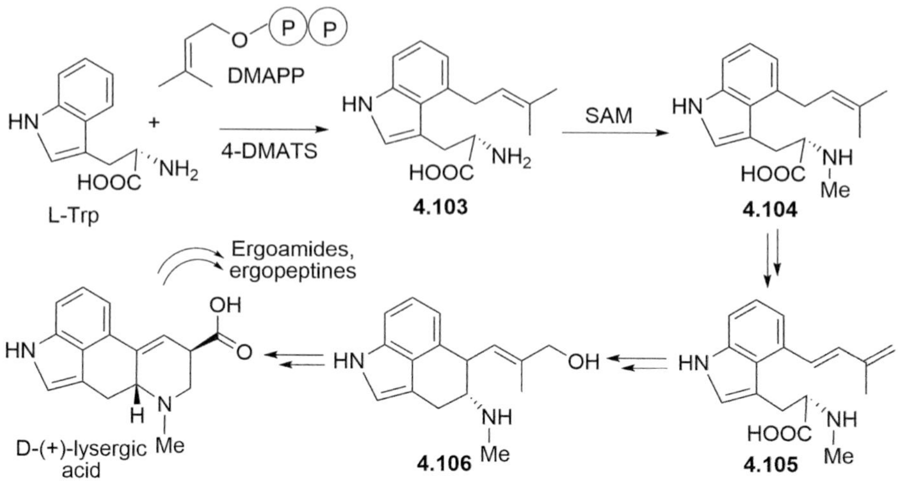

Figure 4.38. Proposed biosynthetic pathway of lysergic acid with ergoline ring formation in *Claviceps* sp.

As regards the biosynthesis of ergot alkaloids, being the amide derivatives of lysergic acid, which contain the ergoline ring, experiments with isotopically labeled precursors indicated that the fundamental building blocks for the lysergic acid skeleton are the amino acid L-tryptophan (L-Trp) and the dimethylallyl diphosphate (DMAPP). Although almost all of the early steps in this biosynthesis have been clarified, we will simplify this process by omitting biological details [102]. Thus, the latter, acting as a prenyl donor, attacks the indole nucleus under prenyltransferase catalysis (in this case, 4-dimethylallyltryptophan synthase 4-DMATS), giving 4-γ,γ-dimethylallyltryptophan **4.103** (Figure 4.38). Its enzymatic N-methylation in the presence of the S-adenosylmethionine (SAM) results in the formation of 4-dimethylallyl-L-abrine **4.104**, where the isoprenyl chain is hydroxylated quickly for the elimination of water, creating a diene **4.105**. Then, a series of successive oxidations mediate the intramolecular cyclization of the prenyl and tryptophan moieties, and decarboxylation should occur to generate the tricyclic ring system of chanoclavine-I **4.106**, a lysergic acid biosynthetic intermediate [102].

Subsequent multiple biotransformation steps conducive to the formation of D-lysergic acid have been postulated to involve cytochrome P-450 monooxygenases. Four NRPS genes control the late steps of the biogenesis of ergopeptines and ergoamides [102, 103].

Regarding the chemical synthesis of these alkaloids, it is considered that if the D-lysergic acid can be prepared quickly, we will have them since they possess the fundamental core of this class of alkaloids and serve as the key intermediate for some of the semi-synthetic drugs. Therefore, since 1954, when Kornfeld and his co-workers accomplished the first total synthesis of lysergic acid, lysergic acid remains a challenging synthetic target. Up to now, about 20 total syntheses have been carried out [104–106]. Retrosynthetic analysis of the lysergic acid skeleton allowed us to find the 3-indole-propionic acid as a helpful starting material related to biogenic tryptophan (Figure 4.39A). Employing this reagent, 3,4-dihydro-benzo[*cd*]

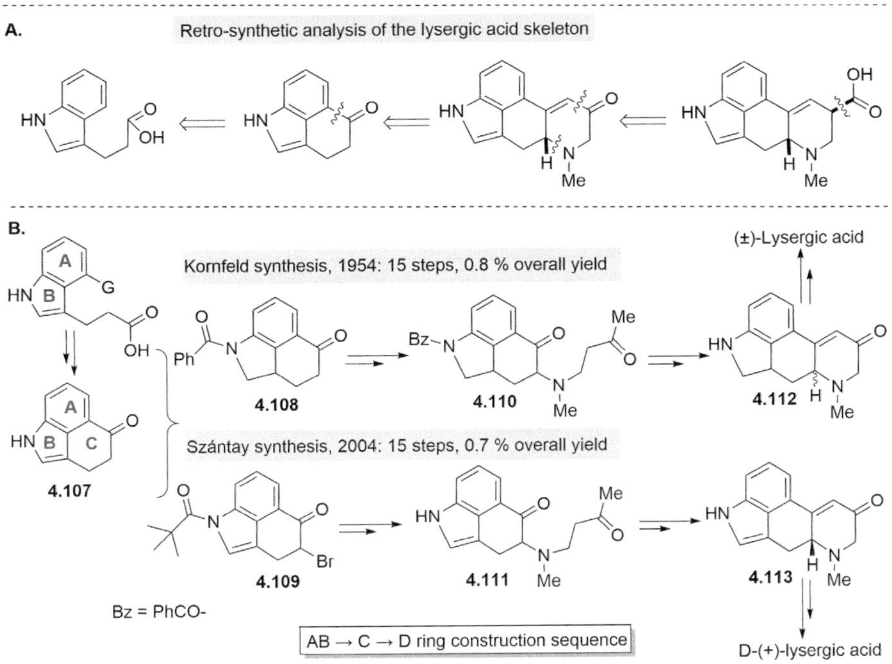

Figure 4.39. A. Retrosynthetic analysis of the lysergic acid skeleton. B. Total syntheses with essential step D-ring cyclization: Kornfeld synthesis of racemic lysergic acid and the first asymmetric synthesis of (+)-lysergic acid via resolution of an intermediate (Szántay synthesis).

indol-5-one **4.107**, known in the literature as Uhle's ketone, can be prepared through an intramolecular Friedel–Crafts acylation [107]. Its N-benzoyl derivative **4.108** (named Kornfeld's ketone) and N-pivaloyl derivative **4.109** were used to construct the lysergic acid skeleton (Figure 4.39B). The key aim of these syntheses was to use the carbonyl group reactivity of **4.108–4.109** for forming the respective amino derivatives **4.110** and **4.111** and take advantage of the presence of appropriate functional groups in the obtained products to undergo intramolecular cyclizations yielding tricyclic skeletons **4.112** and **4.113**, i.e., accomplishing the AB → C → D ring construction sequence. It is noteworthy that Kornfeld synthesis presents a landmark synthesis of the target compound (±)-lysergic acid and Szántay synthetic approach to D ring formation offers a route to enantiomerically pure (+)-lysergic acid (Figure 4.39B).

Synthesis of (±)-lysergic acid based on this retrosynthetic disconnection (i.e., the AB → C → D ring construction sequence) represents a significant part of the organic synthesis development. Currently, a different sequence of construction of the rings, AB → D → C, seems to be particularly attractive. Thus, total syntheses with the crucial step of C-ring cyclization are now known [106]. In 2023, the Smith group described a concise synthesis of (±)-lysergic acid over six synthetic steps, starting with the two commercially available halogenated pyridine and indole precursors **4.114** and **4.115** (Figure 4.40) [108]. This synthesis, which was finalized through the key intramolecular Mizoroki-Heck reaction, represents efficient, scalable,

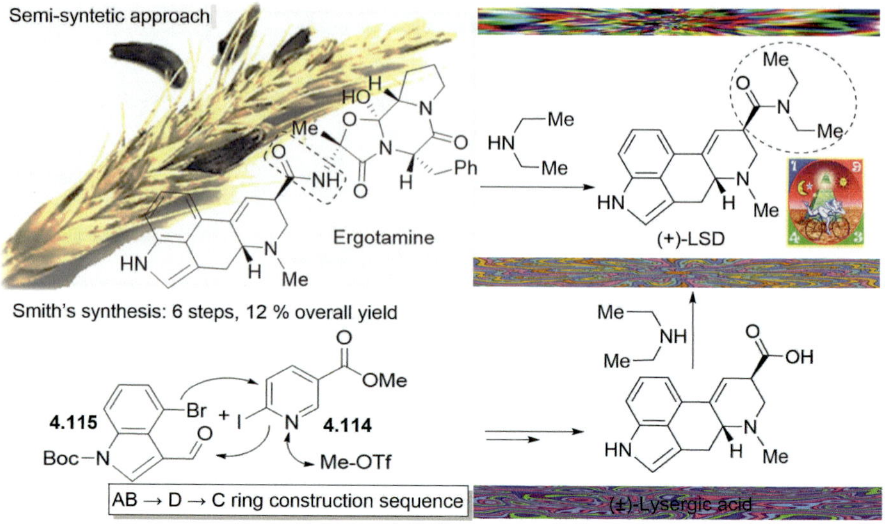

Figure 4.40. Semi-synthetic LSD synthesis realized by Hofmann and Smith's synthesis of racemic lysergic acid.

and potentially modular access to lysergic acid and thus offers more practical opportunities for its derivatization; among them, lysergic acid diethylamide (LSD) remains one of the essential derivatives.

LSD is a molecule created in 1943 by the chemist A. Hofmann (1906–2008). Working at Sandoz laboratories (now called Novartis), he studied the properties of lysergic acid derivatives. He came up with the idea of adding diethylamine to ergomertine, one of the compounds extracted from ergot, which later made it possible to obtain the white crystals that turned out to be lysergic acid diethylamide (Figure 4.40). As it produces alterations in mental functioning – mainly illusions and hallucinations in sub-milligram ($\geq$ 20 μg) oral doses – it caused enthusiasm in the study of mental illnesses. Later, it became a drug of abuse by young people in search of new sensations where "colors are heard, sounds are seen". Also, because of its euphoric effects, a very picturesque term arose among artists – "psychedelic art" [109]. In the 1960s and 1970s, the use of this molecule in research significantly decreased as it did not yield the results initially envisioned. However, as one of the most potent psychoactive agents known, LSD is back in fashion [109]. LSD and other lysergamide hallucinogens mediate behavioral responses through partial activation of the serotonin 5-HT_{2A} receptor [110], which is a G protein-coupled receptor (GPCR). Clinically, this receptor is associated with several psychiatric disorders, including schizophrenia, depression, and obsessive-compulsive disorder. Therefore, synthetic efforts in LSD core construction could accelerate the discovery of more selective drugs for the treatment of a variety of neuropsychiatric disorders [111]. Several species of the *Claviceps* genus, including *Claviceps* strain SD25 and *Claviceps fusiformis* 139/2/1G, metabolize clavicipitic acids, another type of nitrogenous heterocycles. Formally, this alkaloid can be considered to be constructed of indole and perhydroazepine rings. The natural product combines *cis*- and *trans*-

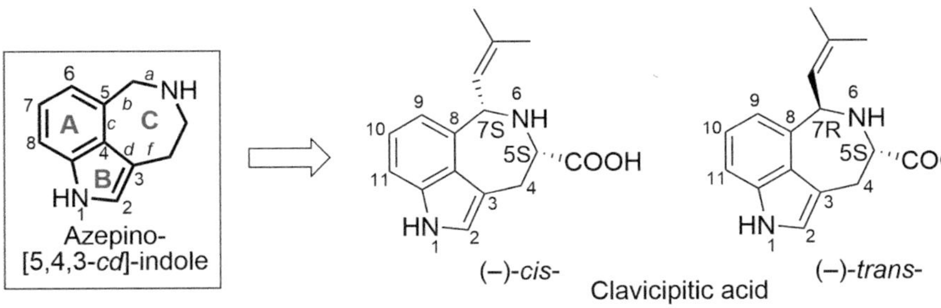

Figure 4.41. Tricyclic azepinoindole structure of clavicipitic acids and its stereochemistry.

diastereomers to the carboxy and 2-methylpropenyl groups (Figure 4.41), and their proportions depend on the specific microorganism from which it was isolated. This alkaloid with the unique tricyclic azepinoindole structure and the cyclic α-amino acid nature was isolated in 1969 by chemists J.E. Robbers and H.G. Floss [112]. Still, its tricyclic structure was revealed only in 1980, when the same authors, with the participation of J. Clardy, revised it using modern physical-chemical techniques (^{1}H NMR, single-crystal X-ray diffraction, high-resolution EM, etc.) [113]. However, 10 years later, Y. Jai and co-workers refined their configurations through intensive studies, including total synthesis, 2D-NMR study, X-ray diffraction analysis, and R_f value comparison [114].

Clavicipitic acid has high antibacterial activity. It was found that its synthetic analogs exhibit potent activity in the central nervous system. Although clavicipitic acid has the only tricyclic skeleton of 1*H*-azepino[5,4,3-*cd*]indole, which is not typical for alkaloids produced by *Claviceps* fungi, both types of alkaloids are biosynthesized from the same precursor, 4-(γ,γ-dimethylallyl)-L-tryptophan, as discussed above.

Due to its unique structure and biological properties, clavicipitic acid has received considerable attention from synthetic chemists over the past three decades, and many novel synthetic strategies have been developed, which were periodically reviewed [115–117]. Some key synthetic points could be noted: (i) functionalization of 4-substituted indole at the 3-position and functionalization of 3-substituted indole at the 4-position are usually used to prepare a needed 3,4-disubstituted indole (AB core); (ii) the construction of the azepinoindole skeleton (ABC skeleton) is realized at a later stage employing catalyzed intermolecular cyclizations.

For example, we examine Jia's synthesis, in which functionalization and indole construction are simultaneously made [118]. The linear synthesis starts with the initial reaction between 3-chloro-2-iodoaniline **4.117**, prepared from 1-chloro-2-iodo-3-nitrob enzene **4.116**, and chiral N-Boc-amino aldehyde **4.118**, and the subsequent N-acetylation of the obtained 3,4-disubstituted indole derivative **4.119** (Boc protection) allowed the formation of the N-Boc-tryptophan derivative **4.120** (Figure 4.42).

Next, a Pd-catalyzed Heck reaction of **4.120** with 2-methyl-3-buten-2-ol **4.121** proceeded efficiently in the presence of palladium acetate-tricyclohexylphosphine ($Pd(OAc)_2$-Cy_3P) catalyst to give alcohol **4.122**. Its cyclization in the presence of $Mg(ClO_4)_2$ in acetonitrile led to the construction of the crucial azepinoindole nucleus (tricyclic system ABC **4.123**) as a diastereoisomeric mixture (*cis-trans*, 5:1).

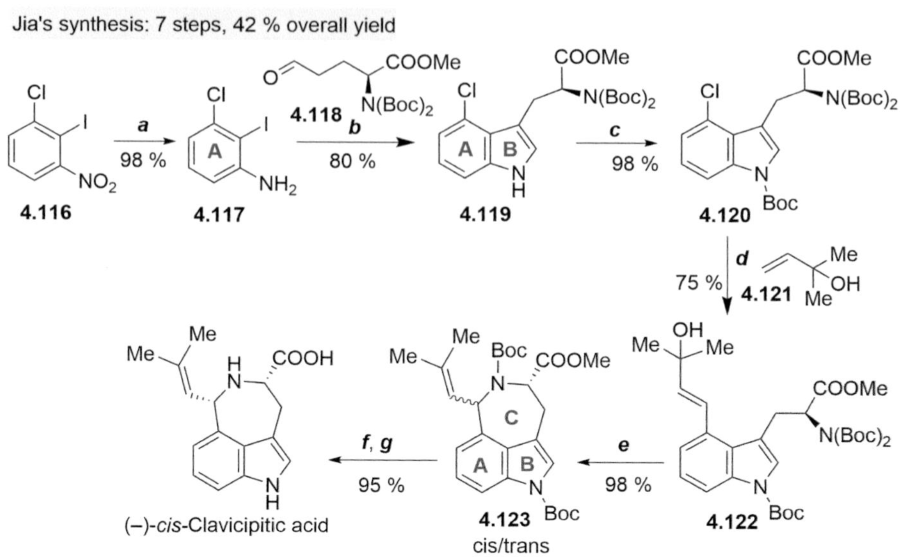

Figure 4.42. Jia's synthesis (2009): (a) $SnCl_2·2H_2O$, EtOH, 70°C; (b) $Pd(OAc)_2$, DMAP, MeCN, rt, 12 hours; (c) $Pd(OAc)_2$-Cy_3P, K_2CO_3, dioxane, 120°C, 12 hours; (d) $Mg(ClO_4)_2$, MeCN, reflux, 2 hours; (e) TMSOTf, 2,6-lutiduine, CH_2Cl_2, 0°C; (f) K_2CO_3, MeOH/H_2O, reflux, 2 d.

The Boc group on the azepine ring was deprotected by trimethylsilyl triflate (TMSOTf), while deprotection of Boc group on the indole ring and hydrolysis of methyl ester was completed simultaneously by using potassium carbonate to obtain optically active alkaloid. This synthesis was efficient: the total yield of (–)-*cis*-clavicipitic acid was 42%, performing 7 steps starting from 1-chloro-2-iodo-3-nitrobenzene [118].

4.2.4 Aflatoxins and Ochratoxins, Oxygenated Polycyclic Molecules from the Genera Aspergillus

The mycotoxins produced by the filamentous molds, especially *Aspergillus flavus* and *Aspergillus parasiticus* fungi, are called aflatoxins (AFTs). Aflatoxins, which show potent carcinogenic, mutagenic, estrogenic, tremorogenic, teratogenic, and immunosuppressive effects on the health of animals and humans, are present in food products. They, thus, are a grave threat to human health [119–121]. The discovery of AFTs began immediately after an epidemic of a disease of turkeys of unknown etiology (Turkey "X" disease) that first appeared in England in 1961 [122]. During extensive efforts to find the cause of this disease, it was elucidated that groundnut meal, commonly food for turkeys, was contaminated with a mold called *A. flavus*, which metabolizes hepatotoxic products. These result in particular polyoxygenated heterocyclic derivatives. Some of the biological importance are aflatoxin B1 (AFB1), aflatoxin M1 (AFM1), and aflatoxin G1 (AFG1), which are based on the tricyclic ABC fragment of dihydrofuro[2,3-*d*]benzo[*b*]furan. They are mainly responsible for the toxic action. The less toxic aflatoxins B2, G2, and M2 are analogs of the tetrahydrofuro[2,3-*b*]benzofuran skeleton (Figure 4.43).

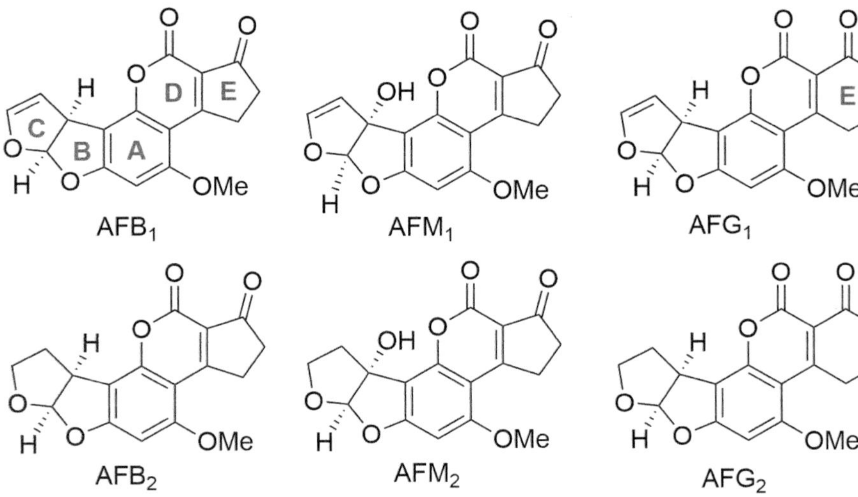

Figure 4.43. Structures of aflatoxins of biological importance.

AFTs are found in various cereals, oilseeds, spices, and nuts. Fungal contamination can occur in the field or during harvest, transport, and storage. AFTs can also infect humans after consumption of aflatoxins-contaminated foods such as eggs, meat and meat products, milk and milk products.

Currently, around 20 different types of AFTs are known. It is important to note that there is some specialization of the fungi: while the species *A. flavus* metabolizes only aflatoxins B1 and B2, *A. parasiticus* can secrete aflatoxins B and G. Aflatoxins M1 and M2 are hydroxylated products of aflatoxins B1 and B2, respectively, modified in the digestive tract of some animals and isolated from milk, urine, and feces. Some other species that produce aflatoxins are *A. nomius*, *A. pseudotamarii*, and *A. ochraceoroseus*.

AFTs are stable crystals ranging in color from white to yellow, odorless, tasteless, and colorless. These molecules are insoluble in water and soluble in conventional organic solvents. These organic solutions are stable when stored in the refrigerator and dark, but they are sensitive to light and decompose in air and oxygen. The abbreviations B and G of the main AFTs come precisely from this characteristic and mean the color of the fluorescence that emanates with UV light, B (Blue) or G (Green). Due to this property, AFTs are easily detectable in thin-layer chromatography. The abbreviations of two other familiar aflatoxins, M1 and M2, are metabolites of aflatoxins B1 and B2 found in milk of animals.

AFTs bind or adhere to nucleic acids, DNA, RNA, and proteins such as albumin, ovalbumin, and hemoglobin and can block protein synthesis, forming an irreversible complex (adducts) that leads to hepatotoxicity and liver cancer (hepatocellular carcinoma). In experimental animals, they are potent carcinogens, mutagens, and teratogens. Aflatoxin B1 is the most powerful and dangerous due to its carcinogenic action. The killing action of AFB1 has been established by tests with *Salmonella typhimurium* in the Ames test, which revealed that AFB1 requires activation to produce mutations. This metabolic activation consists of the transformation of AFB1 into its epoxide-8,9, a labile and reactive compound, which initiates the formation

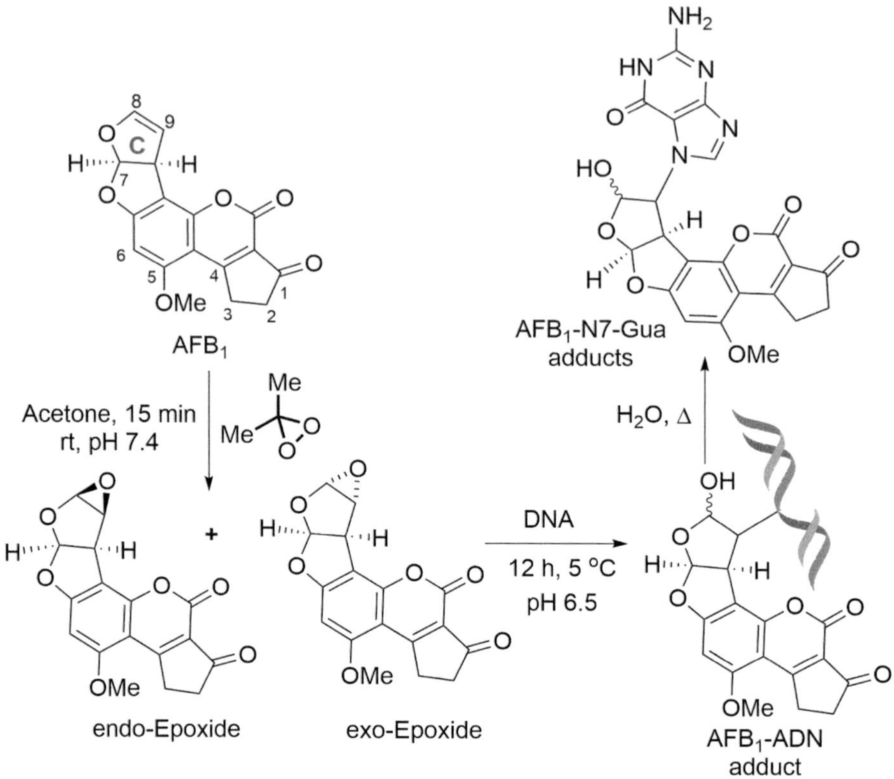

Figure 4.44. Chemical reactivity of AFB1 and its epoxides.

of adducts that cause cancer. The formation of this epoxide is conditioned by the presence of the dihydrofuro[2,3-*d*]benzo[*b*]furan fragment with a double bond between carbons C-8 and C-9. For this reason, aflatoxins B2 and G2, which do not contain this fragment, are much less toxic than aflatoxins B1 and G1. As the metabolite of AFB1, aflatoxin M1 is two orders of magnitude less toxic than AFB1, although it contains the C8=C9 double bond.

These conclusions were confirmed with the preparation of epoxide-8,9-AFB1, carried out by the group of Prof. T.M. Harris in 1988 [123–125]. The synthesis of this epoxide, which consisted of the oxidation of AFB1 by dimethyloxirane, allowed it to be characterized and study its chemical reactivity against various nucleophiles (H_2O, MeOH) and DNA (Figure 4.44). It was shown that (1) two *exo*- and *endo*-isomers of the epoxide are formed, but the *exo*-isomer is the majority; (2) The latter reacts with DNA regio- and stereo-specifically at the N7 position of deoxyguanosine (dGua), forming covalent adducts (AFB1-N7-Gua). Subsequently, this activation and toxic action model of AFTs was refined and completed.

In summary, it can be said that the toxic mode of action of AFTs is due to the presence of a particular double bond in their C-ring, the terminal dihydrofuranic cycle. This double bond is oxidized in the body, in the hepatic microsomal P450 system, through the 1,2 addition of an activated oxygen atom. The epoxides thus formed have high reactivity and intercalate in the DNA chains, forming cationic

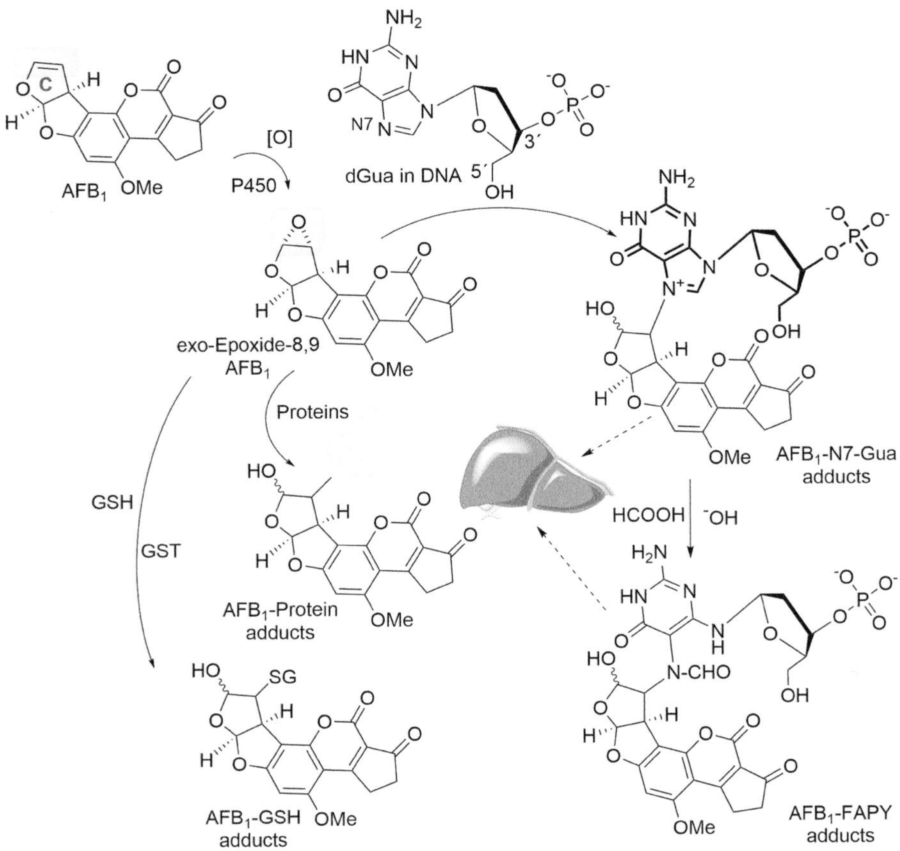

Figure 4.45. Toxic mode of action of AFTs: an example with AFB1.

adducts AFB1-N7-Gua (Figure 4.45). Their formation depends on the bases that flank the guanine; there is a low binding efficiency when the guanine is flanked by sequences rich in adenine (A)-thymine (T), while when it is flanked by another guanine, the binding efficiency ligature increases. These unstable adducts give rise to another product, AFB1-FAPY, which is mainly responsible for genotoxicity.

According to NMR studies, this product presents α- and β-anomeric forms caused by deoxyribose's interconversion. It is known that in duplex DNA, only the β-anomer is found. In contrast, in single-stranded DNA, there is an equilibrium mixture of the α/β anomers with the ratio 2:1. These adducts accumulate for years and cause mutations in the p53 gene and are frequently associated with neoplasms that tend to be more invasive and often cause metastasis. They can also react with glutathione (GSH) through a mechanism mediated by a glutathione-S-transferase (GST), representing the detoxification step of AFTs. Furthermore, exo-epoxide-AFB1 can form adducts with proteins, primarily with albumin, whose lysine residues favor the formation of these adducts. We remember that albumin is an abundant protein in the blood recognized as a meter of ingested dose and that it is synthesized in hepatocytes. On this site, many carcinogens are metabolized to their reactive forms.

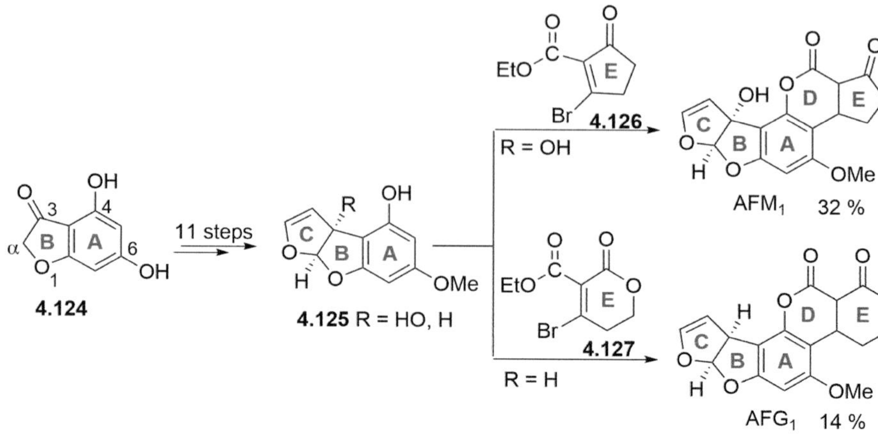

Figure 4.46. First syntheses of (±)-aflatoxins according to the Büchi group.

Aflatoxins do not have toxic properties *in vitro*. Still, *in vivo*, they transform into dangerous poison when they encounter cytochrome P450, which is located in the mitochondria of the adrenal cortex and liver microsomes. This process is called bioactivation. AFTs also act as immunosuppressive molecules and, indirectly, through immunosuppression, can impair reproduction in mammals.

The Büchi group proposed the first synthesis of the aflatoxins AFM1 and AFG1 in the late 1960s and early 1970s [126]. 4,6-Dihydroxybenzo[*b*]furan-3one **4.124** was chosen as the starting product. Through a long path and 11 steps, its transformation into the tricyclic derivatives of 4-hydroxy-6-methoxydihydrofuro[2,3-b]benzofuran **4.125** was achieved. Its condensation with ethyl 2-bromo-5-oxocyclopent-1-ene-1-carboxylate **4.126** in the presence of zinc and magnesium carbonates in dichloromethane formed AFM1 (Figure 4.46).

According to Pechmann, the last stage of this complex synthesis is modifying the well-known method for obtaining coumarins. In the same way, the toxin AFG1 is obtained from dihydrofuro[2,3-*b*]benzofuran. In this case, instead of the cyclopentene derivative **4.126**, they used ethyl 4-bromo-2-oxo-5,6-dihydro-2*H*-pyran-3-carboxylate **4.127**. The cyclocondensation was carried out in the presence of zinc carbonate and lithium iodide. Synthesis of heterocyclic structures **4.125** is crucial in obtaining these mycotoxins.

The median lethal dose of AFB1 0.36 mg/kg body weight is an exceptional range of highly toxic poison (aflatoxin animal half of the lethal dose is found in the most potent carcinogens). Its carcinogenicity is 900 times more than dimethyl nitrosamine-induced liver cancer in the large capacity, 75 times higher than the 3,4-benzopyrene, a large 4,000-fold. The fungi *Aspergillus ochraceus*, *A. westerdijkiae*, and *A. carbonarius* produce other toxic heterocyclic compounds that are less complex than AFTs. These mycotoxins contain the dihydro-isocoumarin (benzo[*c*]pyranone) structure and are called ochratoxins (OTs) A, B, and C (Figure 4.47). OTs are also found in *Penicillium* fungi.

Among three molecules characterized as amides of isocoumarinic acid and L-phenylalanine, ochratoxin A (OTA, 7-carboxyl-5-chloro-8-hydroxyl-3,4-dihydro-

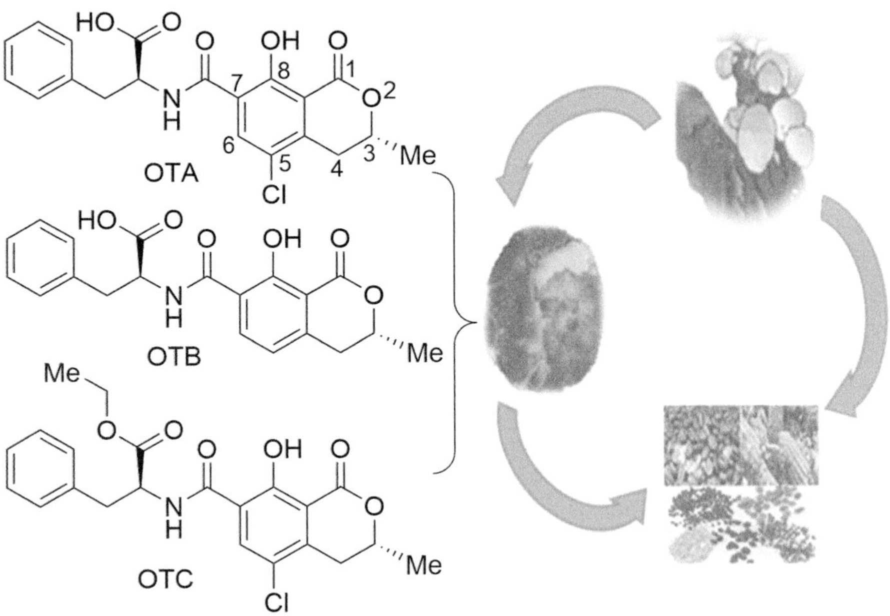

Figure 4.47. Chemical structures of ochratoxins, main mycotoxins of *A. ochraceus* fungi.

3*R*-methyl-isocoumarin-7-L-β-phenylalanine), isolated in 1965, is the most toxic [127]. According to the toxic action, OTA is similar to AFTs. *In vitro* and animal studies, it is genotoxic, carcinogenic, teratogenic, nephrotoxic, and neurotoxic and thereby is a potential risk to human health [128–130].

The mold *A. ochraceus*, the main producer of OTA, frequently infects cereals (corn, wheat, barley, rye, sorghum, and rice) and their processed products. It can also contaminate green coffee beans, meat and its derivatives, grapes, wine, and raisins, among other foods [130].

OTA is a stable white crystalline solid with poor aqueous solubility. Under UV light, this molecule has intense green fluorescence in an acidic medium and blue fluorescence in alkaline conditions. Having two acidic groups (the carboxyl group, pK_a 4.2–4.4, and the phenolic hydroxyl group, pKa 7.0–7.3) in its structure, at physiological pH, OTA is present in monoanionic and anionic forms. For this reason, in the human circulatory system, OTA is almost completely bound to albumin, which considerably affects its toxic-kinetic properties. However, OTA is an inhibitor of protein synthesis in both *in vivo* and *in vitro* models [128].

This mycotoxin blocks the Phe-tRNA synthetase enzyme that encodes phenylalanine. Noting that OTA is the amide derived from L-phenylalanine, it was first hypothesized that the phenylalanine fragment in OTA plays an important role as a competitor between the amino acid phenylalanine and the mycotoxin. However, other studies showed that the dihydro-isocoumarin structure is more important in this tRNA synthetase-OTA interaction. OTA also acts as an inhibitor of phenylalanine hydroxylase, which catalyzes the conversion of the amino acid phenylalanine to tyrosine.

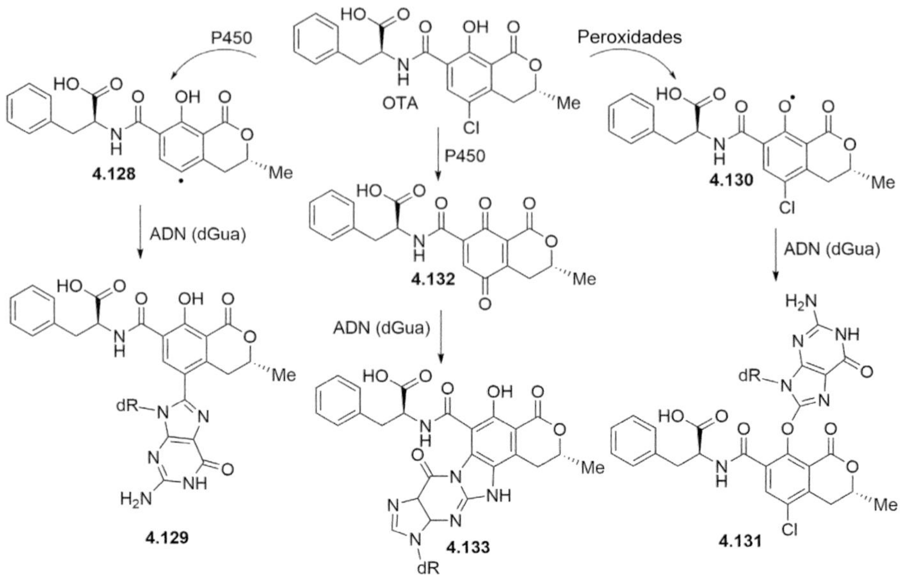

Figure 4.48. Chemical structures of OTA-nucleoside adducts.

Like ATB1 in an animal organism, OTA is activated by cytochrome P450 and peroxidases, forming highly reactive electrophilic products that can covalently bind to DNA, causing mutations and subsequent formation of malignant tumors. Due to the different chemical nature of these products, OTA adducts are different from AFT-DNA adducts.

Despite many *in vitro* and *in vivo* experimental data obtained by isotopic techniques and by LC-MS analysis, the molecular details of these interactions are still under discussion. Some scientists believe that of the three possible adducts, the formation of the C-OTA-dGua adduct **4.129**, which is linked by the C_{5OTA}-C_{8Gua} bond, is most likely. Still, other researchers seriously question the presence of this adduct that comes via the formation of aryl radical **4.128**. The O-OTA-dGua adduct **4.131** proceeds by forming phenoxy radicals **4.130**, while the third possible OTHQ-dGua adduct **4.133** comes after quinone derivative **4.132** (Figure 4.48) [128].

The acute toxicity of OTA in animals demonstrates interspecific variations. The LD_{50} orally ranges between 20–50 mg/kg in rats and mice and 0.2–1.0 mg/kg in dogs, pigs, and chickens, the most sensitive species.

Although there is abundant information on the various toxigenic properties of OTA, its biosynthetic pathway is not completely established. However, both parts of the OTA structure, the dihydro-isocoumarin ring, and the L-phenylalanine moiety, are known to be derived from the polyacetate and shikimate pathways, respectively. It means that we have an example of a biohybrid molecule. Apparently, (*R*)-7-methylmellein **4.134** (3,4-dihydro-3,7-dimethyl-8-hydroxyisocoumarin, typical representative of important PKS-derived fungal metabolites) is an initial cyclic intermediate that is oxidized to β-OT by cytochrome P450 monooxygenase (otaC). Gene encoding NRPS (otaB) catalyzes the formation of an amide bond between β-OT and L-Phen, leading to the biosynthesis of OTB. Then, OTB is chlorinated by

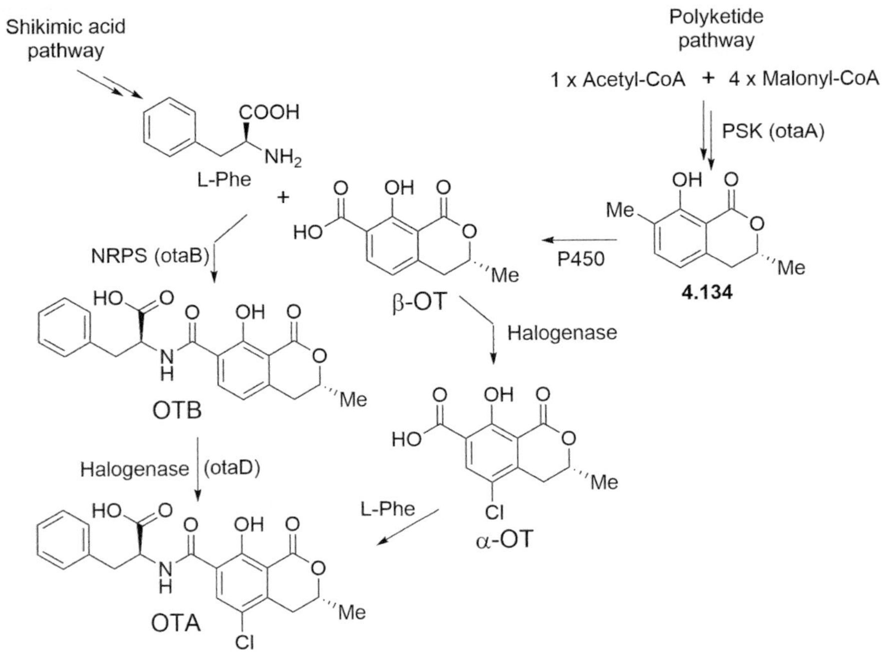

Figure 4.49. Schematic representation of the pathway of OTA biosynthesis.

halogenase (otaD) to produce OTA. Chlorination of β-OT under the same conditions can lead to the formation of α-OT, which may serve as a bioprecursor of OTA (Figure 4.49) [131].

Chemical methods for the total synthesis of OTA usually consist of the α-OT ring construction. Numerous syntheses of α-OT have been published, beginning with Steyn and Holzapfel in 1967 (10% yield in nine steps) [132], followed by Kraus et al. (20% yield in four steps) [133], Snieckus et al. (6% yield in five steps) [134], Gabriele et al. (19% yield in five steps) [135], and finally Cramer et al. (8% yield in five steps) [136]. All racemic syntheses conduce to the formation of the mixture of (R/S)-α-OT and require enantiomeric separation in this step or posterior separation of the two OTA diastereoisomers formed by the coupling of (R/S)-α-OT with L-phenylalanine. The diastereoisomeric separation usually is done on a small scale by high-performance liquid chromatography (HPLC) or preparative thin-layer chromatography (PTLC).As a good example, OTA synthesis by Gabriele et al. (Figure 4.50) is discussed here. Starting materials were dimethyl 3-oxopentanedioate **4.135** and but-2-ynal **4.136**, which were condensed to give dimethyl 2-hydroxy-4-methylbenzene-1,3-dicarboxylate **4.137** in 48% yield. Its treatment with lithium diisopropylamide at –78°C in THF and the addition of acetaldehyde and acidic workup provided lactone derivative **4.138**, which was chlorinated with sulfuryl chloride in dichloromethane at room temperature, followed by hydrolysis of the ester group with lithium hydroxide in methanol finally gave α-OT in 69% yield. Its condensation with protected L-phenylalanine **4.139** in the presence of EDC·HCl and BtOH as coupling agents gave a mixture of protected diastereomers **4.140**. Deprotection of this crude mixture with trifluoroacetic acid led to a mixture of OTA, which was successfully separated by PTLC (Figure 4.50).

Interestingly, following this tactic, Cramer and co-workers obtained all four stereoisomers of ochratoxin A, derived from the racemic α-OT and L- or D-phenylalanine derivatives, and studied the cytotoxicity and apoptotic effects observed in a cultured human liver cell line (Hep G2) [136] (Figure 4.51).

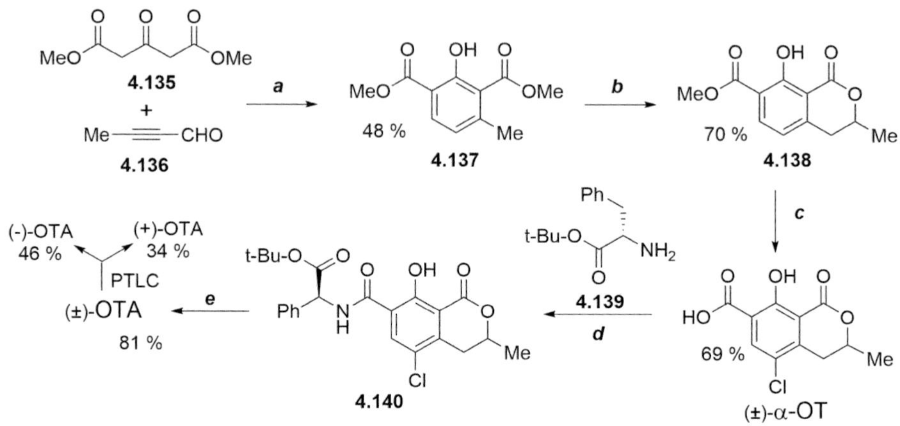

Figure 4.50. Preparation of racemic OTA and separation of the diastereomeric mixture by PTLC: (a) NaH, THF, –10°C, 12 hours; (b) 1. LDA, –75°C, 30 minutes, 2. MeCHO, 3. AcOH, H_2O; (c) 1. SO_2Cl_2, dry CH_2Cl_2, 24 hours, 2. LiOH, MeOH, reflux, 5 hours, 3. 1N aq HCl; (d) 1. EDC·HCl, BtOH, $CHCl_3$, rt, 20 hours, 2. 1N HCl-H_2O; (e) TFA, CH_2Cl_2, rt, 6 hours. EDC·HCl = N-ethyl-N′-(3-dimethylaminopropyl) carbodiimide hydrochloride, BtOH = N-hydroxybenzotriazole.

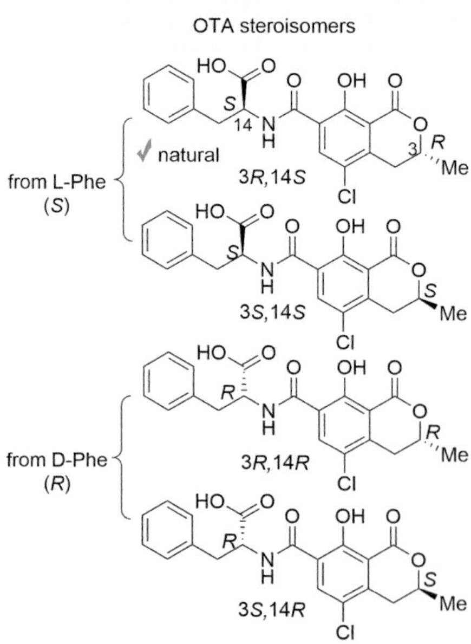

OTA steroisomers	Cytotoxicity IC_{50}, μM	Cytotoxicity ED_{50}, μM	Apoptotic potential ED_{50}, μM
3R,14S (natural)	0.3	0.06	0.32
3S,14S	0.4	0.08	0.26
3R,14R	> 10	> 10	> 10
3S,14R	> 10	3.8	> 10

Figure 4.51. Chemical structures of the ochratoxin A stereoisomers and their cytotoxicity and apoptosis data.

They found that the configuration of the phenylalanine moiety (*S* or *R*) of OTA is responsible for the high cytotoxicity and apoptotic activity of OTA. Namely, only two stereoisomers, having the L-configuration (*S*) at C-14 (from the phenylalanine moiety), displayed a very high cytotoxic potential in all assays applied to determine the toxicity in the *in vitro* model of Hep G2 cells. That indicates that the specificity of enzymes could be the critical basis for the OTA-induced cytotoxicity, and the configuration at C-3 has little importance in toxic effects [136].

To conclude this section, it can be noted that apart from causing health issues, exposure to AFTs and OTs results in enormous economic losses by decreasing the productivity of livestock and farm crops and increasing the medical fees of humans and animals and the mortality rate of animals.

4.3 Conclusion

To summarize this chapter, we must note that an attempt was made to categorize molecules from fungal metabolism, grouped according to the structural and bio-origin aspects; we should remember that the members of the kingdom fungi have particular characteristics that make them different from plants, such as they do not make their own food through photosynthesis like them, but instead they live at the expense of other organisms, living or dead. So, to live in this way, fungi metabolize different molecules, which are functional for their survival.

When fungi attack plants, the latter suffer biological, chemical, or physical stress that induces biochemical and physiological changes in plants. So, in response to fungal infection, plants defend by generating particular metabolites, phytoalexins, that play an essential role in interspecific relationships with plants. Phytoalexins, "invisible" protective molecules of plants, are abundant compounds considered molecular markers of disease resistance. Structurally diverse but united with the unique aim of stopping pathogen invasion, phytoalexins of Vitaceae, Fabaceae, and Cruciferae plants and oat grains were examined, analyzing their biogenesis and chemical synthesis. Among them, phytoalexins of grapes based on monomeric stilbenes, resveratrol, and pterostilbene stand out as potential pharmacological agents in the treatment of neurodegenerative disorders like Parkinson's, Alzheimer's, and Huntington's diseases, amyotrophic lateral sclerosis. Resveratrol is also a promising agent for cancer prevention and treatment, especially prostate cancer. Phytoalexins from Cruciferae plants containing indole and thiazole rings also exhibited anticancer activity.

Fungi also differ from animals because they cannot move or move on the medium or surface in which they grow, but they can interact with them via metabolism. Thus, to live in this way, fungi metabolize different molecules that are functional for their survival; among these are toxic molecules, phytotoxins, and/or mycotoxins. This is the flip side of interspecific plant-fungal relationships. Phytotoxins are the pathogenic fungi's response to the plants that develop toxic substances. They are principally low-molecular-weight secondary metabolites capable of disrupting the vital activity of plant cells or causing their death. Besides the term phytotoxins, more correctly, fungal phytotoxins, mycotoxins are generally used, especially in modern times. The latter substances are also low-molecular-weight fungal metabolites, toxic to vertebrates, exerting severe adverse effects on the health of farm animals and humans who eat contaminated agricultural products. Thus, many fungal metabolites are phytotoxic (toxic to plants) and mycotoxic (toxic to animals). Nevertheless, the limited division between these last groups is elusive. It is noteworthy that among such molecules, there were found highly toxic compounds as potentially harmful food contaminants, e.g., "alternaria toxins", based on the dibenzo-α-pyrone and 4-hydroxy-3-pyrrolin-2-one skeletons from a black mold of the genus *Alternaria*; ophiobolins, phytotoxic sesterterpenes having dicyclopenta[*ad*]cyclooctane system; cytochalasins based on the perhydro-isoindolone ring linked with a macro-lactone system, or aflatoxins, pentacyclic molecules with a characteristic dihydrofuro[2,3-*d*]benzo[*b*]furan ring and drug candidates, e.g., ergot alkaloids possessing the indole[4,3-*fg*]quinoline skeleton, such as nicergoline, bromocriptine and its semi-synthetic analogs pergoline and carbergolide, or methylergonovine, a synthetic ergonovine analog.

Like plants, animals defend themselves against intruder attacks using many methods and tactics, including chemical tactics. All animals are divided into those with a spine (vertebrates) and those without (invertebrates). How do invertebrates and vertebrates use their toxins to survive? It can be seen in subsequent chapters.

4.4 References

[1] Ward, H.M. 1905. Recent research on the parasitism of fungi. Ann. Botan. 19(73): 1–54.

[2] Cobb, N. 1894. Contributions to an economic knowledge of *Australian rusts* (Uredineae). Agric Gaz N S Wales 5(2): 239–251.

[3] Pagán, I. and García-Arenal, F. 2018. Tolerance to plant pathogens: Theory and experimental evidence. Inter J. Mol. Sci. 19(3): 810.

[4] Wu, S., Shan, L. and He, P. 2014. Microbial signature-triggered plant defense responses and early signaling mechanisms. Plant Sci. 228: 118–126.

[5] Oros, G. and Kállai, Z. Phytoanticipins. 2019. The constitutive defense compounds as potential botanical fungicides. pp. 179–229. *In*: Jogaiah S. and Abdelrahman M. [Eds.]. Bioactive Molecules in Plant Defense. Springer, Cham.

[6] Pedras, M.S.C. and Yaya, E.E. 2015. Plant chemical defenses: are all constitutive antimicrobial metabolites phytoanticipins? Nat. Prod. Comm. 10(1): 209–218.

[7] Müller, K.O. and Börger, H. 1940. Experimentelle untersuchungen über die *Phythophthora*-resistenz der Kartoffel. Zugleich ein Beitrag zum Problem der 'erworbenen Resistenz' im Pflanzenreich. Arb. Biol. Reich Land-Forstwirtschaft; 23: 189–223.

[8] Ahuja, I., Kissen, R. and Bones, A.M. 2012. Phytoalexins in defense against pathogens. Trends Plant Sci. 17(2): 73–90.

[9] Chan, E.W.C., Wong, C.W., Tan, Y.H., Foo, J.P.Y., Wong, S.K. and Chan, H.T. 2019. Resveratrol and pterostilbene: A comparative overview of their chemistry, biosynthesis, plant sources and pharmacological properties. J. Appl. Pharm. Sci. 9(7): 124–129.

[10] Shrestha, A., Pandey, R.P. and Sohng, J.K. 2019. Biosynthesis of resveratrol and piceatannol in engineered microbial strains: achievements and perspectives. Appl. Microbiol. Biot. 103(7): 2959–2972.

[11] Huang, H., Liu, R. and Ou, W. 2020. A mini review on the chemical synthesis of resveratrol. Mini-Rev. Org. Chem. 17(5): 546–558.

[12] Khan, Z.A., Iqbal, A. and Shahzad, S.A. 2017. Synthetic approaches toward stilbenes and their related structures. Mol. Divers 21(2): 483–509.

[13] Richard, F. Heck (1931–2015), an American chemist, winner of the Nobel Prize in Chemistry in 2010.

[14] Jeffery, T. and Ferber, B. 2003. One-pot palladium-catalyzed highly chemo-, regio-, and stereoselective synthesis of trans-stilbene derivatives. A concise and convenient synthesis of resveratrol. Tetrahedron Lett. 44(1): 193–197.

[15] Goel, A., Kumar, A. and Raghuvanshi, A. 2013. Synthesis, stereochemistry, structural classification, and chemical reactivity of natural pterocarpans. Chem. Rev. 113(3): 1614–1640.

[16] Selvam, C., Jordan, B.C., Prakash, S., Mutisya, D. and Thilagavathi, R. 2017. Pterocarpan scaffold: A natural lead molecule with diverse pharmacological properties. Eur. J. Med. Chem. 128: 219–236.

[17] Jiménez-González, L., Álvarez-Corral, M., Muñoz-Dorado, M. and Rodríguez-García, I. 2008. Pterocarpans: interesting natural products with antifungal activity and other biological properties. Phytochem. Rev. 7(1): 125–154.

[18] Morimoto, M., Fukumoto, H., Hiratani, M., Chavasiri, W. and Komai, K. 2006. Insect antifeedants, pterocarpans and pterocarpol, in heartwood of Pterocarpus macrocarpus Kruz. Biosci. Biotechnol. Biochem. 70(8): 1864–1868.

[19] Jiménez-González, L., Hernández-Cervantes, C., Álvarez-Corral, M., Muñoz-Dorado, M. and Rodríguez-García, I. 2011. Synthesis of pterocarpans. Nat. Prod. Commun. 6(4): 537–554.

[20] Horino, H. and Inoue, N. 1976. A new route to chromanocoumarans. Synthesis of (±)-pterocarpin. J. Chem. Soc. Chem. Commun. pp. 500–501.

[21] Larock, R.C. Synthesis of substituted pterocarpans. U.S. Patent No. 5,721,371. February 24, 1998.

[22] Yang, X., Zhao, Y., Hsieh, M.T., Xin, G., Wu, R.T., Hsu, P.L. et al. 2017. Total Synthesis of (+)-Medicarpin. J. Nat. Prod. 80(12): 3284–3288.

[23] Pedras, M.S.C., Okanga, F.I., Zaharia, I.L. and Khan, A.Q. 2000. Phytoalexins from crucifers: synthesis, biosynthesis, and biotransformation. Phytochemistry 53(2): 161–176.

[24] Ahuja, I., Kissen, R. and Bones, A.M. 2012. Phytoalexins in defense against pathogens. Trends Plant Sci. 17(2): 73–90.

[25] Takasugi, M., Katsui, N. and Shirata, A. 1986. Isolation of three novel sulphur-containing phytoalexins from the Chinese cabbage *Brassica campestris* L. ssp. *pekinensis* (cruciferae). J. Chem. Soc., Chem. Commun. (14): 1077–1078.

[26] Sabol, M., Kutschy, P., Siegfried, L., Miroššay, A., Suchý, M., Hrbková, H. et al. 2000. Cytotoxic effect of cruciferous phytoalexins against murine L1210 leukemia and B16 melanoma. Biologia (Bratislava) 55(6): 701–707.

[27] Chripkova, M., Zigo, F. and Mojzis, J. 2016. Antiproliferative effect of indole phytoalexins. Molecules 21(12): 1626.

[28] Mezencev, R., Galizzi, M., Kutschy, P. and Docampo, R. 2009. *Trypanosoma cruzi*: antiproliferative effect of indole phytoalexins on intracellular amastigotes *in vitro*. Exp. Parasitol. 122(1): 66–69.

[29] Kutschy, P., Dzurilla, M., Takasugi, M., Török, M., Achbergerová, I., Homzová, R. et al. 1998. New syntheses of indole phytoalexins and related compounds. Tetrahedron 54(14): 3549–3566.

[30] Glawischnig, E., Hansen, B.G., Olsen, C.E. and Halkier, B.A. 2004. Camalexin is synthesized from indole-3-acetaldoxime, a key branching point between primary and secondary metabolism in Arabidopsis. Proc. Natl. Acad. Sci. USA 101(21): 8245–8250.

[31] Pedras, M.S. and Okinyo, D.P. 2008. Remarkable incorporation of the first sulfur-containing indole derivative: Another piece in the biosynthetic puzzle of crucifer phytoalexins. Org. Biomol. Chem. 6(1): 51–54.

[32] Klein, A.P. and Sattely, E.S. 2015. wo cytochromes P450 catalyze S-heterocyclizations in cabbage phytoalexin biosynthesis. Nat. Chem. Biol. 11(11): 837–839.

[33] Dai, H.G., Li, J.T. and Li, T.S. 2006. Efficient and practical synthesis of Mannich bases related to gramine mediated by zinc chloride. Synth. Commun. 36(13): 1829–1835.

[34] Yamada, F., Kobayashi, K., Shimizu, A., Aoki, N. and Somei, M. 1993. A synthesis method of indole-3-Methanamine and/or Gramine from Indole-3-Carboxaldehyde, and Its Application for the Synthesis of Brassinin, Its 4-Substituted Analogues, and 1,3,4,5-Tetrahydropyrrolo [4,3,2-de] quinoline. Heterocycles 36(12): 2783–2804.

[35] Browne, L.M., Conn, K.L., Ayer, W.A. and Tewari, J.P. 1991. The camalexins: new phytoalexins produced in the leaves of *Camelina sativa* (Cruciferae). Tetrahedron 47(24): 3909–3914.

[36] Ayer, W.A., Craw, P.A., Ma, Y.T. and Miao, S. 1992. Synthesis of camalexin and related phytoalexins. Tetrahedron 48(14): 2919–2924.

[37] Moody, C.J., Roffey, J.R., Stephens, M.A. and Stratford, I.J. 1997. Synthesis and cytotoxic activity of indolyl thiazoles. Anti-Cancer Drugs 8(5): 489–499.

[38] Mezencev, R., Updegrove, T., Kutschy, P., Repovská, M. and McDonald, J.F. 2011. Camalexin induces apoptosis in T-leukemia Jurkat cells by increased concentration of reactive oxygen species and activation of caspase-8 and caspase-9. J. Nat. Med. 65: 488–499.

[39] Dzurilla, M., Ružinský, M., Kutschy, P., Tewari, J.P. and Kováčik, V. 1999. Application of 2-substituted ethyl isothiocyanates and 2-aminothiols in the synthesis of the analogs of indole phytoalexin camalexin. Collect. Czechoslov. Chem. Commun. 64(9): 1448–1456.

[40] Pilatova, M., Ivanova, L., Kutschy, P., Varinska, L., Saxunova, L., Repovska, M. et al. 2013. *In vitro* toxicity of camalexin derivatives in human cancer and non-cancer cells. Toxicol. *In Vitro* 27(2): 939–944.

[41] Takasugi, M., Monde, K., Katsui, N. and Shirata, A. 1987. Spirobrassinin, a Novel Sulfur-containing Phytoalexin from the Daikon *Raphanus sativusim* L. var. hortensis (Cruciferae). Chem. Lett. 16(8): 1631–1632.

[42] Suchý, M., Kutschy, P., Monde, K., Goto, H., Harada, N., Takasugi, M. et al. 2001. Synthesis, absolute configuration, and enantiomeric enrichment of a cruciferous oxindole phytoalexin, (S)-(−)-spirobrassinin, and its oxazoline analog. J. Org. Chem. 66(11): 3940–3947.

[43] Liu, L., Zhang, S., Xue, F., Lou, G., Zhang, H., Ma, S. et al. 2011. Catalytic enantioselective henry reactions of isatins: application in the concise synthesis of (*S*)-(−)-Spirobrassinin. Chem. Eur. J. 17(28): 7791–7795.

[44] Kutschy, P., Salayová, A., Čurillová, Z., Kožár, T., Mezencev, R., Mojžiš, J. et al. 2-(Substituted phenyl) amino analogs of 1-methoxyspirobrassinol methyl ether: synthesis and anticancer activity. Bioorg. Med. Chem. 17(10): 3698–3712.

[45] Budovská, M., Tischlerová, V., Mojžiš, J., Harvanová, M., Kozlov, O., Gondová, T. et al. 2017. 2′-Aminoanalogues of the cruciferous phytoalexins spirobrassinin, 1-methoxy-spirobrassinin and 1-methoxyspirobrassinol methyl ether: synthesis and anticancer properties. Tetrahedron 73(44): 6356–6371.

[46] Pedras, M.S.C. and Zaharia, I.L. 2001. Unprecedented Vilsmeier formylation: expedient syntheses of the cruciferous phytoalexins sinalexin and brassilexin and discovery of a new heteroaromatic ring system. Org. Lett. 3(8): 1213–1216.

[47] Suchý, M., Kutschy, P., Dzurilla, M., Kováčik, V., Andreani, A. and Alföldi, J. 2001. 1,3-Thiazino [6,5-*b*]indol-4-one derivatives. The first synthesis of indole phytoalexin cyclobrassinon. Tetrahedron Lett. 42(39): 6961–6963.

[48] Kutschy, P., Suchý, M., Andreani, A., Dzurilla, M., Kováčik, V., Alföldi, J. et al. 2002. A new approach to the synthesis of rare thiazino[6,5-*b*]indol-4-one derivatives. First total synthesis of the indole phytoalexin cyclobrassinon. Tetrahedron 58(44): 9029–9039.

[49] Takasugi, M., Monde, K., Katsui, N. and Shirata, A. 1988. Novel sulfur-containing phytoalexins from the Chinese cabbage *Brassica campestris* L. ssp. *pekinensis* (Cruciferae). Bull Chem. Soc. Jpn. 61(1): 285–289.

[50] Budovská, M., Kudličková, Z., Kutschy, P., Pilátová, M. and Mojžiš, J. 2015. A facile method for the synthesis of indole phytoalexin rutalexin. Tetrahedron Lett. 56(25): 3945–3947.

[51] Miyagawa, H., Ishihara, A., Nishimoto, Ueno, T. and Mayama, S. 1995. Induction of avenanthramides in oat leaves inoculated with crown rust fungus, *Puccinia coronata* f. sp. *avenae*. Biosci. Biotechnol. Biochem. 59(12): 2305–2306.

[52] Collins, F.W. and Mullin, W.J. 1988. High-performance liquid chromatographic determination of avenanthramides, N-aroylanthranilic acid alkaloids from oats. J. Chromatogr. A 445: 363–370.

[53] Collins, F.W. 1989. Oat phenolics: avenanthramides, novel substituted N-cinnamoylanthranilate alkaloids from oat groats and hulls. J. Agric. Food Chem. 37(1): 60–66.

[54] Peterson, D.M., Hahn, M.J. and Emmons, C.L. 2002. Oat avenanthramides exhibit antioxidant activities *in vitro*. Food Chem. 79(4): 473–478.

[55] Bratt, K., Sunnerheim, K., Bryngelsson, S., Fagerlund, A., Engman, L., Andersson, R.E. et al. 2003. Avenanthramides in oats (*Avena sativa* L.) and structure−antioxidant activity relationships. J. Agric. Food Chem. 51(3): 594–600.

[56] Fagerlund, A., Sunnerheim, K. and Dimberg, L.H. 2009. Radical-scavenging and antioxidant activity of avenanthramides. Food Chem. 113(2): 550–556.

[57] Yang, J., Ou, B., Wise, M.L. and Chu, Y. 2014. *In vitro* total antioxidant capacity and anti-inflammatory activity of three common oat-derived avenanthramides. Food Chem. 160: 338–345.

[58] Turrini, E., Maffei, F., Milelli, A., Calcabrini, C. and Fimognari, C. 2019. Overview of the anticancer profile of avenanthramides from oat. Int. J. Mol. Sci. 20(18): 4536.

[59] Wise, M.L., Sreenath, H.K., Skadsen, R.W. and Kaeppler, H.F. 2009. Biosynthesis of avenanthramides in suspension cultures of oat (*Avena sativa*). Plant Cell, Tissue and Organ Culture (PCTOC) 97: 81–90.

[60] Li, Z., Chen, Y., Meesapyodsuk, D. and Qiu, X. 2019. The biosynthetic pathway of major avenanthramides in oat. Metabolites 9(8): 163.

[61] Pretorius, C.J. and Dubery, I.A. 2023. Avenanthramides, distinctive hydroxycinnamoyl conjugates of oat, *Avena sativa* L.: an update on the biosynthesis, Chemistry, and Bioactivities. Plants 12(6): 1388.

[62] Eckroth, D.R. 1972. Benzoxazinone synthesis. An organic chemistry experiment. J. Chem. Educ. 49(1): 66.

[63] Mayama, S., Tani, T., Ueno, T., Hirabayashi, K., Nakashima, T., Fukami, H. et al. 1981. Isolation and structure elucidation of genuine oat phytoalexin, avenalumin I. Tetrahedron. Lett. 22(22): 2103–2106.

[64] Eudes, A., Baidoo, E.E., Yang, F., Burd, H., Hadi, M.Z., Collins, F.W. et al. 2011. Production of tranilast [N-(3′,4′-dimethoxycinnamoyl)-anthranilic acid] and its analogs in yeast *Saccharomyces cerevisiae*. Adv. Appl. Microbiol. 89: 989–1000.

[65] Lee, S.J., Sim, G.Y., Kang, H., Yeo, W.S., Kim, B.G. and Ahn, J.H. 2018. Synthesis of avenanthramides using engineered *Escherichia coli*. Microbial Cell Factories 17(1): 1–12.

[66] Keller, N.P. 2019. Fungal secondary metabolism: regulation, function and drug discovery. Nat. Rev. Microbiol. 17(3): 167–180.

[67] Raistrick, H., Stickings, C.E. and Thomas, R. 1953. Studies in the biochemistry of microorganisms. 90. Alternariol and alternariol monomethyl ether, metabolic products of *Alternaria tenuis*. Biochem. J. 55(3): 421–433.

[68] Wang, H., Guo, Y., Luo, Z., Gao, L., Li, R., Zhang, Y. et al. 2022. Recent advances in Alternaria phytotoxins: a review of their occurrence, structure, bioactivity, and biosynthesis. J. Fungi 8(2): 168.

[69] Pero, R.W., Posner, H., Blois, M., Harvan, D. and Spalding, J.W. 1973. Toxicity of metabolites produced by the "Alternaria". Environ Health Perspect. 4: 87–94.

[70] Bensassi, F., Gallerne, C., el Dein, O.S., Hajlaoui, M.R., Bacha, H. and Lemaire, C. 2012. Cell death induced by the *Alternaria* mycotoxin alternariol. Toxicol *In Vitro* 26(6): 915–923.

[71] Fernandez-Blanco, C., Juan-Garcia, A., Juan, C., Font, G. and Ruiz, M.J. 2016. Alternariol induce toxicity via cell death and mitochondrial damage on Caco-2 cells. Food Chem. Toxicol. 88: 32–39.

[72] Aly, A.H., Edrada-Ebel, R., Indriani, I.D., Wray, V., Müller, W.E.G., Totzke, F. et al. 2008. Cytotoxic metabolites from the fungal endophyte *Alternaria* sp. and their subsequent detection in Its host plant polygonum senegalense. Nat. Prod. 71: 972–998.

[73] Aichinger, G., Del Favero, G., Warth, B. and Marko, D. 2021. Alternaria toxins-Still emerging? Compr. Rev. Food Sci. Food Saf. 20(5): 4390–4406.

[74] Schäberle, T.F. 2016. Biosynthesis of α-pyrones. Beilstein. J. Org. Chem. 12(1): 571–588.

[75] Koch, K., Podlech, J., Pfeiffer, E. and Metzler, M. 2005. Total synthesis of alternariol. J. Org. Chem. 70(8): 3275–3276.

[76] Won, M., Kwon, S. and Kim, T.H. 2015. An efficient synthesis of alternariol. J. Korean Chem. Soc. 59(5): 471–474.

[77] Sebald, M.A., Gebauer, J. and Koch, M. 2022. Concise syntheses of alternariol, alternariol-9-monomethyl ether and their D3-isotopologues. Synthesis 54(19): 4285–4293.

[78] Schobert, R. and Schlenk, A. 2008. Tetramic and tetronic acids: an update on new derivatives and biological aspects. Bioorg. Med. Chem. 16(8): 4203–4221.

[79] Mikula, H., Horkel, E., Hans, P., Hametner, C. and Fröhlich, J. 2013. Structure and tautomerism of tenuazonic acid–a synergetic computational and spectroscopic approach. J. Hazard Mater 250: 308–3317.

[80] Zaghouani, M. 2016. Nay B. 3-Acylated tetramic and tetronic acids as natural metal binders: myth or reality? Nat. Prod. Rep. 33(4): 540–548.

[81] Chen, S. and Qiang, S. 2017. Recent advances in tenuazonic acid as a potential herbicide. Pesticide Biochemistry and Physiology 143: 252–257.

[82] Sotelo-Cerón, N.D. and Martínez-Alvarez, J.C. and Maldonado-Mendoza, I.E. 2023. Recent advances in the bioherbicidal potential of tenuazonic acid, an *Alternaria* spp. mycotoxin. J. Mex Chem. Soc. 67(3): 266–274.

[83] Schobert, R., Jagusch, C., Melanophy, C. and Mullen, G. 2004. Synthesis and reactions of polymer-bound $Ph_3P{=}C{=}C{=}O$: a quick route to tenuazonic acid and other optically pure 5-substituted tetramates. Org. Biomol. Chem. 2(23): 3524–3529.

[84] Rico-Yuste, A., Abouhany, R., Urraca, J.L., Descalzo, A.B., Orellana, G. and Moreno-Bondi, M.C. 2021. Eu(III)-Templated molecularly imprinted polymer used as a luminescent sensor for the determination of tenuazonic acid mycotoxin in food samples. Sens Actuators B Chem. 329: 129256.

[85] Poliseno, V., Chaves, S., Brunetti, L., Loiodice, F., Carrieri, A., Laghezza, A. et al. 2021. Derivatives of tenuazonic acid as potential new multi-target anti-Alzheimer's disease agents. Biomolecules 11(1): 111.

[86] Peukert, S., Sun, Y., Zhang, R., Hurley, B., Sabio, M., Shen, X. et al. 2008. Design and structure-activity relationships of potent and selective inhibitors of undecaprenyl pyrophosphate synthase (UPPS): tetramic, tetronic acids and dihydropyridin-2-ones. Bioorg. Med. Chem. Lett. 18(6): 1840–1844.

[87] Nakamura, M. and Ishibashi, K. 1958. On the new antibiotics "Ophiobolin", produced by *Ophiobolus miyabeanus*. J. Agric. Chem. Soc. Jpn. 32: 739–744.

[88] Masi, M., Dasari, R., Evidente, A., Mathieu, V. and Kornienko, A. 2019. Chemistry and biology of ophiobolin A and its congeners. Bioorg. Med. Chem. Lett. 29(7): 859–869.

[89] Scherlach, K., Boettger, D., Remme, N. and Hertweck, C. 2010. The chemistry and biology of cytochalasans. Nat. Prod. Rep. 27(6): 869–886.

[90] Stork, G. and Nakamura, E. 1983. A simplified total synthesis of cytochalasins via an intramolecular Diels-Alder reaction. J. Am. Chem. Soc. 105(16): 5510–5512.

[91] Stork, G., Nakahara, Y., Nakahara, Y. and Greenlee, W.J. 1978. Total synthesis of cytochalasin B. J. Am. Chem. Soc. 100(24): 7775–7777.

[92] Valli, M., Souza, J.M., Chelucci, R.C., Biasetto, C.R., Araujo, A.R., Bolzani, V.D.S. et al. 2022. Identification of natural cytochalasins as leads for neglected tropical diseases drug discovery. PloS ONE 17(10): e0275002.

[93] Tian, W., Deng, Z. and Hong, K. 2017. The biological activities of sesterterpenoid-type ophiobolins. Mar Drugs 15(7): 229.

[94] Tsuna, K., Noguchi, N. and Nakada, M. 2011. Convergent Total Synthesis of (+)-Ophiobolin A. Angew. Chem. 123(40): 9624–9627.

[95] Tsuna, K., Noguchi, N. and Nakada, M. 2013. Enantioselective Total Synthesis of (+)-Ophiobolin A. Chemistry–A European Journal 19(17): 5476–5486.

[96] Rowley, M., Tsukamoto, M. and Kishi, Y. 1989. Total synthesis of (+)-ophiobolin C. J. Am. Chem. Soc. 111(7): 2735–2737.

[97] Truax, N.J. and Romo, D. 2020. Bridging the gap between natural product synthesis and drug discovery. Nat. Prod. Rep. 37(11): 1436–1453.

[98] Tao, Y., Reisenauer, K.N., Masi, M., Evidente, A., Taube, J.H. and Romo, D. 2020. Pharmacophore-directed retrosynthesis applied to ophiobolin A: simplified bicyclic derivatives displaying anticancer activity. Org. Lett. 22(21): 8307–8312.

[99] Grzybowski, A., Pawlikowska-Łagód, K. and Polak, A. 2021. Ergotism and Saint Anthony's fire. Clin. Dermatol. 39(6): 1088–1094.

[100] Sharma, N., Sharma, V.K., Manikyam, H.K. andKrishna, A.B. 2016. Ergot alkaloids: A review on therapeutic applications. Eur. J. Med. Plants 14(3): 1–17.

[101] Schiff, Jr. P.L. 2006. Ergot and its alkaloids. Am. J. Pharm. Educ. 70(5): 98.

[102] Wallwey, C. and Li, S.M. 2011. Ergot alkaloids: structure diversity, biosynthetic gene clusters and functional proof of biosynthetic genes. Nat. Prod. Rep. 28(3): 496–510.

[103] Gerhards, N., Neubauer, L., Tudzynski, P. and Li, S.M. 2014. Biosynthetic pathways of ergot alkaloids. Toxins 6(12): 3281–3295.

[104] Liu, H. and Jia, Y. 2017. Ergot alkaloids: synthetic approaches to lysergic acid and clavine alkaloids. Nat. Prod. Rep. 34(4): 411–432.
[105] Jastrzębski, M.K., Kaczor, A.A. and Wróbel, T.M. 2022. Methods of lysergic acid synthesis-the key ergot alkaloid. Molecules 27(21): 7322.
[106] Nutt, M.J., Woolf, N. and Stewart, S.G. 2023. Overview of the synthetic approaches to lysergic acid as a precursor to the psychedelic LSD. Aust. J. Chem. 76(5): 279–287.
[107] Bartoccini, F. and Piersanti, G. 2021. Synthesis and reactivity of Uhle's ketone and its derivatives. Synthesis 53(08): 1396–1408.
[108] Knight, B.J., Harbit, R.C. and Smith, J.M. 2023. Six-step synthesis of (±)-lysergic acid. J. Org. Chem. 88(4): 2158–2165.
[109] Nichols, D.E. 2018. Dark classics in chemical neuroscience: lysergic acid diethylamide (LSD). ACS Chem. Neurosci. 9(10): 2331–2343.
[110] Kim, K., Che, T., Panova, O., DiBerto, J.F., Lyu, J., Krumm, B.E. et al. 2020. Structure of a hallucinogen-activated Gq-coupled 5-HT2A serotonin receptor. Cell 182(6): 1574–1588.
[111] Fuentes, J.J., Fonseca, F., Elices, M., Farré, M. and Torrens, M. 2020. Therapeutic use of LSD in psychiatry: a systematic review of randomized-controlled clinical trials. Front Psychiatry 10: 943.
[112] Robbers, J.E. and Floss, H.G. 1969. Clavicipitic acid, a new 4-substituted indolic amino acid obtained from submerged cultures of the ergot fungus. Tetrahedron Lett. (23): 1857–1858.
[113] Robbers, J.E., Otsuka, H., Floss, H.G., Arnold, E.V. and Clardy, J. 1980. Clavicipitic acid: its structure, biosynthesis, and role in ergot alkaloid formation. J. Org. Chem. 45(6): 1117–1121.
[114] Xu, Z., Hu, W., Liu, Q., Zhang, L. and Jia, Y. 2010. Total synthesis of clavicipitic acid and aurantioclavine: stereochemistry of clavicipitic acid revisited. J. Org. Chem. 75(22): 7626–7635.
[115] McCabe, S.R. and Wipf, P. 2016. Total synthesis, biosynthesis and biological profiles of clavine alkaloids. Org. Biomol. Chem. 14(25): 5894–5913.
[116] Ito, M., Tahara, Y.K. and Shibata, T. 2016. Strategies for the total synthesis of clavicipitic acid. Chem. Eur. J. 22(16): 5468–5477.
[117] Ji, C.B., Hu, X., Jiang, S.Z. and Yang, J.H. 2022. Efforts and strategies for the syntheses of clavine-type ergot alkaloid aurantioclavine. Eur. J. Org. Chem. (44): e202201000.
[118] Xu, Z., Li, Q., Zhang, L. and Jia, Y. 2009. Efficient Total Synthesis of (–)-cis-Clavicipitic Acid. J. Org. Chem. 74(17): 6859–6862.
[119] Kumar, P., Mahato, D.K., Kamle, M., Mohanta, T.K. and Kang, S.G. 2017. Aflatoxins: A global concern for food safety, human health and their management. Front Microbiol. 7: 2170.
[120] Awuchi, C.G., Amagwula, I.O., Priya, P., Kumar, R., Yezdani, U. and Khan, M.G. 2020. Aflatoxins in foods and feeds: A review on health implications, detection, and control. Bull Environ. Pharmacol. Life Sci. 9: 149–155.
[121] Abrar, M., Anjum, F.M., Butt, M.S., Pasha, I., Randhawa, M.A., Saeed, F. et al. 2013 Aflatoxins: biosynthesis, occurrence, toxicity, and remedies. Crit. Rev. Food Sci. Nutr. Crit. Rev. Food Sci. 53(8): 862–874.
[122] Richard, J.L. 2008. Discovery of aflatoxins and significant historical features. Toxin Reviews 27(3-4): 171–201.
[123] Baertschi, S.W., Raney, K.D., Stone, M.P. and Harris, T.M. 1988. Preparation of the 8, 9-epoxide of the mycotoxin aflatoxin B1: the ultimate carcinogenic species. J. Am. Chem. Soc. 110(23): 7929–7931.
[124] Brown, K.L., Bren, U., Stone, M.P. and Guengerich, F.P. 2009. Inherent stereospecificity in the reaction of aflatoxin B1 8,9-epoxide with deoxyguanosine and efficiency of DNA catalysis. Chem. Res. Toxicol. 22(5): 913–917.
[125] Stone, M.P., Banerjee, S., Brown, K.L. and Egli, M. 2011. Chemistry and biology of aflatoxin-DNA adducts. Frontiers in Nucleic Acids. ACS Symposium Series 1082: 147–166.
[126] Yang, L. and Wang, Z. 2021. Advances in the total synthesis of aflatoxins. Front Chem. 9: 779765.
[127] Van der Merwe, K.J., Steyn, P.S., Fourie, L., Scott, D.B. and Theron, J.J. 1965. Ochratoxin A, a toxic metabolite produced by *Aspergillus ochraceus* Wilh. Nature 205(4976): 1112–1113.
[128] Ringot, D., Chango, A., Schneider, Y.J. and Larondelle, Y. 2006. Toxicokinetics and toxicodynamics of ochratoxin A, an update. Chem. Biol. Interact. 159(1): 18–46.
[129] Kőszegi, T. and Poór, M. 2016. Ochratoxin A: molecular interactions, mechanisms of toxicity and prevention at the molecular level. Toxins 8(4): 111.

[130] Liu, W.C., Pushparaj, K., Meyyazhagan, A., Arumugam, V.A., Pappuswamy, M. and Bhotla, H.K. 2022. Ochratoxin A as an alarming health threat for livestock and human: A review on molecular interactions, mechanism of toxicity, detection, detoxification, and dietary prophylaxis. Toxicon. 213: 59–75.
[131] Gallo, A., Ferrara, M. and Perrone, G. 2017. Recent advances on the molecular aspects of ochratoxin A biosynthesis. Curr. Opin. Food Sci. 17: 49–56.
[132] Steyn, P.S. and Holzapfel, C.W. 1967. The synthesis of ochratoxins A and B: metabolites of *Aspergillus ochraceus* Wilh. Tetrahedron 23(11): 4449–4461.
[133] Kraus, G.A. 1981. A facile synthesis of ochratoxin A. J. Org. Chem. 46(1): 201–202.
[134] Sibi, M.P., Chattopadhyay, S., Dankwardt, J.W. and Snieckus, V. 1985. Combinational O-aryl carbamate and benzamide directed ortho metalation reactions. Synthesis of ochratoxin A and ochratoxin B. J. Am. Chem. Soc. 107(22): 6312–6315.
[135] Gabriele, B., Attya, M., Fazio, A., Donna, L.D., Plastina, P. and Sindona, G. 2009. Synthesis 11: 1815–1820.
[136] Cramer, B., Harrer, H., Nakamura, K., Uemura, D. and Humpf, H.U. 2010. Total synthesis and cytotoxicity evaluation of all ochratoxin A stereoisomers. Bioorg. Med. Chem. 18(1): 343–347.

CHAPTER 5

Toxic Heterocyclic Molecules Utilized by Terrestrial Arthropods

5. INTRODUCTION: CHEMICAL "WEAPONS" OF ARTHROPODS ARE A KEY FACTOR IN THE FIGHT FOR EXISTENCE

Arthropods, the leading group of terrestrial invertebrate animals (phylum Arthropoda), have undoubtedly achieved the best biological result. They are the most numerous (comprising about 75–85% of all animals), are found in the most remarkable diversity of ecological habitats, and can eat a wide variety of foods [1–3]. They are characterized by containing an external skeleton (exoskeleton), a segmented body, and, most notably, articulated appendages, which gives them their name derived from Greek.

Like plants and fungi, terrestrial invertebrate animals, including insects, myriapods, and arachnids, can produce multiple toxic substances to defend themselves from attack by other animals or kill the prey they feed. Although spiders and scorpions (arachnids), whose composition of venoms is often dominated by proteins and peptides, are probably the animals that first come to mind when considering venomous arthropods, we will examine more insect species (e.g., ants, wasps, and beetles) that possess venoms and toxic substance based on small molecules. However, we can see harmful small molecules from spiders.

Toxic substances, therefore, play an important role in the struggle for existence and constitute a fundamental element for the survival of the species that possess them. How do these toxic substances originate? Are they synthesized in the animal's body or from the food chain? In many cases, the compounds present descend from the secondary metabolites of plants. However, in several species, the biosynthesis of their toxins was verified.

Over millions of years, insects, masters of survival, have developed numerous effective strategies, such as feeding, reproduction, dispersal and migration, and defense mechanisms. The latter includes the use of toxic substances, which are the focus of this chapter. Among insects, many species, such as ants, bees, and beetles, know how to use their chemical weapons. Although they are small and appear to be fragile, they produce complex mixtures of toxic alkaloid molecules and toxic oxygen-containing heterocycles. Ants and beetles have undoubtedly been the most prolific producers of repellent and/or toxic compounds [4].

5.1 Toxic Nitrogen-Containing Compounds

Nitrogen-containing compounds are important metabolites that generate plants, fungi, and animals. We already know some examples from the previous chapter. In this section, we have selected toxic alkaloids from ants, bees, and beetles. Insect alkaloids are frequently used as interspecific repellents and insecticides. Most are toxic; therefore, it is unsurprising that ant venom alkaloids are employed to overcome competitors, nest assailants, and prey. Notably, alkaloid-rich venoms can also serve as nicotinic antagonists, inducing rapid prey paralysis and death [5].

5.1.1 Tetraponerins and Solenopsins from Ants

The vast majority of the 13,000 species of ants can produce their venom. Ant venom can be dangerous to many insects on which they feed. In addition, many species of ants use it as a defensive weapon. Ants can use three defense or attack mechanisms. The first is biting with their jaws (soldier ants). The second is expelling through their anal region. The third is inoculating toxic substances using an inoculating device similar to those of wasps and bees. The bites of some species produce a painful and itchy vesicle that, after a few days, turns into an eschar.

The ants *Tetraponera allaborans*, *T. rufinigara*, *T. binghami*, and *T. clypeata* of the genus *Tetraponera* are specialized in trees, have shapes similar to wasps, and feed alone. Although the ability to sting is common for ants, these species do not sting but instead smear a contact poison, a mixture of toxic alkaloids, upon their enemies. This serves as a defensive weapon against other invading insects, displaying potent, selective insecticide activity.

A group of compounds in the venom of these ants, the constituents called tetraponerins (from (+)-T-1 to (+)-T-8), belong to a rare class of alkaloids. These alkaloids form a unique family of heterocycles, which share a quinolizidine or indolizidine fragment with another indolizidine ring. These ABC tricyclic alkaloids have a rare amino structure in alkaloids. They can be divided into two groups depending on the size of the rings: 5-6-5 (tetraponerins T-1, T-2, T-5 and T-6) or 6 -6-5 (tetraponerins T-3, T-4, T-7 and T-8) with the structures of decahydropyrido[1,2-*c*] pyrrolo[1',2'-*a*]pyrimidine and decahydrodipyrrolo[1,2-*a*:1',2'-*c*]pyrimidine, respectively (Figure 5.1) [6].

In each group, tetraponerins differ in the length of the alkyl chain located at C-5 (propyl, Pr, or pentyl, Pent) and in the configuration of this stereocenter. In general, the 6-6-5 group is more abundant where the principal component is tetraponerin (+)-T-8 with the (5*S*,6a*R*,11a*R*) configuration, while tetraponerins from the 5-6-5 group are less common and tetraponerins T-1 and T-2 were identified as a trace. Isolating each component of this mixture is not an easy task. In 1987, the first isolation and structural elucidation of these natural products were reported. The extraction (CH_2Cl_2/MeOH) of 1,500 Tetraponera ants, followed by chromatographic separation

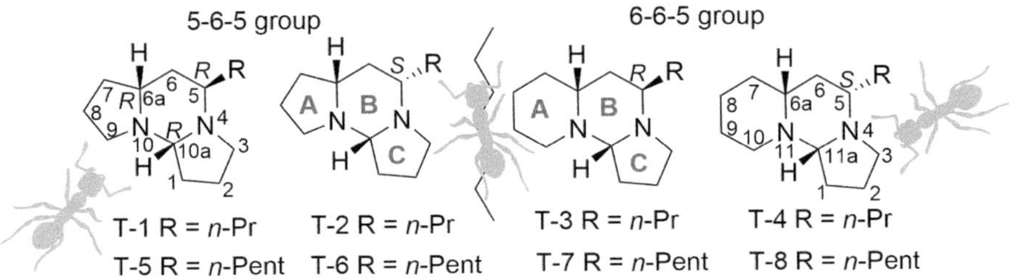

Figure 5.1. Defensive alkaloids from *Tetraponera* ants.

processes of the alkaloid fraction, yielded small quantities of each chiral molecule: 1.6 mg of (+)-T-3, 3.6 mg of (+)-T-4, 4.6 mg of (+)-T-5, 3.0 mg of (+)-T-6, 6.6 mg of (+)-T-7, and 10.5 mg of (+)-T-8 [6]. The principal component, tetraponerine T-8, is toxic to the ant *Myrmica rubra* on topical application in methanolic solution (LD_{50} = 2.0×10^{-9} mol/ant mg). Its toxicity is high compared to a natural alkaloidal insecticide, nicotine (LD_{50} = $2.5\cdot10^{-8}$ mol/ant mg).Thus, the unusual structure and insecticidal activities of T-8 have made it an attractive target for total synthesis. Indeed, several approaches to obtain this alkaloid and its analogs have been reported. Braekman and his group developed a stereocontrolled total synthesis of this ant alkaloid, starting from 3,4,5,6-tetrahydropyridine-l-oxide **5.1**, which was found valuable reagent for the synthesis of (±)-T-8 after performing retrosynthetic analysis with the opening of the B ring of T-8 (Figure 5.2A) [7]. So, [3+2]-cycloaddition reaction of nitrone **5.1**, prepared by HgO oxidation of 1-hydroxypiperidine, with 1-heptene **5.2** afforded isoxazolidine derivative **5.3** in a 94% yield after chromatography on alumina. Its hydrogenolysis by H_2/Raney nickel furnished 2-(hydroalkyl)piperidine **5.4**, which was converted into 2-(succiniminylalkyl)piperidine **5.5** in two steps; first, the benzylation reaction of NH group of **5.4** and then, the Mitsunobu reaction of N-benzyl protected **5.4** with succinimide as the nucleophile in the presence of diethyl azodicarboxylate (DEAD) in dry THF (Figure 5.2B).

The transformation of **5.5** into 8-oxo-tetraponerin T-8 (**5.7**) was performed using a one-pot catalytic debenzylation-reductive cyclization process under neutral conditions (H_2-Pd/C, MeOH) through the formation of **5.6**. Finally, oxo-T-8 was subjected to $LiAlH_4$ reduction in refluxing THF that allowed to obtain (±)-T-8 (Figure 5.2B). It is noteworthy that an excellent level of stereocontrol was achieved at each stage (d.e. > 95%) by preparing (±)-T-8 in 6 steps with a 27% yield.

Another approach to the synthesis of T-8, proposed by Barluenga and co-workers, starts with cyclic imine (6-methyl-2,3,4,5-tetrahydropyridine, **5.8**), found after executing retrosynthetic analysis with the elimination of the C ring of T-8 (Figure 5.3A) [8]. Successful disconnection of the T-8 skeleton allowed the use of semicyclic azabutadiene **5.10**, a crucial precursor in this synthesis. It was prepared from the readily available imine **5.8**, which was prepared from δ-valerolactam and hexanenitrile **5.9** (Figure 5.3B). Next, azabutadiene **5.10** was allowed to react overnight with 4-bromobutyraldehyde (**5.11**) in refluxing THF in the presence of anhydrous sodium sulfate as a dehydrating agent; the sodium sulfate was then filtered off, and the presumed pyrimidinium salt **5.12** was reduced at –20°C with

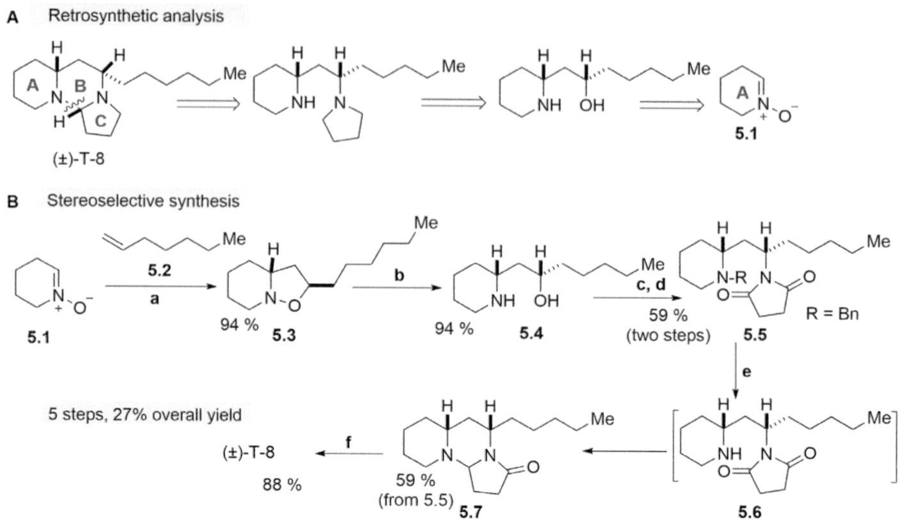

Figure 5.2. A. Retrosynthetic analysis of tetraponerine T-8. B. Its total synthesis starting from piperidine oxide **5.1**: (a) $CHCl_3$, 0–20°C, reflux, 12 hours; (b) H_2, Ra-Ni, MeOH, rt; (c) $PhCH_2Br$, KOH/EtOH-H_2O, reflux, 30 minutes; (d) PPh_3, DEAD (EtOOC-N=N-COOEt), succinimide, THF, rt under Ar, 48 hours; (e) H_2, Pd/C, MeOH, rt, 15 hours; (f) $LiAlH_4$, dry THF, rt, 8 hours.

sodium borohydride in THF/MeOH. After workup and chromatographic purification, racemic tetraponerine-8 was isolated as the sole stereoisomer in 67% yield.

Thus, the latter synthesis is the most straightforward, high-yield synthesis of the alkaloid tetraponerin-8 based on the facile cyclization of 4-amino-1-azadiene. Moreover, this methodology involves only inexpensive, readily available materials (δ-valerolactam, hexanenitrile, and 4-bromobutyraldehyde) and should allow the obtention of the product or its analogs in multigram quantities.

Biological studies indicate tetraponerins are paralyzing poisons segregated by the host ants against their enemies. Therefore, it is not surprising that they exhibit insecticidal and neurotoxic activities. Recent pharmacological research reveals tetraponerins are effective inhibitors in micromolar concentrations of vertebrate neuromuscular, ganglionic, and brain nicotinic acetylcholine receptors [9]. It was also found that T-5, T-8, and their C-5 alkyl analogs **5.13–5.15** are cytotoxic against the HT-29 cell line (colorectal carcinoma), showing lower IC_{50} values (i.e., more active) when the larger hydrocarbonated chain at the C5 are present (Figure 5.4) [10]. These results revealed that the cytotoxicities are highly sensitive to the size of the alkyl substituents.

It is also interesting to observe that the toxic activities are not sensitive to the stereochemistries of the alkyl substituents C-5. Continuing the study of chemistry and biology of tetraponerin alkaloids, Gonzalez-Gomez and co-workers described stereoselective total synthesis of all known tetraponerins, from (+)-T-1 to (+)-T-8, demonstrating the anticancer activity of C-5 pentyl tetraponerins (T-5 to T-8) against four human cell lines (A2780 from ovary carcinoma, NCI-H460 from lung, Caco-2 from colon and MCF-7 from breast) [11]. Surprisingly, in most of the cases, the cellular growth inhibition was not enough to determine the IC_{50} value;

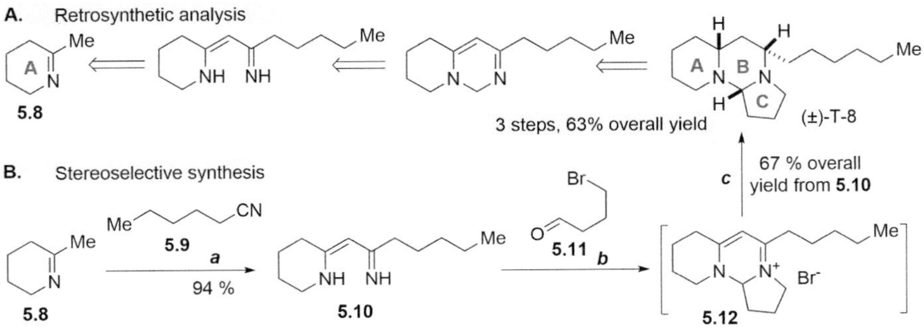

Figure 5.3. A. Retrosynthetic analysis of tetraponerin T-8. B. Its total synthesis starting from 6-methyl-2,3,4,5-tetrahydropyridine **5.8**: (a) LDA, –78°C, 1 hour; (b) Na_2SO_4, THF/MeOH, reflux, 14 hours; (c) $NaBH_4$, THF/MeOH, –20°C to 0°C, 6 hours.

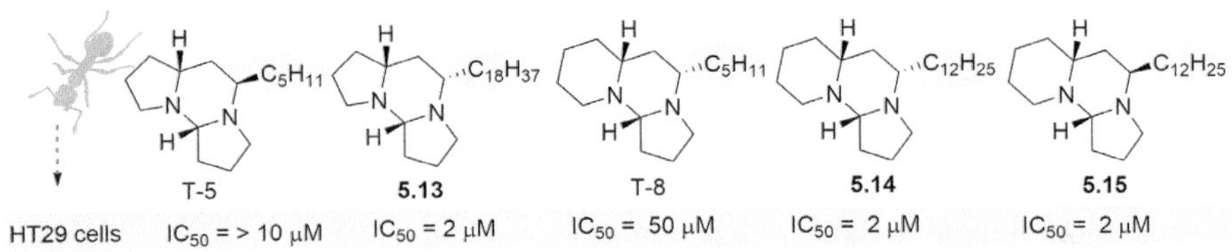

Figure 5.4. Structures and IC_{50} values of T-5, T-8, and their C-5 alkyl analogs **5.13–5.15** against HT29 cancer cells.

nevertheless, (+)-T-7 exhibited encouraging results against A2780, MCF-7 (cellular growth inhibition > 50%), being more efficient against MCF-7 cancer cells.

The biogenesis of tetraponerin T-8, the major component of the venom, appears to have a mixed origin; the pyrrolidine ring of tetraponerin skeleton is formed from ornithine, which gives rise to glutamic acid and then putrecine. The 12-carbon chain of 12-carbon polyacetate precursor **5.16** is derived from combining six acetate units (Figure 5.5) [12].

Interestingly, in contrast with T-8, tetraponerin T-6 is biosynthesized from two molecules of putrescine and a seven-carbon moiety coming from an eight-carbon polyacetate chain through decarboxylation [13].

With their yellowish heads and bodies of a characteristic reddish color, ants of the genus *Solenopsis* (fire ants) are venomous and aggressive. In a dangerous situation, the workers release pheromones as a type of alarm call to warn their sisters and prompt the soldiers to defend the anthill using their chemical weapons. When they sting, they do so to hold on and then inject their venom, producing a sensation similar to the burning caused by fire. The main components of their venom are simple lipophilic 2-methyl-6-alkylpiperidines and 2-methyl-6-alkenylpiperidines (over 95%), known as solenopsin alkaloids [14, 15]. These alkaloids in the venom of fire ants are used to hunt and defeat their opponents or as a defensive measure against pathogens. The successful exploitation of this venom by some invasive fire ants (*S. geminata*, *S. richteri*, and *S. invicta*) may have facilitated their rapid spread to multiple areas, where they establish new populations, replacing the dominant native ant species [16].

The structures of solenopsins differ from each other by the relative configuration of their substituents and the length of the alkyl chain at the C-6 position of the

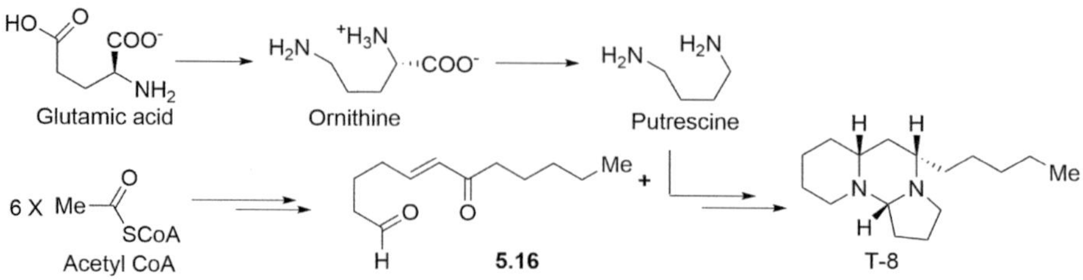

Figure 5.5. Hypothetical biosynthetic scheme of tetraponerin-8.

ring (Figure 5.6). The following structural features can be observed: (i) methyl and alkyl(alkenyl) groups could present *cis*- or *trans*-configuration, and (ii) the two carbon atoms at the C-2 and C-6-positions of the piperidine ring are chiral. Thus, the absolute configuration of an alkaloid component can be (2*R*,6*R*), (2*R*,6*S*), (2*S*,6*R*), and (2*S*,6*S*). In practice, it was found that the absolute configuration of *trans*-alkaloids (solenopsins) is almost always (2*R*,6*R*), while that of *cis*-alkaloids (isolenopsins) is (2*R*,6*S*). In some species of ants, along with these solenopsins, the *trans*- or *cis*- 2-methyl-6-alkenylpiperidines are minor constituents. The first are called dehydrosolenopsins, which contain the C=C double bond with the *Z* (*cis*) configuration in their aliphatic chains [16–19].

It is interesting to note that the solenopsin mixture can be easily extracted from *Solenopsis* ants by directly dissecting the venom glands or dipping the ants in an organic solvent. The piperidine mixture can then be partially purified through traditional thin-layer or silica column chromatography based on the relative affinity of solenopsins for silicates and different solvents.

Fire ant venom showed hemolytic, necrotic, insecticidal, antifungal, and antibiotic activity. The main component of the venom is solenopsin A (*trans*-6-methyl-2-undecanylpiperidine). Isolated from the species *S. invicta*, solenopsin A inhibits angiogenesis through the phosphoinositide-3-kinase (PI3K) signaling pathway in mice and zebrafish embryos, and intercellular communication (quorum sensing) in *Pseudomonas aeruginosa* bacteria. It is capable of producing toxic cardiovascular and central nervous system effects in mice. That shows their capacity to cross the blood-brain barrier [16, 20–22]. These toxic effects benefit these ants as active vertebrate and invertebrate prey predators. Solenopsin alkaloids also proved toxic to *Trypanosoma cruzi* amastigotes and epimastigotes, the etiologic agent of Chagas disease [23].

As previously mentioned, most alkaloids are primarily described from plant extracts, illustrated by the well-studied compounds coniine and nicotine (Section 3.4). Thus, it is often assumed that ants can sequester alkaloids by feeding on plants into their venoms. However, laboratory ant colonies can produce alkaloids without such foods, thus demonstrating they synthesize the compounds directly. The biosynthesis of solenopsins has been studied by feeding 3,000–4,000 ants of *S. geminata* species with labeled sodium acetate ([1-^{14}C] and [2-^{14}C]). This ant species was chosen because its venom is essentially made up of *trans*- and *cis*-solenopsin A, unlike other *Solenopsis* species, whose venoms are much more complex. Studies indicated that the majority component is generated from 9 units of acetyl-CoA to the 18-carbon polyketides **5.17** and/or **5.18**. The formation of two isomeric forms

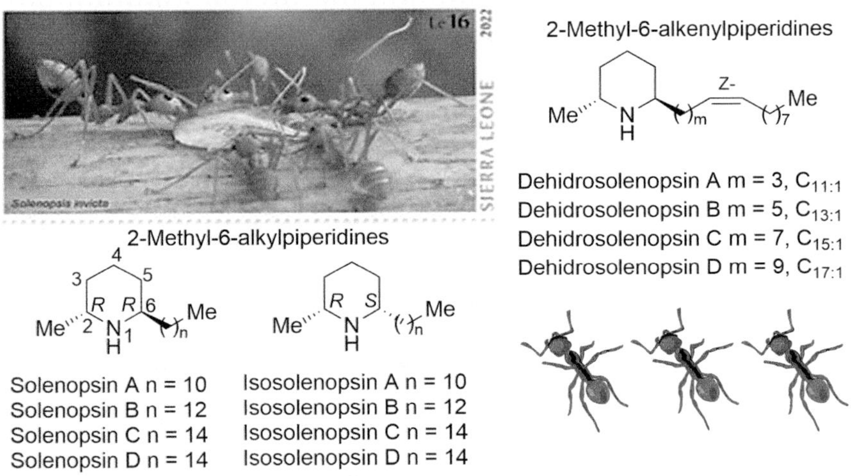

Figure 5.6. Structures of piperidine alkaloids from *Solenopsis* fire ant venom. *Trans*-2-methyl-6-alkenylpiperidines are only shown, and *cis*-isomers and dehydroisosolenopsins are not present here.

of solenopsins suggests that there are two alternative routes (route A and route B) where the proposed cyclized precursors are of piperideine nature, $\Delta^{1,2}$-piperideines **5.19** and/or $\Delta^{1,6}$-piperideines **5.20** (Figure 5.7). The subsequent identification of new metabolites of these structures isolated from the species *S. richteri* corroborates this proposed biogenesis [24].

Thus, solenopsins are oxygen-free polyketide alkaloids, as their nitrogen atom is inserted into a polyketide carbon skeleton. They are not metabolically derived from amino acids. However, the origin of the nitrogen atom in solenopsins has not yet been clarified.

The preparation of solenopsins in the laboratory was the subject of many studies [25]. As a result, numerous routes are found to obtain these alkaloids, including main components, (iso)solenopsins A. These piperidine alkaloids can be easily prepared in racemic form from 2,6-dimethylpyridine (2,6-lutidine, **5.21**) by reacting it with the appropriate alkyl bromide **5.22** in the presence of a strong base, followed by the catalyze reduction (5% rhodium on charcoal) of the obtained pyridine **5.23** to afford isosolenopsin A [25]. Alterantively, they can be prepared from 4-chloropyridine **5.24** in four steps (steps c–f), using the reagent of Grignard, undecylmagnesium bromide **5.25** in tetrahydrofuran (THF) at –78°C and phenyl chloroformate (PhOCOCl) to afford 1,2-dihydropyridine **5.26**. Then, the treatment of **5.26** with potassium *tert*-butoxide in THF gives derivative **5.27**, which is methylated (MeI/BuLi) and hydrogenated (H_2/Pd-C) into tetrahydropyridine **5.28**. The desired solenopsin A alkaloid was then prepared in six steps from 4-chloropyridine with an overall yield of 35% [26] (Figure 5.8).

These routes of obtaining racemic solenopsin alkaloids were the first but still the most convenient synthesis. Notably, both initial pyridines are inexpensive materials, which is an essential factor in planning the synthesis of a natural product.

Different methodologies carried out the asymmetric synthesis of solenopsin A; almost all start with optically active α-amino acids or use chiral auxiliaries [25].

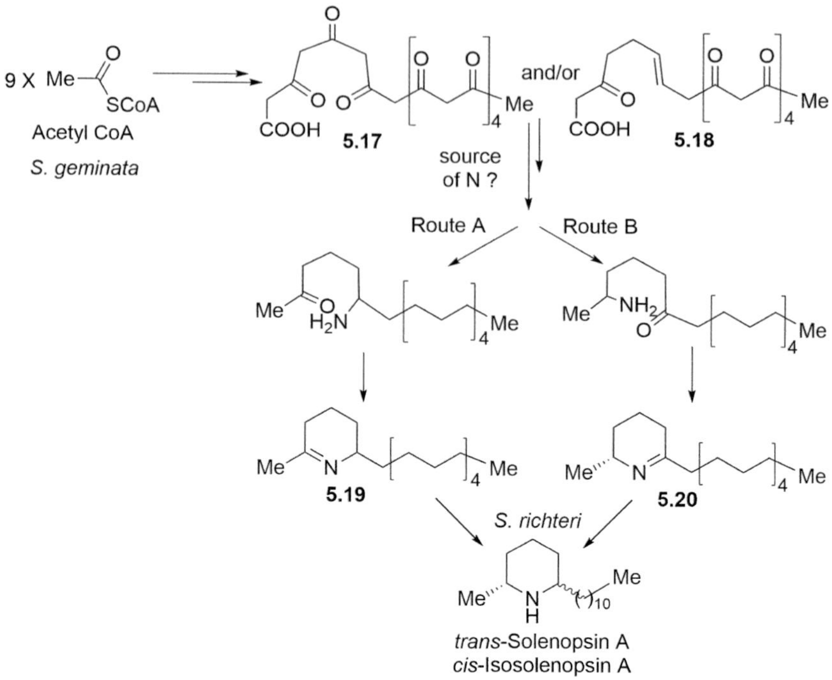

Figure 5.7. Proposed biosynthesis of isomeric solenopsins A.

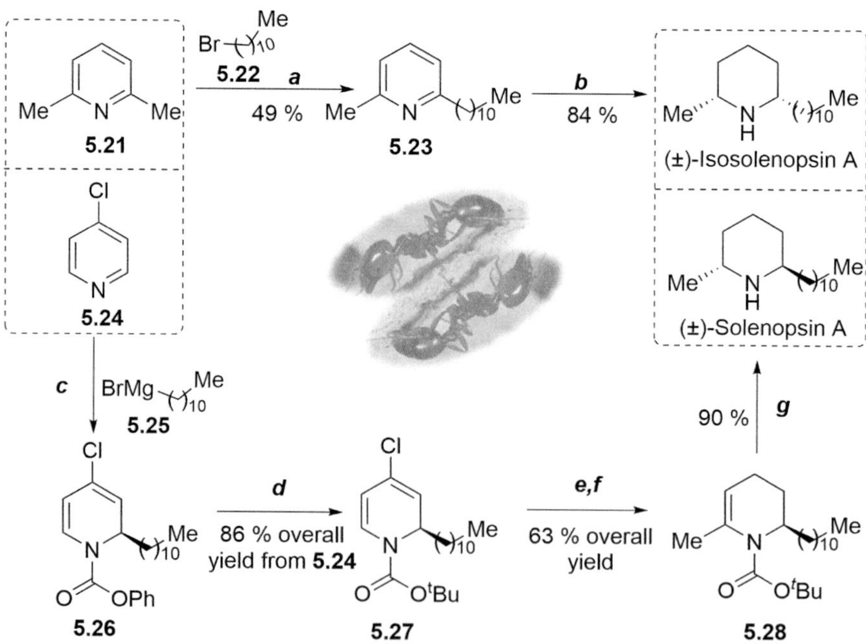

Figure 5.8. Synthesis of (±)-solenopsin A and (±)-isosolenopsin A using available pyridine reagents: (a) LDA, –78°C or PhLi, ether, rt; (b) H_2, 5% rhodium on charcoal, EtOH; (c) THF, –78°C, PhOCOCl, 30 minutes; (d) THF, –42°C, *tert*-BuOK, 1 hour; (e) MeI, BuLi; (f) H_2, Pd-C, Li_2CO_3, MeOH, 0°C; (g) $NaBH_3CN/CF_3COOH$.

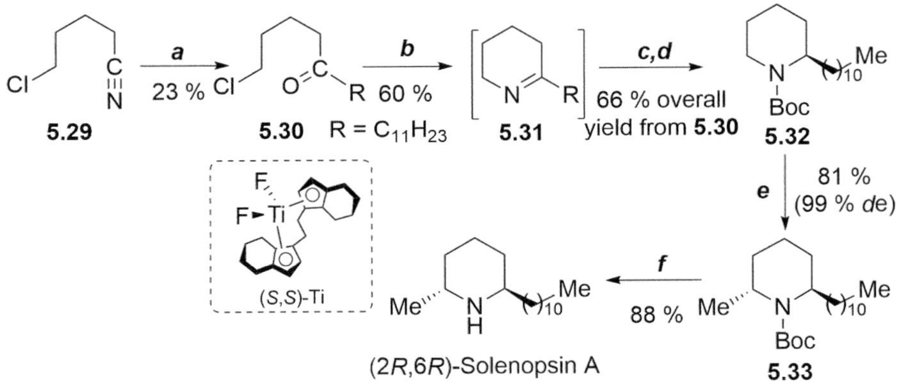

Figure 5.9. Short enantioselective total synthesis of (2*R*,6*R*)-*trans*-solenopsin A via catalytic asymmetric imine hydrosilylation: (a) $C_{11}H_{23}$-MgBr, Et_2O, rt, 2 hours; (b) 1. NaN_3, benzene, Bu_4NBr, 80°C, 12 hours, 2. PPh_3, 12 hours; (c) THF, $PhSiH_3$, cat. (*S*,*S*)-Ti; (d) $(Boc)_2O$, Et_3N, Et_2O; (e) *sec*-BuLi, TMEDA, Me_2SO_4; (f) CF_3COOH, CH_2Cl_2, rt, 12 hours. TMEDA = 1,2-bis(dimethylamino)ethane, $(CH_3)_2NCH_2CH_2N(CH_3)_2$.

The enantioselective syntheses of natural (2*R*,6*R*)-*trans*-solenopsin A, performed by Prof. Buchwald and his student Reding in 1998, uses 5-chloro-valeronitrile **5.29** as an acyclic precursor to construct the piperidine ring **5.31** via the formation of the chloro-ketone **5.30** [27] (Figure 5.9).

The enantioselective imino reduction reaction of **5.31**, a key step called the asymmetric hydrosilylation process, is carried out with 2 equivalent phenylsilane ($PhSiH_3$) in the presence of 1 mol% chiral titanium catalyst at room temperature. The obtained chiral piperidine product was directly converted to the BOC-derivative **5.32**, which then was methylated in the C-6 position in a stereocontrolled manner to give **5.33**. Its deprotection afforded (2*R*,6*R*)-*trans*-solenopsin A in 88% yield. Although the route examined is short (six steps), the method requires special catalysts and reaction conditions, which are difficult to acquire and limit its synthetic use.

5.1.2 Defensive Coccinellidae Alkaloids

The funny beetles of the Coccinellidae family, also known as ladybird beetles, use their chemical weapons in a common way: to protect themselves from eternal enemies such as birds, ants, or amphibians. They produce several types of heterocycles with a very bitter taste. These tiny insects with a round or oval shape are usually brightly colored, with black spots on an orange, yellow, or red background. Their colors keep away predators, who usually associate these bright colors with poison, whose main components are repulsive alkaloids [28–30]. In a dangerous situation or predator attack, these beetles can emit orange fluid droplets of hemolymph from the tibiofemoral joints of their legs. This behavior is referred to as reflex-bleeding. Interestingly, the red coloration of many ladybirds is associated with the presence of 2-methoxy-3-alkylpyrazines (also alkaloids), which give rise to the strong smell detected when these insects undergo reflex-bleeding.

Due to their wide habitat, these insects are known to many nationalities. Interestingly, most of their traditional names are associated with religion or beliefs in good luck and happiness, such as ladybird, ladybug, Mary Beetle, bishop-that-burneth, "bishy barnabee", etc. However, the exact etymology of these names is not known.

Ladybird beetles generate a wide variety of fatty acid-derived alkaloids, most of which are considered to be of autogenous origin and many of which closely resemble the ant venom alkaloids described in the previous section. Some characteristic alkaloids from these beetles are displayed in Figure 5.10. It is noteworthy that these beetles and their larvae are very voracious, destroying large quantities of aphids, psyllids, mealybugs, citrus bugs, scale insects, and mites. They play a central ecological role in regulating populations of agricultural pests. For example, the harlequin ladybird (*H. axyridis*) has been introduced into many countries as a biological control agent but has become an invasive species that threatens the biodiversity of native ladybirds. Its invasive success has been attributed to its vigorous resistance against various pathogens.

Despite their structural diversity, they have a common biogenetic origin. In the *in vitro* incorporation experiments and isotopic labeling of [1-^{14}C] and [2-^{14}C] acetates, it was proven the three common alkaloids precoccinelline, coccinelline, and harmonine are produced from stearic acid rather than a polyketide pathway [28]. In contrast, the alkaloid adaline can be generated using myristic acid. Thus, their biogenesis starts from fatty acid precursors undergoing straightforward oxidation and amination reactions [31]. It should be noted that coccinelline and precoccinelline possess a tricyclic perhydro-9b-azaphenalene skeleton, which can be considered decahydropyrido-quinolizine ring system. In contrast, adaline and harmonine are based on a 9-azabicyclo[3.3.1]nonane ring system (so-called homotropanes system) and a 1,17-diaminooctadec-9-ene backbone in *Z*-form, respectively (Figure 5.10). These alkaloids are less common in nature but not less important. Nine so-called monomeric azaphenalene alkaloids were found, which differ in their stereochemistries and the presence of a C=C bond or N-oxide. Precoccinelline and coccinelline, devoid of optical activity, were the first of these alkaloids to be characterized in 1971 from

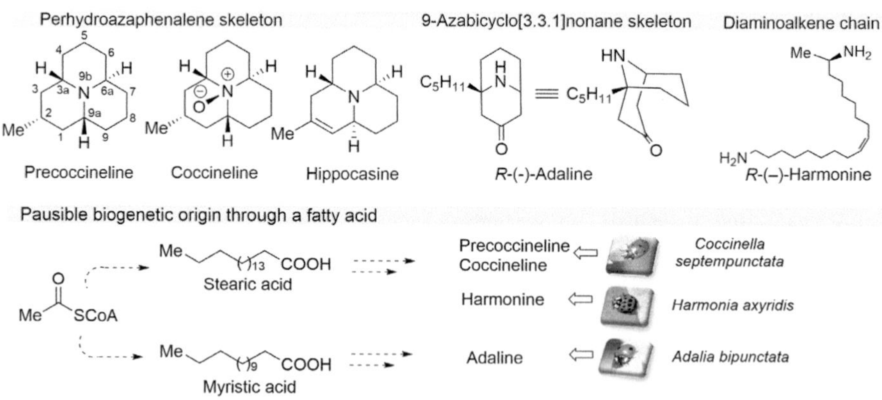

Figure 5.10. Structures of characteristic alkaloids from ladybird beetles and their shortened biosynthetic scheme. Ladybird beetles, including *Coccinella septempunctata*, *Harmonia axyridis*, and *Adalia bipunctata*, produce their respective toxic alkaloids to protect against potential predators. Ladybird beetles are not poisonous to humans but have a foul odor that deters predators.

the methanol extract of *C. septempunctata*. Natural adaline and harmonine have been assigned the *R* configuration [29]. Interestingly, precoccinelline was also reported as a minor alkaloid in a Panamanian dendrobatid frog (*Dendrobates auratus*) in 1992 [32] (Section 6.1).

Some azaphenalene alkaloids have high biological activity. Coccinelline, in addition to its unpleasant bitter taste, has a powerful repellent effect on ants (*Formica rufa*) even in the form of a 0.1–0.5% aqueous solution, which is significantly less than its concentration in the excreted hemolymph. In addition, coccinelline has a serious toxic effect on quails and blue-tit chicks, destroying the birds' liver. Also, these alkaloids probably affect the nervous system of higher animals. There is evidence that perhydroazaphenalene derivatives are powerful antagonists of serotonin receptors, acting also as non-competitive inhibitors of mammalian nicotinic acetylcholine receptors (nAchRs) [33]. Adaline and harmonine are no less interesting biomodels in agrochemical and drug research. Adaline is a potent antagonist of insect and specific mammalian nAchRs and acts as an open-channel blocker of these receptors [34]. Harmonine showed a degree of selectivity for the insect nAChR [35]. Thus, these ladybird alkaloids may be of interest as potential sources for lead candidates in the search for novel insecticides. Moreover, it displays antibacterial and antimalarial activity against mycobacteria (*Mycobacterium tuberculosis*) and protozoan parasites (*Plasmodium falciparum*), which cause tuberculosis and malaria diseases, respectively [36]. It also reveals cytotoxicity against various human tumor cell lines, including A549 (human lung cancer), SK-OV3 (human ovarian cancer), SK-MEL2 (human skin cancer), XF498 (human CNS cancer), and HCT15 (human colon cancer) [37].

The small amount of these alkaloids isolated from ladybirds (e.g., for (−)-adaline, 35 mg from 800 specimens, and coccinelline, 135 mg from 1,600 specimens) and their interesting pharmacological properties emphasize the desirability of practical syntheses for further biological studies [29, 38, 39].

The first synthesis of coccinelline, described by Ayer and Furuichi in 1976, uses 2,6-lutidine **5.21** as an inexpensive starting material, also employed for the isosolenopsin A synthesis. The route consists of eight consecutive steps using special reactivity of both methyl groups in **5.21** (Figure 5.11). Its functionalization in a strong base (PhLi) and electrophilic reagents allows it to reach pyridine derivative **5.36**, which is reduced to piperidine derivative **5.37** (formation of the ring A). Its acid hydrolysis gives the hemiketal **5.38** (construction of the AB skeleton), whose cyclization using pyrrolidine and acetic acid in refluxing tetrahydrofuran provides tricyclic ketone **5.39** [29].

Precoccinelline is prepared from this ketone via a one-pot method: treatment of the ketone with methyllithium, followed by dehydration of the resulting carbinol with thionyl chloride, yields the air-sensitive olefin, which is hydrogenated to give precoccinelline. This is then converted to coccinelline by reaction with *m*-chloroperbenzoic acid [29].

Another synthesis of coccinelline based on the Robinson-Schöpf methodology consists of using amino dialdehyde or its dimethyl acetal **5.40** as starting materials that must be prepared from malonate and acrolein (Figure 5.12) [40]. This is an entirely different synthetic approach to the azaphenalene skeleton of the ladybird

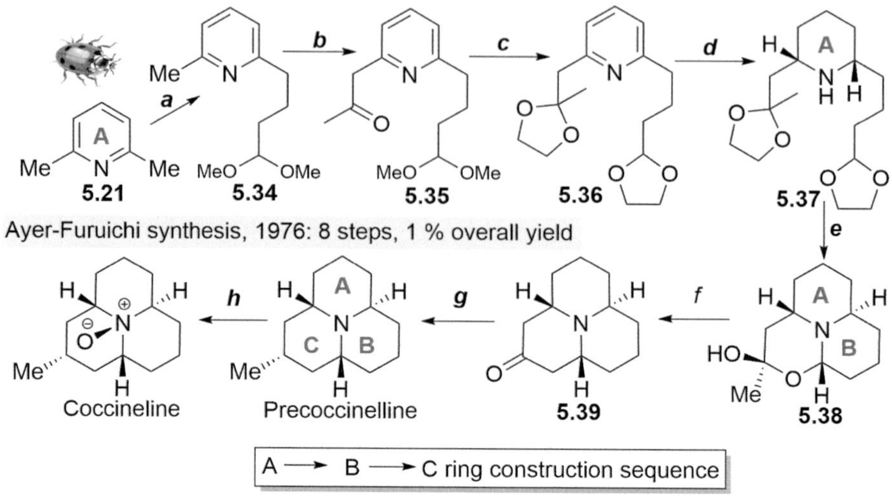

Figure 5.11. The simplified scheme of the first synthesis of coccinelline: (a) PhLi, $BrCH_2CH_2CH(OEt)_2$, Et_2O, 76%; (b) PhLi, MeCN, Et_2O, H^+; (c) $HOCH_2CH_2OH$, TSOH, benzene, reflux, 48% from **5.34**; (d) Na/*i*-AmOH/Δ (86%); (e) 5% aq. HCl; (f) 1 equivalent. pyrrolidine, 2 equivalent AcOH, THF, reflux; (g) 1. CH_3Li, 2. $SOCl_2/CH_2Cl_2$, 3. H_2/Pt, MeOH, rt, 27%; (h) *m*-chloroperbenzoic acid, CH_2Cl_2.

beetles. Condensation of acyclic derivative **5.40** with dimethyl acetonedicarboxylate ($MeO_2CCH_2C(O)CH_2CO_2Me$) followed by acid hydrolysis (pH = 1–2) produces only one stereoisomer of the tricyclic ketoesters, ABC system **5.41**. Its decarbomethoxylation yields ketone **5.39**. The treatment of the latter with methylenetriphenylphosphorane ($PhP{=}CH_2$) generates olefin **5.42**, from which precoccinelline was made by catalytic hydrogenation. This is then promptly transformed into coccinelline.

The first enantioselective synthesis of precoccinelline/coccinelline pair was developed by Snyder and collaborators in 2014, starting from the commercially available chiral N-Boc-(*S*)-(−)-piperidine-2-ethanol [41].

Various synthetic strategies have been employed in synthesizing another alkaloid type from ladybird beetles, namely adaline, an azabicyclo[3.3.1]nonane alkaloid [29, 39]. The early preparation of racemic adaline is based on the Robinson-Schöpf biomimetic strategy that involves crucial Mannich-type tandem reactions between 5-oxodecanal (β-ketoaldehyde, **5.47**) and ketoglutaric acid **5.48** in the presence of chloride ammonium as a nitrogen source (Figure 5.13). A crucial reagent **5.47** is synthesized in two steps: (i) using the hetero-Diels-Alder cycloaddition reaction of 1-octen-3-one **5.44** with methoxyethene (methyl vinyl ether, **5.45**) to afford 2-methoxy-6-pentyl-3,4-dihydro-2*H*-pyran **5.46** and (ii) acidic cleavage of the generated dihydropyran ring [42].

The stereoselective synthesis of this alkaloid, realized by Yu and co-workers in 2009, also starts from commercially available dihydropyran **5.49** and involves an intramolecular allylation of a cyclic imine **5.52** to give bridged bicyclic system **5.53** as a critical step (Figure 5.14) [43]. Note that the first step leads to the formation of the same ketoaldehyde **5.47** that Tursch's team used in its biomimetic synthesis.

Thus, the allylic transfer reaction of the latter with tributylstannyl-silane derivative, $Bu_3SnCH_2\text{-}C(=CH_2)\text{-}CH_2SiMe_3$, in the presence of a chiral Brønsted acid

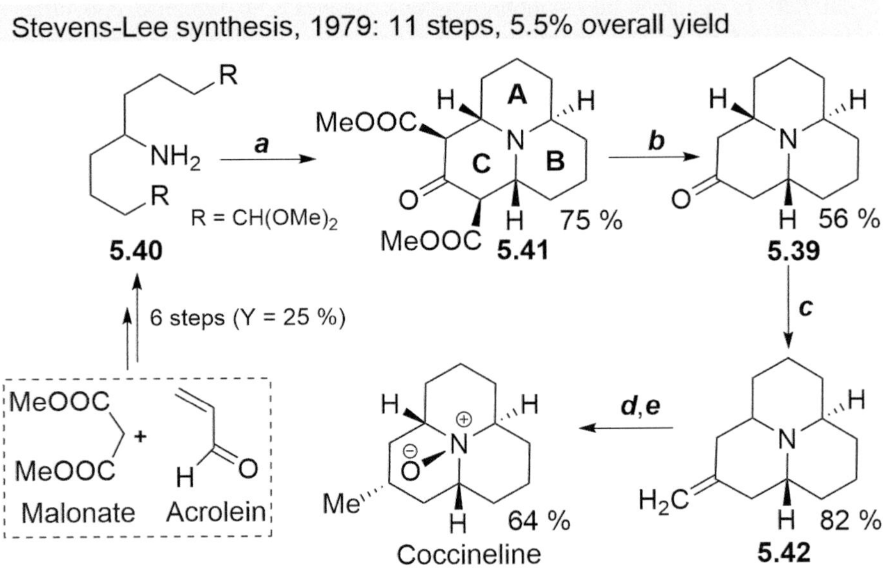

Figure 5.12. The Stevens-Lee synthesis of coccinelline using dimethyl acetal of amine dialdehyde **5.40**: (a) dilute HCl, $MeO_2CCH_2C(=O)CH_2CO_2Me$, 24 hours; (b) NaCl, DMF, reflux, under N_2, 4 hours; (c) Ph_3P/CH_3-Br, *n*-BuLi, Et_2O, under N_2, rt, 1 hour to reflux, 30 minutes; (d) H_2, 10% Pd/C, MeOH, rt; (e) *m*-chloroperbenzoic acid, CH_2Cl_2.

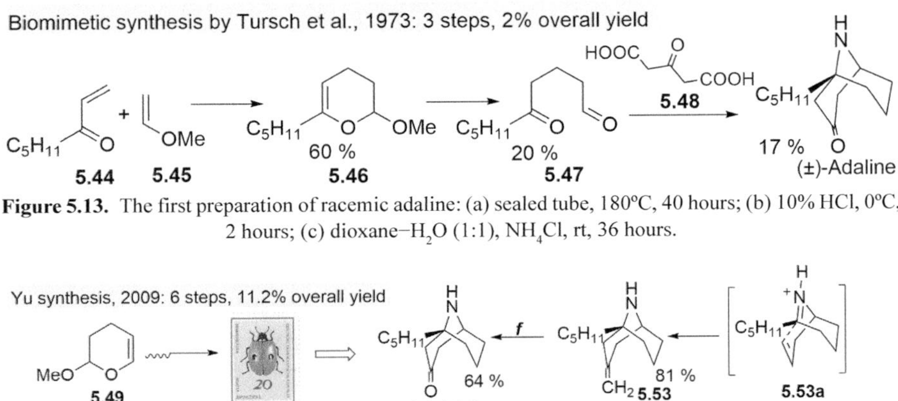

Figure 5.13. The first preparation of racemic adaline: (a) sealed tube, 180°C, 40 hours; (b) 10% HCl, 0°C, 2 hours; (c) dioxane–H_2O (1:1), NH_4Cl, rt, 36 hours.

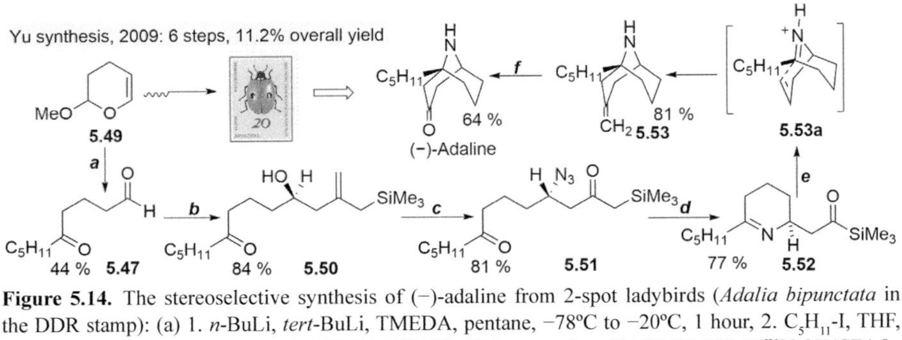

Figure 5.14. The stereoselective synthesis of (−)-adaline from 2-spot ladybirds (*Adalia bipunctata* in the DDR stamp): (a) 1. *n*-BuLi, *tert*-BuLi, TMEDA, pentane, −78°C to −20°C, 1 hour, 2. C_5H_{11}-I, THF, −20°C, 4 hours, 3. H_3O^+; (b) $Bu_3SnCH_2C(=CH_2)CH_2SiMe_3$, cat. 5 mol% (*S*)-BINOL-$Ti^{IV}[OCH(CF_3)_2]_2$, $PhCF_3$, −20°C, 12 hours; (c) DIAD, HN_3 (benzene solution), THF, rt, 2 hours; (d) Ph_3P, Et_2O, rt, 4 hours; (e) 1. CF_3COOH, toluene, 0°C, 5 minutes, 2. Bu_3SnF, 0°C, 1 hour, then rt, 12 hours; (f) 3 mol% OsO_4, KIO_4, THF–H_2O (3:1), rt. DIAD = diisopropyl azodicarboxylate, $(CH_3)_2CHOOC-N=N-COOCH(CH_3)_2$.

(*S*)-BINOL-Ti^{IV} catalyst allows obtaining keto alcohol **5.50**, which is converted to azido ketone **5.51** by Mitsunobu reaction. Then, this azido ketone is cyclized into cyclic imine **5.52**. Its treatment with TFA and tributyltin fluoride (Bu_3SnF) conduces to the formation of bicyclic alkene **5.53** via an intermediate **5.53a**. Finally, this alkene is oxidized in the presence of osmium tetroxide and potassium periodate, to provide (−)-adaline (Figure 5.14). Homotropane-based compounds like adaline continue attracting the attention of researchers involved in developing new synthetic methodologies to reproduce these natural products and synthesize their analogs [44].

Recalling the aliphatic, long-chain diamine harmonine (Figure 5.10), we should note that it was first found in 1985 in five lady beetle species [45], and only in 2002 was it discovered to be the major defense compound of *H. axyridis* [37]. Its complete structural confirmation and its first racemic synthesis were reported by Braconnier et al. [46]. Using 1,6-hexanediol and 1,8-octanediol as initial commercially available materials, they achieved the total synthesis of (±)-harmonine in 15 steps with an overall yield of 6.5%. Six years later, for the first time, Enders and Bartzen described their enantioselective synthesis of (*R*)-(−)-harmonine that yielded a chiral product in 13 steps and an overall yield of 5% (97%, ee) [47]. These long and costly syntheses have been the only source of synthetic harmonine for years. However, the design and synthesis of harmonine and its analogs have recently gained renewed attention due to their promising properties in clinical and agricultural use. In search for short and efficient approaches to this alkaloid, the harmonine structure was retrosynthetically envisioned using two different strategies: (i) the classical Wittig reaction and (ii) the modern *Z*-selective (*cis*-selective) cross-metathesis (Figure 5.15). As a result, the essential building blocks A–D could be found [48, 49]. All these can be prepared from readily available starting materials. For example, in the harmonine synthesis by the Indian researchers, a Wittig reaction was used as the critical step to build up the carbon chain using amino aldehyde, available from D-alanine, as the chiral reagent for completing the block A construction [48].

Thus, the interaction of chiral amino aldehyde **5.54** and (5-hydroxyhexyl)-triphenylphosphonium bromide **5.55** in a strongly basic medium (*tert*-BuOK) (i.e., Witting reaction) results in olefin **5.56**, which is hydrogenated immediately to give the saturated amino alcohol **5.57** and its Swern oxidation allows to obtain the key block A. On the other hand, N-Boc-aminodecene **5.58**, available from undecylinic acid, is converted into block B via another amino alcohol, **5.59**, on the right side of the carbon chain backbone (Figure 5.16). Therefore, having both blocks A and B, a two-step procedure involving the formation of ylide from crude phosphonium salt B, followed by the Witting reaction with reagent A (*tert*-BuOK, THF, rt, 2 hours) and subsequent acid hydrolysis (TFA, CH_2Cl_2, rt, 4 hours), affords (*R*)-Z-harmonine as a green oil.

After six steps, the authors prepared the main blocks A and B with good overall yields and obtained the desired alkaloid product with a good 12% overall yield following the Witting reaction approach [48]. However, Australian researchers surpassed the abovementioned synthesis by exploiting a Z-selective cross-metathesis reaction approach [49]. They modified olefin blocks C and D, replacing N-Boc-amino functional groups with alcohol and ester ones in blocks E and F (Figure 5.17A). To achieve harmonine synthesis, both blocks and their cross-metathesis product needed to be synthesized. To this end, 7-bromohept-1-ene **5.60**, 9-decenoic acid **5.62**, and

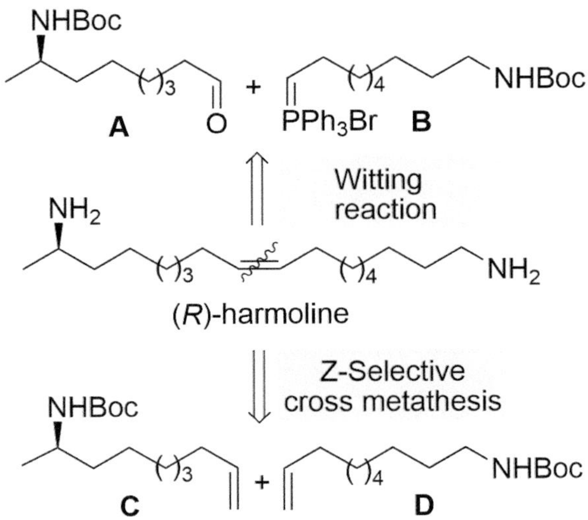

Figure 5.15. The retrosynthetic strategies to access harmonine conducing to the vital building blocks A–D. Blocks A and C belong to the left-side carbon chain backbone, while blocks B and D fit in the right-side carbon chain backbone of harmonine alkaloid.

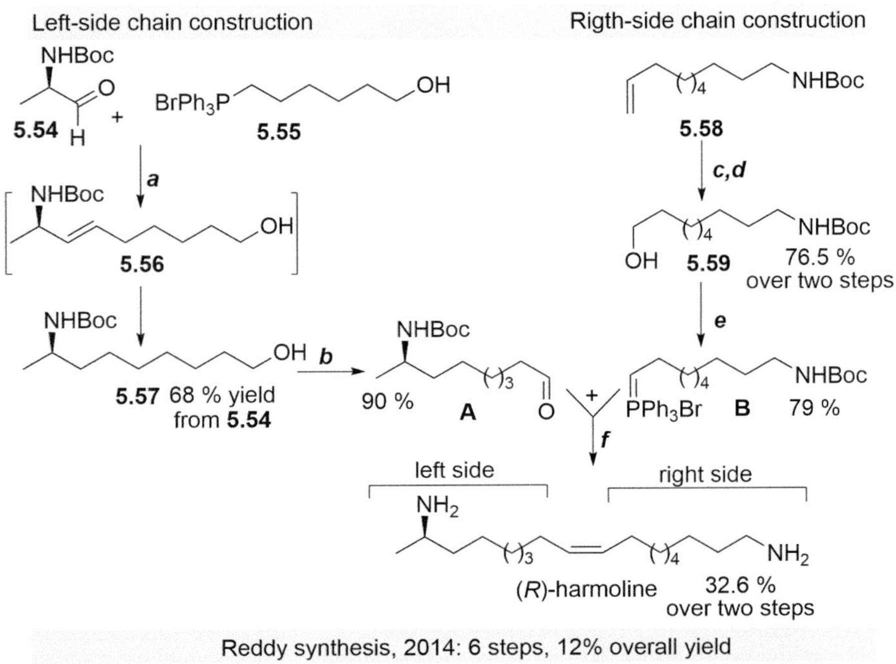

Figure 5.16. Convergent synthesis of the harmonine skeleton based on the Witting reaction: (a) 1. *tert*-BuOK, THF, 0°C to rt giving **5.56** (68%), 2. H_2, 10% Pd/C, EtOH affording **5.57** (99%); (b) $(COCl)_2$, DMSO, Et_3N, CH_2Cl_2, −70°C to 0°C; (c) OsO_4 (2.5% solution in *tert*-BuOH), $NaIO_4$, dioxane–H_2O, rt, 4 hours (85%); (d) $NaBH_4$, EtOH, 0°C to rt, 12 hours (90%); (e) 1. CBr_4, PPh_3, CH_2Cl_2, 0°C, 3 hours (79%), 2. PPh_3, MeCN, reflux, 48 hours; (f) 1. *tert*-BuOK, THF, 0°C to rt, 2 hours (46%), 2. 20% CF_3COOH–CH_2Cl_2, rt, 4 hours (71%).

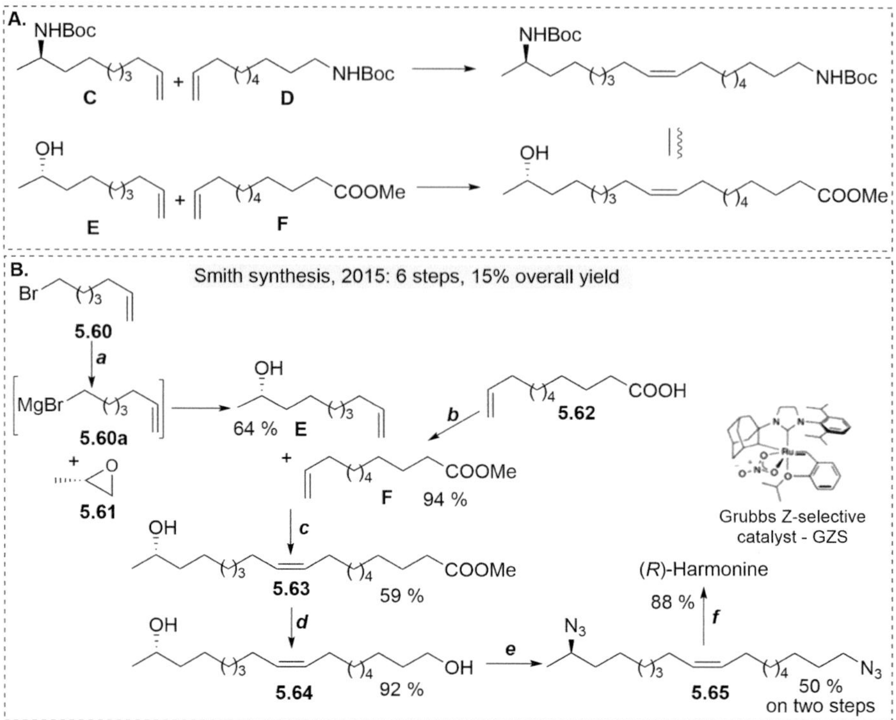

Figure 5.17. A. A new strategy for synthesizing harmonine is using crucial olefin reagents containing alcohol and ester groups. B. Convergent synthesis of chiral harmonine alkaloid based on the *Z*-selective cross-metathesis reaction: (a) 1. Mg, Et_2O, cat. I_2; 2. $CuBrSMe_2$, THF, −78°C to rt, 16 hours; (b) MeOH, H_2SO_4, rt, 24 hours; (c) 2.6 mol% GZS cat., CH_2Cl_2, rt, 24 hours; (d) $LiAlH_4$, THF, 0°C, 1 hour; (e) 1. MsCl, DIPA, CH_2Cl_2, 0°C, 3 hours, 2. NaN_3, DMF, rt, 23 hours; (f) $LiAlH_4$, THF, rt, 1.5 h. DIPA = N,N-diisopropylamine.

Grubbs Z-selective catalyst (GZS) were selected. Thus, crucial reagent E, (*S*)-dec-9-en-2-ol, was readily prepared from **5.60** through the Grignard reaction of **5.60a** and enantiopure propylene oxide **5.61**. At the same time, another alkene partner, F, was obtained using the straightforward Fischer esterification of **5.62** with no trouble (Figure 5.17B).

Thus, the obtained chiral enol E with the opposite configuration to that of the block A and methyl 9-decenoate F were subjected to the crucial cross-metathesis reaction in the presence of a CZS catalyst in dichloromethane; the ester *cis*-alkene product **5.63** was prepared in a 59% yield. The latter was reduced to endiol **6.64**, which was then converted into the corresponding mesylate (MsCl), which was immediately reacted with NaN_3 to afford diazide **5.65** with the needed configuration. Finally, reduction with $LiAlH_4$ provided (*R*)-harmonine in 15% yield over six steps (Figure 5.17B) [49]. Notably, methyl-17-hydroxy oleate **6.64** is available from bio-renewable feedstock sophorolipids [50] that represents green access to harmonine, suitable for scaling up the multicolored Asian lady beetle.

The described synthesis of harmonine provided larger quantities for bioassays. For example, Röhrich and co-workers reported the efficiency of (*R*)-harmonine against *Escherichia coli* and *Bacillus subtilis*, determining the minimal inhibitory concentration for *E. coli* to be 89 μM and 44 μM for *B. subtilis* [38]. Additionally, Boland et al. [51] studied its antiparasitic activity against the causative agent of cutaneous leishmaniasis, *Leishmania major*, showing that the half maximal inhibitory concentration (IC_{50}) of harmonine was 14.2 μM against *L. major* promastigotes and 2.4 μM against the intracellular and clinically relevant *L. major* amastigote form. Notably, these data surpass the activity of the antileishmanial drug Miltefosine (IC_{50} = 36.2 μM and 33.0 μM, respectively). Moreover, harmonine shows a good selectivity index (SI) of 15.2 toward *L. major*, considerably higher than the SI of 2.0 for Miltefosine [51].

5.1.3 Polyamines and Acyl Polyamines from Spiders and Wasps

Carnivorous and dangerous spiders are some of the most famous and studied animals. In the past, legends proliferated about spider venoms and their chemical properties. Venom is produced in specialized venom glands and is used to fight competitors or aggressors and/or for predation. Spider venom is efficient against a broad spectrum of prey groups, primarily invertebrates, mainly insects; it also affects humans [52]. Like spider venom, wasp venom impacts the physiological aspects of the human body and could also lead to an allergic reaction [53]. Interestingly, social wasps normally use their venom for defense and self-preservation, while solitary wasps tend to use their venom for predation rather than defense. Moreover, the social wasp's venom has evolved over myriad years to make its bites more painful and its immune and allergic responses more intense than those of solitary wasps [54].

The chemical study of spider and wasp venom is difficult because its concentration is too small. Spider venom is obtained by electrical milking, and wasp venom is milked by presenting an agar block to a female wasp for stinging. Currently, it is known that the chemical constitution of these venoms is a complex mixture of different classes of compounds, among them are inorganic salts, low molecular weight organic molecules (< 1 kDa: amino acids, amines, polyamines, acylpolyamines, and purines) and macromolecules of a protein nature (polypeptides of 3−6 kDa with 3–5 disulfide bonds and proteins with more than 10 kDa). Some of the most essential venom components spiders produce are polyamines and acylpolyamines. These polyamine toxins are also found in the wasp venom. The polyamine components of venoms had been identified as a new class of neurotoxins in the early 1980s in connection with the newly found property to block the neuromuscular impulse in the glutamatergic synapses of insects. The structural diversity of such polyamide toxins is too enormous (more than 100 structures of these toxins are known) [55, 56].

Let's look at the structures of these toxins (Figure 5.18). Acylpolyamines are a diverse class of molecules built from an aromatic acyl group and a polyamine backbone (some with amino acids in the backbone). They possess masses ranging from 350 to 1,000 Da. The structural organization of these venoms consists of four parts: part I – a lipophilic, aromatic acyl moiety; part II – a linker amino acid residue; part III – the polyamine backbone chain; and part IV – the backbone tail (Figure 5.18A). Thus, such acylpolyamines are polyamines conjugated to amino acids, such as Tyr, Arg, Asn, β-Ala, Gly, Lys, etc. [57]. The aromatic acyl group and the

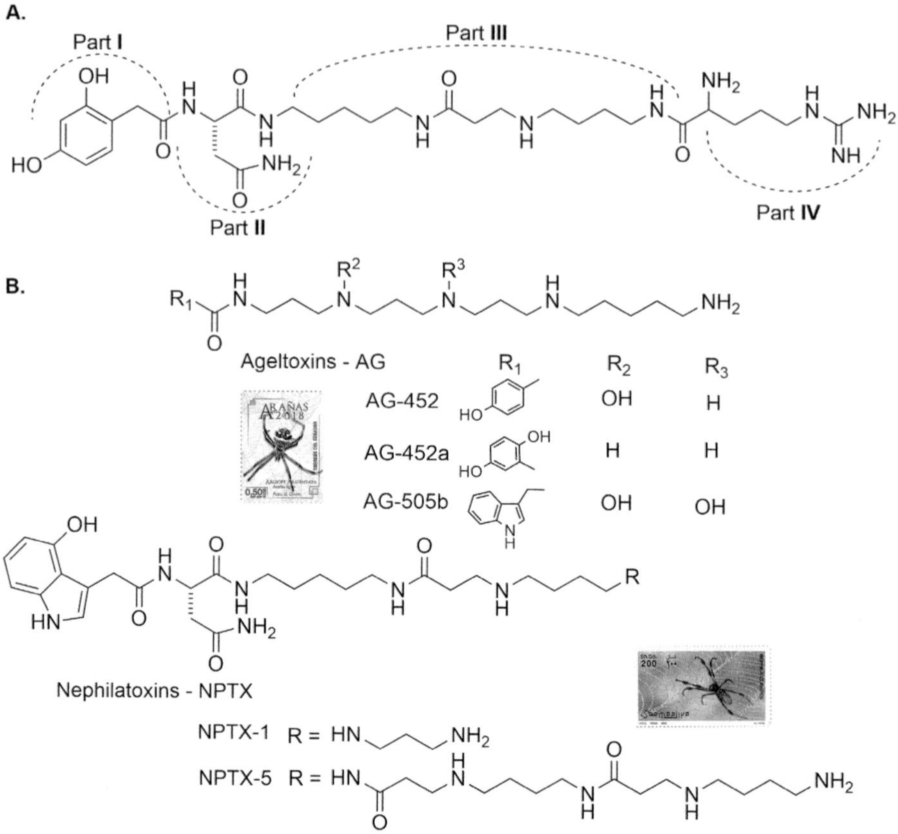

Figure 5.18. A. Structural parts of the acylpolyamine toxins from spider and wasp venoms. B. Examples of acylpolyamine toxins isolated from spider venoms. Sometimes, these molecules from both groups are found as the quaternium ammonium salt of the terminal nitrogen (-N^+Me_3), like acetylcholine. Note that ageltoxins were later renamed to α-agotoxins and are more widely used in modern literature. The stamps show funnel-web spiders *Argiope argentata* and *Nephila clavipes*, which produce such acylpolyamine toxins.

polyamine backbone constitute the essential parts of these compounds shared by all known toxins of this class. Meanwhile, the linker amino acid(s) and the tail constitute the nonessential parts and are present only in some toxins. These acylpolyamines are hydrophilic molecules that provoke the fast paralysis of their prey. Historically, acylpolyamines have been classified into two fairly numerous groups (in total, more than 30 toxins of this type were identified): the ageltoxins (AG), which are the N-aroylpolyamines of five nitrogen atoms (pentamines) with the hydroxybenzene or (hydroxy)indolic fragment, and nephilatoxins (NPTX), polyamides with the methylcarbamide and (hydroxy)indolic functions (Figure 5.18B).

Ageltoxins are major constituents of the venom of the spiders *Agelenopsis* spp. (Agelenidae spiders), while nephilatoxins are found in the venoms of Nephilinae spiders (*Nephila clavata*, *Nephila clavipes*, etc.). Venoms of other orb-weaver spiders, *Argiope lobata and Argiope aurantia* (*Agelenidae spiders*), contain similar polyamine toxins, i.e., argiotoxin 636 (ArgTX-636) and argiotoxin 659 (ArgTX-659).

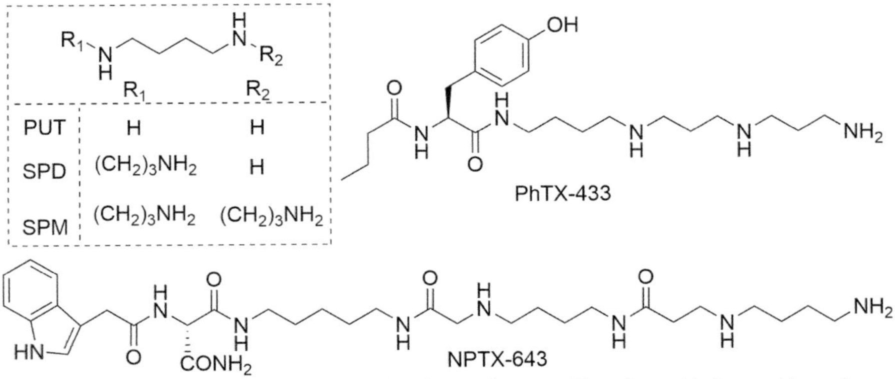

Figure 5.19. Biogenetic polyamines and polyamine conjugates with amino acids from spider and wasp venoms to paralyze insect prey. NPTX-643 shows subunit-specific activity on the glutamate receptors, while PhTX-433 inhibits both nicotinic acetylcholine and ionotropic glutamate receptors.

They are skillful builders of spider webs with their strange funnel shape. *Argiope* spider venom is still used as indigenous arrow poison for hunting in southern Africa.

Structural dissimilarity of spider and wasp toxins can be seen in Figure 5.19, where nephilatoxin-643 (NPTX-643, also JSTX-3), found in the venom of the Joro spider, *Nephila clavata*, and philanthotoxin-433 (PhTX-433), isolated from the venom of the Egyptian digger wasp (female) *Philanthus triangulum*, present. Additionally, we can observe the structures of primary biogenetic polyamines, putrescine (PUT), spermidine (SPD), and spermine (SPM), which are involved in a variety of fundamental biological functions of animals, plants, and microorganisms.

Interacting strongly with the neuromuscular junctions of insects [glutamate receptors, especially N-methyl-D-aspartate (NMDA)], acylpolyamines from spider and wasp venoms represent an exciting molecular basis for the design of novel pharmaceutical drugs. These drugs have potential applications in treating conditions such as hypoxia/ischemia and trauma of the central nervous system, seizures, hypoglycemia, and the pathogenesis of chronic neurodegenerative disorders, including amyotrophic lateral sclerosis, Huntington's, Alzheimer's and Parkinson's disease [54, 58]. However, one of the shortcomings of native spider and wasp acylpolyamine toxins is their lack of selectivity for the family of ionotropic glutamate receptors. An exception is that JSTX-3 is commercially available and has subunit-specific activity on the glutamate receptors.Therefore, developing specific agonists and antagonists for ionotropic glutamate receptors has potential. To obtain native spider and wasp acylpolyamines and their analogs, various synthetic protocols have been devised involving either the selective functionalization/protection of the amino functions of polyamines or the assembly of the polyamine skeleton from suitably protected amino building blocks [57–61].

A synthetic pathway for the preparation of acylpolyamine toxin NPTX-473B from the *N. borbonica* venom could be achieved starting with the coupling of the N-α-Boc-L-asparagine *p*-nitrophenyl ester **5.66** and mono-Ns-putrescine **5.67** to give triamine intermediate **5.68** (Figure 5.20). On the other hand, mono-Ns-1,3-diaminopropane precursor **5.69** converts into bromoethyl N-protected diamine intermediate **5.70**, which combines with **5.68** to afford the desired polyamine

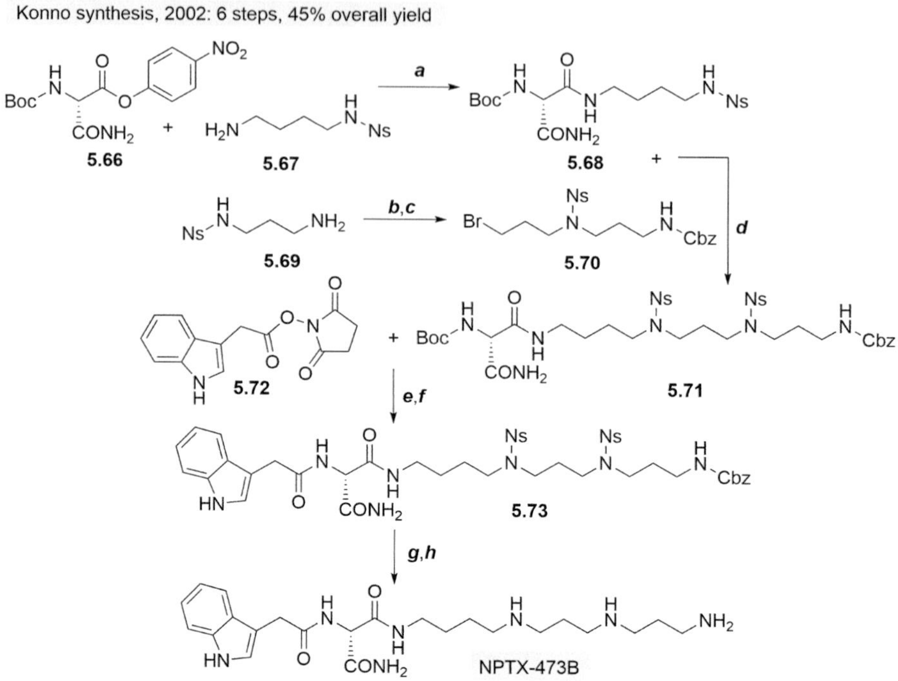

Figure 5.20. Solution-phase synthesis of NPTX-473B: (a) DMF (95%); (b) CbzCl, Et_3N/CH_2Cl_2 (99%); (c) 1,3-dibromopropane, Cs_2CO_3, DMF (91%); (d) Cs_2CO_3, TBAI, DMF (94%); (e) TFA, $CHCl_3$; (f) Et_3N, DMF (84%, two steps); (g) $HSCH_2CH_2OH$, DBU, DMF; (h) H_2-Pd-OH)$_2$, AcOH (60%, two steps). CBzCl = benzyl chloroformate, TBAI = tetrabutylammonium iodide.

backbone **5.71**. Note that a 2-nitrobenzenesulfonamide (Ns) group is a protecting and activating group.

The obtained polyamine **5.71** was later deprotected and condensed with indoleacetic acid N-hydroxysuccinimide ester **5.72** to afford the fully protected toxin **5.73**, which was deprotected to lead to the formation of NPTX-473B [62]. It is one of the most effective and short syntheses of acylpolyamines; for example, the solution-phase synthesis of argiotoxins from venoms of *Argiope* spiders contains 10−14 steps.

Regarding a wasp toxin, philanthotoxin-433, limited chemical syntheses of PhTX-433 have been reported to the present day, always using a multistep solution-phase strategy. In 2000, Hall and co-workers described the solid-phase synthesis of PhTX-433 achieved on a trityl diaminopropyl derivatized resin **5.74** (Figure 5.21) [63]. From **5.74**, two successive steps of amide coupling with N-(9-fluorenylmethoxycarbonyl) L-glutamic acid $HO_2C(CH_2)_2$NHFmoc and deprotection steps afforded dipeptide **5.75**, whose reduction followed by oxidative workup gave triamine **5.76**.

Its treatment with Boc provided the intermediate **5.77**, which was subsequently coupled with Fmoc-L-Tyr(*tert*-Bu), treated with piperidine, and coupled with butyric acid to afford N-Boc-protected toxin **5.78**. Then, its cleavage from the resin provided crude PhTX-433 as a tris(trifluoroacetate) ammonium salt. Since this first synthesis, other solid-phase syntheses have been described to successfully obtain PhTX-433 in

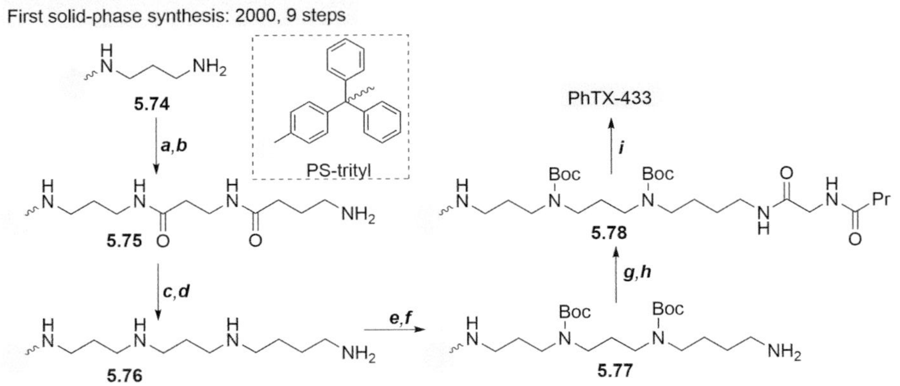

Figure 5.21. Solid-phase synthesis of PhTX-433: (a) 1. $HO_2C(CH_2)_2NHFmoc$, HBTU, HOBt, *i*-Pr_2EtN, DMF, 2. 20% Piperidine, DMF; (b) Repeat step (a) with $HO_2(CH_2)_3NHFmoc$; (c) 1. 1M BH_3, THF, 65°C, 48 hours, 2. Iodine, THF, *i*-Pr_2EtN-AcOH, rt, 4 hours; (d) 2-Ac-dimedone, DMF; (e) Boc_2O, *i*-Pr_2EtN, CH_2Cl_2; (f) 2% NH_2NH_2, DMF; (g) 1. HO-Tyr(*tert*-Bu)-Fmoc, HBTU, HOBt, *i*-Pr_2EtN, DMF, 2. 20% Piperidine, DMF; (h) $CH_3CH_2CH_2CO_2H$, HBTU, HOBt, *i*-Pr_2EtN, DMF; (i) TFA-H_2O, *i*-Pr_3SiH. Fmoc = 9-fluorenylmethyl chloroformate, HBTU = 2-(1*H*-benzotriazole-1-yl)-1,1,3,3-tetramethyluronium hexafluorophosphate, HOBt = 1-hydroxybenzotriazole.

nine steps. The desired toxin was obtained with an increased purity to > 90% with a 55% overall yield from dipeptide **5.75**.

5.2 Toxic Oxygen-Containing Heterocycles

Oxygen-containing heterocycles are ubiquitous in biologically active natural products. They are mainly found in carbohydrates, polyketides, peptides, and terpenoids. In previous sections of Chapter 2, we discussed some of them from plants, which can be a great source of inspiration in drug discovery. Now, we will examine ones from ants and beetles. In addition to defensive, toxic alkaloids, ants and beetles also use diverse oxygen-containing heterocycles, which are involved in insects' communication systems, either as pheromones or allomones, as well as their chemical weapons.

5.2.1 Iridoid Molecules from Meat Ants and Leaf Beetles

As seen before (Section 2.4), iridoids are oxygenated cyclic molecules abundant in plants. This name indicates that they derive from the volatile monoterpenes iridodial and iridomyrmecin, which are the components of the defensive secretion in the ants from the genus *Iridomyrmex*. In addition, dolicoidal and isoiridomyrmecin should be mentioned as their chemical defense against attackers. These ants and many others spray their defensive monoterpenoids, such as these toxic monoterpene aldehydes (Figure 5.22).

Structurally, iridodial and dolicoidal are tri-substituted pentane derivatives with three chiral centers and can present eight possible stereoisomers. As stereoisomers, iridomyrmecin and isoiridomyrmecin possess a cyclopentane-pyran bicycle condensed in the *cis*-configuration and belong to the non-glycosidic iridoids, specially

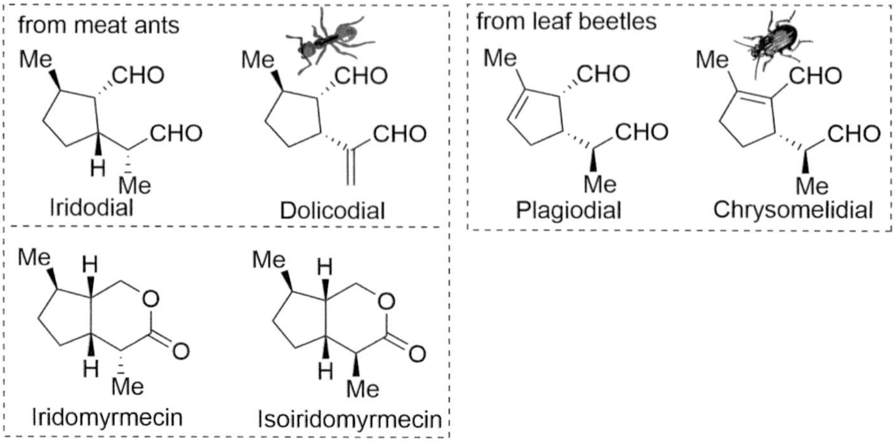

Figure 5.22. Structures of the monoterpene aldehydes from meat ants and leaf beetles.

iridoid lactones of type-I, which have a carbonyl group at C-3. Notably, iridodial can be thaumerized in solutions into cyclic form, corresponding to iridomyrmecin. Iridomyrmecin and isoiridomyrmecin are secreted as a toxic mixture from the anal glands of the Australian meat ant *Iridomyrmex detectus* or the Argentine ant *Iridomyrmex humilis*. These anal gland constituents represented approximately 4% and 1% of body weight, respectively.

The larvae of many chrysomelid beetles or leaf beetles also produce similar volatile methylcyclopentanoid molecules, such as plagiodial and chrysomelidial. The latter are isomeric pentene derived from the double C=C bond and might possess 8 and 4 diastereoisomers, respectively. However, the structural similarity between the plagiodial/chrysomelidial and iridodial/dolicoidal pairs is obvious (Figure 5.22). Interestingly, it was first isolated from a chrysomelid beetle, *Plagiodera versicolora*, in 1977, but in 2008, it was also discovered in glandular secretions from the oribatid mite, *Oribotritia berlesei*.

These toxic monoterpene aldehydes are found in many known plants where ants and beetles co-inhabit. However, it has been proven that insects do not "sequester" toxic substances from plants, but rather are biosynthesized *de novo* from dimethylallylpyrophosphate (geranyl diphosphate) via the mevalonic acid pathway, similar to that found for analogous compounds in plants [64–66]. For example, the larvae of *Gastrophysa viridula* and *Phaedon brassicae* beetles synthesize their molecules, chrysomelidial and plagiodial, in the abdominal glands and store them in the special vesicle (defensive section) where several enzymes, β-glucosidase, and oxidase, can be activated when these insects are in danger. The biogenesis of these molecules begins in the abdominal glands with geraniol via 8-hydroxy-geraniol **5.79** to give its conjugate **5.80** with the glucose transferred to the defensive section (Figure 5.23). Through enzymatic action, this is converted into 8-oxo-citral **5.81**, and then, depending on the specific species, the chrysomelidial (*G. viridula*) or/and the plagiodial (*P. brassicae*) are formed.

It is easy to see that these substances are biosynthetically related to citronellal and iridodial. In the laboratory, the three-step synthesis of natural (+)-iridodial from

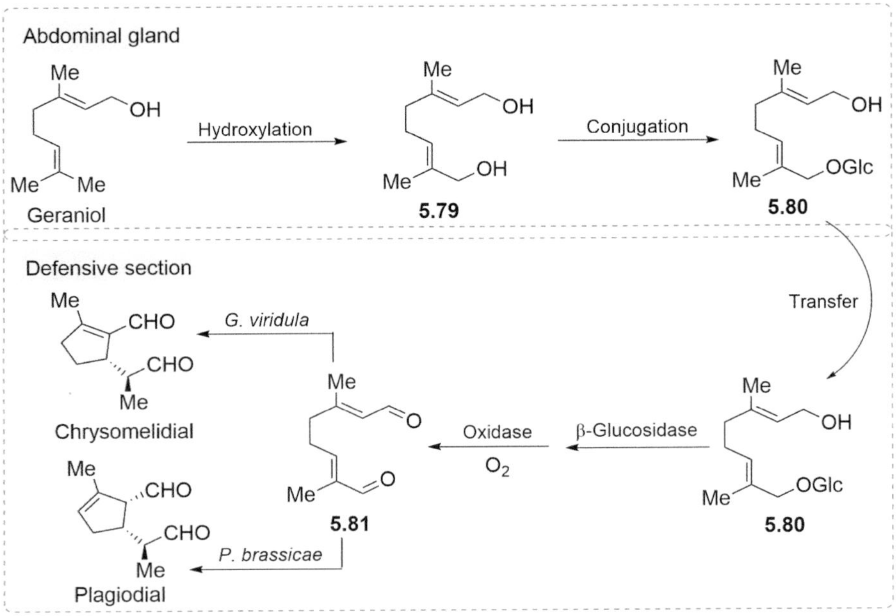

Figure 5.23. Biogenesis of monoterpenes from chrysomelid larvae, which are strictly phytophagous. Like other specialist herbivores, they are expected to use toxins from their host plants for their chemical defense.

(−)-citronellal via the 8-oxo-citral (**5.81**) formation in a solution of dilute HCl and acetone was realized by Prof. Robert Robinson et al. in 1959 [67], with the ring formation being a normal Michael reaction. However, the process was ineffective [68]. Pagnoni and co-workers also used 8-oxo-citral **5.81** in the base-catalyzed synthesis of (±)-chrysomelidial. Under specific base conditions (0.01 N NaOH in MeOH-H_2O, 98:2), the Cannizzaro reaction cyclization was smoothly achieved, producing a mixture of chrysomelidial (3*R*,8*R* and 3*S*,8*S*) and its diastereomer, dehydroiridodial (3*S*,8*R* and 3*R*,8*S*) in a 1.2:1 ratio in 82% yield. The mixture of two pairs of enantiomers was separated using silica gel column chromatography [69]. This mixture was also prepared from (*R*)-limonene [70].

To elucidate the absolute configuration of chrysomelidial, optically active (5*S*,8*S*)-chrysomelidiol **5.86** was synthesized in four steps, as summarized in Figure 5.24 [71]. The synthesis begins with the enantioselective Mukaiyama-Michael reaction between cyclopentenone derivative **5.82** and propionate silylketene acetal **5.83** in the presence of *in situ* formed a chiral bis(oxazoline)-Cu(II) ligand to provide a key chiral cyclopentanone **5.84**. This was then treated with NaH and diethyl chlorophosphate, $(Et_2O)_2P(O)Cl$, followed by the Gilman reagent (Me_2CuLi) to afford methylcyclopentene diester **5.85**. Finally, by reducing it with $LiAlH_4$, the desired diol **5.86** was formed (Figure 5.24).

Citronellal, an acyclic monoterpene, is also a close precursor in the construction of the bicyclic system of these iridoids with the hexahydro-4,7-dimethylcyclopenta[*c*]pyran-3(1*H*)-one structure. Therefore, its diastereoselective synthesis from citronellal is logical and well justified [72, 73]. Thus, we can observe the divergent synthesis

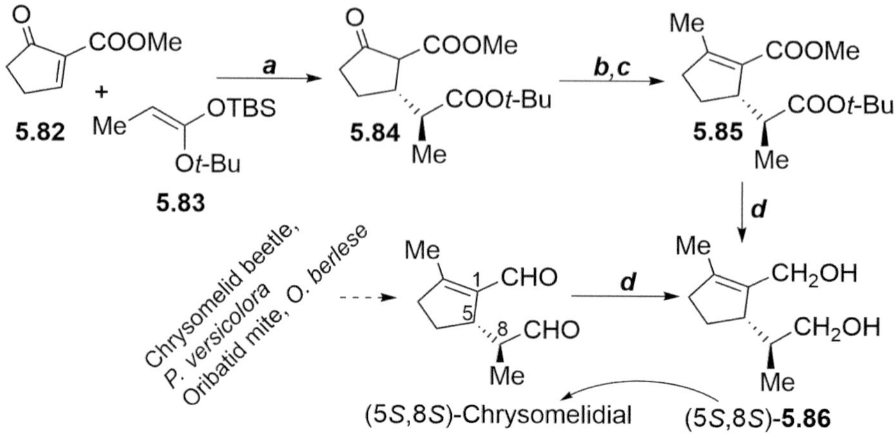

Figure 5.24. Elucidation of the configuration of chrysomelidial through the synthesis of optically active chrysomelidial: (a) (*R*,*R*)-2,2'-isopropylidene-bis(4-phenyl-2-oxazoline), $Cu(OTf)_2$, toluene (45%); (b) NaH, $ClP(O)(OEt)_2$, Et_2O (83%); (c) MeLi, CuI, Et_2O (80%); (d) $LiAlH_4$, Et_2O (52%).

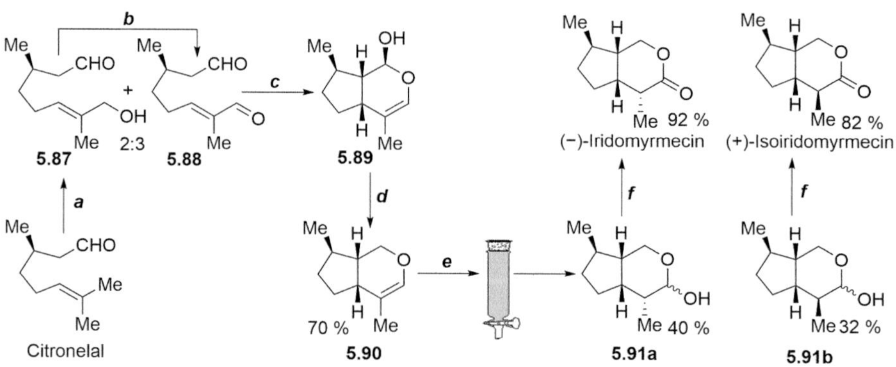

Figure 5.25. Six-step synthesis of iridomyrmecin and isoiridomyrmecin: (a) cat SeO_2, *tert*-BuOOH, salicylic acid, CH_2Cl_2, rt; (b) IBX, DMSO; (c) 1. NHMePh, Et_2O, 2. *p*-TsOH, wet THF; (d) Et_3SiH, $BF_3{\cdot}OEt_2$, CH_2Cl_2, −35°C; (e) $Hg(OAc)_2$, $NaBH_4$, THF, H_2O, 0°C; (f) TPAP, NMO, CH_2Cl_2. IBX = 2-Iodoxybenzoic acid, NMO = 4-Methylmorpholine 4-oxide, TPA= Tetrapropylammonium perruthenate ($N^+(C_3H_7)_4RuO_4^-$).

realized by Hofferberth et al. that allows the preparation of iridomyrmecin and isoiridomyrmecin in good quantities (Figure 5.25) [72]. Citronellal is first converted to nepetalactol **5.89** (see Figure 2.30) via a mixture **5.87/5.88** formation in three stages. A key process involves the treatment of 1,8-enedial **5.88** with N-methylaniline in ether, followed by acid-catalyzed hydrolysis of the resulting aminal. Synthesis of **5.88** is achieved in 30–45% yield over two steps from citronellal.

Thus, the obtained **5.89** with 71% yield for two steps is then hydrogenated with the $BF_3{\cdot}OEt_2$–Et_3SiH system to give bicyclic enol ether **5.90**, which is subjected to regioselective mercuration-reductive demercuration process providing diastereomers of acetals **5.91a–b**. This mixture is partially resolved into two pairs of two diastereomers (**5.91a** and **5.91b**) using silica gel chromatography. Separated like this, they are oxidized with a catalytic mild oxidizing agent, TPAP, to provide iridomyrmecin and isoiridomyrmecin in quantitative yield (Figure 5.25).

All these oxygenated cyclic molecules have potent antifungal, insecticidal, and antibiotic properties. Iridodial has high lachrymatory activity on vertebrates. As a result, total synthesis and synthetic studies of such arthropod iridoids are still active in synthetic organic chemistry.

Beetles from other families generate more exciting and complex oxygenated heterocycles than iridoids, which will be examined in the following sections.

5.2.2 Cantharidin, Tricyclic Terpenoid from Blister Beetles

Some beetles of the Meloidae family of diverse genera, i.e., *Lytta (= Cantharis) vesicatoria* and *Mylabris quadripunctata* or *M. phalerata*) belong to vesicant insects, producing a bright bluish-green liquid containing a toxic and irritant substance called cantharidin (CTD) [74]. The beetles use it as a defensive chemical weapon to protect them from predators. Although CTD is toxic to many arthropods, some frogs, toads, birds, and mammals consume CTD-containing beetles. For example, the great bustards, *Otis tarda*, a vulnerable and protected European bird species, ingest highly toxic beetles, *Berberomeloe majalis*, habituated the Mediterranean area.

CTD exhibited harsh effects on the gastrointestinal tract, kidney, and ureter in mammals, including humans. Nevertheless, despite these poisonous effects, historically, CTD has been used as a medicine for centuries [75–77]. The better studied among them are beetles of the *Lytta* and *Mylabris* genus. They are commonly known as blister beetles, where *Lytta vesicatoria*, a historical example of the cantharides, stands out. Initially, this species was only gathered in Spain. Because of that, this beetle was called *Musca hispanica*. In subsequent years, it was found in many European countries. Crude cantharidin is known as a Spanish fly.

These tiny insects (up to 2.5 cm long), variable in color, do not bite or sting. Their principal defensive armament, CTD, is in all male and female blister beetle body fluids and is biosynthesized by the male beetle. However, the female acquires CTD as a copulatory gift from her mate. Moreover, males accumulate large amounts of this molecule in their reproductive organs, mainly in male accessory glands. Nevertheless, there is still a gap in knowledge on the specific site of its *in vivo* production. The CTD concentration in the blister beetles differs between species; however, it is known that the CTD content for a blister beetle is about 5.0–5.4% in dry-weight bodies [75]. Thus, as a defensive means, cantharidin with a high concentration in the hemolymph is discharged from the leg joints.

From a chemical point of view, cantharidin is an oxygenated bridged bicyclic molecule of the 7-oxabicyclo[2.2.1]heptane (7-oxanorbornane) system with the stereochemistry of the *exo,exo*-2,3-dimethyl-2,3-dicarboxylic acid anhydride (Figure 5.26). Figure 5.26 shows that CTD is an adduct formed by 2,5-dimethyl-tetrahydrofuran with maleic anhydride, an unusual monoterpenoid C10. Noteworthy, its structure is symmetrical, and it is not optically active.

Subsequently, 10 carbon molecules of cantharidin were derived from mevalonate or farnesol precursor. However, the complete details of its biosynthesis are not known until now. The production of CTD seems to be a complex of several oxidation processes from farnesol, which comes from the mevalonate pathway via either linkage of two isoprene units in tail-to-tail or head-to-tail configuration (Figure 5.27) [78–80].

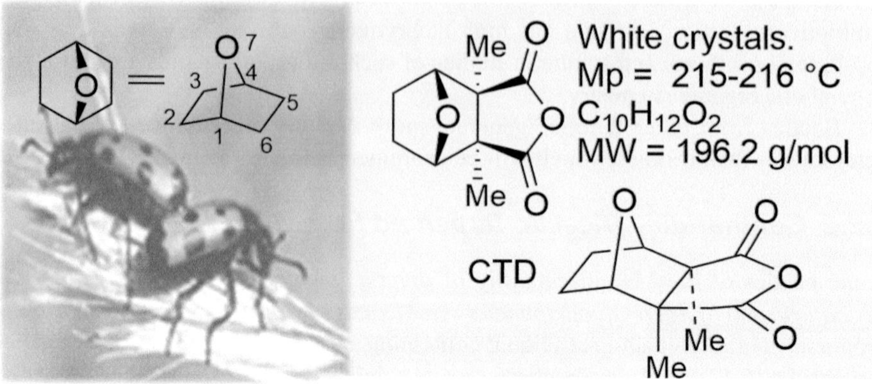

Figure 5.26. Structural representation of CTD with its 7-oxabicyclo[2.2.1]heptane system and an image of the *M. quadripunctata* blister beetles. The dried body of similar Chinese blister beetles (*M. caragnae* and *M. cichorii*) has been used medicinally in China for over 2,000 years. It is still used in folk medicine today in Asia. CTD was first obtained from *L. vesicatoria* in Spain (1810, Robiquet), and its molecular structure was identified by Gadamer in 1914.

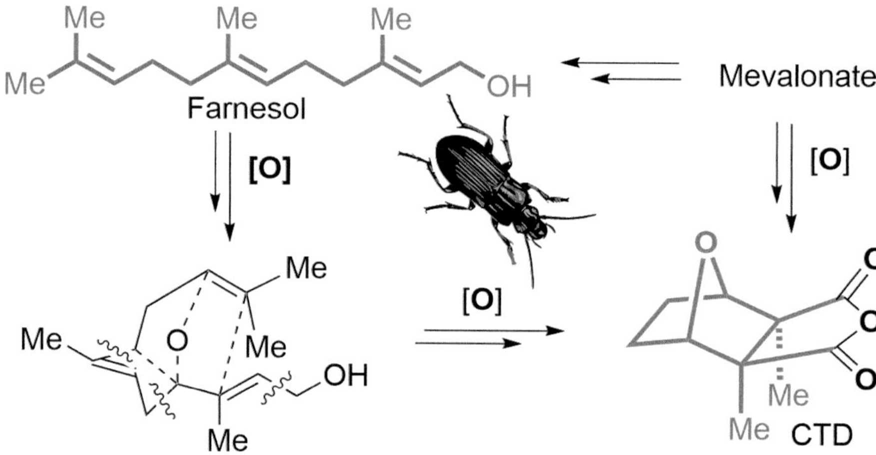

Figure 5.27. Possible biosynthesis of cantharidin through multiple oxidation processes on farnesol, which occurs via the mevalonate (MVA) pathway and cantharidinimide formation via an enzymatic insertion of the nitrogen atom.

It should be noted that *Mylabris phalerata* Pallas, habituated principally in Korea and China, is the only blister beetle in which the natural cantharidimide **5.92** and its several derivatives, like the 5-substituted cantharidinimide **5.93**, have been isolated as yellow oils in low concentrations, ≤ 0.002% [81–83]. Possible biosynthesis of these cantharidinimides occurs through a biochemical transformation in which the oxygen is substituted by the nitrogen atom, involving an enzymatic natural source of nitrogen (Figure 5.27).

CTD has a long history in popular and traditional medicine. In ancient times, the substance isolated from Spanish flies was wrongly used as a diuretic and aphrodisiac. Nowadays, it is known that pure CTD, given orally, can cause the death of a man.

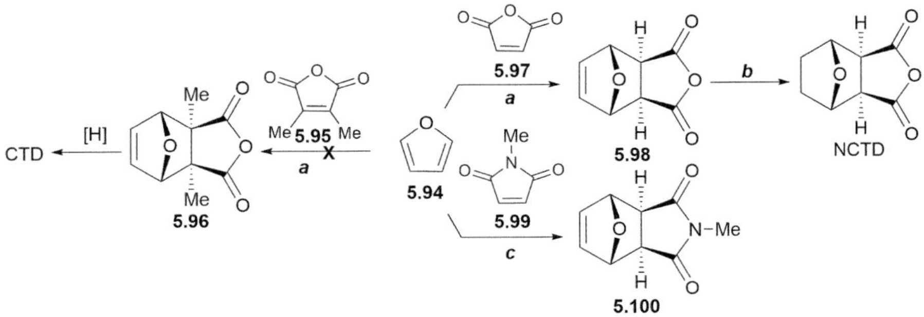

Figure 5.28. Diels-Alder cycloaddition of furan and maleimide derivatives for the CTD and NCTD synthesis: (a) toluene, 80°C, 24 hours; (b) H_2, 10% Pd/C, THF, rt, 4 hours; (c) THF, rt, 42 hours.

Its toxicity is high (LD_{50} = 0.5–1.0 mg/kg in mice). Therefore, in Asia, cantharidin is used diluted only for external use, in the form of oil or ointment; it is helpful as a medicine to treat certain rheumatic forms, warts, and skin molluscum contagiosum disease caused by the virus *Molluscum contagiosum*. There are two available liquid preparations for dermatologic use, Canthacur® (0.7% cantharidin) and Canthacur PS® (1% cantharidin 30%/salicylic acid/2% podophylotoxin); however, neither is approved by the Food and Drug Administration (FDA) [84].

Cantharidin can inhibit various tumor cell lines, and thus, it and its derivatives, especially cantharidinimide's analogs, are promising compounds for anticancer research [85]. They also display potent antiparasitic and insecticidal properties [86, 87].

Given these important properties, the practical synthesis is considered a relevant issue. The most apparent synthesis of cantharidin would be the direct cycloaddition of furan **5.94** and dimethylmaleic anhydride **5.95** to afford the respective cycloadduct (5,6-dehydronorcantharidin **5.96**) and its catalytic hydrogenation. However, this simple cycloaddition reaction under natural conditions failed, while this two-step process, using furan and maleic anhydride **5.97**, led to the formation of its closest analog, norcantharidin (NCTD). The Diels-Alder of furan and N-methylmaleimide **5.97** to obtain norcantharimide **5.100** occurs easier than the NCTD formation. (Figure 5.28).

The first total synthesis of cantharidin, consisting of 11 steps from furan **5.94** and dimethyl ester of acetylenedicarboxylic acid **5.101** as raw materials, was carried out by the American chemist G. Stork in 1951, but complete details were published in 1953 [88]. The synthesis is entirely linear and uses only the classical reactions within which the Diels-Alder reaction ([4+2] cycloaddition was used twice: diene **5.94** and dienophile **5.101** to give 3,6-epoxy-3,6-dihydrophthalate **5.102** and diene **5.102** and dienophile **5.103** to afford *exo*-5,8-epoxy-decalin derivative **5.105**) (Figure 5.29).

In the same year, Schenck and Wirtz improved this method. They used 1,3-butadiene **5.102** and 3,4-dimethyl maleic anhydride **5.95** as raw materials through linear seven-step synthesis to obtain the final product cantharidin [89]. These syntheses have long been the main method of cantharidin synthesis. However, using the Diels-Alder cycloaddition methodology was still attractive as a straightforward approach to CTD molecules. Only in 1980, Dauben et al. [90] were able to carry out the Diels-Alder reaction between furan **5.94** (diene) and 2,5-dihydrothiophene-

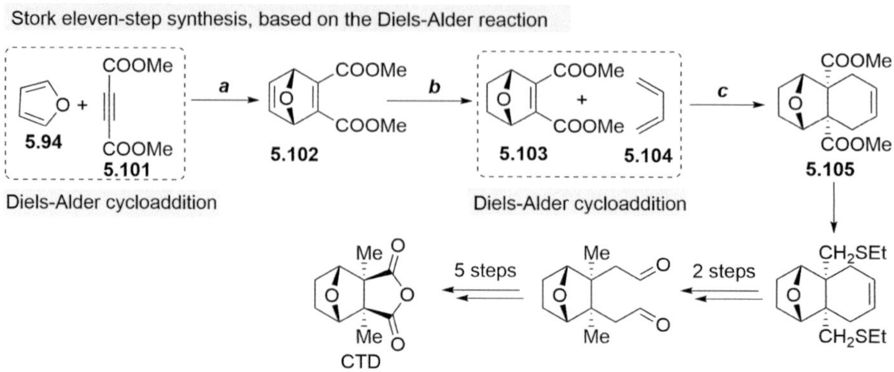

Figure 5.29. Simplified scheme of the first total synthesis of CTD realized by Prof. Gilbert Stork (1921–2017): (a) benzene, reflux, 14 hours (77%); (b) H_2, 10%-Pd/C, acetone, rt (78%); (c) EtOH, an autoclave, 100°C, 24 hours (88%).

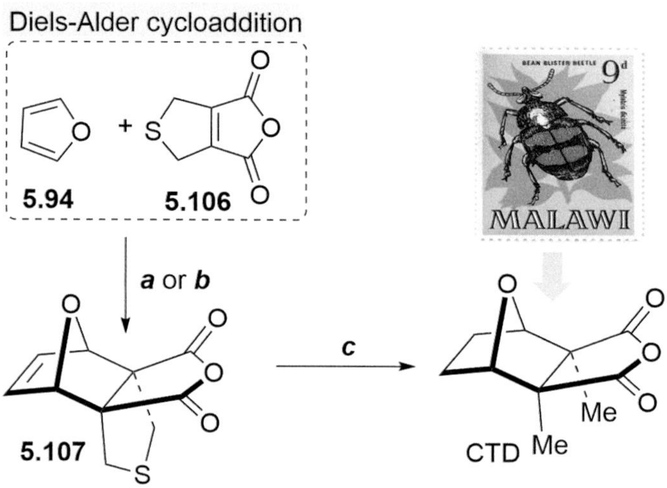

Figure 5.30. Straightforward synthesis of CTD using a high-pressure [4+2] cycloaddition reaction of furan and 2,5-dihydrothiophene-3,4-dicarboxylic anhydride or employing this cycloaddition with lithium perchlorate in diethyl ether medium at rt: (a) CH_2Cl_2, 15 Kbar, 5 hours (~ 75%); (b) 5.0 M $LiClO_4$, Et_2O, rt, 9.5 hours (~ 70%); (c) Ni-Raney, EtOAc, reflux (63%).

3,4-dicarboxylic anhydride **5.106** (dienophile) that requires high pressures (15 Kbar, 1 kbar = 986.9 atm) and sophisticated equipment (Figure 5.30).

The cycloadduct **5.107**, obtained in this way, was a major *exo*-isomer adduct, which could be separated by chromatography on silica gel. Having all the elements of the desired CTD molecule, this cycloadduct was directly converted to cantharidin, reducing the double bond and desulfurizing tetrahydrothiophene fragment in one step over Raney nickel in refluxing ethyl acetate (Figure 5.30). Thus, the synthesis appears simple and effective, but using high pressures in the chemical laboratory requires the appropriate infrastructure. Ten years later, these problems were solved by condensing **5.94** and **5.106** in 5.0 M lithium perchlorate in diethyl ether.

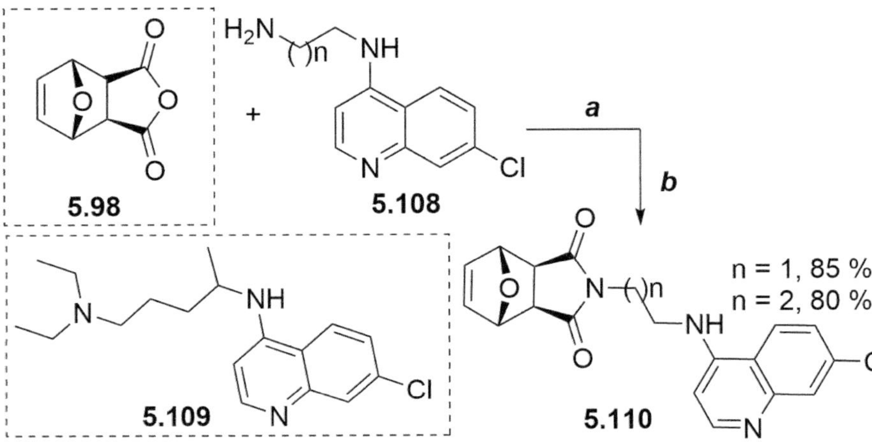

Figure 5.31. Simple synthesis of dehydronorcantharidinimide-chloroquine hybrids: (a) CH_2Cl_2, rt, 2 hours; (b) AcOANa, Ac_2O, 60°C, 5 hours.

Under these conditions, the cycloadduct formation occurs smoothly at room temperature [91].

Readily available 5,6-dehydronorcantharidin **5.98** is an excellent raw material for generating new norcantharidinimide analogs, functional in drug research. For example, a combination of the dehydronorcantharidin ring with potentially bioactive hetero-molecules **5.108**, similar to chloroquine **5.109**, could yield the novel 5,6-dehydronorcantharidinimide-chloroquine hybrids **5.110** in excellent yields, which are potential antitumor and antimalarial agents (Figure 5.31) [92].

In conclusion, we mention that biologically, cantharidin and cantharidinimide are easily absorbed into the lipid layers of the epidermal cell membrane. At the molecular level, they are excellent inhibitors of protein phosphatase 1 and protein phosphatase 2A (PP1 and PP2A), which play an essential role in the control of the cell cycle and apoptosis, and that is why their analogs exhibit potent antitumor activity *in vitro* and *in vivo* against various types of cancer cells, primarily hepatocellular carcinoma cells. However, cantharidinimides have better inhibition of PP1 and PP2A.

5.2.3 Pederin, Vesicant Toxic Amide with two Tetrahydropyran Rings from Paederus Beetles

The beetles of the Staphylinidae family (rove beetles) belonging to the genus *Paederus* (*Paederus fuscipes*, *P. brasiliensis*, and *P. ferus*), widely distributed worldwide, also have as a chemical weapon. This vesicant poison serves as a defense against their predators (spiders, frogs, lizards, etc.). Most species are small and do not exceed 1 cm in length. The first description of the species of the genus *Paederus* was made in 1775 by the Danish entomologist Fabricius. At this time, he only included two species. In the following two centuries, more than 600 species were identified, including *Paederus fuscipes*, the most famous species of southern Europe [93]. They inhabit riverbanks, swamps, and irrigated fields, feeding mainly on insects, mites, and soil nematodes, behaving like actual predators. Laboratory tests confirmed that it effectively protects *Paederus* sp. larvae against wolf spiders.

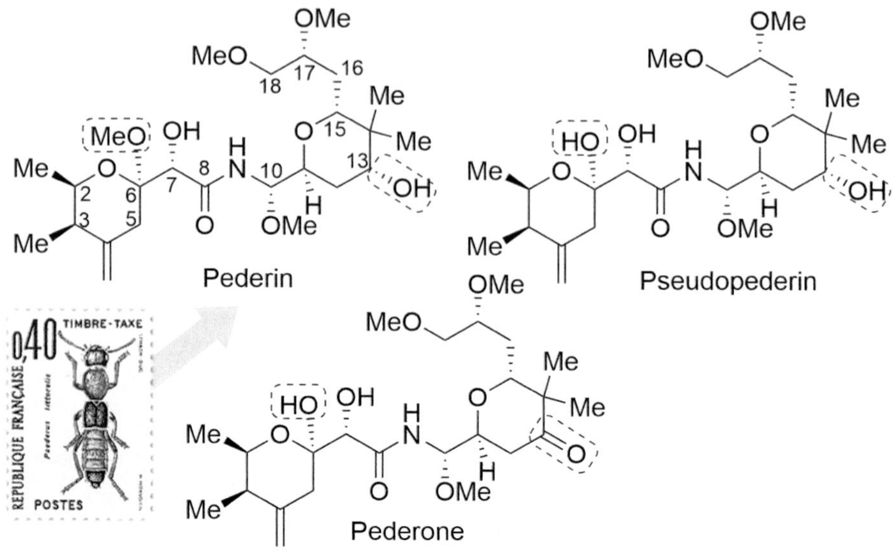

Figure 5.32. Structures of pederin, a principal venom constituent, and pseudopederin and pederone, minor components of toxic secretions by *Paederus* beetles. One species, *Paederus littovalis* is present on the stamp image.

Like meloid beetles, the paederus do not sting or bite; their defensive venom is in their hemolymph. Pederin (or pederine, $C_{25}H_{45}O_9N$, MW = 503.63 g/mol, mp = 122–122.5°C) (Figure 5.32), the main component of the venom of the beetle *P. fuscipes*, was isolated in 1949, but its structure was established in 1965 and then revised (by X-ray analysis) in 1968. Interestingly, 25 million of these beetles (approximately 100 kg) had to be sacrificed to establish the provisional structure of this venom component. Pederin is highly toxic, with an LD_{50} of 1 mg/kg or less for humans [93].

From the studies carried out, it is known that (1) each adult beetle weighs about 4 mg, with males containing about 0.025% of pederin by weight, while females contain up to 10 times more of the substance than males; (2) the minor components, pseudopederin and pederone, differ only from the functional groups. All three molecules are amides with the acetal function of two tetrahydropyran rings (Figure 5.32).

Throughout these years, *Paederus* beetles and their toxin, pederin, were in the spotlight, mainly due to their impressive and varied biological activities. In addition to its vesicant, necrotic, and toxic properties, the pederin toxin demonstrated potent *in vitro* cytotoxicity against a wide range of cancer cell lines (growth inhibition, IC_{50} 1.0–2.0 ng/mL) [94]. These antitumor effects are due to an inhibition of protein biosynthesis caused by binding to the ribosome. This mechanism causes a ribotoxic stress response and, ultimately, apoptosis.

Additionally, it acts as a powerful inhibitor of plant cell growth: The 0.5 μg/mL concentration of pederin strongly inhibits the germination of field lupine (*Lupinus albus*) seeds. It also produces a metaphase block in the mitosis of the meristems from the common onion (*Allium cepa*) roots [95].

Complete lysis of cells caused by pederin is observed at 20 nM concentration. In humans, it can cause peeling of the skin (dermatitis). This molecule's small concentration (1.5 ng/mL) is sufficient to generate a severe skin abscess. Interestingly, simple contact with the beetle on the skin is not enough to trigger the condition. Only releasing toxic hemolymph when acting on the epidermis causes vesicant dermatitis. This occurs by alarming the insect or crushing it. The "toxic juices" of a *Paederus sabaeus*, an African rove beetle, known as the "Nairobi fly", if they accidentally reached the cornea, would produce painful keratoconjunctivitis or periorbital dermatitis, popularly called "Nairobi eye" [96]. As mentioned before, the genus *Paederus* includes several hundred species, but only 30 species are recognized as responsible for human dermatoses.

The biosynthesis of pederin with the incorporation of radiolabeled acetate and glycine is consistent with its biosynthesis using an unusual polyketide pathway. The biosynthesis of pederin and its homologs is accomplished by hybrid polyketide/nonribosomal peptide synthases (PKS/NRPS) employing an unusual trans-acyltransferase (AT) PKS mechanism [97, 98]. However, the detailed biogenesis of pederin is not known so far.

Due to the pharmacological interest of this toxin, many synthetic efforts have been made to achieve a practical total synthesis of pederin and generate its analogs, which are necessary in pharmacological research. Chemically, the total synthesis of pederin presents a challenge to synthetic organic chemists. Its intricate structure, decorated with nine chiral centers, comprises the pederic acid unit **5.111** (A ring) and the pedamide part **5.112** (B ring) connected through the N-acylation of an imidate ester bridge. The first synthesizes of pederin were carried out in the years 82–90 [99–104]. The routes proposed in these historical syntheses and other more modern ones include the preparation of both functionalized heterocyclic A and B rings in parts and their subsequent union according to the retrosynthetic analysis of pederin illustrated in Figure 5.33.

The disconnection of the peptide bond, a logical step in the synthesis of pederin, reveals the two crucial precursors (pederic acid and pedamide), whose subsequent retrosynthetic analyses allow the total synthesis of these molecules to be better designed using the appropriate synthons. Matsumoto and his team reported the first total stereoselective synthesis of (+)-pederin in 1982 [99–101]. The crucial coupling reaction to afford an intermediate **5.115** was accomplished by briefly treating pederic acid derivative **5.113** with thionyl chloride in the presence of pyridine in dichloromethane, followed by adding methyl pedimidate **5.114** (Figure 5.34). The N-acyl aminal bridge was then constructed by reducing **5.115** with sodium borohydride in ethanol, generating a diastereoisomer mixture of N-acyl aminals **5.116** (dr 7:2). The subsequent O-benzylation process of the latter allowed the preparation of the needed product **5.117**, which was chromatographically separated as a minor component. Then, the requisite terminal alkene fragment in A ring was introduced in a two-step manner (oxidation of **5.117** and thermolysis of the obtained selenoxide derivative). Upon hydrolysis of the C-7 and C-13 benzoates, pederin was afforded.

Nakata and his group reported the synthesis of pederin three years later, employing different starting materials for the essential precursors and intermediates to improve synthetic procedures [102, 103]. It is worth noting that these syntheses

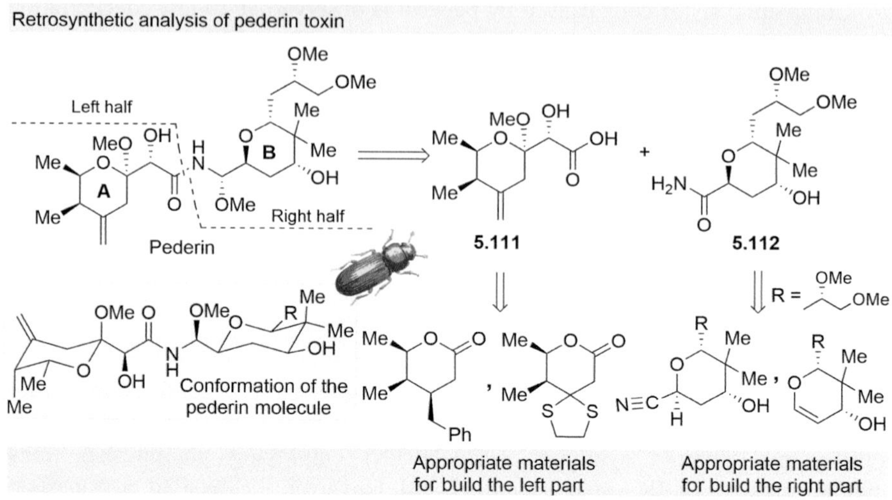

Figure 5.33. Retrosynthetic analysis of pederin, leading to the key precursor, pederic acid and pedamide derivatives.

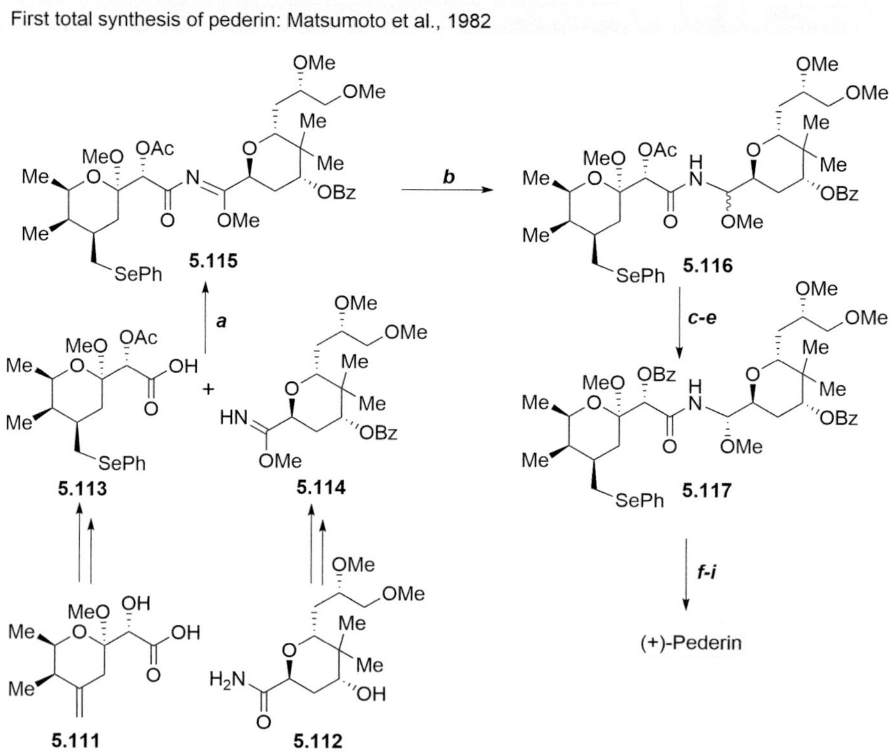

Figure 5.34. First total synthesis of pederin via forming N-acyl aminal bridge. Simplified scheme: (a) 1. $SOCl_2$, Py, CH_2Cl_2, rt, 5 minutes, 2. Et_3N, CH_2Cl_2, rt, 2 hours; (b) $NaBH_4$, –20°C, 30 minutes; (c) LiOH, MeOH, rt, 3 hours; (d) BzCl, DMAP, Py, rt, 12 hours; (e) chromatographic separation; (f) $NaIO_4$, MeOH, rt, 1 hour; (g) benzene, Et_3N, reflux, 30 minutes; (h) LiOH, MeOH, rt, 3 hours.

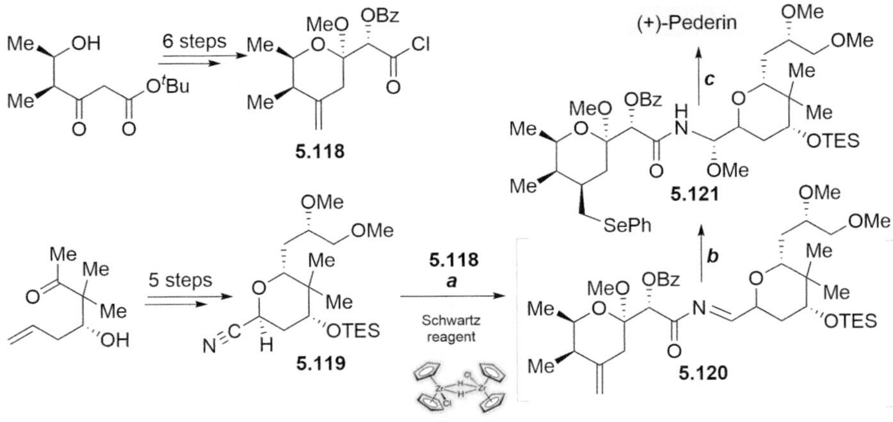

Figure 5.35. Floreancig synthesis of pederin, simplified scheme: (a) $Cp_2Zr(H)Cl$, CH_2Cl_2, rt, then –78°C, **5.118**; (b) MeOH, $Mg(ClO_4)_2$ (43%); (c) Bu_4NF, THF, then LiOH, MeOH, H_2O (86%). TES = triethylsilyl.

have the longest linear sequence (more than 20 steps). In 2007, Jewett and Rawal completed a concise total diastereoselective synthesis of pederin. The synthetic route proceeds in 12 steps and has an 11% overall yield [105]. Four years later, the Floreancig team also disclosed their finding on the total of pederin with an original and effective linear sequence from commercially available materials [106]. Like other strategies, the tactic carried out presents a condensation reaction between a pederic acid derivative **5.118** (A ring) and a tetrahydropyranyl nitrile **5.119** (B ring) formed in previous stages. This coupling reaction proceeded in the presence of the Schwartz reagent (zirconocene hydrochloride, $Cp_2Zr(H)Cl$ or $(C_5H_5)_2Zr(H)Cl$), forming the acylimine intermediate **5.120** *in situ* (Figure 5.35) and then, adding methanol and magnesium perchlorate to the formed intermediate led to the formation of acyl aminal precursor **5.121** as a single stereoisomer in 43% yield. The final step of the total synthesis of pederin was completed through an efficient one-pot cleavage of both protecting groups.

The aforementioned syntheses provide adequate amounts of pederin for pharmacological testing and access to its analogs, which are useful in developing structure-activity relationships and, therefore, in medicinal chemistry. This is also important because several close analogs of pederin are found in marine sponges of the genera *Theonella*, *Mycale*, and *Stylinos*, exhibiting potent antitumor activities (Section 7.3).

5.3 Conclusion

The unique structures and pharmacology of alkaloidal toxins from insects, mainly ants, wasps, and beetles, continue attracting many scientists' attention. Tetraponerins and solenopsins, the main toxic components of ant venoms, and defensive structurally diverse alkaloids (coccinelline, adaline, and harmonine) from fragile ladybird beetles are cyclic molecules containing at least one piperidine ring.

However, these insects "learned" to generate and use their chemical weapons, which were useful in human activity. These alkaloids function primarily as defensive compounds but possess antibacterial, antifungal, insecticidal, hemolytic, CNS, and anticancer properties. Notably, the antibacterial activity of harmonine is most pronounced against fast-growing mycobacteria, *Mycobacterium tuberculosis*, the growth of both chloroquine-sensitive and -resistant *Plasmodium falciparum* strains, and *Leishmania major* promastigotes.

Acylpolyamines are one of the main non-peptide compounds present in spider and wasp venom and also represent a promising alternative in the search for new molecules with antimicrobial action.

Iridoid-related chemistry continues to grow in the fields of biosynthesis, synthetic organic chemistry, medicinal chemistry, and chemical biology. Iridoids from meat ants and leaf beetles are also no exception. All these oxygenated cyclic molecules have strong antifungal, insecticidal, and antibiotic properties.

Blister beetles and their toxins (cantharidin and pederin) have inspired a notable amount of research in natural product isolation, structure determination, total synthesis, biological studies, and biosynthetic investigations. From these investigations, many interesting toxic compounds have been synthesized. Their studies demonstrate the importance of the symbiotic relationship between natural product chemistry, analysis (structure determination), pharmacology, synthesis, and design that can lead to the creation of new medicinal agents.

5.4 References

[1] Giribet, G. and Edgecombe, G.D. 2012. Reevaluating the arthropod tree of life. Annu. Rev. Entomol. 57: 167–186.

[2] Stork, N.E., McBroom, J., Gely, C. and Hamilton, A.J. 2015. New approaches narrow global species estimates for beetles, insects, and terrestrial arthropods. Proc. Natl. Acad. Sci USA 112(24): 7519–7523.

[3] Stork, N.E. 2018. How many species of insects and other terrestrial arthropods are there on Earth? Annu. Rev. Entomol. 63: 31–45.

[4] Laurent, P., Braekman, J.C. and Daloze, D. 2005. Insect chemical defense. pp. 168–225. *In*: Schulz S. [Ed.]. The Chemistry of Pheromones and other Semiochemicals II. Topics in Current Chemistry. Springer, Berlin, Heidelberg. 240.

[5] Fox, E.G.P. and Adams, R.M. 2022. On the biological diversity of ant alkaloids. Annu. Rev. Entomol. 67: 367–385.

[6] Braekman, J.C., Daloze, D., Pasteeis, J.M., Hecke, P.V., Declercq, J.P., Sinnwell, V. et al. 1987. Tetraponerine-8, an alkaloidal contact poison in a Neoguinean pseudomyrmecine ant, *Tetraponera* sp. Z Naturforsch C 42(5): 627–630.

[7] Merlin, P., Braekman, J.C. and Daloze, D. 1991. Stereoselective total synthesis of (±)-tetraponerine-8. Tetrahedron 47(23): 3805–3816.

[8] Barluenga, J., Tomas, M., Kouznetsov, V. and Rubio, E. 1994. An extremely short, stereoselective synthesis of (±)-tetraponerine-8. J. Org. Chem. 59(13): 3699–3700.

[9] Kem, W.R., Wildeboer, K., LeFrancois, S., Raja, M.A., Marszalec, W. and Braekman, J.C. 2004. Nicotinic receptor inhibition by *Tetraponera* ant alkaloids. Cell Mol. Neurobiol. 24: 535–551.

[10] Rouchaud, A. and Braekman, J.C. 2009. Synthesis of new analogues of the tetraponerines. Eur. J. Org. Chem. 2009(16): 2666–2674.

[11] Bosque, I., Gonzalez-Gomez, J.C., Loza, M.I. and Brea, J. 2014. Natural tetraponerines: a general synthesis and antiproliferative activity. J. Org. Chem. 79(9): 3982–3991.

[12] Renson, B., Merlin, P., Daloze, D., Braekman, J.C., Roisin, Y. and Pasteels, J.M. 1994. Biosynthesis of tetraponerine-8, a defence alkaloid of the ant *Tetraponera* sp. Can. J. Chem. 72(1) 105–109.

[13] Devijver, C., Braekman, J.C., Daloze, D. and Pasteels, J.M. 1997. Biosynthesis of tetraponerine-6. Evidence that two different pathways are operating in the biosynthesis of the two tetraponerine skeletons. Chem. Comm. (7): 661–662.

[14] MacConnell, J.G., Blum, M.S. and Fales, H.M. 1970. Alkaloid from fire ant venom: identification and synthesis. Science 168(3933): 840–841.

[15] MacConnell, J.G., Blum, M.S. and Fales, H.M. 1971. The chemistry of fire ant venom. Tetrahedron 27(6): 1129–1139.

[16] Xu, G. and Chen, L. 2023. Biological activities and ecological significance of fire ant venom alkaloids. Toxins 15(7): 439.

[17] Chen, J., Cantrell, C.L., Shang, H.W. and Rojas, M.G. 2009. Piperideine alkaloids from the poison gland of the red imported fire ant (Hymenoptera: *Formicidae*). J. Agric Food Chem. 57 (8): 3128–3133.

[18] Chen, L. and Fadamiro, H.Y. 2009. Re-investigation of venom chemistry of Solenopsis fire ants. II. Identification of novel alkaloids in *S. invicta*. Toxicon. 53(5): 479–486.

[19] Araújo, F.D., Botelho, M.A., Trigo, J.R. and Marsaioli, A.J. 2018. Absolute configuration of solenopsis piperidines is a tool to classify fire ants (Formicidae: Myrmicinae). J. Braz. Chem. Soc. 29(2): 398–403.

[20] Touchard, A., Aili, S.R., Fox, E.G.P., Escoubas, P., Orivel, J., Nicholson, G.M. et al. 2016. The biochemical toxin arsenal from ant venoms. Toxins 8(1): 30.

[21] Arbiser, J.L., Kau, T., Konar, M., Narra, K., Ramchandran, R., Summers, S.A. et al. 2007. Solenopsin, the alkaloidal component of the fire ant (*Solenopsis invicta*), is a naturally occurring inhibitor of phosphatidylinositol-3-kinase signaling and angiogenesis. Blood 109(2): 560–565.

[22] Howell, G., Butler, J., Farley, J.M., Liu, H.L., Nanayakkara, N.P.D., Yates, A. et al. 2005. Cardiodepressant and neurologic actions of *Solenopsis invicta* (imported fire ant) venom alkaloids. Ann. Allerg. Asthma. Im. 94(3): 380–386.

[23] Silva, R.C.C., Fox, E.G., Gomes, F.M., Feijó, D.F., Ramos, I., Koeller, C.M. et al. 2020. Venom alkaloids against Chagas disease parasite: search for effective therapies. Sci. Rep. 10(1): 10642.

[24] Leclercq, S., Braekman, J.C., Daloze, D., Pasteels, J.M. and Van der Meer, R.K. 1996. Biosynthesis of the solenopsins, venom alkaloids of the fire ants. Naturwissenschaften 83: 222–225.

[25] Leclercq, S., Daloze, D. and Braekman, J.C. 1996. Synthesis of the fire ant alkaloids, solenopsins. A review. Org. Prep. Proc. Inter. 28(5): 499–543.

[26] Comins, D.L. and Weglarz, M.A. 1991. Stereocontrolled preparation of cis-and trans-2,6-dialkylpiperidines via 1-acyldihydropyridine intermediates. Synthesis of (±)-solenopsin A and (±)-dihydropinidine. J. Org. Chem. 56(7): 2506–2512.

[27] Reding, M.T. and Buchwald, S.L. 1998. Short enantioselective total syntheses of the piperidine Alkaloids (*S*)-Coniine and (2 *R*, 6*R*)-*trans*-Solenopsin A via Catalytic Asymmetric Imine Hydrosilylation. J. Org. Chem. 63(18): 6344–6347.

[28] Daloze, D., Braekman, J.C. and Pasteels, J.M. 1994. Ladybird defence alkaloids: Structural, chemotaxonomic and biosynthetic aspects (Col.: Coccinellidae). Chemoecology 5: 173–183.

[29] Glisan King, A. and Meinwald, J. 1996. Review of the defensive chemistry of coccinellids. Chem. Rev. 96(3): 1105–1122.

[30] Housecroft, C.E. 2019. Nature's Chemical Weapons: Beetle Defenses: Chemical Education. Chimia 73(5): 420–421.

[31] Haulotte, E., Laurent, P. and Braekman, J.C. 2012. Biosynthesis of defensive coccinellidae alkaloids: incorporation of fatty acids in adaline, coccinelline, and harmonine. Eur. J. Org. Chem. 2012(10): 1907–1912.

[32] Daly, J.W., Secunda, S.I., Garraffo, H.M., Spande, T.F., Wisnieski, A., Nishihira, C., Cover, Jr. J.F. et al. 1992. Variability in alkaloid profiles in neotropical poison frogs (Dendrobatidae): genetic versus environmental determinants. Toxicon. 30(8): 887–898.

[33] Leong, R.L., Xing, H., Braekman, J.C. and Kem, W.R. 2015. Non-competitive inhibition of nicotinic acetylcholine receptors by ladybird beetle alkaloids. Neurochem. Res. 40: 2078–2086.

[34] Richards, D.P., Patel, R.N., Duce, I.R., Khambay, B.P., Birkett, M.A., Pickett, J.A. et al. 2022. (−)-Adaline from the *Adalia* Genus of ladybirds is a potent antagonist of insect and specific mammalian nicotinic acetylcholine receptors. Molecules 27(20): 7074.

[35] Patel, R.N., Richards, D.P., Duce, I.R., Birkett, M.A., Sattelle, D.B. and Mellor, I.R. 2020. Actions on mammalian and insect nicotinic acetylcholine receptors of harmonine-containing alkaloid extracts from the harlequin ladybird *Harmonia axyridis*. Pestic. Biochem. Physiol. 166: 104561.
[36] Röhrich, C.R., Ngwa, C.J., Wiesner, J., Schmidtberg, H., Degenkolb, T., Kollewe, C. et al. 2012. Harmonine, a defence compound from the harlequin ladybird, inhibits mycobacterial growth and demonstrates multi-stage antimalarial activity. Biol. Lett. 8(2): 308–311.
[37] Alam, N., Choi, I.S., Song, K.S., Hong, J.K., Lee, C.O. and Jung, J.H. 2002. A new alkaloid from two coccinellid beetles *Harmonia axyridis* and *Aiolocaria hexaspilota*. Bull. Korean Chem. Soc. 23(3): 497–499.
[38] Alujas-Burgos, S., Bayón, P. and Figueredo, M. 2018. Recent advances in the synthesis of azaphenalene alkaloids: First enantioselective approaches. Org. Biomol. Chem. 16(37): 8218–8229.
[39] Lima, D.J., Santana, A.E., Birkett, M.A. and Porto, R.S. 2021. Recent progress in the synthesis of homotropane alkaloids adaline, euphococcinine and N-methyleuphococcinine. Beilstein J. Org. Chem. 17(1): 28–41.
[40] Stevens, R.V. and Lee, A.W. 1979. Stereochemistry of the robinson-schoepf reaction. a stereospecific total synthesis of the ladybug defense alkaloids precoccinelline and coccinelline. J. Am. Chem. Soc. 101(23): 7032–7035.
[41] Sherwood, T.C., Trotta, A.H. and Snyder, S.A. 2014. A Strategy for complex dimer formation when biomimicry fails: Total synthesis of ten coccinellid alkaloids. J. Am. Chem. Soc. 136(27): 9743–9753.
[42] Tursch, B., Chome, C., Braemkan, J.C. and Daloze, D. 1973. Chemical ecology of arthropods. VIII. Synthesis and absolute configuration of adaline. Bull Soc. Chim. Belg. 82(9-10): 699–703.
[43] Lee, B., Kwon, J. and Yu, C.M. 2009. Asymmetric sequential allylic transfer strategy for the synthesis of (-)-Adaline and (-)-Euphococcinine. Synlett 2009(9): 1498–1500.
[44] Hernández-Ibáñez, S., Sirvent, A., Yus, M. and Foubelo, F. 2023. Stereoselective synthesis of 1-substituted homotropanones, including natural alkaloid (−)-adaline. Molecules 28(5): 2414.
[45] Braconnier, M.F., Braekman, J.C., Daloze, D. and Pasteels, J.M. 1985. (Z)-1, 17-diaminooctadec-9-ene, a novel aliphatic diamine from Coccinellidae. Experientia 41: 519−520.
[46] Braconnier, M.F., Braekman, J.C. and Daloze, D. 1985. Synthesis of the racemic form of (Z)-1, 17-Diaminooctadec-9-Ene, an aliphatic diamine from coccinellidae, determination of the absolute configuration of the (+)-Naturally-occurring antipode. Bull. Soc. Chim. Belg. 94(8): 605−613.
[47] Enders, D. and Bartzen, D. 1991. Enantioselective total synthesis of harmonine, a defence alkaloid of ladybugs (Coleoptera: Coccinellidae). Liebigs Ann. Chem. 1991(6): 569−574.
[48] Philkhana, S.C., Dhasaiyan, P., Prasad, B.L.V. and Reddy, D.S. 2014. Access to harmonine, a chemical weapon of ladybird beetles. RSC Adv. 4(58): 30923–30926.
[49] Abel, S.A.G., Olivier, W.J., Pederson, R.L., Bissember, A.C. and Smith, J.A. 2015. Complete stereocontrol in the synthesis of harmonine and novel analogues facilitated by a grubbs z-selective cross-metathesis catalyst. Aust. J. Chem. 68(12): 1815−1820.
[50] Zerkowski, J.A. and Solaiman, D.K. 2006. Synthesis of polyfunctional fatty amines from sophorolipid-derived 17-hydroxy oleic acid. J. Am. Oil Chem. Soc. 83(7): 621−628.
[51] Nagel, N.C., Masic, A., Schurigt, U. and Boland, W. 2015. Efficient synthesis of (*R*)-harmonine–the toxic principle of the multicolored Asian lady beetle (*Harmonia axyridis*). Org. Biomol. Chem. 13(18): 5139−5146.
[52] Langenegger, N., Nentwig, W. and Kuhn-Nentwig, L. 2019. Spider venom: components, modes of action, and novel strategies in transcriptomic and proteomic analyses. Toxins 11(10): 611.
[53] Abd El-Wahed, A., Yosri, N., Sakr, H.H., Du, M., Algethami, A.F., Zhao, C. et al. 2021. Wasp venom biochemical components and their potential in biological applications and nanotechnological interventions. Toxins 13(3): 206.
[54] Luo, L., Kamau, P.M. and Lai, R. 2022. Bioactive peptides and proteins from wasp venoms. Biomolecules 12(4): 527.
[55] Long, Q., Zhou, W., Zhou, H., Tang, Y., Chen, W., Liu, Q. and Bian, X. 2024. Polyamine-containing natural products: structure, bioactivity, and biosynthesis. Nat. Prod. Rep.; https://doi.org/10.1039/D2NP00087C.

[56] Rogoza, L.N., Salakhutdinov, N.F. and Tolstikov, G.A. 2006. Polymethyleneamine alkaloids of animal origin: II. Polyamine neurotoxins. Russ. J. Bioorg. Chem. 32:23–36.
[57] Karigiannis, G. and Papaioannou, D. 2000. Structure, biological activity and synthesis of polyamine analogues and conjugates. Eur. J. Org. Chem. 2000(10): 1841–1863.
[58] Estrada, G., Villegas, E. and Corzo, G. 2007. Spider venoms: a rich source of acylpolyamines and peptides as new leads for CNS drugs. Nat. Prod. Rep. 24(1): 145–161.
[59] Frølund, S., Bella, A., Kristensen, A.S., Ziegler, H.L., Witt, M., Olsen, C.A. et al. 2010. Assessment of structurally diverse philanthotoxin analogues for inhibitory activity on ionotropic glutamate receptor subtypes: discovery of nanomolar, nonselective, and use-dependent antagonists. J. Med. Chem. 53(20): 7441–7451.
[60] Négrel, S. and Brunel, J.M. 2021. Synthesis and biological activities of naturally functionalized polyamines: An overview. Curr. Med. Chem. 28(17): 3406–3448.
[61] Vassileiou, C., Kalantzi, S., Vachlioti, E., Athanassopoulos, C.M., Koutsakis, C., Piperigkou, Z. et al. 2022. New analogs of polyamine toxins from spiders and wasps: liquid phase fragment synthesis and evaluation of antiproliferative activity. Molecules 27(2): 447.
[62] Nihei, K.I., Kato, M.J., Yamane, T., Palma, M.S. and Konno, K. 2002. An efficient and versatile synthesis of acylpolyamine spider toxins. Bioorg. Med. Chem. Lett. 12(3): 299–302.
[63] Wang, F., Manku, S. and Hall, D.G. 2000. Solid phase syntheses of polyamine toxins HO-416b and PhTX-433. Use of an efficient polyamide reduction strategy that facilitates access to branched analogues. Org. Lett. 2(11): 1581–1583.
[64] Laurent, P., Braekman, J.C., Daloze, D. and Pasteels, J. 2003. Biosynthesis of defensive compounds from beetles and ants. Eur. J. Org. Chem. 2003(15): 2733–2743.
[65] Burse, A. and Boland, W. 2017. Deciphering the route to cyclic monoterpenes in Chrysomelina leaf beetles: source of new biocatalysts for industrial application? Z Naturforsch C. 72(9–10): 417–427.
[66] Kouda, R. and Yakushiji, F. 2020. Recent advances in Iridoid chemistry: biosynthesis and chemical synthesis. Chem. Asian J. 15(22): 3771–3783.
[67] Robert Robinson (1886–1975) was a British organic chemist and Nobel laureate acknowledged in 1947 for his research on plant dyestuffs (anthocyanins) and alkaloids.
[68] Clark, K.J., Fray, G.I., Jaeger, R.H. and Robinson, R. 1959. Synthesis of D-and L-isoiridomyrmecin and related compounds. Tetrahedron 6(3): 217–224.
[69] Bellesia, F., Ghelfi, F., Pagnoni, U.M. and Pinetti, A. 1986. The base-catalyzed cyclization of 10-oxocitral synthesis of chrysomelidal and dehydroiridodial. Tetrahedron Lett. 27(3): 381–382.
[70] Meinwald, J. and Jones, T.H. 1978. Synthesis and stereochemistry of chrysomelidial and plagiolactone. J. Am. Chem. Soc. 100(6): 1883–1886.
[71] Shimizu, N., Yakumaru, R., Sakata, T., Shimano, S. and Kuwahara, Y. 2012. The absolute configuration of chrysomelidial: a widely distributed defensive component among oribotririid mites (Acari: Oribatida). J. Chem. Ecol. 38: 29–35.
[72] Fischman, C.J., Adler, S. and Hofferberth, J.E. 2013. Divergent diastereoselective synthesis of iridomyrmecin, isoiridomyrmecin, teucrimulactone, and dolicholactone from citronellol. J. Org. Chem. 78(14): 7318–7323.
[73] Tammisetti, R., Chaudhari, P.D., Hong, B.C. and Chien, S.Y. 2023. From citronellal to iridoids: asymmetric synthesis of iridoids and their analogues via organocatalytic intramolecular Michael reactions. Org. Biomol. Chem. 21(20): 4200–4205.
[74] Carrel, J.E., McCairel, M.H., Slagle, A.J., Doom, J.P., Brill, J. and McCormick, J.P. 1993. Cantharidin production in a blister beetle. Experientia 49(2): 171–174.
[75] Puerto Galvis, C.E., Vargas Méndez, L.Y., Kouznetsov, V.V. 2013. Cantharidin-based small molecules as potential therapeutic agents. Chem. Biol. Drug Des. 82(5): 477–499.
[76] Wang, G., Dong, J. and Deng, L. 2018. Overview of cantharidin and its analogues. Curr. Med. Chem. 25(17): 2034–2044.
[77] Moed, L., Shwayder, T.A. and Chang, M.W. 2001. Cantharidin revisited: a blistering defense of an ancient medicine. Archiv. Dermatol. 137(10): 1357–1360.
[78] McCormick, J.P., Carrel, J.E. and Doom, J.P. 1986. Origin of oxygen atoms in cantharidin biosynthesized by beetles. J. Am. Chem. Soc. 108(25): 8071–8074.

[79] Lü, S., Jiang, M., Huo, T., Li, X. and Zhang, Y. 2016. 3-hydroxy-3-methyl glutaryl coenzyme A reductase: an essential actor in the biosynthesis of cantharidin in the blister beetle *E picauta chinensis* Laporte. Insect. Mol. Biol. 25(1): 58–71.

[80] Huang, Y., Wang, Z., Zha, S., Wang, Y., Jiang, W., Liao, Y. et al. 2016. *De novo* transcriptome and expression profile analysis to reveal genes and pathways potentially involved in cantharidin biosynthesis in the blister beetle *Mylabris cichorii.* PLoS One 11(1): e0146953.

[81] Nakatani, T., Konishi, T., Miyahara, K. and Noda, N. 2004. Three Novel Cantharidin-Related Compounds from the Chinese Blister Beetle, *Mylabris phalerata* PALL. Chem. Pharm. Bull. 52(7): 807–809.

[82] Nakatani, T., Jinpo, K. and Noda, N. 2007. Cantharimide Dimers from the Chinese Blister Beetle, *Mylabris phalerate* PALLAS. Chem. Pharm. Bull. 55(1): 92–94.

[83] Zeng, Y.B., Liu, X.L., Zhang, Y., Li, C.J., Zhang, D.M., Peng, Y.Z. et al. 2016. Cantharimide and its derivatives from the blister beetle *Mylabris phalerata* Palla. J. Nat. Prod. 79(8): 2032–2038.

[84] Vakharia, P.P., Chopra, R., Silverberg, N.B. and Silverberg, J.I. 2018. Efficacy and safety of topical cantharidin treatment for molluscum contagiosum and warts: a systematic review. Am. J. Clin. Dermatol. 19: 791–803.

[85] Naz, F., Wu, Y., Zhang, N., Yang, Z. and Yu, C. 2020. Anticancer attributes of cantharidin: involved molecular mechanisms and pathways. Molecules 25(14): 3279.

[86] Whitman, D.W., Andrés, M.F., Martínez-Díaz, R.A., Ibáñez-Escribano, A., Olmeda, A.S. and González-Coloma, A. 2019. Antiparasitic properties of cantharidin and the blister beetle *Berberomeloe majalis* (Coleoptera: Meloidae). Toxins 11(4): 234.

[87] Khan, R.A., Rashid, M. and Naveed, M. 2022. Cantharidin: a chemical precursor for the development of novel bioinsecticides. Bulg. Chem. Comm. 54(1): 19–28.

[88] Stork, G., Tamelen, E.E.V., Friedman, L.J. and Burgstahler, A.W. 1953. A stereospecific synthesis of cantharidin. J. Am. Chem. Soc. 75(2): 384–392.

[89] Schenck, G.O. and Wirtz, R. 1953. Eine weitere Synthese des Cantharidins. Naturwissenschaften 40(22): 581–581.

[90] Dauben, W.G., Kessel, C.R. and Takemura, K.H. 1980. Simple, efficient total synthesis of cantharidin via a high-pressure Diels-Alder reaction. J. Am. Chem. Soc. 102(22): 6893–6894.

[91] Grieco, P.A., Nunes, J.J. and Gaul, M.D. 1990. Dramatic rate accelerations of Diels-Alder reactions in 5 M lithium perchlorate-diethyl ether: the cantharidin problem reexamined. J. Am. Chem. Soc. 112(11): 4595–4596.

[92] Rojas Ruiz, F.A. and Kouznetsov, V.V. 2011. Property based design and synthesis of new chloroquine hybrids via simple incorporation of 2-imino-thiazolidin4-one or 1H-Pyrrol-2,5-dione fragments on the 4-amino-7-chloroquinoline side chain. J. Braz. Chem. Soc. 22: 1774–1781.

[93] Frank, J.H. and Kanamitsu, K. 1987. Paederus, sensu lato (Coleoptera: Staphylinidae): natural history and medical importance. J. Med. Entomol. 24: 155–191.

[94] Brega, A., Falaschi, A., De Carli, L. and Pavan, M. 1968. Studies on the mechanism of action of pederine. J. Cell Biol. 36: 485–496.

[95] Tiboni, O., Parisi, B. and Ciferri, O. 1968. The mode of action of pederin, a drug inhibiting protein synthesis in eucaryotic organisms. Plant Biosystem. 102(4): 337–345.

[96] Sadhasivamohan, A., Palaniappan, V. and Karthikeyan, K. 2022. Nairobi eye–"Wake and see" disease. Oman. J. Ophthal. 15(3): 417–418.

[97] Piel, J. 2010. Biosynthesis of polyketides by trans-AT polyketide synthases. Nat. Prod. Rep. 27(7): 996–1047.

[98] Hemmerling, F. and Hahn, F. 2016. Biosynthesis of oxygen and nitrogen-containing heterocycles in polyketides. Beilstein J. Org. Chem. 12(1): 1512–1550.

[99] Matsuda, F., Yanagiya, M. and Matsumoto, T. 1982. Total synthesis of (+)-pederine. A simple synthetic method for N-(1-methoxyalkyl) amides. Tetrahedron Lett. 23(39): 4043–4046.

[100] Matsuda, F., Tomiyoshi, N., Yanagiya, M. and Matsumoto, T. 1983. A new stereocontrolled synthesis of (+)-pederine. Unusual conformation around c-10—c-11 bond in pederine derivatives. Tetrahedron Lett. 24(12): 1277–1280.

[101] Matsuda, F., Tomiyoshi, N., Yanagiya, M. and Matsumoto, T. 1988. Stereocontrolled total synthesis of (+)-pederine. Tetrahedron 44(23): 7063–7080.

[102] Nakata, T., Nagao, S., Mori, N. and Oishi, T. 1985. Total synthesis of (+)-pederin. 1. Stereocontrolled synthesis of (+)-benzoylpedamide. Tetrahedron Lett. 26(52): 6461–6464.
[103] Nakata, T., Nagao, S. and Oishi, T. 1985. Total synthesis of (+)-pederin. 2. Stereocontrolled synthesis of (+)-benzoylselenopederic acid and total synthesis of (+)-pederin. Tetrahedron. Lett. 26(52): 6465–6468.
[104] Jarowicki, K., Kocieński, P., Marczak, S. and Willson, T. 1990. A synthesis of (+)-pederin. The metallated dihydropyran approach. Tetrahedron Lett. 31(24): 3433–3436.
[105] Jewett, J.C. and Rawal, V.H. 2007. Total synthesis of pederin. Angew Chem. 119(34): 6622–6624.
[106] Wu, F., Green, M.E. and Floreancig, P.E. 2011. Total synthesis of pederin and analogues. Angew Chem. Int. Ed. 50 (5): 1131–1134.

Chapter 6

Toxic Heterocyclic Molecules used by Tetrapods

6. INTRODUCTION: CHEMICAL WEAPONS FROM THE SKIN AND FEATHERS OF VERTEBRATES TO SURVIVE

The superclass Tetrapoda, a fascinating group of terrestrial vertebrates with four limbs, includes reptiles, birds, mammals, and amphibians. In this chapter, we delve into the latter with their unique survival strategies. Among these, we encounter the intriguing world of toads and frogs, some mammals, and birds. These creatures provide us with a captivating subject of study with their ability to use toxic and noxious molecules to survive.

The toads and frogs belong to *Anura*, the most significant order of amphibians, generally referred to by the common word "frogs". This order includes nearly 8,000 species divided into 33 families. However, there are some noticeable differences in physical aspects between toads and frogs: (i) the first is warty, while the second possesses smooth, slimy skin; (ii) frogs are more likely to be brightly colored than toads; and (iii) frogs have long legs for hopping, while toads sport stumpy bodies and shorter legs for walking. For our interest, it is more important that their skin secretions are a rich source of bioactive molecules, which allow them to distract predators and escape [1–3].

Some wild birds and rats are added to the list of toxic vertebrate animals, which already includes toads and frogs. It is strange to read this. Do these tetrapods really exist? Until 1992, we had no idea. Now, we know that some avian species are toxic to humans and other animals when touched or consumed. Poisonous birds sequester or obtain toxic substances from the plants or animals they feed on, most commonly from insects [4].

6.1 Amphibian Neurotoxins: Alkaloidal Molecules Coming from Frog Skins

The skin of amphibians exercises a broad spectrum of functions, which are essential for their homeostasis and interaction with the environment. Diverse multicellular skin (mucous and granular) glands produce milky fluid secretions containing a real cocktail of toxic, complex mixtures of alkaloids, biogenic amines, peptides, and more. These secretions are involved in both vital processes and defense strategies against potential aggressors, including large predators or pathogenic microorganisms, such as fungi, bacteria, and viruses. When captured, many frogs respond immediately by increasing their body lubrication, making it challenging to keep them under a human hand or in the mouth of a predator.

The Dendrobatidae family comprises more than 180 species, generally divided among various genera. This section discusses frogs of the genera *Dendrobates, Oophaga, Phyllobates*, and *Epipedobates*.

The most striking characteristic of this frog family is its shiny, colorful skin, which can be observed in a wide range of colors, such as bright yellow and orange to dark blue and red. The presence of toxic molecules in the skin and its intense and striking colors are closely related. These colors alert the frogs' predators that they contain dangerous substances that can considerably affect the "well-being" of the aggressors. This color function is known as aposematic coloration, and the phenomenon is called aposematism, which is almost always defensive. So different and strikingly tiny poison-arrow frogs belong to the family Dendrobatidae [5, 6].

Initially, it was thought that the presence of alkaloids in frogs resulted from biosynthesis. However, the complete and meticulous research of scientist John W. Daly (1933–2008) discovered that most of these alkaloids are sequestered unchanged from their diet, primarily arthropods [7, 8]. The fact that captive-bred poison dart frogs had no alkaloids was a turning point in the biosynthetic and sequestration hypothesis studies.

The frogs of *Oophaga, Dendrobates,* and *Epipedobates* genera contain various defensive chemical weapons based on different heterocyclic systems: indolizidine, decahydroquinoline, 1-azaspiro[5.5]undecane, and piperidine. However, these are formed from the same probable biogenetic predecessor found in ants or beetles, which possess a piperidine derivative A (Figure 6.1).

All these lipophilic alkaloid molecules have several chiral centers and are found as racemates (two isomers) or chiral forms (one of the two enantiomers). The typical piperidine ring for these alkaloids is usually in the more energetically stable chair conformation. Notably, the designation of some dendrobatid skin alkaloids as "toxins" is misleading. Some compounds are ineffective in our tests, not toxic *in vivo*, or weak in commonly used *ex vivo* assays [9].

Note also that in the scientific literature of the 80s–90s, some species, such as *D. pumilio*, *D. speciosus, D. histrionicus*, and *D. tricolor*, were the subjects of many chemical-pharmacological studies and were described as members of the genus *Dendrobate*s. These species were then reclassified. The first three belong to the genus *Oophaga* and are known as *O. pumilio* (red and blue arrow frog), *O. speciosus* (or speciosa,- splendid poison frog) and *O. histrionicus* (harlequin poison frog), respectively. The same thing happened with the species *Dendrobates tricolor*: now, it belongs to the genus *Epipedobates* and is called *Epipedobates tricolor*.

This chapter gives old and new names (in parentheses) to avoid confusion and discrepancies with these biological classifications. With this matter clarified, we can proceed with describing some species of the Dendrobatidae family.

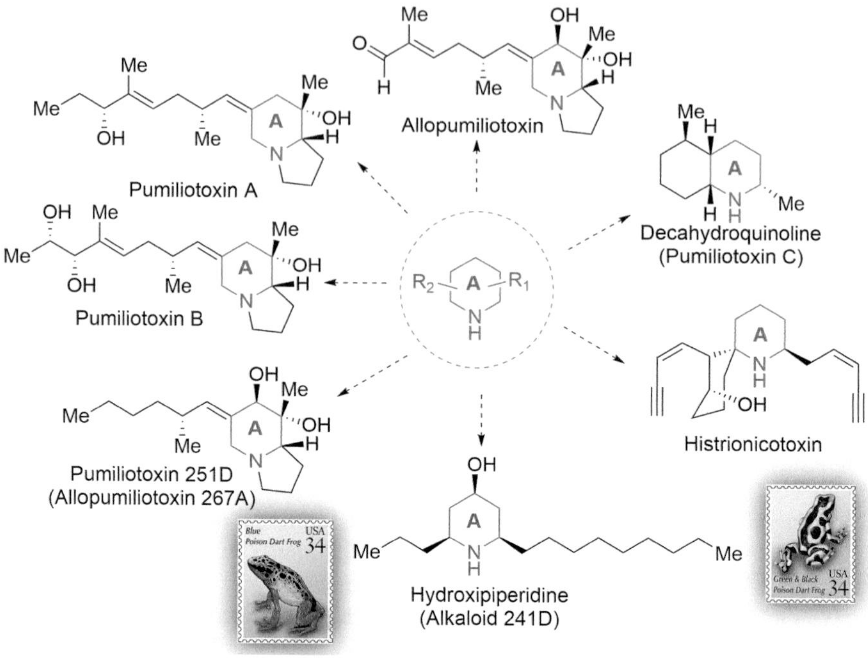

Figure 6.1. Structural diversity of toxic alkaloids isolated from the skin of frogs of the *Oophaga*, *Dendrobates,* and *Epipedobates* genera. These alkaloids are possibly generated from the identical biogenetic predecessor, i.e., are acquired via the food chain (synthesized *de novo*). The stamps show images of the poison dart frogs.

6.1.1 Pumiliotoxin Alkaloids: Indolizidine Compounds from the Frogs Oophaga Pumilio

The skin of miniature and beautiful Panamanian frogs, *Dendrobates* (now *Oophaga*) *pumilio*, so-called poison-arrow (dart) frogs, provides characteristic indolizidine alkaloids known as pumiliotoxins (PTXs) and allopumiliotoxins (aPTXs). Among these groups (~ 80 alkaloids), we will mention only the most famous, which are major and present in the skin of dendrobatid frogs at levels of up to 100–200 μg per frog [10]. It is necessary to slowly observe the structures of these toxins to see the structural differences. First, alloPTXs contain a hydroxyl group at the C-7 position of the indolizidine ring; second, both PTX and aPTXs differ in the chemical nature of the side chain, R. Thus, the pumiliotoxin alkaloids are characterized by the bicyclic 8-hydroxy-8-methyl-6-(*E*)-alkylidene-indolizidine ring system with the natural (8*S*,11*R*)-configuration. In contrast, the allopumiliotoxins contain an additional hydroxy group at C-7. Both are alkylidene-indolizidine derivatives (Figure 6.2). However, they can be named l-azabicyclo[4.3.0]nonane derivatives. Interestingly, PTX-A and -B were the second and third dendrobatid alkaloids to be isolated and were initially obtained in 1967 from the skin of the Panamanian poison frog *D. pumilio*. Elucidation of their structure was problematic due to their instability in acid, which was likely due to their allylic hydroxyl group. The structure of these toxins remained unknown until 1980, when a simpler alkaloid, pumiliotoxin 251D,

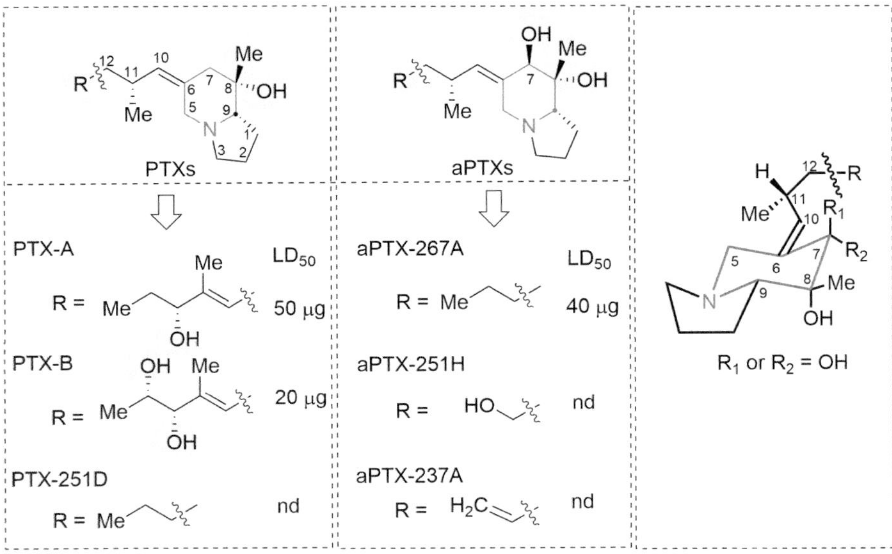

Figure 6.2. Structural difference between the selected pumiliotoxins/allopumiliotoxins and their toxicity. The figure shows the stereochemistry and conformation of these indolizidine alkaloids. Note that PTX analogs with the natural (8*S*,11*R*)-configuration are, in all cases, more active than diastereomers epimeric at one or the other of these positions. Nd = not determined.

was isolated as the major alkaloid component of skin extracts of the Ecuadorian poison frog, *D.* (now *Epipedobates*) *tricolor*. Interestingly, the abbreviation in the name of pumiliotoxin 251D indicates the molecular mass of this toxin, as in other cases of alkaloids and peptides present in animal venoms.

These valuable molecules are found in small quantities in the epidermis of arrow frogs. The studies carried out by Daly in the 1960s show interesting and unique data (Table 6.1): the average concentration of the mentioned alkaloid substances in the skin of a specimen is at the microgram level, while the average weight of the skin of a frog does not exceed 300 milligrams.

From these data, it can be deduced that, unfortunately, many of these diminutive frogs must be sacrificed for each toxin's isolation and structural and biological characterization. For example, alcoholic extraction of the skin of 250 *O. pumilio* (*D. pumilio*) specimens and chromatographic isolation of alkaloidal products gave 17 mg of PTX-A, 20 mg of PTX-B, and 16 mg of PTX-C. The original name "pumiliotoxin C" (PTX-C), which is a derivative of decahydroquinoline (5-methyl-

Table 6.1. Some characteristics of PTX alkaloids and their amounts in the skin of the specific anuran.

Toxin	Empirical Formula	Molecular Mass, g/mol	Average Quantity, μg Per Frog	Sources	Average Skin Weight of a Specimen
PTX-A	$C_{19}H_{33}NO_2$	307	80	*O. pumilio*	~ 90 mg
PTX-B	$C_{19}H_{33}NO_3$	323	120	*O. pumilio*	~ 90 mg
PTX-215D	$C_{16}H_{29}NO$	251	37	*E. anthonyi*	nd

2-propyl-decahydroquinoline), is no longer used due to possible confusion with the true pumiliotoxins PTX-A, PTX-B which are derivatives of indolizidine (Figure 6.1). Several substituted decahydroquinolines, such as PTX-C, act as non-competitive blockers of nicotinic receptors. PTX-C has local anesthetic activity acting on sodium channels and exhibits less toxicity: its LD_{50} is 400 μg per mouse of 20 g.

The members of this group are quite toxic, showing LD_{50} values are about 50 μg per mouse of 20 g. For example, PTX-A causes death in mice at subcutaneous (s.c.) doses as low as 50 μg per mouse, and alloPTX-267A showed 40 μg LD_{50} values. The most toxic is the PTX-B; its LD_{50} = 20 μg per mouse. It is noteworthy that an isomer of alloPTX-267A with the (8*R*,11*S*)-configuration is inactive. These data are comparable with a minimum lethal dose for mice of about 2 mg/kg, while a 10 mg/kg dose of PTX (+)-251D administered s.c. to mice occasioned convulsions, hyperalgesia, and death [11].

These alkylidene-indolizidine alkaloids display particularly significant pharmacological and pesticidal activities. Pumiliotoxins are highly toxic and are believed to protect frogs against predators. Thus, some of them have been investigated against ectoparasitic arthropods. For example, PTX-251D, with a simple alkyl side chain, showed good insecticidal activity against larvae of the tobacco budworm (*Heliothis virescens*), a highly polyphagous pest of many economically important crops worldwide. When injected into budworm larvae, (+)-PTX-251D causes convulsions with only a dose of 10 ng per larva and exhibits an LD_{50} of 150 ng per larva [12]. It was about 100-fold less active than the commercial pyrethroid fenvalerate in the injection assay and about 30-fold less active in the topical assay. Interestingly, PTX-A (now designated PTX-(+)-307A) and PTX-B (now PTX-(+)-323A), which are highly toxic to mice, were weakly toxic to budworms, being 20-fold less active than PTX-251D. When female yellow fever mosquitoes (*Aedes aegypti*), a hematophagous and pathogen vector, were tested for responses to PTX molecules, it was found that both enantiomers, natural PTX (+)-251D and unnatural PTX (-)-251D presented reduced landing and feeding by these mosquitoes. However, the first did so at lower concentrations and provoked contact toxicosis, as showed by impaired flight and leg autotomy [13].

PTX molecules were also tested against red fire ants (*Solenopsis invicta*), one of the world's most damaging and invasive pest species. Studies of ants' responses to animal-derived alkaloids are focused on feeding deterrents in the defensive secretions of arthropods. Examining an alkaloid PTX-251D as a contact deterrent of these ants, Meer and coworkers observed the efficacy of PTX-251D against *S. invicta*, which was expressed in the reduced ambulation and elicited convulsions in ants at a low concentration (386.5 ng/cm^2) that is lower than its estimated cutaneous concentration (6–9 μg/cm^2) [14].

Thus, penetrating the insect cuticle, lipophilic PTX-251D molecule induces convulsions, leg autotomy, immobility, and death. The mentioned above contact toxicity is consistent with a function in anuran chemical defense. In addition to this toxicity, PTX molecules are positive modulators of voltage-gated sodium channels (VGSCs), showing cardiotonic activity [15]. However, unlike other PTXs (PTX-A, PTX-B, etc.), PTX-251D does not share this property but can cause convulsions and death in mammalians and insects. It inhibited the Na^+ influx through the mammalian

VGSCs and affected the steady-state activation and inactivation. Additionally, PTX-251D inhibited the K^+ efflux through voltage-gated potassium channels (VGPCs) [16]. Thus, the PTX-251D molecule inhibits both VGSCs and VGPCs, which is why this toxin is more active against arthropods in the abovementioned tests.

In 2007, Peter Kovacic suggested that the highly conjugated iminium form **6.1** may be responsible for the toxicity of the PTX-251D alkaloid [17] (Figure 6.3). The PTX-251D structure could be converted into toxic iminium **6.1** during a metabolic oxidation reaction, which may start with the facile dehydration of the tertiary alcohol, promoted by conjugation with the exocyclic alkene, generating a quinoid-like intermediate. Tertiary amine, a common moiety in alkaloids (salts or *N*-oxides), is readily converted to iminium.

Due to the scarcity of the mentioned alkaloids and natural sources, the chemical literature describes different preparation methods. Prof. Larry E. Overman and colleagues were responsible for the early syntheses of pumiliotoxins and allopumiliotoxins [18]. They first analyzed the structural peculiarity of these toxins, considering the difficulty of stereochemical control of the exo-(*Z*)-alkylidene side, and developed an original synthetic strategy based on the iminium ion-vinylsilane cyclization process. It should be noted that a stereoselective introduction of the alkylidene moiety at the C-6 position still poses a considerable challenge. Their strategy of assembly of the target alkaloids is summarized in Figure 6.4.

This retrosynthetic analysis allowed us to obtain pyrrolidine epoxide **6.3** and various alkynylsilane precursors **6.4**. Both precursors are readily available from L-proline **6.5** and trimethylsilyl-acetylene derivatives **6.6.** To obtain PTX 251D, the simplest alkaloid of this series, (*R*)-silylalkyne **6.10** and *N*-(benzyloxycarbonyl)-

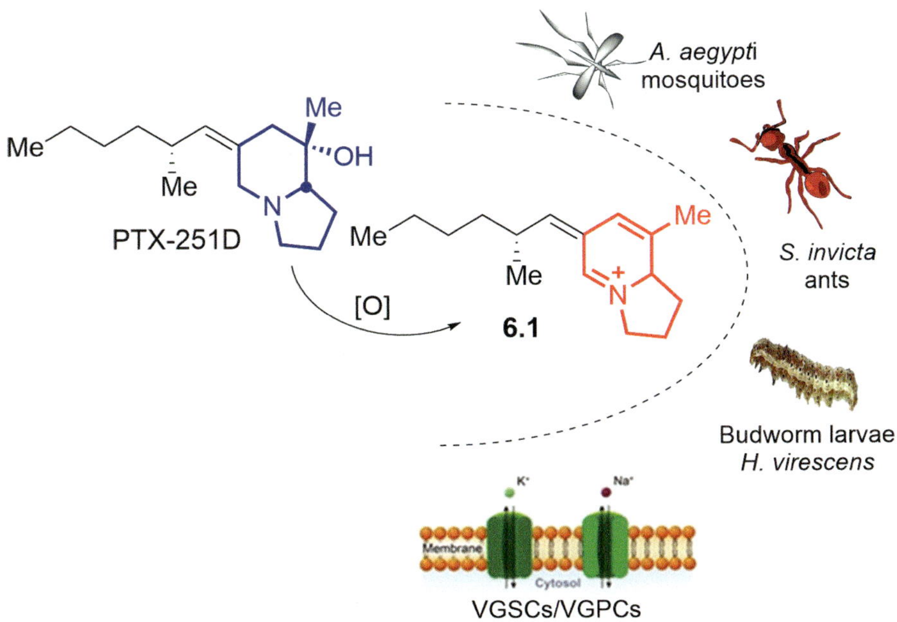

Figure 6.3. Iminium form of PTX-251D **6.1**, responsible for toxicity against arthropods and the inhibition of ion channels.

L-proline epoxide **6.11** were prepared from the respective (*S*)-1-heptyn-3-ol (**6.8**) and *N*-(benzyloxycarbonyl)-L-proline (**6.9**) (Figure 6.5) [19]. A crucial step of its synthesis is the generation of vinylalanate **6.12** from **6.10**. Then, it was reacted with epoxide **6.11** to provide bicyclic carbamate **6.13a** in 38% yield. The latter was converted into cyclopenta-oxazolidine **6.13b**, a key molecule. Its subsequent heating in ethanol in the presence of 1 equivalent of camphor sulfonic acid (CSA) smoothly yielded (+)-PTX-251D through iminium intermediate **6.14** (compared with **6.2**). After its chromatographic purification, followed by crystallization of the

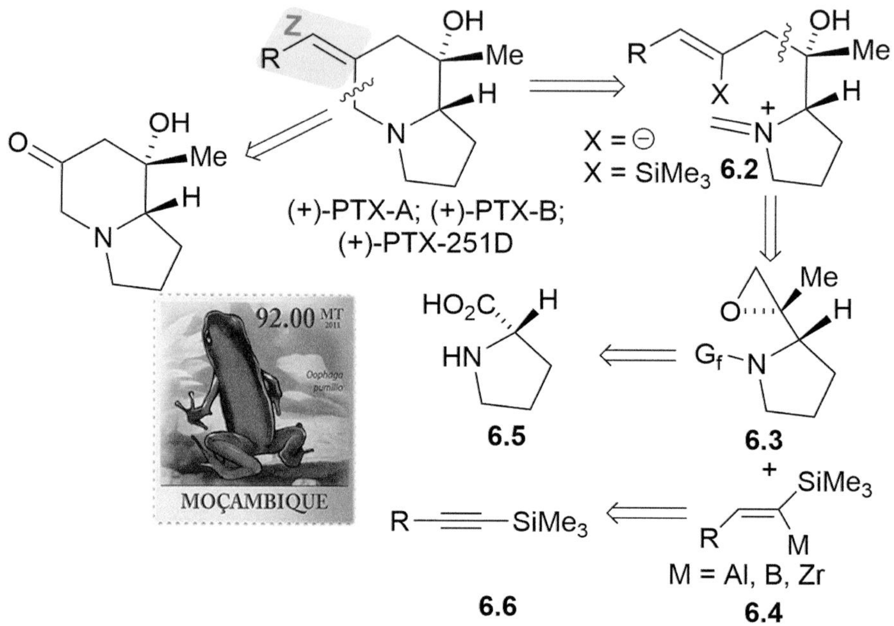

Figure 6.4. First retrosynthetic analysis and synthesis strategy for pumiliotoxins 251D (Prof. Overman and coworkers) and an image of *O. pumilio* frog in the Mozambican stamp.

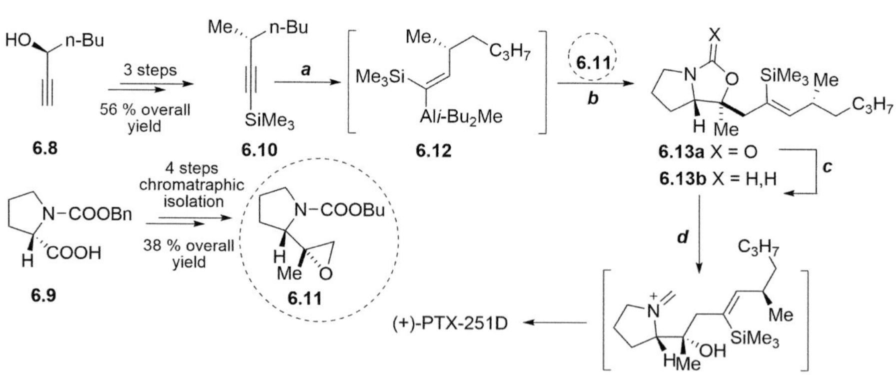

Figure 6.5. First enantiospecific total synthesis of (+)-PTX-251D was realized by Overman and Bell in 1981: (a) DIBALH, MeLi; (b) Et_2O, reflux, 48 hours, 38%; (c) 1. 3M KOH, 2. $(CH_2O)_n$, EtOH, 80%; (d) CSA, EtOH, reflux, 60% from **6.13a**. DIBALH = Diisobutylaluminium hydride, i-Bu_2AlH.

hydrochloride salt, (+)-PTX-251D hydrochloride was obtained in 60% overall yield from cyclopentaoxazolidinone **6.13a**.

Notable, the synthesis of this alkaloid was accomplished in 13% overall yield from (*S*)-1-heptyn-3-ol (**6.8**) and 4.7% yield from *N*-(benzyloxycarbonyl)-L-proline (**6.9**).

Synthetic efforts targeting (+)-PTX-B, a problematic goal due to the side *syn* diol chain configuration, were more labor intensive [20]. Retrosynthesis of this alkaloid allowed us to two main precursors, phosphorene derivative A and alkylidene-indolizidine aldehyde B (Figure 6.6). Preparation of an equivalent reagent **6.17** to precursor A was achieved in six steps from available ethyl L-lactate (ethyl (*S*)-2-hydroxypropanoate, **6.15**). Another key precursor B, indolizidine aldehyde **6.23**, was assembled in a long linear route from acrylonitrile (**6.18**), where the reaction of (*R*)-

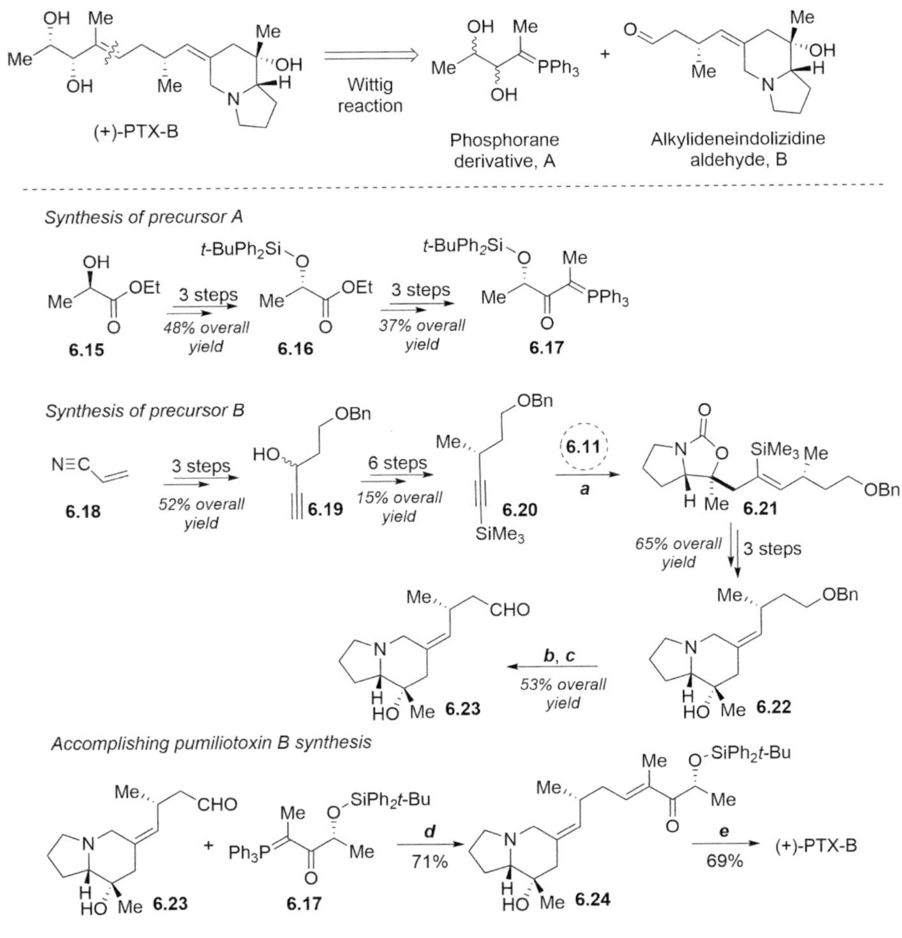

Figure 6.6. Simplified scheme of total synthesis for (+)-(15*R*,16*R*)-PTX-B (1984): (a) 1. DIBALH, hexane, MeLi, 2. **6.11**, THF, 56°C; (b) Li, NH_3; (c) Swern oxidation; (d) CH_2Cl_2, 40°C, 5 days; (e) $LiAlH_4$, THF, –20°C, 30 minutes.

alkynylsilane **6.20** and epoxide **6.11** to afford bicyclic carbamate **6.21** is similar to that employed in the first synthesis of PTX-251D.

Then, this bicyclic carbamate **6.21** was converted into the needed indolizidine aldehyde **6.23** in 5 steps. Finally, there are two essential precursors, **6.17** and **6.23**, and their condensation under mild reaction conditions (refluxing dichloromethane) is readily provided (*Z*)-alkylidene-indolizidine **6.24**. Fortunately, the threo-selective reduction of the latter with excess $LiAlH_4$ was accompanied by desilylation, directly affording (+)-PTX-B in a 69% yield. This synthesis unambiguously established the complete stereo-structure and provided 150 mg of this potentially important cardiac agent for further pharmacological investigations. Notably, this first synthesis of (+)-PTX-B was performed in 10 steps, and 4.2% yield from *N*-(benzyloxycarbonyl)-L-proline (**6.9**) and 17 steps and 1.6% yield from acrylonitrile (**6.18**) [20].

In 1985, Overman and Linn used the iminium ion-vinylsilane cyclization described above in the enantioselective total synthesis of (+)-PTX-A [21], which was isolated from dendrobatid frogs as a mixture of C-15 epimers. The major epimer has the (*S*)-configuration at C-15. Thus, the synthesis started with alcohol, (*S*)-(–)-2-methyl-1-penten-3-ol (**6.25**) as a chiral starting substrate that was converted into silylalkyne **6.26** after nine consecutive steps (Figure 6.7). Coupling of the vinylalanate derivative of the latter with epoxide **6.11** in refluxing THF provided cyclopentaoxazolidinone **6.27** in 52% yield.

Iminium ion-vinylsilane cyclization of **6.28** via intermediate **6.29** was best completed in methanol buffered with pyridine-pyridinium tosylates and provided alkylidene-indolizidine **6.30** in 71% overall yield from **6.27**. Pure (+)-PTX-A was isolated and purified by column chromatography (SiO_2, $CHCl_3$-MeOH-12N NH_4OH, 50:1:1). Therefore, this synthesis was accomplished in 13 steps with an 8.2% overall

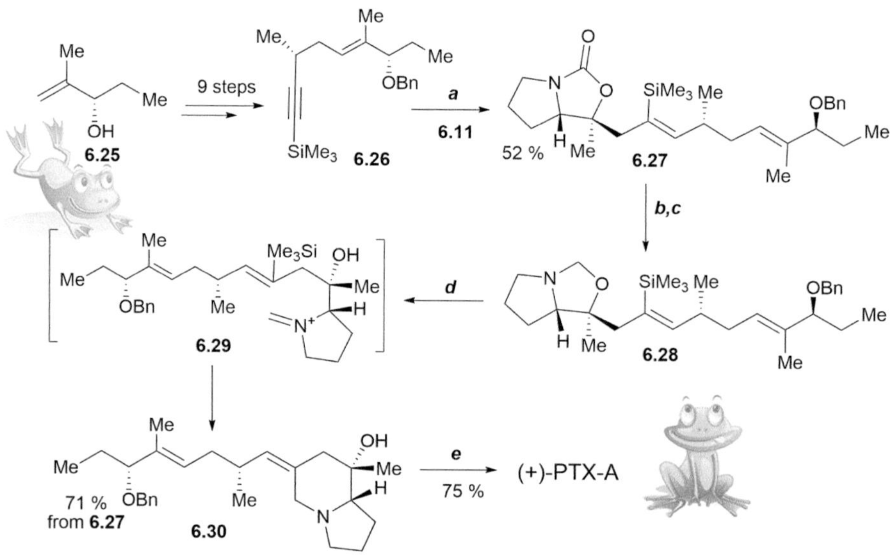

Figure 6.7. Simplified scheme of total synthesis for (+)-(15*S*)-PTX-A (1985): (a) 1. DIBALH, hexane, MeLi, 2. **6.11**, THF, 56°C; (b) 3M KOH, reflux, (c) $(CH_2O)_n$, EtOH; (d) pyridinium *p*-toluenesulfonate, MeOH, pH = 4.5, 80°C; (e) Li/NH_3-THF, –78°C.

yield from (*S*)-2-methyl-1-penten-3-ol (**6.25**) and in nine steps with a 5.7% overall yield from *N*-(benzyloxycarbonyl)-L-proline (**6.9**) [21].

A major shortcoming of the iminium ion-vinylsilane cyclization route to this pumiliotoxin group is the coupling of the vinylsilane side chain and pyrrolidine epoxide fragments. This step involves a particular optimization for each side chain nucleophile and is often low-yielding.

Thus, another cyclization route based on the iminium ion-alkyne strategy was developed in 1988. These Mannich-type cyclizations of alkynes need the presence of reactive external nucleophiles. Iodide is particularly effective, and cyclizations with this promoter can occur under various conditions [22–24]. An example of a cyclization of **6.32b** to obtain solely the (*Z*)-1-iodobuthylidene stereoisomer **6.33a** is depicted in Figure 6.8. The reaction of epoxide **6.9** with diethyl(l-hexynyl) alane (**6.31**) provided the suitable coupled product **6.32a** in 95% yield. Cleavage of the benzyloxycarbonyl group of this product was achieved under basic reaction conditions to afford the desired alkynyl pyrrolidine **6.32b** in 75% yield. Its treatment with paraformaldehyde (5 equivalent) and camphorsulfonic acid (1 equivalent) in the presence of NaI gave the iodo-alkylidene indolizidine **6.33a** via the formaldiminium ion intermediate.

Treatment of **6.33a** with *n*-butyl-lithium by quenching with methanol provided nor-11-methylpumiliotoxin 237A (**6.33b**) in 83% yield. Similarly, PTX-A was prepared in five steps with a 43% overall yield from the readily available epoxide

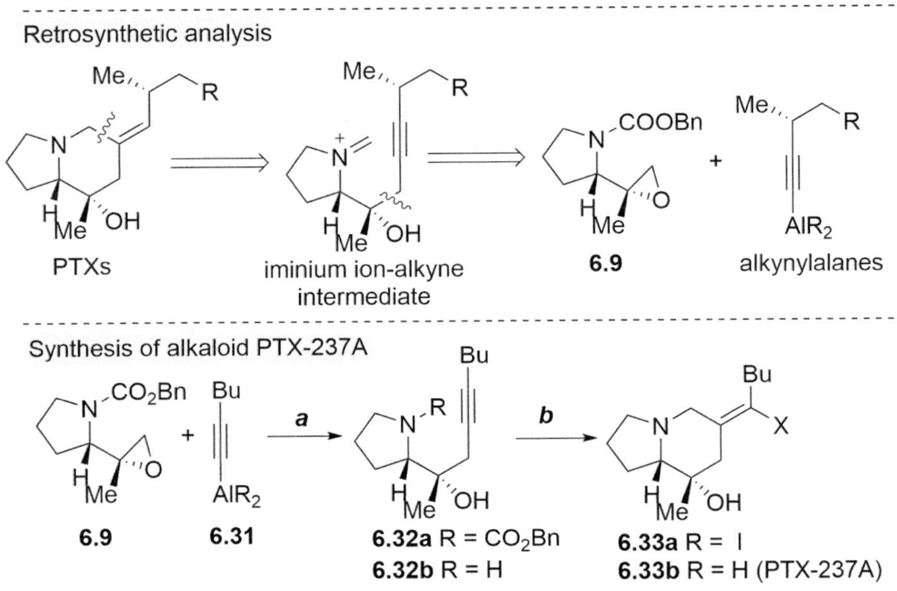

Figure 6.8. Retrosynthetic analysis of PTXs based on the iminium ion-alkyne strategy allows us to find an iminium intermediate conducive to starting materials, proline derivative (**6.9**), and alkynylalanes. The synthesis of nor-11-methylpumiliotoxin 237A (**6.33b**) as an example: (a) 1. Toluene-hexane, 7:1, 0°C, 95% (**6.32a**), 2. 0.15 M $Ba(OH)_2$, glyme-H_2O, 3:2, reflux, 40 hours, 75% (**6.32b**); (b) 1. $(CH_2O)_n$, CSA, MeCN, NaI, 100°C, 1 hour, 82% (**6.33a**), 2. *n*-BuLi (2.2 equivalent), –78°C, Et_2O, 83% (**6.33b**).

6.9 and appropriately substituted alkyne [23]. This route to PTXs remains a highly practical method for preparing them and related analogs.

Over time, several different strategies have been developed to synthesize the ubiquitous (*Z*)-alkene. One is based on highly selective nickel (0)-catalyzed epoxide-alkyne reductive cyclizations [25]. Nickel maintains a unique place among the transition metals in terms of reactivity trends, functional group tolerance, and catalytic activity. Based on the retrosynthetic analysis of the PTXs (Figure 6.9), Jamison and Woodin used a stereospecific intramolecular nickel-catalyzed reductive cyclization of epoxy *N*-alkynyl pyrrolidine molecule to install the (*Z*)-alkene in the synthesis of pumiliotoxin 251D [26]. The second step of retrosynthetic analysis allowed us to define epoxy *N*-alloc-proline derivative and propargyl bromide.

Thus, epoxide **6.36** was prepared from carbamate *N*-alloc-protected proline methyl ester **6.34** in a three-step procedure through ketone **6.35** formation. At the same time, propargyl bromide **6.38** ((*S*)-1-bromo-4-methyloct-2-yne) was synthesized from (*R*)-2-methylhexanal in a two-step protocol. Having key reagents **6.37** and **6.38**, synthesis of epoxy alkyne **6.39**, a PTX-251D precursor, was performed in a two-step manner first removal of the allyloxycarbonyl (Alloc) protection group was effected with a catalytic $Pd(PPh_3)_4$ and diethylamine to give and free amine **6.37**, which was treated with propargyl bromide **6.38**, setting the stage for the critical nickel-catalyzed step ($Ni(cod)_2$/PMe_2Ph in the presence of triethylborane). Thus, pumiliotoxin 251D was synthesized from commercial materials with a 17% overall yield in nine linear steps. Among over 800 lipophilic alkaloids detected in the skin extracts of poison frogs, pumiliotoxins are a large indolizidine class of poison frog

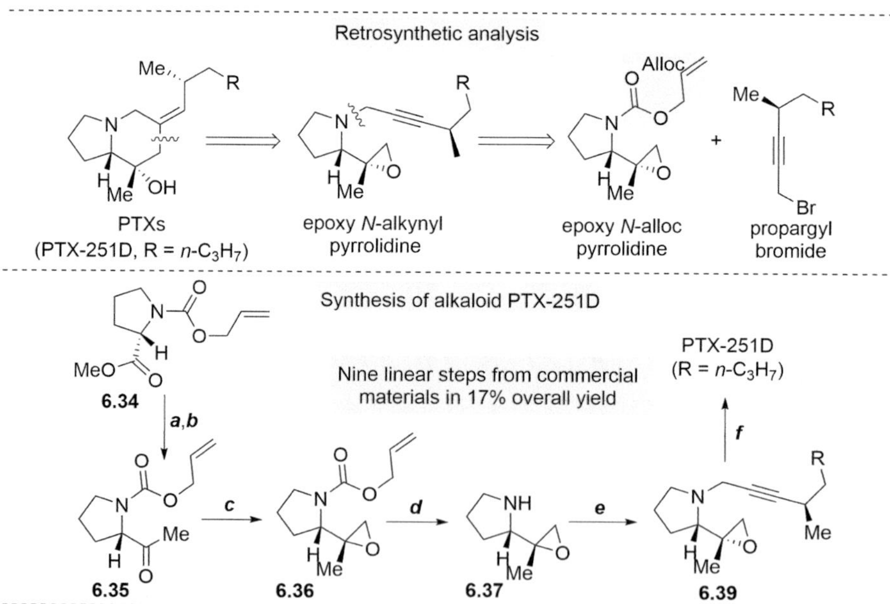

Figure 6.9. Retrosynthetic analysis of PTX-251D and its synthesis based on nickel-catalyzed epoxide-alkyne reductive cyclization (2007): (a) MeONHMe·HCl, Me_3Al, CH_2Cl_2, 0°C to rt; (b) MeMgBr, THF, 0°C to rt; (c) Me_3SiCl, *n*-BuLi, THF, –20°C; (d) 5 mol% $Pd(PPh_3)_4$, Et_2NH, THF; (e) **6.38**, Na_2CO_3, acetone; (f) 20 mol% $Ni(cod)_2$, 40 mol% PMe_2Ph, Et_3B, 65°C. $Ni(cod)_2$ = bis(cyclooctadiene)nickel (0).

alkaloids. As other alkaloid classes have been described in diverse book chapters, we leave you to investigate them independently. In the next section, we will learn about an interesting frog alkaloid similar to our well-known plant alkaloid nicotine (Section 3.4).

6.1.2 Frog Alkaloid Epibatidine: A Toxic Molecule Isolated as a "Nicotinic Alkaloid"

One of the most interesting frogs is the Ecuadorian frog *Epipedobates anthonyi* (old name: *E. tricolor*). This species lives at an altitude of 150 m to 1,400 m on the western slope of the Western Cordillera in Ecuador and Peru. This species is named in honor of mammalogist Harold Elmer Anthony, who discovered it. Studies of the venom of these dart frogs provided other alkaloid toxins with pharmacological properties. One is epibatidine, a toxic molecule (LD_{50} = 0.4 µg per mouse of 20 g) with high analgesic activity (ED_{50} = 2.5 mg/kg).

Interestingly, epibatidine was found in 1974 and isolated in 1980 [27] as a trace alkaloid together with the major alkaloid (21 mg) pumiliotoxin 251D mentioned earlier, but its molecular structure was confirmed only in 1992 [28, 29]. Due to its unique structure, interesting physiological properties, and rarity in nature (a few micrograms as an *N*-acetyl derivative is obtained from 750 *E. anthonyi* frogs), epibatidine became a crucial medical research model and has stimulated extensive synthetic efforts to develop new agents as potential analgesics. Since then, different methods for its total synthesis have appeared, and from 1993 to 1994, this alkaloid became available in sufficient quantities for broad pharmacological studies. Now, it is well-known that it possesses an *exo*-2-(6-chloro-3-pyridyl)-7-azabicyclo[2.2.1] heptane structure and its analgesic potency (ED_{50} = 0.005 mg/kg, hot plate method and 0.020 mg/kg using tail flick method) surpasses morphine (ED_{50} = 1.0 mg/kg and 10 mg/kg) by 100–200 times, and this analgesic activity mediated by non-opioid receptors, nicotinic acetylcholine receptors (nAChR) [30, 31]. This molecule contains two nitrogenous rings: azanorbornane (7-azabicyclo[2.2.1]heptane) and 2-chloropyridine. The chloropyridinyl fragment is linked to azanorbornane by the *exo* bond with the configuration (1*R*,2*R*,4*S*) [32] (Figure 6.10).

Thus, epibatidine is a non-opioid analgesic and agonist of neuronal nAChRs subtype α4β2 [31, 33]. Epibatidine binds to nAChR in the brain and at other neuromuscular junctions throughout the body that cause seizures and respiratory and digestive problems. Remember that the neuronal nAChRs are pentameric structures combining only two classes of subunits (α and β). In the mammalian central nervous system (CNS), the heterodimeric nAChR subtype α4β2 is the most expressed. Epibatidine has high affinity to nAChRs for this subtype, a potent agonist with an inhibitory constant K_i = 40 pM. Moreover, epibatidine toxicity could be related to its ability to activate not only this central neuronal subtype but also the ganglionic nAchR subtype α3β4 and α-7 nicotinic receptors (α-7nACh) present both peripherally and centrally in the nervous system. The latter is involved in various neurological conditions, such as schizophrenia, Parkinson's, and Alzheimer's disease.

Extensive modification of this alkaloid would be necessary to provide more SAR information for designing a helpful therapeutic agent. How is epibatidine made an accessible biological object? One year after publishing its structure, four groups developed their routes for epibatidine synthesis in racemate form [34–37]. For the first time, one of these groups, Fletcher and coworkers, showed that the natural alkaloid possessed the (1*R*,2*R*,4*S*)-stereochemistry through its synthesis [35]. These initial syntheses are based on the construction methodologies of the 7-azabicyclo[2.2.1]heptane system: the Diels-Alder reaction of the *N*-substituted pyrrole (acting as a diene) with activated dienophiles or the intramolecular cyclization of the 1,4-bifunctional hexane derivatives, which are laborious and lengthy, i.e., not quite effective. However, these allowed preliminary pharmacological studies of epibatidine and its analogs to be carried out. More than 100 synthetic routes have been developed since the epibatidine structure was assigned in 1992–93 [38–40]. Despite the sizable synthetic variety of methods for this alkaloid, its commercial availability, according to current prices in 2024 (5 mg US$ 219, Sigma-Aldrich), is scarce.

As examples of epibatidine preparation, we want to examine its synthesis based on the Diels-Alder cycloaddition reaction employing an *N*-protected pyrrole as the diene and substituted acetylene as the dienophile to create the rapid construction of N-acyl 7-azanorbornan-2-one, a key synthetic intermediate in the total synthesis of epibatidine. Thus, Trudell and his Ph.D. student Chen reported a short and efficient total synthesis of (±)-epibatidine using the following retrosynthetic analysis (Figure 6.11) [41].

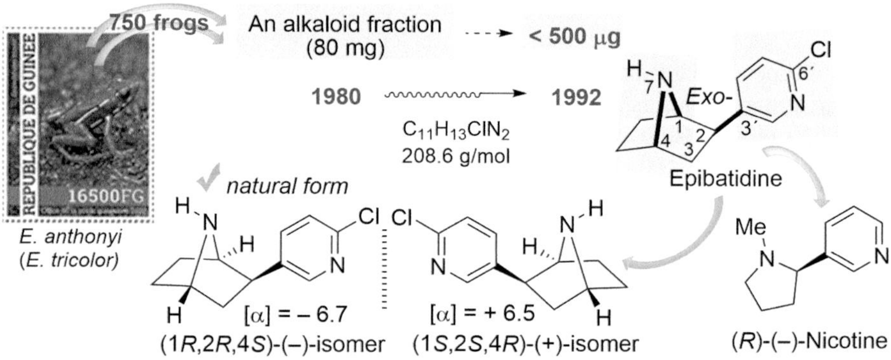

Figure 6.10. Pictorial diagram of the epibatine alkaloid discovery and its structural features. Epibatine is the *exo*-isomer of the two possible geometric isomers and can exist as two enantiomers: (1*S*,2*S*,4*R*)-(+)-epibatidine and (1*R*,2*R*,4*S*)-(–)-epibatidine. The natural compound is the (–)-isomer, although there is little difference in pharmacological activity between (+)- and (–)-isomers.

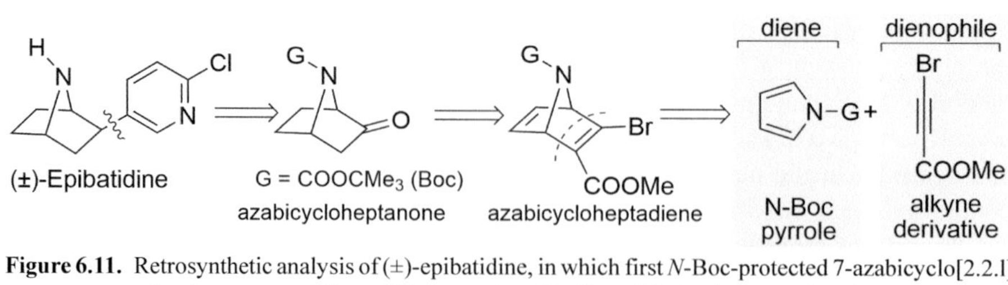

Figure 6.11. Retrosynthetic analysis of (±)-epibatidine, in which first *N*-Boc-protected 7-azabicyclo[2.2.1] heptan-2-one and then *N*-Boc-pyrrole with alkynyl derivative were found.

With a clean synthetic plan, the key azabicycloheptanone **6.42** was prepared in four steps with an overall yield of 40%. Thermic treatment of *N*-Boc pyrrole (**6.39**) with methyl 3-bromopropiolate (**6.40**) resulted in bromoalkenyl cycloadduct **6.41** as a sole product (Diels-Alder cycloadduct) with 60% yield (Figure 6.12). The latter was converted in the needed *N*-Boc-azabicycloheptanone **6.42**, transforming methyl bromovinyl carboxylate fragment in an β-ketoester group, which was then successfully decarboxylated. Note that the *N*-Boc (*tert*-butyloxycarbonyl) protecting group was preferred to other acyl groups because it is removed under milder conditions.

Adding 5-lithio-2-chloropyridine (**6.43** and *n*-BuLi, –78°C) to the obtained ketone **6.42** gave stereoselectively alcohol **6.44**. This tertiary alcohol was deoxygenated by a radical reaction via its methyl oxalyl ester with Bu_3SnH in the presence of AIBN. Epimerization of *endo*-pyridinyl derivative **6.45** to the requisite *exo*-pyridinyl isomer was achieved by refluxing in the presence of potassium *tert*-butoxide. The last step, deprotection of the *N*-Boc group, provided epibatidine in racemic form (Figure 6.12). Reviewing this multistep synthesis, we can conclude that the racemic epibatidine was obtained in 18% overall yield using eight steps.When another important key synthetic intermediate, *N*-Boc-7-azanorbornene (**6.49**), was proposed for use in the preparation of epibatidine, considerable improvements were made in the epibatidine synthesis. For example, Carroll and coworkers developed a suitable procedure to prepare **6.49** on a multigram scale from *N*-Boc pyrrole (**6.39**) and *p*-toluenesulfonyl acetylene (**6.46**) (Figure 6.13) [42]. The initial Diels-Alder reaction of the latter reagents gave cycloadduct **6.47**, which was selectively reduced using *in situ* prepared nickel boride ($Ni(OAc)_4$ + $NaBH_4$) to provide 2-sulfonyl 7-azanorbornene **6.48**.

Then, a two-step desulfonation process of **6.48** yielded the desired 7-azanorbornen **6.49**. Its reductive Heck arylation using 2-amino-5-iodopyridine (**6.50**) exclusively afforded the *exo*-pyridinyl azabicycloalkane **6.51**. Racemic epibatidine was obtained

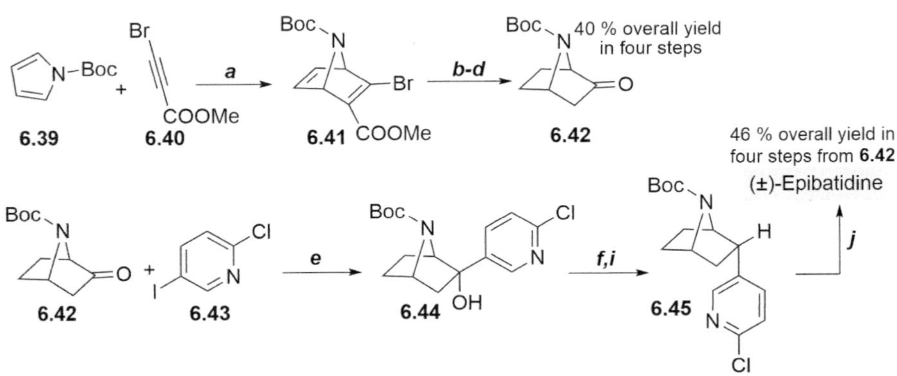

Figure 6.12. Total synthesis of (±)-epibatidine (Zhang and Trudell, 1996): (a) 90–95°C under argon, 30 hours; (b) 1. Et_2NH, Et_3N, MeCN, rt, under argon, 1.5 hours, 2. 10% HCl, 4 hours; (c) H_2, 10% Pd/C, MeOH, 10 hours; (d) 1. 10% HCl, 100–110°C, under argon, 3 hours, 2. $(Boc)_2O$, rt, 24 hours; (e) *n*-BuLi, THF, –78°C, 3 hours, then warmed to –50°C, 30 minutes; (f) 1. $ClCOCO_2Me$, DMAP, MeCN, rt, under argon, 10 minutes, 2. Bu_3SnH, AIBN, toluene, reflux, 1 hour; (i) *tert*-BuOK, *tert*-BuOH, reflux, under argon, 45 hours; (*j*) CF_3CO_2H, rt, under argon, 3 hours. AIBN = 2,2′-azobis(isobutyronitrile), DMAP = 4-(dimethylamino)pyridine.

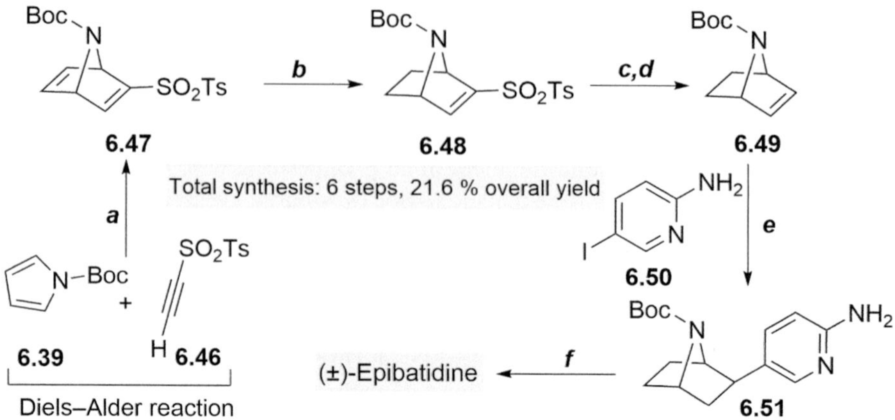

Figure 6.13. The improved synthesis of (±)-epibatidine using *N*-Boc-7-azanorbornene as a close precursor (Carroll and coworkers, 1998): (a) 80°C, 65%; (b) $Ni(OAc)_4$ + $NaBH_4$, MeOH, 86%; (c) Bu_3SnH, AIBN, benzene, 78–91%; (d) Bu_4NF, THF, 93–98%; (e) $Pd(OAc)_2$, *n*-Bu_4NCl, HCOOK, DMF, 100°C, 68%; (f) $NaNO_2$, HCl, CuCl, 76%.

by diazotizing the latter, followed by treatment with cuprous chloride in hydrochloric acid (Figure 6.13). This efficient methodology is well-suited for multigram scale synthesis: a short route (six steps) generates the alkaloid in 21.6% overall yield and allows for pyridinyl analogs to be prepared. For example, various diazotization reactions of **6.51** with simple transformation generated an excellent series of epibatidine analogs **6.52** (Figure 6.14) for their SAR studies on affinity, efficacy, and subtype selectivity of neuronal nAChRs [43–45]. Unfortunately, these studies' results indicated modest differences in functional potency and efficacy, suggesting that these analogs are unlikely to be selective enough for CNS neuronal nAChRs over peripheral nAChRs to avoid peripheral toxicity [45]. Some derivatives of epibatidine have been described to produce fewer toxicities in rodents. Nevertheless, the decrease in toxic effects observed in these studies was insufficient to ensure safe therapeutic/toxicity ratios.

In search of reducing the toxicity of epibatidine, the chloropyridinyl fragment was also replaced by the methylisoxazolyl ring. Epiboxidine, a methylisoxazole analog of epibatidine (Figure 6.14), has also been reported to produce fewer toxicities (*ca.* 20-fold less toxic) than epibatidine in mice [46]. However, epiboxidine was 10-fold less potent than epibatidine as an antinociceptive agent and, in sodium flux assays, was not significantly less potent than epibatidine at ganglionic nAChRs. Thus, the development of CNS-selective epibatidine analogs will require continuous effort.

The first synthetic route to (±)-epiboxidine is based on the work of Dongly Bai and coworkers, which describes building a 7-azabicyclo[2.2.1]heptane ring system by contracting the tropinone skeleton via Favorkii rearrangement [47]. This is an original and interesting approach to epibatidine and epiboxidine from tropinone, a precursor of alkaloid atropine, a close analog of the well-known local anesthetic cocaine. Let us examine this route. Commercially available tropinone (**6.52**) was converted to its *N*-ethoxycarbonyl derivative **6.53** by treatment with ethyl chloroformate (ClCOOEt)

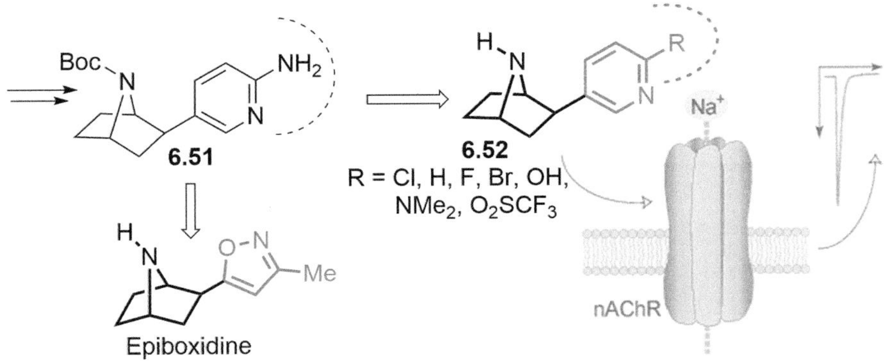

Figure 6.14. Structure of epibatidine and analogs used in competition binding assays (affinity and efficacy).

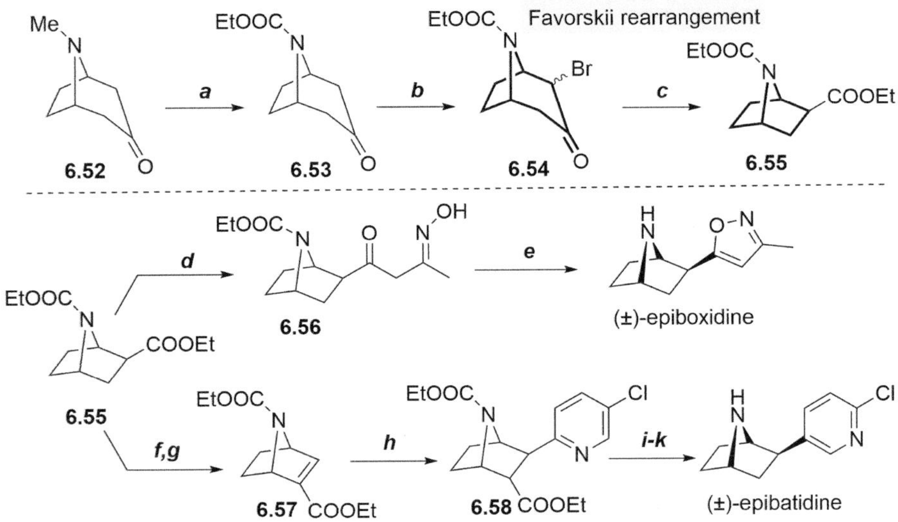

Figure 6.15. Synthetic approach to construct 7-azabicyclo[2.2.1]heptanone skeleton from 8-azabicyclo[3.2.1]octanone ring using the Favorskii rearrangement. Syntheses of (±)-epiboxidine and (±)-epibatidine: (a) $ClCO_2Et$ (2 equivalent), K_2CO_3 (cat.), toluene, reflux, 3 hours, 86%; (b) $CuBr_2$ (2 equivalent), $CHCl_3$, EtOAc, reflux, 1 hour; (c) MeONa (3 equivalent), DME, rt, 0.5 h, 56%; (d) Acetone oxime (Me_2C=N-OH), THF, *n*-BuLi; (e) conc. HCl, 80°C; (f) LDA, THF, –78°C, 20 minutes, then PhSeBr; (g) 30% H_2O_2, CH_2Cl_2, rt, 15 minutes, 68% from **6.55**; (h) **6.43**, cat. $(Ph_3P)_2Pd(OAc)_2$, Et_3N, HCO_2H, DMF, rt, 4 d, 56%; (i) $LiOH{\cdot}H_2O$, MeOH-H_2O (3:2), 45°C, 1 hour, 92%; (j) $(COCl)_2$, sodium 2-thioxopyridin-1(2*H*)-olate, *tert*-BuSH, benzene, reflux, 3 hours, 75%; (k) Me_3SiI, $CHCl_3$, reflux, 4 hours, 83%.

(Figure 6.15). Subsequently, its bromination reaction with cupric bromide afforded the monobrominated tropinone **6.54**. The latter, without purification, was subjected to the Favorskii rearrangement using sodium methoxide that allowed to give the 7-azabicyclo[2.2.1]heptane skeleton **6.55** in three steps with 48% overall yield.

This crucial intermediate **6.55** was treated with acetone oxime in the presence of *n*-butyl lithium to afford azabicyclo[2.2.1]heptane derivative **6.56**, which was quickly converted into the desired racemic epiboxidine under acid reaction conditions. At the same time, **6.55** could be transformed into racemic epibatidine using a step sequence to obtain the alkaloid product with a 10.5% overall yield from tropinone **6.52** (Figure 6.15).

With this, we finish our story about the toxins found in the skin of graceful frogs and move on to study the toxic molecules contained in nasty toads.

6.2 Amphibian Neurotoxic Molecules Coming from Toad Skin Glands

The common toad can differ in color from olive brown to green, dark brown, or gray, sometimes with dark patterns. Its skin is typically dry and "warty". Toads are active at night when they hunt for various invertebrates, including slugs, spiders, worms, aphids, and ants. They sneak up on prey and use their sticky tongues to catch it. Toads, like frogs, have no weapons for aggression. The poisonous secretions of their glands constitute the best and, in many cases, the only measure to protect themselves from their enemies. When toads are disturbed, they inflate their lungs and assume a stereotyped posture, presenting their parotoid glands to the aggressor. The paratoids then become turgid and ready to release poisonous shots in case any external mechanical pressure is exerted, such as the bite of an aggressor. Predator's self-poisoning is characteristic of amphibian defense, which is classified as a "passive defense" [48].

From the point of view of poisonous secretions, the most important tailless amphibians are the true toads of the genus *Bufo* in the family Bufonidae. The parotoid or skin gland secretion of bufonid toads is a rich source of toxic metabolites against parasites, pathogens, and natural enemies, such as snakes, birds, or mammals. Depending on the species, these skin secretions contain various compounds, such as biogenic amines, water-soluble steroids, lipid-soluble alkaloids, peptides, and proteins. The difference in chemical composition is also based on differences in the methods of collecting and processing. Interestingly, the main composition was approximately 103 toad venom lactone compounds and 12 indole alkaloids as chemical constituents in toads [49]. Such mixture compounds isolated from toads have significant therapeutic value for many diseases. Medicinal raw materials based on these toxins are widely used clinically worldwide, especially in China, Japan, and South Korea [50–52]. Although different constituents may have diverse functions, it is well-known now that steroidal bufadienolides and bufotoxins are the main bioactive compounds in toad toxins. Therefore, we first examine toxic molecules from toads with this group.

6.2.1 Bufadienolides and Bufotoxins: Toxic Steroid α-Pyranones Produced by Toad Skin Glands

As mentioned above, the most distributed components in the venom of toads *Bufo* and *Rhinella* are steroid molecules called bufogenins and bufotoxins. The latter are C-3 glycosides bound to suberylarginine. These molecules, together with biogenic amines and alkaloids, play a crucial role in the defense of toads against predators and microorganisms. Steroid compounds contain 24 carbon atoms (aglycone unit) and are part of the group of steroidal cardiac poisons, many of which are found in plants

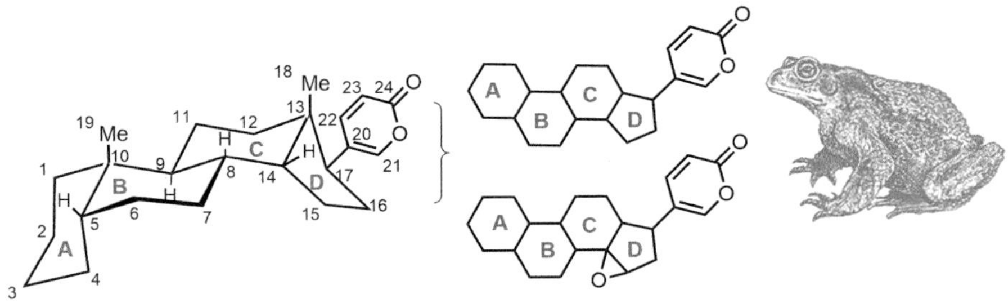

Figure 6.16. Carbon skeleton of bufogenins (bufadienolides) and their most stable conformation.

(Section 2.5; Figure 3.32). Most bufogenins, i.e., bufadienolides, have the *cis*-fusion of the A/B and C/D rings and the *trans*-junction of the B/C rings, forming the unique *cis-trans-cis*-configuration (Figure 6.16).

Another structural feature of bufogenins is the β-oriented heterocyclic 2-pyranone (α-pyrone) ring (penta-2,4-dien-5-olide ring or δ-valerolactone) at the C-17 position. Bufogenins differ from each other primarily in the number and position of the hydroxy groups. Each bufogenin contains at least one hydroxy group in the C-3 position, essential for glycoside formation (in certain plants) and bufotoxin (conjugates) formation. Additional HO groups can be found at positions C-11, C-12, or C-16. Some bufogenins contain fused oxirane cycles at carbons C-14 and C-15 of the steroid skeleton or carbons C-5 and C-6 of the 2-pyranone ring [53].

As mentioned before (Section 2.5), compounds of this type in small doses are stimulants of cardiac activity. However, experimentally inoculated in large doses, they exert a toxic action on the heart and determine its arrest in systole. They are 50–100 times stronger than digitalis extract. To kill a rabbit weighing 1 kg, an injection of 28 mg of this poison is enough. It is estimated that the contents of the granular glands of 10 toads can cause the death of a man. However, it was found that some bufogenins have an anesthetic potency estimated to be 90 times greater than that of cocaine. It is also interesting to note that the toad resists its poison.

Similar compounds, like gamabufotalin, arenobufagin, telocinobufagin, bufotalin, cinobufotalin, bufalin, cinobufagin, and resibufogenin (Figure 6.17), are present in the processed and dried venom from parotoid glands of the different *Bufo* species (*Bufo bufo gargarizans* Cantor, *B. melanosticus* Schneider, *B. marinus (= Rhinella marina)*, and *B. viridis*).

These toad toxins are identified as major constituents of ChanSu, a traditional Chinese medicine used for treating pain and inflammatory diseases [54, 55]. On the other hand, an aqueous extract from the skin and parotid venom glands of *B. b. gargarizans* Cantor is a traditional Chinese medicine called Cinobufacini (HuachanSu), which has been widely used in clinical cancer therapy in China [55, 56]. Note that its chemical composition is almost the same (Table 6.2).

This is not surprising since the link between inflammation and cancer is now evident. The complete analysis of the complex activation or deactivation mechanisms comes from our book discussion. Let us emphasize that the microenvironment of tumors is typically rich in pro-inflammatory mediators, such as tumor necrosis factor-alpha (TNF-α), interleukin (IL)-1, IL-6, IL-8, nuclear factor-kappa B

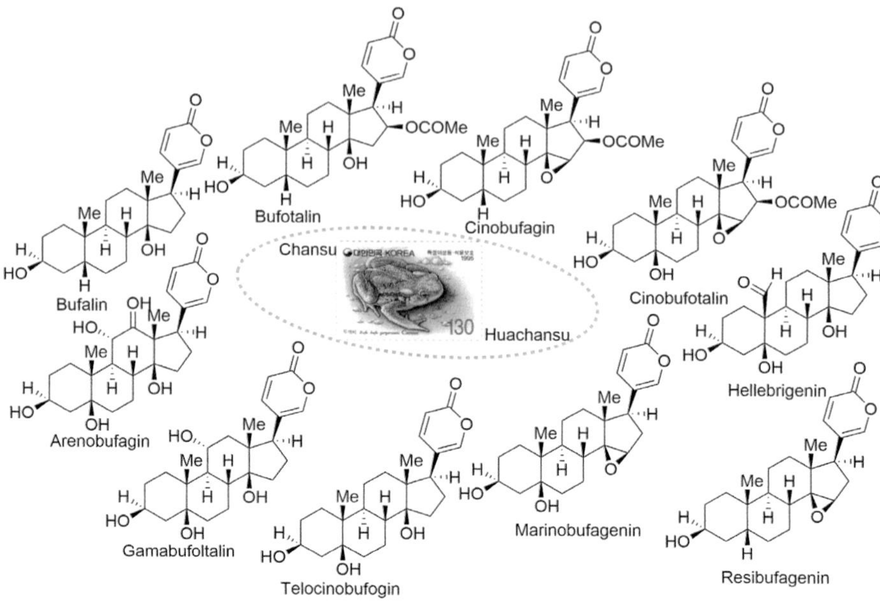

Figure 6.17. Chemical structures of the main bufadienolide components in ChanSu and HuachanSu. These are liposoluble components of toad venom.

Table 6.2. Chemical properties of ChanSu and HuachanSu.

Drug	Chemical Composition[a]	Method
Chansu	Arenobufagin, bufalin, bufotalin, cinobufagin, cinobufotalin, gamabufotalin, resibufogenin, telocinobufagin, and marinobufagenin.	HPLC and LC-DAD-MS/MS[b]
Huachansu	Arenobufagin, bufalin, bufotalin, cinobufagin, cinobufotalin, hellebrigenin, gamabufotalin, recinobufagin, and telocinobufagin.	HPLC and triple quadrupole mass spectrometry

[a] Placed in alphabetical order; [b] Liquid chromatography (LC) coupled to diode array detection (DAD) and tandem mass spectrometry (MS/MS).

(NF-κB), cyclooxygenases-2 (COX-2), and inducible nitrogen synthesis (iNOS), which are essential components of the tumor initiation and promotion [57].

Bufadienolides bufalin and cinobufagin are recognized as the major bioactive components in ChanSu and HuachanSu. At the same time, marinobufagenin is a major cardiotonic steroid component in the venom of *Bufo marinus* (now *Rhinella marina* L.). However, the exact chemical composition depends on the analytical method used [58, 59]. Moreover, bufadienolide levels and secretion compositions depend on environmental salinity and climate [60].

Isolation, purification, and characterization of each component of a mixture of toad toxins are always laborious. In 1985, Nambara and coworkers left us a description of this work [61]. The crushed and pulverized one hundred dried bodies (1.7 kg) of the Korean toad *B. b. gargarizans* Cantor were extracted with ethanol (50 L) for three months at room temperature. After removing insoluble materials by filtration through a layer of Celite, the filtrate was concentrated in a vacuum below 50°C and partitioned with the hexane-water system and the AcOEt-H_2O system. The latter contains the needed bufadienolides and their conjugate components, separated into the AcOEt and aqueous layers. After *vacuum* concentration, the AcOEt layer afforded a brown residue (1 g), chromatographed on silica gel with benzene containing stepwise-increasing concentrations of AcOEt. Then, further purification of dried eluate obtained with benzene-AcOEt (2:1) by reversed-phase HPLC gave finally the following bufadienolides: resibufogenin (< 1 mg), cinobufagin (26 mg), bufalin (4 mg), bufotalin (4 mg), cinobufotalin (< 1 mg), marinobufagenin (< 1 mg), and gamabufotalin (5 mg) [61].

In 2010, dos Santos and coworkers described the isolation of cytotoxic bufadienolides from the parotoid glands of the Brazilian toad *Rhinella schneideri* [62]. They showed that manual compression of 16 adult toads can yield 1.28 g of the crude extract. Using first-column chromatography (aluminum oxide, stepwise mixing of hexane/ethyl acetate) and then preparative TLC chromatography (silica gel, hexane: ethyl acetate, 1:9) of an isolated fraction, six stable crystalline compounds with well-defined melting points were obtained. These components are marinobufagin, bufalin, telocinobufagin, hellebrigenin (Figure 6.17), and β-sitosterol, which lacks the 2-pyranone ring (Table 6.3). The latter is the most abundant phytosterol.

The cytotoxic evaluation showed that these bufadienolides exhibited moderate to strong activity against human HL-60 (leukemia), SF-295 (nervous system), MDA-MB-435 (melanoma), and HCT-8 (colon) cancer cell strains without hemolysis of mouse erythrocytes. Bufalin, telocinobufagin, hellebrigenin, and β-sitosterol presented potent inhibitory activity against all cancer cell strains tested (IC_{50} from 0.0007 to 0.16 μM), while marinobufagin revealed moderate activity (IC_{50} from 1.07 to 3.52 μM). Notably, β-sitosterol displayed the highest anti-proliferative activity against all human tumor cells screened (IC_{50} 0.0007 to 0.00019 μM). Moreover, this phytosterol promoted apoptosis in human breast cancer cells MDA-MB-231. Telocinobufagenin and hellebrigenin showed good activity against the

Table 6.3. Bufogenins isolated from the skin of the toad *Rhinella schneideri*.

Bufogenins	**Isolated Weight, mg**	**Yield, %**[a]	**Melting Point, °C**
Marinobufagenin	705	55	215-217
Bufalin	72	5.6	231–234
Telocinobufagenin	116	9.1	166–170
Hellebrigenin	22	1.7	154–160
β-Sitosterol 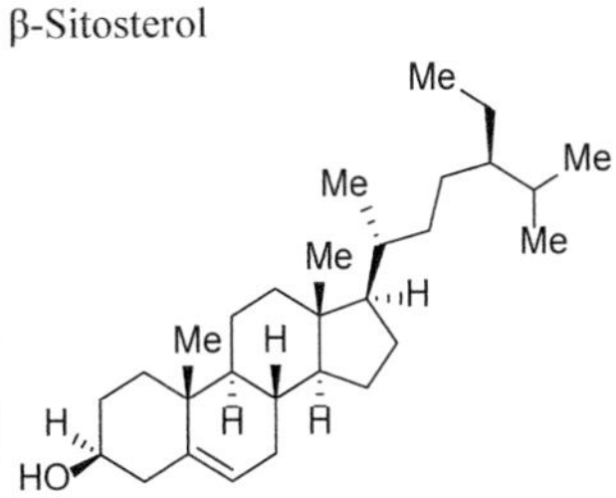	28	2.2	142–144

[a] The yields were calculated based on the crude extract mass.

promastigotes of *Leishmania chagasi*, which causes leishmania disease; the latter bufadienolide is also active against *Trypanosoma cruzi* promastigotes, which cause Chagas disease. Therefore, the structure-activity relationship of bufadienolides remains the focus of modern investigations, which aim to gain a better understanding of many physiological processes.

Knowing how to synthesize these bufadienolides separately is necessary to study their biology. Total syntheses of these steroid molecules developed only in the 1970s and 1980s have been well discussed in various reviews [63–65]. Here, we will not discuss in detail these works. Let us only note that (1) the construction of the 2-pyrone ring in these syntheses involves tedious and harsh stepwise manipulations from a linear or saturated precursor, (2) the choice of the steroid scaffold as the available starting material is also an essential synthetic task in the planation synthesis, and (3) generally, these sequences are linear and somewhat lengthy.

In 1969, Sondheimer and coworkers reported the first total synthesis of bufalin via the resibufogenin formation using 14α-hydroxycortexolone **6.59** as an available starting material (Figure 6.18). The desired bufalin molecule was obtained in less than 1% yield in more than 28 steps of transformation, which was unsuitable for large-scale industrial production. In 1970, Petti and his group developed another approach to these bufadienolides toxins. They developed the first total synthesis of bufalin from digitoxigenin **6.60** as available material from digitoxin (Section 2.5). Both approaches involve the formed 14-dehydrobufalin acetate **6.61**, a close bufalin precursor. Thus, its late-stage oxidation with *meta*-chloroperbenzoic acid (*m*-CPBA) allowed the preparation of resibufogenin acetate **6.62**. Selective saponification of this acetate using basic alumina afforded the product resibufogenin, and its simple

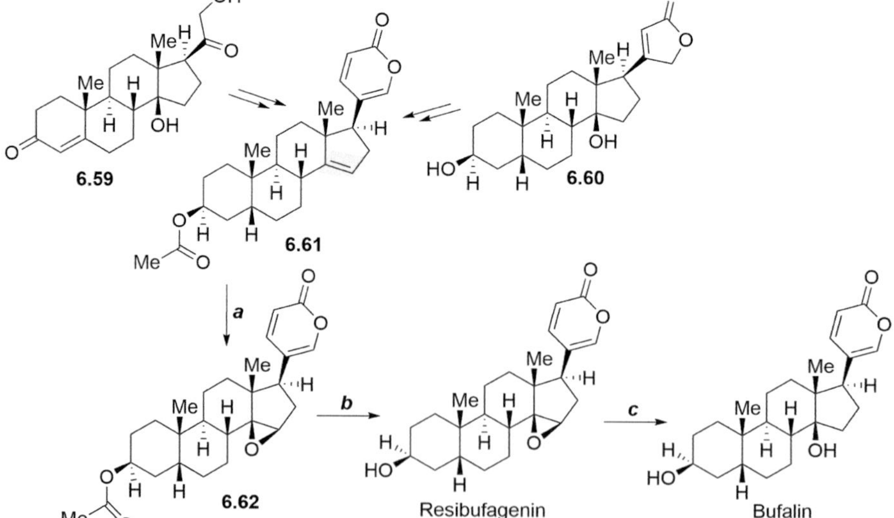

Figure 6.18. Simplified approaches to bufalin synthesis (1969–1970): (a) *m*-CPBA, CH_2Cl_2 or NBS, acetone; (b) alkaline alumina ether; (c) $LiAlH_4$, Et_2O.

reduction ($LiALH_4$, ether) led to the cleavage of the epoxide ring and the formation of the bufalin molecule (Figure 6.18).

Pettit's group also completed the synthesis of telocinobufagin from digitoxigenin and the synthesis of cinobufagin from bufotalin. However, these pioneering studies are now considered to be antiquated [63].

Modern and efficient bufadienolide routes recently developed are based on the direct Stille coupling of the 2-pyrone ring and commercially available 4-androstene-3,17-dione skeletons [66,67]. Starting from **6.63** as the building block, Inoue and coworkers developed an eight-step synthesis of C17-iodo containing intermediate **6.64**, an indispensable precursor to building C17β-2-pyrone substituted skeleton **6.66** (Figure 6.19) [66]. This skeleton was assembled using the Stille coupling reaction with 5-(trimethylstannyl)-2*H*-pyran-2-one (**6.65**). Then, the most electron-rich C17-double bond of **6.66** was epoxidized chemoselectively over the potentially more exposed diene of the 2-pyrone ring to afford **6.67** with the *cis*-fused CD-ring. Its epoxide rearrangement to **6.68** was not easily accomplished because of the unusual reactivity of 2-pyrone. This was possible using a combination of trimethylsilyl trifluoromethanesulfonate (TMSOTf) (2.4 equivalent) and 2,6-lutidine (5.1 equivalent) in CH_2Cl_2. $NaBH_4$ reduction of **6.68** (via an S_N2 reaction) allowed for the stereoselectively obtaining of C16β-alcohol **6.69**, and finally, its reductive

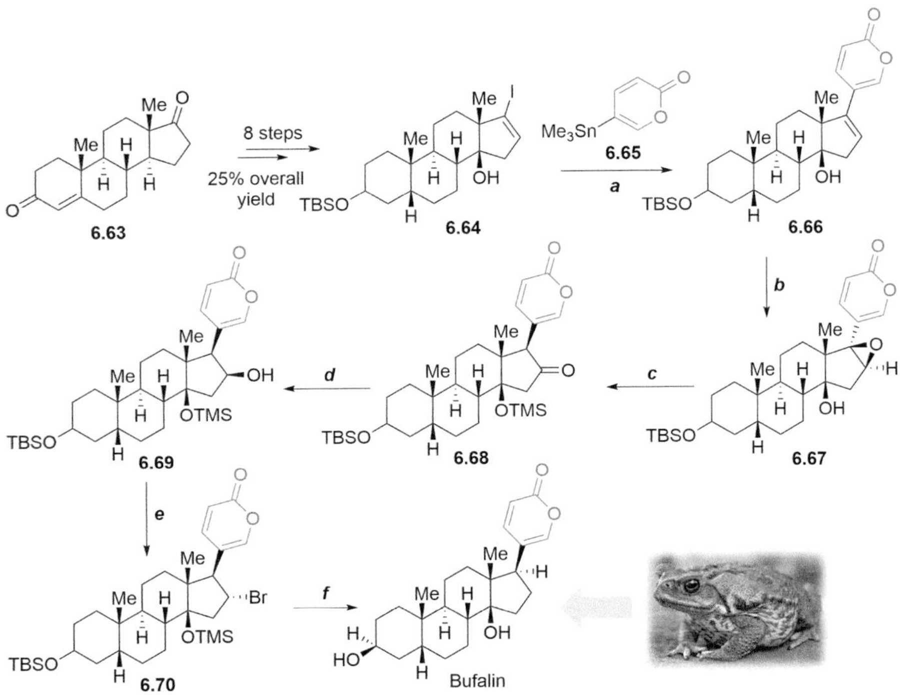

Figure 6.19. Total synthesis of bufalin developed by Inoue and coworkers (2020): (a) $Pd(PPh_3)_4$, CuCl/LiCl, DMSO/THF, 60°C, 67%; (b) *m*-CPBA, Na_2CO_3, CH_2Cl_2, –78°C to –20°C, 77%; (c) TMSOTf, 2,6-lutidine, CH_2Cl_2; –78°C to rt, 68%; (d) $NaBH_4$, MeOH, THF, 0°C, 93%; (e) 1. CBr_4, imidazole, PPh_3, CH_2Cl_2, 40°C, 68%; (f) *n*-Bu_3SnH, Et_3B, under an O_2 atmosphere, 2. HCl–MeOH, 75%.

removal via C16-bromide formation (comp. **6.70**) produced bufalin, which exhibited high activity (GI_{50} = 13.3 nM) against MCF-7 human breast cancer.

Two years later, interested in exploring the medicinal chemistry of cinobufagin, Nagorny and coworkers developed an efficient synthesis of cinobufagin using Inoue's approach from readily available dehydroepiandrosterone **6.71** (Figure 6.20) [67]. The latter was converted into a key enone **6.72** in five steps with a good 55% overall yield. The vinyl triflate intermediate **6.73** was then generated by treating **6.72** with 2,6-di-*tert*-butyl-4-methylpyridine (DTBMP) and triflic anhydride at 0°C in CH_2Cl_2 and used without purification as the coupling partner of the known stannane **6.65** in a Stille reaction to afford diene **6.74** containing a C17-pyrone ring. The photooxidation of this latter with the singlet oxygen in the presence of methylene blue and cobalt(II) tetraphenyl porphyrin (CoTPP) catalyst resulted in 64% formation of **6.76** containing β-**6.76** as the major diastereomer (α/β = 1:1.6), which could be separated chromatographically. Its House-Meinwald rearrangement into keto-epoxide **6.77** was realized using scandium(III) trifluoromethanesulfonate $Sc(OTf)_3$ (10 mol%) in anhydrous dichloromethane molecular sieves. Reduction with sodium borohydride followed by acetylation with an acetic anhydride/DMAP/pyridine system allowed to obtain TBS-protected cinobufagin **6.78** in 61% yield over three steps. Finally, its deprotection with HF·pyridine in THF at room temperature gave cinobufagin (Figure 6.20).

Thus, this first synthesis of cinobufagin was realized in 12 steps and 7.6% overall yield from readily available dehydroepiandrosterone **6.71** as the starting material.

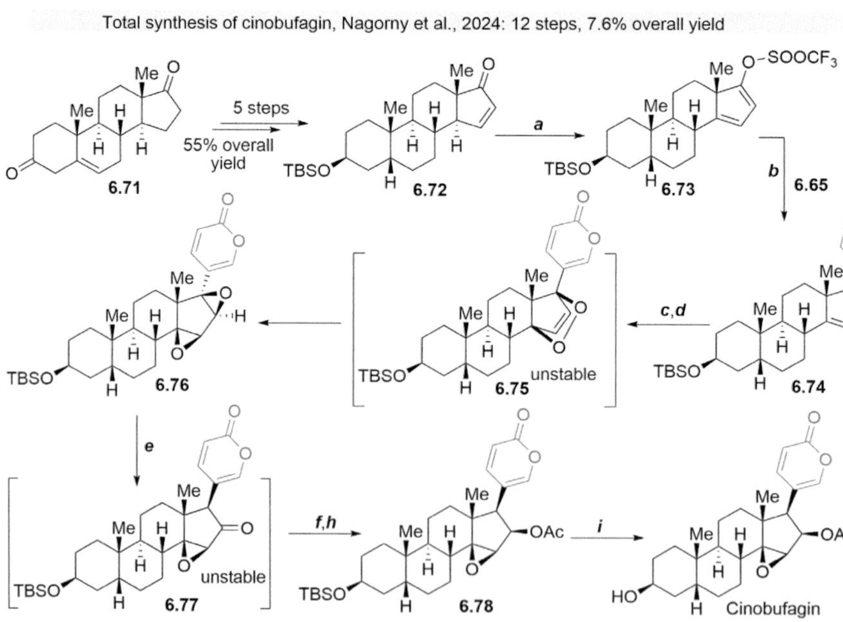

Figure 6.20. Total synthesis of cinobufagin developed by Nagorny and coworkers (2024): (a) Tf_2O, DTBMP; (b) $Pd(PPh_3)_4$, LiCl/CuCl, DMSO, 96% from **6.72**; (c) $^1O_2$, hv, methylene blue, $NaHCO_3$, CH_2Cl_2, −78°C; (d) CoTPP, −78°C, CH_2Cl_2, 64% over two steps; (e) $Sc(OTf)_3$, CH_2Cl_2, 4 Å MS; (f) $NaBH_4$; (h) Ac_2O, DMAP, 61% over three steps; (i) HF·Pyridine, 94%.

These synthetic studies of both groups could be used as the platform for preparing butalin and cinobufagin analogs. However, perhaps the most significant sources of bufadienolides for medicinal chemistry applications remain the toads of the *Bufo* genus. Interestingly, unlike Dendrobates frogs, toads themselves biosynthesized these bufadienolides toxins, but how they do this is poorly understood. Various cholesterol derivatives have been proposed as precursors in butadienolides synthesis.

Bufotoxins (bufotoxin, cinobufotoxin, gammabufotoxin, etc.) are conjugates of bufogenins with suberylarginine or other acids. Since suberylarginin is the diacid, two types of esters of C-3 substituted bufogenins can theoretically be generated. Thus, it happens in practice: these esters were identified. For example, the binding between bufotalin and suberilarginine can generate the more studied and characterized bufotoxin or another bufotoxin T12 (Figure 6.21).

Bufotoxin ($C_{40}H_{60}N_4O_{10}$, MW = 756.93 g/mol) was found in the skin of the European toad *Bufo vulgaris* and the Asian toad *Bufo gargarizans*. However, this is not a unique conjugated molecule. So far, other bufotoxins have been found, derived from C-3 substituted bufogenins and amino acids. For example, bufotoxin T11 is an ester between telocinobufagenin and succinoylarginine (Figure 6.22).

Exposure to these toxins detected only in toads can also cause cardiac abnormalities. Other symptoms include itching, depression, weakness, collapsed lungs, cardiac arrest, and seizures. Dizziness, diarrhea, and vomiting may also occur. If adequate treatment is not provided, cardiovascular symptoms can cause the

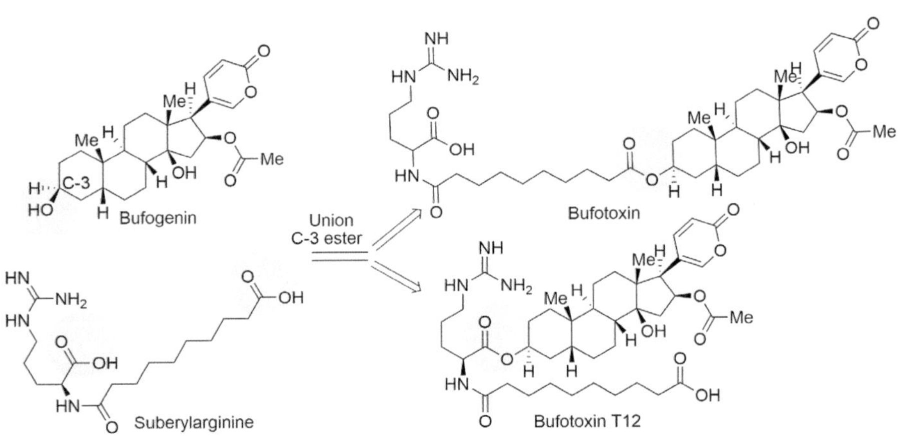

Figure 6.21. Structures of bufotoxin and bufotoxin T12.

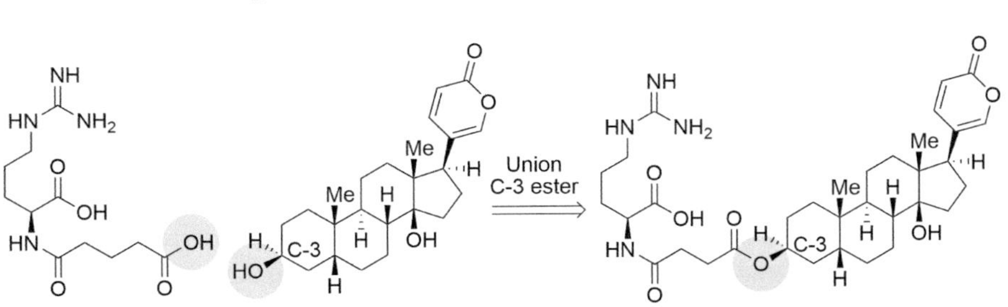

Figure 6.22. Structure of bufotoxin T11.

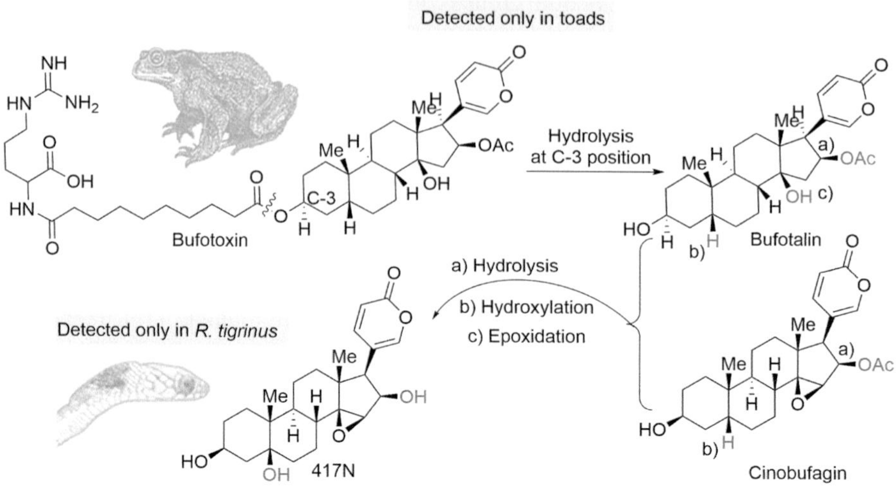

Figure 6.23. Hypothesized chemical reactions of toad bufadienolides and biochemical transformations by the snakes to own bufadienolides like 417N. The moieties involved in the reaction steps are shown in gray.

animal's death that ingests the toad venom. However, this defense is not sufficient against some snakes. Japanese natricine snakes *Rhabdophis tigrinus* ingest toads (*B. fowleri* and *B. terrestris*), sequester them, accumulate toad bufadienolides in the nuchal glands, and use them for their chemical defense. Moreover, these snakes can modify some [68–70]. Studies indicated three types of modifications by the snakes: (1) hydrolytic cleavage of the conjugated forms at the C-3 position, changing to a hydroxyl group; (2) epimerization of the hydroxyl group at the C-3 position; and (3) hydroxylation at various positions of the steroid skeleton (Figure 6.23). It is believed that *R. tigrinus* can convert dietary bufotoxin to various bufadienolides before or during accumulation in the nuchal gland.

The accumulation of sequestered toxins in the specialized defensive structures of *R. tigrinus* is unique among terrestrial vertebrates. It represents a rare case in which defensive toxins are acquired from vertebrate prey.

Another toxin group of toads is simple indole derivatives known as indolealkylamines, which contain the 5-hydroxytryptamine (5-HT) skeleton. We will mention these in the next section.

6.2.2 Indole Alkaloids Produced by Toads

More than 14 of these indolealkylamines, including 5-methoxy-N,N-dimethyltryptamine (5-MeO-DMT), have been identified in the skins of toad species. Among these, bufotenine, bufotenidine, or bufoviridine are simple 5-substituted indole derivatives, i.e., 5-hydroxytryptamine (5-HT) or serotonin derivatives [52]. At the same time, dehydrobutotenine and bufothionine are cyclic derivatives, which are based on the tricyclic structure (ABC) of 1,3,4,5-tetrahydropyrrolo[4,3,2-*de*] quinoline ring system (Figure 6.24).

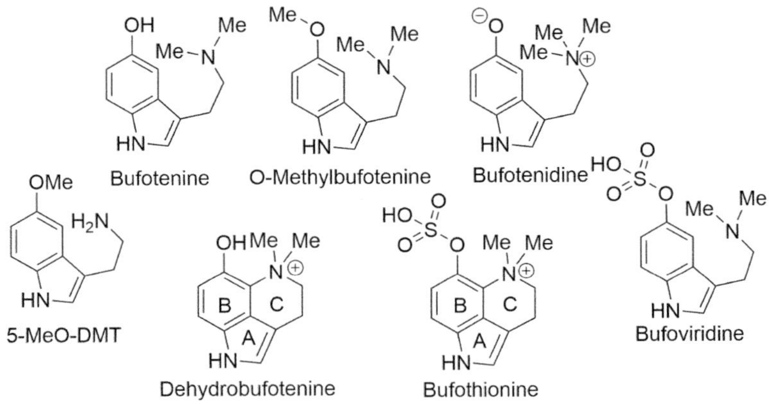

Figure 6.24. Structure of major indolealkylamines found in *Bufo* toad species. These are water-soluble ingredients of toad venom.

Simple 5-substituted indole derivatives, especially 5-MeO-DMT and bufotenine, could be potential therapeutic compounds in developing new agents for treating several neurologic disorders due to their potential 5-HT2A receptor selectivity in the central nervous system. Still, they are one of the most potent psychedelics like LSD (Section 4.2.3; Figure 4.40), behaving 5-HT2A agonists or partial agonists. It is interesting to note that 5-MeO-DMT was first synthesized in the laboratory in 1936 and then found in plants and fungi, and also in the gland secretions of the Sonoran Desert toad, *Incilius* (formerly *Bufo*) *alvarius*, in 1967. In animals, 5-MeO-DMT is converted into bufotenine via an O-demethylation reaction by the cytochrome P450 2D6 enzyme (CYP2D6). However, the principal metabolite is 5-methoxyindoleacetic acid [71]. According to toxicity studies, the LD_{50} of 5-MeO-DMT in mice ranges from 48 to 278 mg/kg, depending on the route of administration. In cats, it is 15 mg/kg (intramuscular); in sheep, it is 1–2 mg/kg (subcutaneous) and 1–5 mg/kg (intravenous). In humans, a high dose, which can lead to marked serotonin syndrome or other serious adverse effects, is > 2–30 mg/kg (depending on the route of administration).

Bufotenine produces considerable LSD psychological effects, i.e., euphoria, color hallucinations, extreme anxiety, and a sense of imminent death, but its toxicity (LD_{50}) in rodents has been estimated at 200 to 300 mg/kg. Death occurs by respiratory arrest [72, 73]. Bufotenine, a tryptamine alkaloid resulting from the methylation of serotonin, also functions as a potent constrictor of blood vessels, causing a rise in blood pressure and analgesic agent, showing synergistic analgesia with morphine's painful effect. Bufotenine and its quaternary ammonium salt, bufotenidine, are readily prepared from 5-benzyloxyindole, an available raw material [74].

In contrast, dehydrobutotenine (DBT) and bufothionine (BT) displayed potent *in vitro* cytotoxicity against various human tumor cell lines that were thought to act as DNA topoisomerase II inhibitors. Moreover, both tetrahydropyrroloquinoline-based alkaloids were used as dry powder inhalers to prepare antitumor drugs for treating lung tumors and pulmonary neoplasm [52]. Notable, the same heterocyclic system was found in several marine products. Compounds of this structure have received considerable attention because they exhibit potent *in vitro* cytotoxicity against human tumor cells.

Therefore, advanced pharmacological studies of these alkaloids and their analogs depend on their synthetic accessibility. All synthetic efforts in the preparation of these alkaloids of the ABC ring system can be divided into two groups according to the strategy employed, generally using aniline derivatives as the suitable starting material: the B → AB → ABC formation of the rings (from the indoles-Ind in the last stage) and the construction B → BC → ABC (from the quinolines-Qu) (Figure 6.25).

Dehydrobufotenine was isolated by Wieland in 1937 from the skin of the South American toad *Bufo marinus* (= *Rhinella marina*). These toads are the most enormous toad in the lowlands of Tropical America. Adult toads vary in body length from 90–200 mm and weigh 80 g to 1.2 kg. They are known as cane toads. The common name "cane toad" comes from the original purpose of using them to eradicate pests from sugar cane crops. Curiously, describing an isolated toxin, Wieland proposed a wrong structure of 5-hydroxy-indole with a 2-(dimethylamino)vinyl substituent at the C-3 position. More than 20 years later, in 1961, the correct structure of DBT was proposed. Only four synthetic methods are available for its preparation [75–78]. All these methods present some synthetic inconveniencies, especially harsh reaction conditions and little synthetic accessibility of raw indole materials. We present briefly the three most effective methods shown in Figure 6.26.

Gannon and coworkers realized the first synthesis of DBT from 5-benzyloxygramine **6.79**, which was nitrated to give the needed 3,4-disubstituted

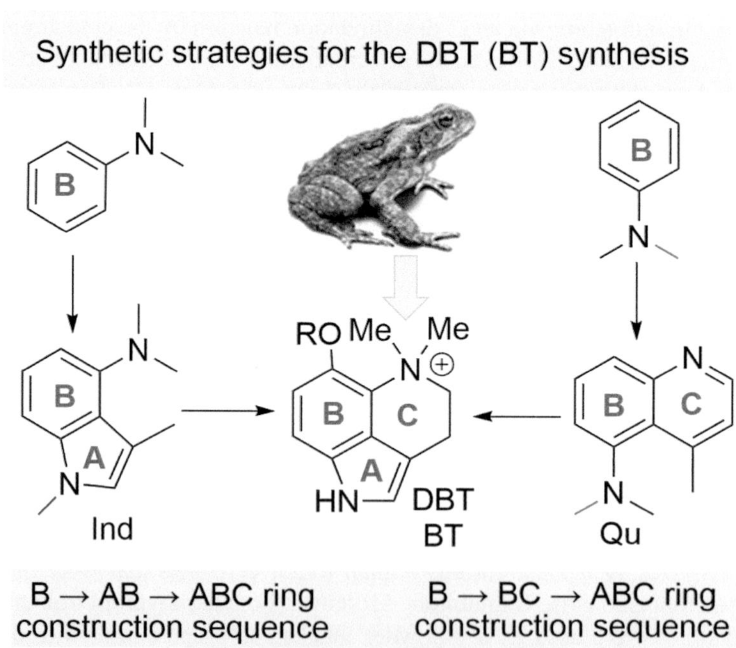

Figure 6.25. Consecutive strategies in synthesizing DBT and BT: 1. The tricyclic heterocycle ABC was constructed from a preexisting 3,4-disubstituted indoles-Ind (AB). 2. The tetrahydropyrrolo[4,3,2-*de*] quinoline ring system was prepared from 5-amino-substituted quinolines-Qu (BC).

indole **6.80**. The subsequent esterification, cyclization, and deprotection reactions allowed us to obtain DBT alkaloid as a hydrochloride with an overall yield of 10% [75]. Buchwald group started the DBT construction with 2-bromo-4-methoxyaniline **6.81** as a raw material to obtain functionalized 3,4-disubstituted indole derivative **6.82**. The ring closure of the latter, an essential step, was realized via the intramolecular Pd-catalyzed aryl amination ($Pd(PPh_3)_4$, K_2CO_3, and NEt_3). In this manner, DBT was synthesized in six steps with a 17% overall yield from **6.81** [76]. In 2021, Wang and coworkers achieved the efficient synthesis of DBT in eight steps with an overall yield of 8% from 5-methoxyindole **6.83**. The preparation of 3,4-disubstituted indole **6.84** and its conversion into tetrahydropyrroloquinoline derivatives were realized under mild reaction conditions [78]. The latter method allowed the preparation of a series of DBT derivatives evaluated for their antiviral and fungicidal activities. Among them, 6-methoxy-DBT analogs **6.85** and **6.86** resulted in potent and promising anti-plant virus agents (tobacco mosaic virus), and acylhydrazone derivative **6.87** exhibited a broad spectrum and efficient fungicidal activity, which can be further optimized as a fungicidal candidate, especially for *Sclerotinia sclerotiorum* fungus.

Joule and Alvarez [79] developed an alternative strategy, the DBT ring construction, starting from a quinoline skeleton. They used 6-methoxy-4-methylqinoline **6.88** as the starting material to produce 8-methoxy pyrroloquinoline **6.93**, an analog of DBT (Figure 6.27).

Regioselective nitration of **6.88** gave its 5-nitro derivative, **6.89**, was the first step in its preparation. Oxidation, acetal formation, and sequential nitro group and pyridine ring C in the quinoline system reduction processes led to the formation of amino-tetrahydroquinoline compound **6.92**, which was quickly converted into

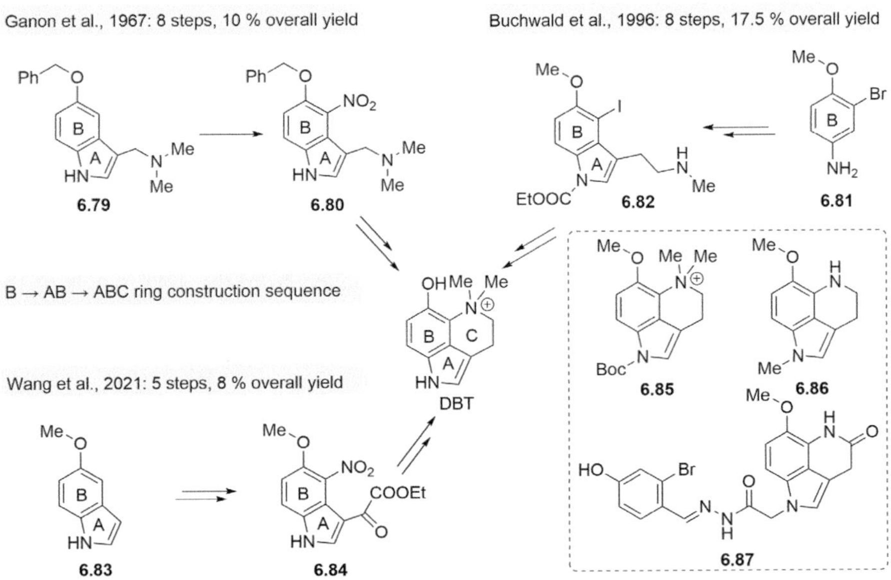

Figure 6.26. Synthetic methods for DBT preparation using raw indole materials **6.79–6.84** and structures of the promising DBT analogs **6.85–6.87** as anti-plant virus and pathogen agents.

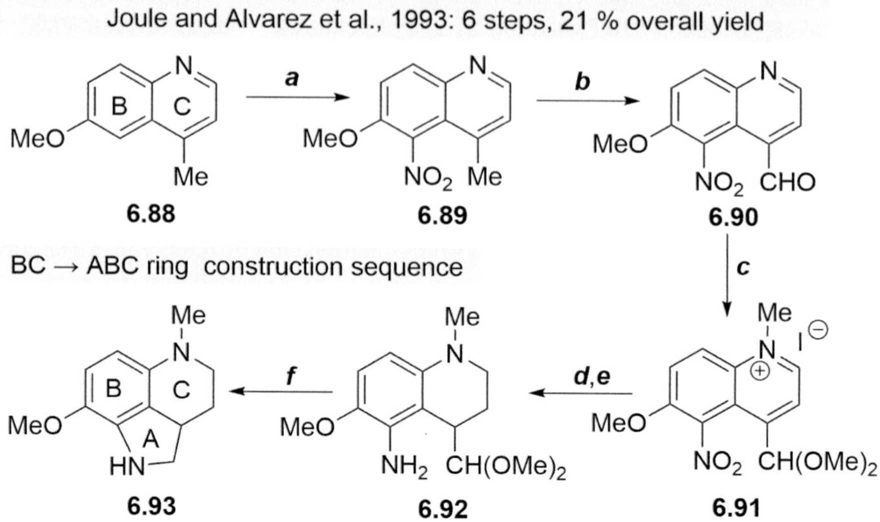

Figure 6.27. Synthesis of 8-methoxy-DBT analog **6.93**, using used 6-methoxy-4-methylqinoline **6.88** as starting material (1993): (a) fumic HNO_3, –40°C to 0°C, 90%; (b) I_2, *tert*-BuI, $FeCl_2$, DMSO, 80°C, 80%; (d) 1. dry HCl/MeOH/reflux, 2. MeI, MeCN, 50°C, 80%; (d) $NaBH_4$, MeOH, rt, 81%; (e) H_2/Pt-C, MeOH, 75%; (f) *p*-TsOH, THF, reflux, 3 hours, 58%.

the 8-methoxy-DBT analog **6.93** in a two-step protocol involving acetal hydrolysis and ring closure. However, this efficient approach was not developed and applied to various analogs.

Our discussion of the toxins isolated from ugly toads ends here. We now turn to the toxic alkaloids in fragile frogs and rare birds.

6.3 Batrachotoxin Alkaloids: Toxic Metabolites in the Skin of Phyllobate Frogs and Passerine Birds' Feathers

Before we talk about bird toxins, we should first tell you a little about the most toxic frogs of the genus *Phyllobates* (*P. aurotaenia*, *P. terribilis,* and *P. bicolor*) that live west of the Andes (Chocó, Colombia). These frogs secrete more active poison than toxins such as curare, aconitine, and veratridine. In the late 1960s, a mixture of toxic substances called batrachotoxins (BTXs) was isolated from the skin glands of the golden dart frog *P. aurotaenia* [80]. The name for these unique steroidal alkaloids was derived from the Greek "batrachos", meaning frog.

The main component of the venom is (−)-batrachotoxin [(−)-BTX] (LD_{50} = 2.0 µg, subcutaneous, per mouse of 20 g), which is a liposoluble steroid alkaloid. This neurotoxin binds specific receptor sites on sodium channels in nerve and muscle membranes, paralyzing predators and prey [81]. The strong agonistic effect of BTX is due to its activation capacity of the dual binding sites of channels [82]. Its steroid structure ABCD consists of the atypical *cis*-fusion of the A/B and C/D rings like in the cardenolide glycosides and bufadienolides. In addition, the toxin contains in its structure the six-membered C3-hemiacetal across the AB-ring, a seven-membered oxazepane on the CD-ring, and the dimethyl-pyrrole carboxy group (Figure 6.28).

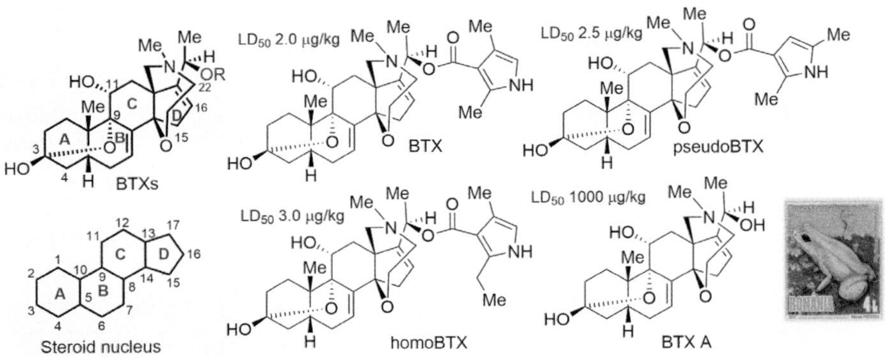

Figure 6.28. Structures of the venom's main component isolated from the frogs' skin *P. aurotaenia*. From the methanolic extracts of 5,000 frogs, a mixture of 11 mg of BTX, 16 mg of homoBTX, 1 mg of pseudoBTX, and 42 mg of BTX-A was obtained. Alkaloids were isolated in pure form using a silica gel chromatographic column with an eluent mixture of cyclohexane, triethyl amine, chloroform, and methanol (16:4:1:1). Notably, since the first isolation in 1963, an estimated 12,300 *P. aurotenia*, 1,290 *P. terribilis*, and 24 *P. bicolor* have been sacrificed to isolate a total of 1.2 grams of BTX, homoBTX, and BTX-A.

The golden poison frog has enough toxin to kill ten men, and the Indigenous Emberá people of Colombia rub their blowgun darts on the frogs' backs when hunting.

Other components, homobatrachotoxin (homoBTX, LD_{50} = 3.0 μg/kg) and pseudobatrachotoxin (pseudoBTX, LD_{50} = 2.5 μg/kg), differ from this one in the position and nature of the alkyl groups on the pyrrole ring. In addition, batrachotoxin A (BTX-A) was isolated and does not contain the pyrrolic fragment. Apparently, due to this fact, this alkaloid is much less toxic (LD_{50} = 1,000 μg/mouse) than the other components. However, it should be noted that the toxicity of BTX-A exceeds that of strychnine, a famous alkaloid isolated from the small tree of the genus *Strychnos* known as the vomica nut (*Strychnos nux vomica* L.). Comparing the lethal toxicities of BTX, which contains a substituted pyrrole ester hetero-fragment, and BTX-A, it can be observed the 2,4-dimethyl-1*H*-pyrrole-3-carboxy group at the C20 position plays a crucial role in ion channel binding affinity. In its absence, BTX-A displays only 1/500th of the toxicity of BTX. It should be noted that the most poisonous frog is the golden dart frog *P. terribilis* since its skin contains nine times more BTXs than the other two species of the genus *Phyllobates*. Interestingly, *P. terribilis* frogs are perhaps the only creatures immune to their poison. The average venom in the skin of a wild *P. terribilis* is estimated at 1 mg, enough to kill about 10,000 mice. This estimate may vary, but most agree that this dose is enough to kill between 10 and 20 humans.

In captivity, frogs of the genus *Phyllobates* feed on fruit flies and small crickets. As a result, they do not produce BTXs and lose their toxicity. The same thing happens with the other frogs of the Dendrobatidae family. All captive poison frogs lose their toxicity when deprived of certain foods, and captive-born golden poison frogs are harmless.

Now, we all know that birds are warm-blooded animals. Their bodies are covered with feathers. Their best defense against terrestrial predators is their wings.

However, some birds contain toxic substances in their feathers. What purpose do these toxins serve for the birds, and which birds have them?

One such bird is the pitohui (family Pachycephalidae), endemic to New Guinea. This bird, described for the first time in 1827, is present in all important ornithological collections worldwide. Pitohuís are brightly colored omnivorous birds. The *Pitohui* genus comprises six species, of which three (*Pitohui dichrous*, *P. kirhocephalus* and *P. ferrugineus*), are significantly more toxic than the others. The substance found in the skin and feathers of three passerine bird species of pitohui is one hundred times more harmful than strychnine. However, until 1992, few suspected that birds' wings could be poisonous. Curiously, the active ingredients of this poison are batrachotoxins (BTX), homoBTX, and BTX-A (Figure 6.28), among others [83]. BTXs were later found in New Guinea toxic birds of the genus *Ifrita* [84]. They are also part of the poison of South American *Phyllobates* frogs we discussed above in the same section. It is believed that such toxins provide some protection against the birds' natural enemies, such as parasites and predators, including humans.

It was found that most of the toxins are concentrated in the skin and feathers. While birds have much more homoBTX than BTX, the amount of BTX in frogs is almost double that of homoBTX. In addition to the BTXs mentioned, the esters of BTX-A (acetate, crotonate, and 3-hydroxypentanate) (Figure 6.29) are also found in bird venom; they were not found in frog venom. The bicolor pitohui (*P. dichrous*) is the most poisonous of the three *Pitohui* species. It is estimated that a 65 g bird concentrates between 15 and 20 μg of homoBTX in the skin, while less than 3 μg is found in the feathers. According to the quantification of the extract of the feathers of *Ifrita* birds, "trash birds," the level of alkaloids is 5.30 μg of BTX-A (crotonate), 0.29 μg of BTX, and 3.40 μg of homoBTX.

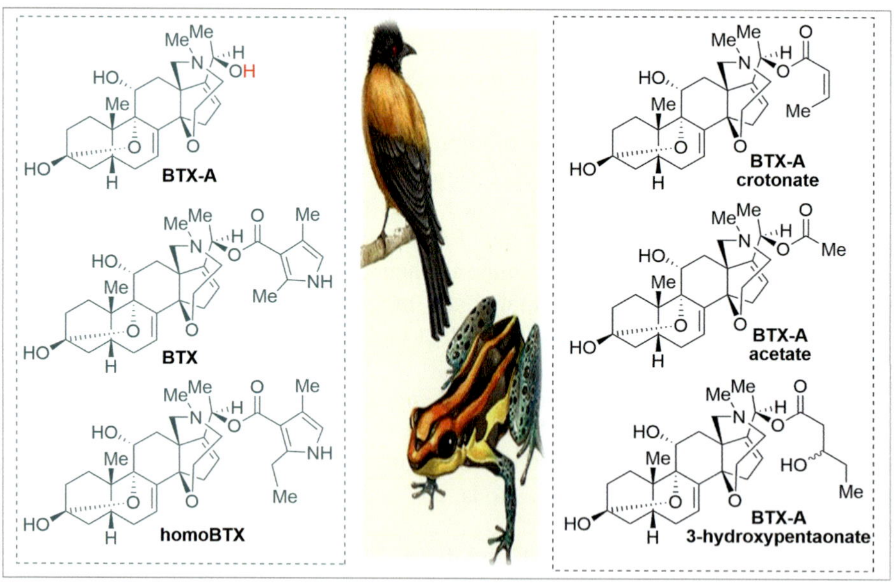

Figure 6.29. Batrachotoxins found in the skin and wings of New Guinea passerine birds. The BTXs marked in gray were previously found in the skin of *Phyllobates* frogs.

It should be noted that these birds, like *Phyllobates* frogs, do not produce batrachotoxins themselves but rather obtain them from some beetles of the *Choresine* genus (family Meryridae), which are part of their diet. It was shown that the amount of BTX found in the *Choresine semiopaca* beetle is 1.8 μg [85]. Although beetles can use the mevalonate pathway to create terpenes, it is unknown if they can create steroidal skeletons such as BTXs. Therefore, these may originate from plant phytosterols and are modified by the beetle by a symbiont or by a set of symbionts.

Exhibiting potent agonistic activity of voltage-gated sodium channels (VGSC), an obligatory class of membrane proteins for bioelectrical signaling, the BTX molecule is considered an optimal "lead" compound for the rational design of small-molecule modulators of VGSC activity. Since its collection from the natural source is restricted, the problem of its accessibility must be solved in order to study this neurotoxin in detail. The synthesis of this alkaloid with such a complex architecture is an arduous synthetic challenge.

Witkop and coworkers, who isolated this toxin, initiated the first synthetic developments. To confirm the structure of BTX, they converted the natural BTX

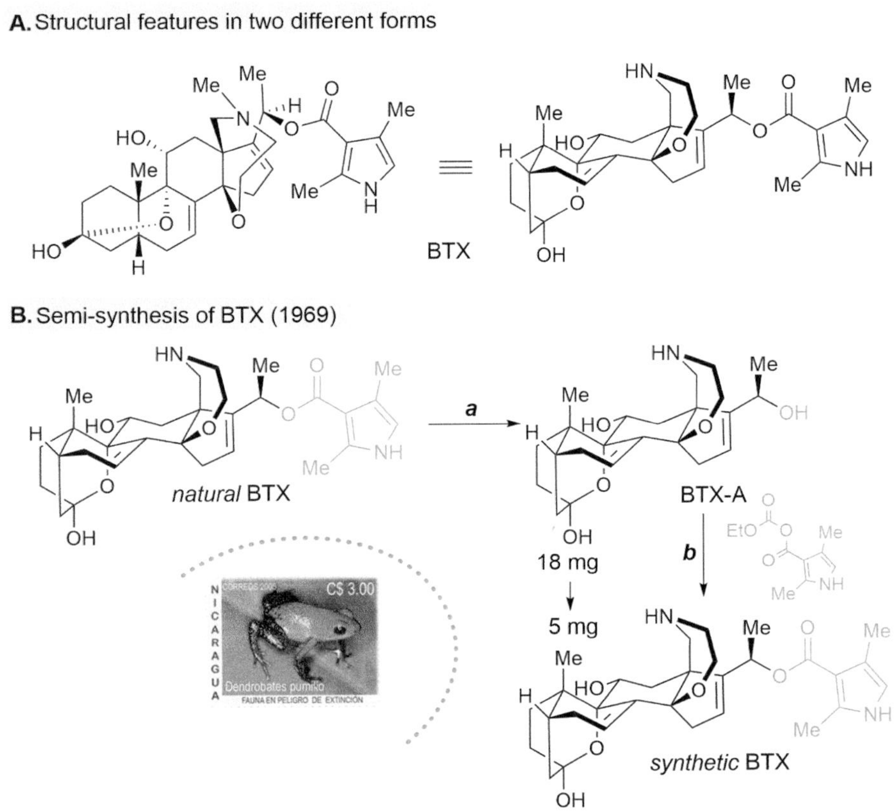

Figure 6.30. Structural analysis of batrachotoxin. A. Different structural representations of the BTX skeleton help us better understand the structural features of BTX and synthesis design. B. One-step semi-synthesis of BTX from BTX-A: (a) 2.0 N KOH (aq)/MeOH (1:1), 60°C, sealed tube under a nitrogen, 16 hours; (b) 2,4-dimethyl-1*H*-pyrrole-3-carboxylic anhydride, 2.0 N KOH (aq), $CHCl_3$, rt, 5 hours, 28%.

molecule into synthetic BTX using a simple two-step hydrolysis-esterification protocol, i.e., semi-synthesis (Figure 6.30) [80].

Then, in 1972, the partial synthesis of (±)-BTX-A, a molecule lacking the C-20 pyrrole ester, was disclosed using 11α-acetoxy-progesterone as the starting material [86]. However, the length of this linear synthesis is laborious; (±)-BTX-A was obtained in 45 steps with 0.0022% overall yield. This example precluded the production of BTX or its analogs, and since then, there have been many synthetic studies on feasible routes to this alkaloid group. Like bufadienolides syntheses, the success of BTX-A and BTX's designed syntheses depends much on the choice of the steroidal or steroidal-like scaffold as an available starting material. The existing total or partial synthesis of BTX-A in racemic and chiral forms illustrates the previous sentence. Therefore, Prof. Yoshito Kishi and his group performed the first total synthesis of (±)-BTX-A [87], starting with racemic Wieland-Hiesche ketone (9-methyl- $\Delta^{5(10)}$-octalin-1,6-dione, **6.94**), which was converted into a key precursor **6.95** with the constructed ABCD skeleton (Figure 6.31). Finally, the target (±)-BTX-A was obtained in 48 steps (0.70% yield), thus also completing the formal synthesis of (±)-BTX.

Another valuable starting material for synthesizing these steroidal alkaloids is the Hajos-Parrish ketone (7a-methyl-2,3,6,7-tetrahydroindene-1,5-dione,

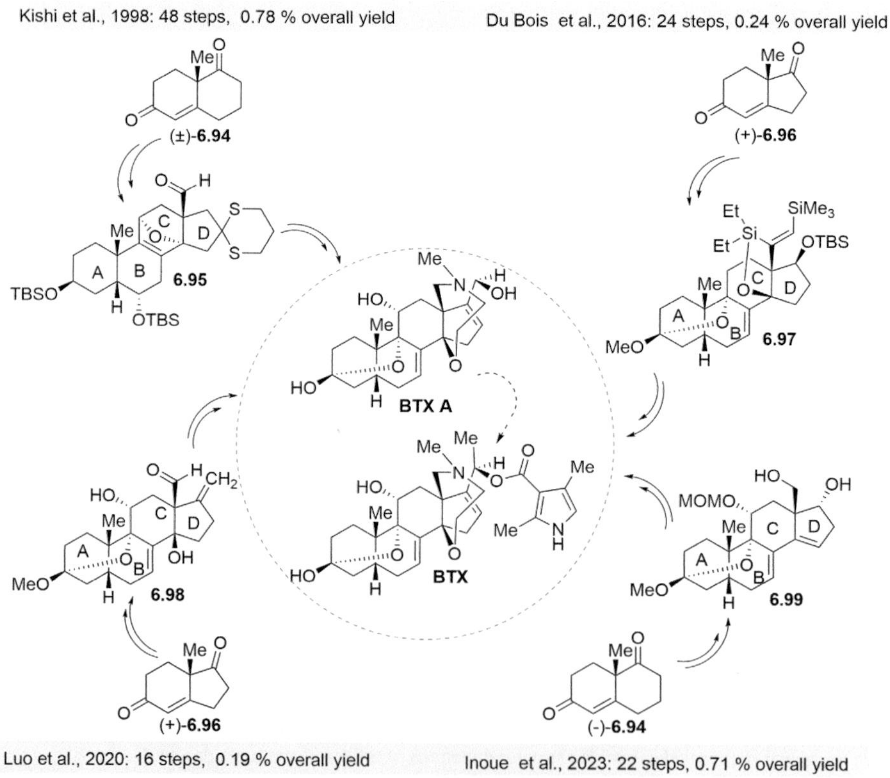

Figure 6.31. General aspects of BTX syntheses and selected starting materials. MOM = methoxymethyl; TBS = *tert*-butyl(dimethyl)silyl.

6.96). Using an (*S*)-form of this ketone, the Du Bois group achieved asymmetric syntheses of the natural (−) and non-natural (+) antipodes of batrachotoxin [88]. They transformed it into an adequate functionalized skeleton ABD **6.97** ready to complete the construction of a complex BTX ring system (Figure 6.31). Thus, BTX in enantiomeric form was prepared in 24 linear steps with a 0.24% overall yield.

The Luo group also used (*S*)-(+)-Hajos-Parrish ketone to synthesize (-)-BTX-A [89]. Modifying his structure by a long-step route, they obtained steroidal ABCD skeleton **6.98** with the six-membered C3-hemiacetal in the AB-ring, which was transformed into the desired chiral BTX-A in four steps (Figure 6.31). Therefore, this new 16-step synthesis provides the natural product with a 0.19% overall yield. In contrast, Inoue and coworkers addressed the Wieland-Hiesche ketone again when planning their design for the BTX construction [90]. Multitransformations of (+)-Wieland-Miescher ketone **6.94** led to the formation of the vital precursor **6.99**, in which a seven-membered oxazepane was installed on the CD-ring (Figure 6.31).

These total syntheses highlight the investigators' creativity and have enriched the science of organic synthesis. However, BTX as a biochemical reagent is still expensive (10 μg, 285 $, Santa Cruz Biotechnology).

Many scientists are permanently interested in the defensive neurotoxins of amphibians and birds that we have just studied. Every year, new information is added to their chemistry and biology that could lead to the creation of various therapeutics. Thus, from observing the toxic effects of plants or animals, in this case, amphibians and birds, to the design of medicines or agrochemicals, there is a long path of chemical and biological research. In contrast, little is known about the toxic substances produced by mammals.

6.4 Toxic Molecules from Mammals

Mammals are vertebrate animals like amphibians and birds and are divided into several groups. However, little is known about mammals that process and use toxic substances to defend against predators. Venomous mammals are rare, and their venoms (mainly of protein origin) have not been well characterized [91]. Some interesting examples have been found regarding small molecules that protect mammals against their depredators.

The defensive secretion of skunks (family Mustelidae) is well-studied. A characteristic of mephitids is that they secrete an odorous liquid from their anal glands. When disturbed, they can expel this liquid substance, sometimes reaching a distance of 2–3 meters from the back of their body. They raise their tail high and launch the jet, which irritates anyone who bothers them. GC-MS analyses of this repulsive-smelling secretion demonstrate the presence of seven major compounds, including thiols, their thioacetates, and 2-thio-substituted quinolines (Figure 6.32) [92]. The major component in skunk spray was erroneously believed to be 1-butanethiol until it was later shown that the actual compound was (*E*)-2-butene-1-thiol. More recently, two studies identified the third principal compound in the secretion as either (*E*)-2-butenyl methyl disulfide or (*E*)-2-butenyl propyl sulfide [93].

Thiols stand out among these due to their horrible smell. The persistence of skunk odor may be attributed to thioacetates in the secretion. Thioacetates' decomposition

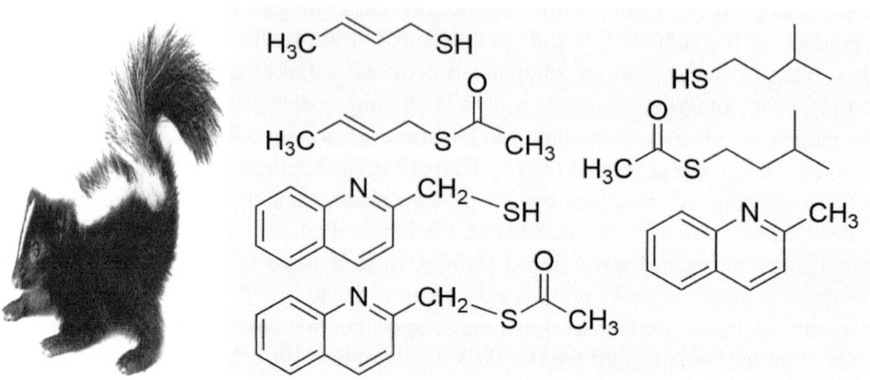

Figure 6.32. Striped skunk and its defensive weapons.

in water is slow, and the most powerful thiols are released, thus causing the odor. To speed up the destruction of thioacetates, "chemical" methods can be used by washing these animals with soda soaps (catalyzing hydrolysis under slightly basic conditions).

The maned rat, African crested rat, or Imhaus's hamster (*Lophiomys imhausi*) is a species of rodent in the family Cricetidae. It is the only species in its subfamily and genus. It measures between 25 cm and 36 cm and weighs 600 g to 900 g. It is a large cricetid with short legs and a long tail. Their coloration is usually black and white or brown and white, with stripes and spots. It is found in East Africa. It lives in forests, semi-deserts, and savannas, frequently in rocky areas where it finds its plant food. Precisely there, the African crested rat has developed an interesting and unique defense mechanism: it chews the root and bark of the highly toxic Acokanthera bush (family Apocynaceae), known as the "African poison-arrow tree" or "ouabaio tree", and then deliberately rubs its fur against this bush, which secretes a cardenolide called ouabain [94]. We member this molecule as a potent sodium pump inhibitor (Section 2.5). More extensive exposures to this cardenolide induce vomiting, dysrhythmia, convulsions, difficulty breathing, and cardiac arrest. Most mammals are naturally susceptible to cardenolides; however, muroid rodents like Imhaus's hamsters are more resistant. Thus, this highly dangerous substance, acquired from Acokanthera bushes, can poison terrestrial predators that attack with their mouths, e.g., wild dog, *Lycaon pictus*; black-backed jackal, *Canis mesomelas*; hyena, *Hyaena hyaena*; and leopard, *Panthera pardus* (Figure 6.33). When alarmed, the crested rats retract their heads and flare out their flanks containing these poisonous hairs to ward off predators.

This is why, without producing said poison, the maned rat acquires toxic molecules and makes incredible use of them. This strange and curious behavior can be said to be artificially poisonous. Paradoxically, the rat is poisonous but strictly vegetarian. The same rat is entirely immune to this poison that can kill an elephant.

We should also remember that the border between poison and drugs is quite illusory. The discussed plant cardenolides and toad bufadienolides belong to the drug-like cardiotonic steroids. Digoxin, bufalin, and marinobufagenin treat cardiac and other ailments in minute concentrations. Ouabain also triggers Na^+/K^+-ATPase-

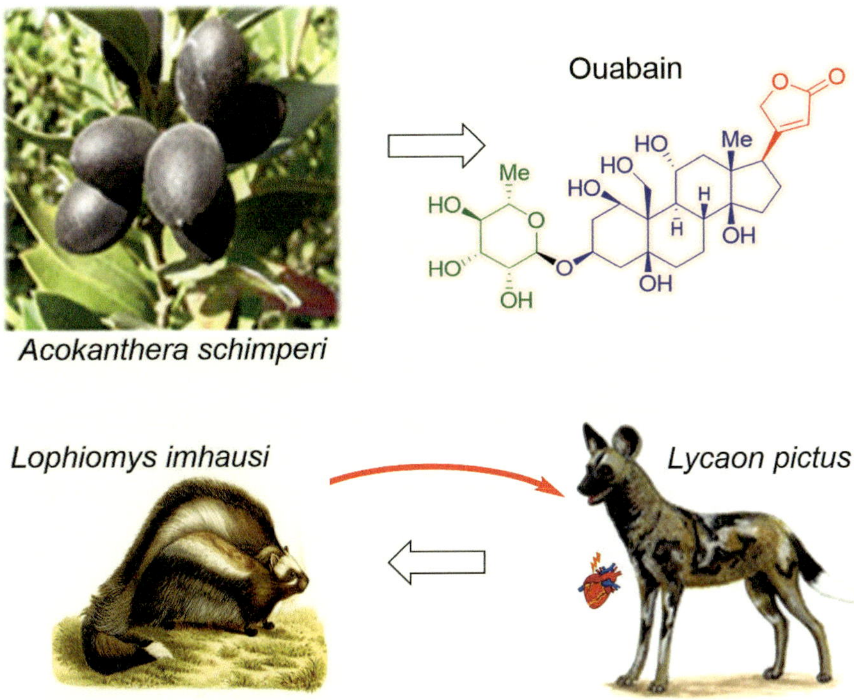

Figure 6.33. Exceptional chemical defense in mammals: *L. imhausi* is unique in its ability to sequester cardenolides from *A. schimperi*.

mediated signaling pathways at low concentrations, inducing several biological regulatory effects, including cell proliferation, hypertrophy, and apoptosis [95]. Thus, it and other aforementioned cardiotonic steroids could be potential anticancer therapeutics.

The structure of ouabain consists of α–L-rhamnoside and an aglycone called ouabagenin. So if the ouabagenin skeleton can be synthesized, ouabain synthesis is in our hands. The highly oxygenated structure of ouabagenin has posed a formidable synthetic challenge. Only five successful syntheses of this cardenolide have been reported [96–100]. As the first total synthesis of ouabain and their aglycone (ouabagenin) realized by the Deslongchamps group was examined in Section 2.5 (Figure 2.42) [96], first we want to comment briefly on present approaches to ouabagenin [97–99] and then discuss the synthesis performed by the Tang group in 2023 [100].

The Baran group utilized cortisone acetate **6.100** as a cheap starting material, which was transformed into a key-protected boronate ester precursor **6.101**. Ouabagenin was obtained from the latter through redox and stereochemical relays. The modified Stille cross-coupling reaction in the last-stage synthesis of ouabagenin allowed the installation of the butenolide moiety (Figure 6.34) [97]. Inoue and coworkers chose highly oxygenated decalone **6.102**, prepared from (*R*)-perillaldehyde in 15 steps, as the starting material for their ouabagenin synthesis. Then, this polycyclic ketone was subjected to multiple transformations to assemble

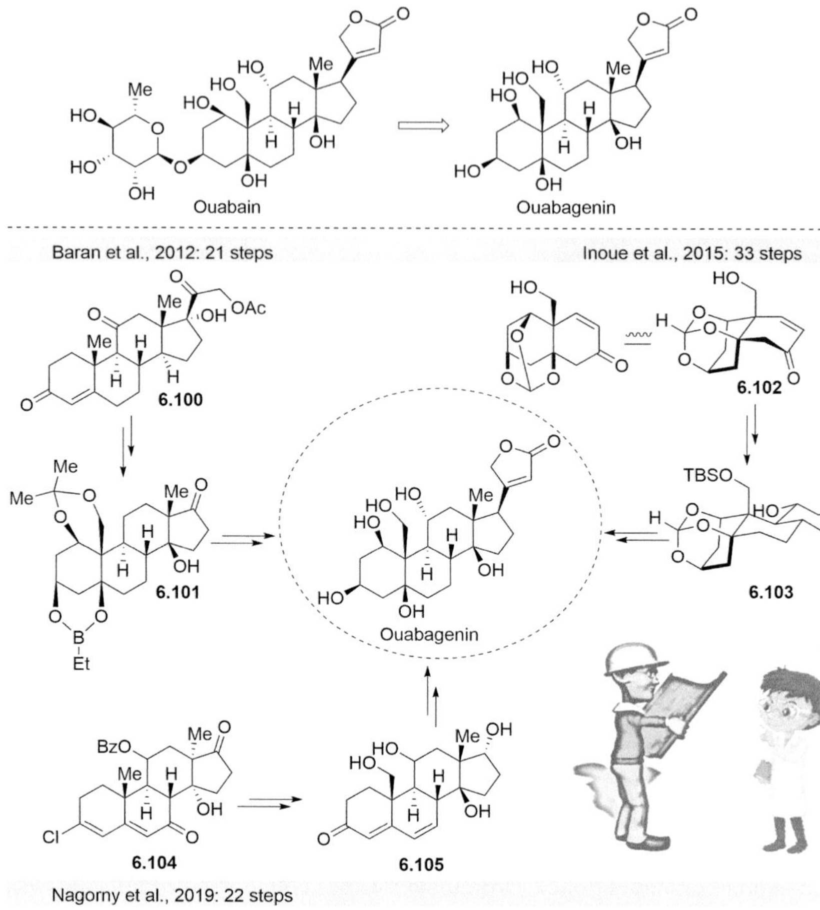

Figure 6.34. General features of ouabagenin syntheses and selected starting materials. Bz = benzoyl; TBS = *tert*-butyl(dimethyl)silyl.

the key skeleton ABCD **6.103**, in which the butenolide moiety was installed through the same process of the Stille cross-coupling reaction (Figure 6.34) [98].

Nagorny and coworkers [99] performed the synthesis of ouabagenin, starting with 2-methylcyclopentane-1,3-dione and 3-ethoxycyclohex-2-en-1-one to construct the ABCD skeleton **6.104** (seven steps, 45% overall yield), which was transmuted into a key precursor **6.105**. A long-route transformation of the latter allowed the preparation of the desired cardenolide (Figure 6.34).

Despite these efforts, rapid and efficient synthesis of highly oxygenated cardiotonic steroids remains in high demand to support biological studies searching for new therapeutic agents and testing new synthetic strategies and tactics.

The Tang group proposed a new synthetic strategy for the construction of the ouabagenin skeleton [100], in which the Hajos-Parrish ketone ketal **6.106**, derived from (+)-Hajos-Parrish ketone **6.96**, was used as an available starting material (Figure 6.35). Its interaction with tosylate **6.107** in toluene with NaH as the base

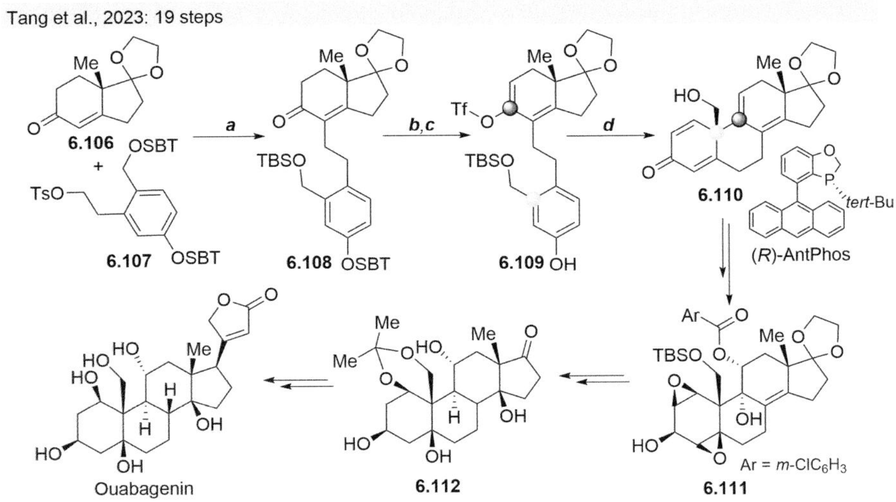

Figure 6.35. Simplified scheme of quabagenin synthesis realized by the Tang group (2023): (a) NaH, toluene, 110°C, 4 hours, 59%; (b) LiHMDS, $PhNTf_2$, THF, −78°C, 20 minutes, then rt, 2 hours, 85%; (c) KF, tetraethylene glycol/EtOAc, rt, 1 hours, 94%; (d) 2 mol% [Pd(cinnamoyl)Cl_2], 5 mol% (*R*)-AntPhos, K_2CO_3, toluene, 85°C, 10 hours, 76%. Ts = 4-toluenesulfonyl; Tf = trifluoromethanesulfonyl; TBS = *tert*-butyl(dimethyl)silyl.

afforded the coupling product **6.108**, which was effectively transformed into the enol triflate **6.109**. Then, its palladium-catalyzed dearomative cyclization led to the formation of the unsaturated steroidal skeleton **6.110**, possessing four olefinic functionalities for installing other hydroxy groups. Its synthesis was achieved in four steps from the Hajos-Parrish ketone ketal **6.106** in 36% overall yield on a decagram scale.

Next, further transformations of this steroid skeleton allowed for the obtaining of an intermediate **6.111**, with all installed hydroxy functions in three steps. A close precursor, so-called protected ouabageninone **6.112**, was prepared from **6.111** in nine other steps. In the final stages, the introduction of the cardenolide moiety and acetal deprotection completed the total synthesis of ouabagenin in 19 steps from the Hajos-Parrish ketone ketal **6.106** (Figure 6.35).

The described synthesis is the shortest and most efficient of the existing total syntheses of this highly oxygenated cardiotonic steroid, which offers synthetic versatility and practicality in the search for new therapeutic agents.

6.5 Conclusion

In summary, amphibians use their venom as a defensive weapon only in an accessory manner. Poison is not for eating but for "not being eaten". A predator that bites an amphibian with active skin secretion will be forced to spit out such a "spicy" morsel and feel a strong burning sensation in the mouth. What molecules help them survive deadly attacks?

We have seen that the frog pumiliotoxin alkaloids contain an indolizidine ring, i.e., azabicyclo[4.3.0]nonane ring with polyfunctional moieties that are diversity-oriented. In contrast, the epibatidine alkaloid possesses a 7-azabicyclo[2.2.1]heptane structure. The first are positive modulators of voltage-gated sodium channels, showing cardiotonic activity, and the latter is a non-opioid analgesic and agonist of neuronal nAChRs subtype α4β2. Both have become important models for drug research as well as toad neurotoxins. Among them, bufadienolides bufalin and cinobufagin are recognized as the major bioactive components in ChanSu and HuachanSu, traditional Chinese medicine used for treating pain and inflammatory diseases and in clinical cancer therapy, respectively. Chemically, they are steroid molecules, similar to plant cardenolides, with an β-oriented heterocyclic 2-pyranone ring instead of the 2-furanone moiety.

Amphibian venoms, especially those of toads, have been used in antitumor therapies. They may also have some biomedical applications, such as analgesics, hemostatics, and, above all, cardiostimulants. On the other hand, toads exude the toxin 5-MeO-DMT from glands within their skin to deter predators. While the substance puts off predators, people who consume 5-MeO-DMT can have a psychedelic experience. Psychedelic substances have gained attention in recent years as treatments for depression and other central nervous system disorders.

Batrachotoxin alkaloids, like BTX, homoBTX, and BTX-A, found in the skin of phyllobate frogs and passerine birds' feathers, are based on the steroid ABCD skeleton. Still, they are more complex structurally, having the six-membered C3-hemiacetal across the AB-ring, a seven-membered oxazepane on the CD-ring, and the dimethyl-pyrrole carboxy group. The main component of the paralyzing venom of golden frogs is (−)-BTX, one of the most toxic nonprotein molecules. While passerine birds have much more homoBTX than BTX, the amount of BTX in frogs is almost double that of homoBTX. It seems that these markedly different animals obtain them from their similar diet. As these neurotoxins in amphibians and birds are in mg-level quantities, developing strategies and tactics for their synthesis is essential. Each total synthesis of each complex molecule is a challenge and a step forward that helps expand knowledge about natural products.

The same goes for the highly toxic steroid glycoside ouabain and its aglycone ouabagenin, which is present in the African ouabaio tree. Imhaus's hamsters sequester this steroid from the bushes of their diet, store it in their fur, and use it in case of danger. This is a unique defense mechanism developed by this mammalian rodent. Skunks also have another defensive tactic: when disturbed, they can expel an odorous liquid, which contains repulsive-smelling thiols, their thioacetates, and 2-thio-substituted quinolines.

In previous chapters, we have analyzed, examined, and studied these and numerous other small toxic molecules. All of them are metabolites of living terrestrial organisms. Many more complex molecules are found beneath ocean water, which we will address in the last chapter. We have a difficult task ahead because studying the behavior of marine organisms is much more complex, and there is less reliable data on chemical defense. However, the selected material will reflect trends in the chemistry and biology of marine metabolites used to protect organisms in their struggle for survival.

6.6 References

[1] Clarke, B.T. 1997. The natural history of amphibian skin secretions, their normal functioning and potential medical applications. Biol. Rev. 72(3): 365–379.

[2] Daly, J.W., Garraffo, H.M. and Spande, T.F. 1993. Alkaloids from amphibians. pp. 185–288. *In*: Cordell, G.A. [Ed.]. The Alkaloids. Vol. 43. San Diego: Academic Press.

[3] Santos, J.C., Tarvin, R.D. and O'Connell, L.A. 2016. A review of chemical defense in poison frogs (Dendrobatidae): ecology, pharmacokinetics, and autoresistance. pp. 305–337. *In*: Schulte, B., Goodwin, T., Ferkin, M. [Eds.]. Chemical Signals in Vertebrates 13. Springer, Cham. https://doi.org/10.1007/978-3-319-22026-0_21.

[4] Ligabue-Braun, R. and Carlini, C.R. 2015. Poisonous birds: A timely review. Toxicon. 99: 102–108.

[5] Daly, J.W. and Myers, C.W. 1967. Toxicity of Panamanian poison frogs (*Dendrobates*): some biological and chemical aspects. Science 156(3777): 970–973.

[6] Saporito, R.A., Donnelly, M.A., Spande, T.F. and Garraffo, H.M. 2012. A review of chemical ecology in poison frogs. Chemoecology 22: 159–168.

[7] Daly, J.W., Secunda, S.I., Garraffo, H.M., Spande, T.F., Wisnieski, A., Cover, J.F. Jr. 1994. An uptake system for dietary alkaloids in poison frogs (Dendrobatidae). Toxicon. 32: 657–663.

[8] Daly, J.W., Garraffo, H.M., Spande, T.F., Jaramillo, C. and Rand, A.S. 1994. Dietary source for skin alkaloids of poison frogs (Dendrobatidae)? J. Chem. Ecol. 20: 943–955.

[9] Weldon, P.J. 2017. Poison frogs, defensive alkaloids, and sleepless mice: critique of a toxicity bioassay. Chemoecology 27(4): 123–126.

[10] Daly, J.W., Spande, T.F. and Garraffo, H.M. 2005. Alkaloids from amphibian skin: a tabulation of over eight-hundred compounds. J. Nat. Prod. 68(10): 1556–1575.

[11] Daly, J.W. and Myers, C.W. 1967. Toxicity of Panamanian poison frogs (*Dendrobates*): some biological and chemical aspects. Science 156(3777): 970–973.

[12] Bargar, T.M., Lett, R.M., Johnson, P.L., Hunter, J.E., Chang, C.P., Pernich, D.J. et al. 1995. Toxicity of pumiliotoxin 251D and synthetic analogs to the cotton pest *Heliothis virescens*. J. Agric. Food Chem. 43(4): 1044–1051.

[13] Weldon, P.J., Kramer, M., Gordon, S., Spande, T.F. and Daly, J.W. 2006. A common pumiliotoxin from poison frogs exhibits enantioselective toxicity against mosquitoes. Proc. Natl. Acad. Sci. USA 103(47): 17818–17821.

[14] Weldon, P.J., Cardoza, Y.J., Vander Meer, R.K., Hoffmann, W.C., Daly, J.W. and Spande, T.F. 2013. Contact toxicities of anuran skin alkaloids against the fire ant (*Solenopsis invicta*). Naturwissenschaften 100: 185–192.

[15] Daly, J.W., Gusovsky, F., McNeal, E.T., Secunda, S., Bell, M., Creveling, C.R. et al. 1990. Pumiliotoxin alkaloids: a new class of sodium channel agents. Biochem. Pharmacol. 40: 315–326.

[16] Vandendriessche, T., Abdel-Mottaleb, Y., Maertens, C., Cuypers, E., Sudau, A., Nubbemeyer, U. et al. 2008. Modulation of voltage-gated Na+ and K+ channels by pumiliotoxin 251D: A "joint venture" alkaloid from arthropods and amphibians. Toxicon. 51(3): 334–344.

[17] Kovacic, P. 2007. Toxicity mechanism of frog mosquito repellent: electron transfer and radical species--useful aspects. Med. Hypotheses 69(1): 225–226.

[18] Franklin, A.S. and Overman, L.E. 1996. Total syntheses of pumiliotoxin A and allopumiliotoxin alkaloids. Interplay of pharmacologically active natural products and new synthetic methods and strategies. Chem. Rev. 96(1): 505–522.

[19] Overman, L.E. and Bell, K.L. 1981. Enantiospecific total synthesis of dendrobatid toxin 251D. A short chiral entry to the cardiac-active pumiliotoxin A alkaloids via stereospecific iminium ion-vinylsilane cyclizations. J. Am. Chem. Soc. 103(7): 1851–1853.

[20] Overman, L.E., Bell, K.L. and Ito, F. 1984. Enantioselective total syntheses of pumiliotoxin B and pumiliotoxin 251D. A general entry to the pumiliotoxin A alkaloids via stereospecific iminium ion-vinylsilane cyclizations. J. Am. Chem. Soc. 106(15): 4192–4201.

[21] Overman, L.E. and Lin, N.H. 1985. Enantioselective total synthesis of (+)-pumiliotoxin A. J. Org. Chem. 50(19): 3669–3670.

[22] Overman, L.E. and Sharp, M.J. 1988. Nucleophile-promoted electrophilic cyclization reactions of alkynes. J. Am. Chem. Soc. 110(2): 612–614.

[23] Overman, L.E. and Sharp, M.J. 1988. Enantioselective total synthesis of the pumiliotoxin A alkaloids via reductive iminium ion-alkyne cyclizations. Total synthesis of (+)-pumiliotoxin A. Tetrahedron Lett. 29(8): 901–904.
[24] Overman, L.E. and Sarkar, A.K. 1992. Iodide-promoted aldiminium ion-alkyne cyclizations. A convenient synthesis of substituted tetrahydropyridines and 3-alkylidenepiperidines. Tetrahedron Lett. 33(29): 4103–4106.
[25] Montgomery, J. 2004. Nickel-catalyzed reductive cyclizations and couplings. Angew. Chem. Int. Ed. 43(30): 3890–3908.
[26] Woodin, K.S. and Jamison, T.F. 2007. Total synthesis of pumiliotoxins 209F and 251D via Late-Stage, Nickel-Catalyzed Epoxide–Alkyne Reductive Cyclization. J. Org. Chem. 72(19): 7451–7454.
[27] Daly, J.W., Tokuyama, T., Fujiwara, T., Highet, R.J. and Karle, I.L. 1980. A new class of indolizidine alkaloids from the poison frog, Dendrobates tricolor. X-ray analysis of 8-hydroxy-8-methyl-6-(2'-methylhexylidene)-1-azabicyclo[4.3.0]nonane. J. Am. Chem. Soc. 102(2): 830–836.
[28] Spande, T.F, Garraffo, H.M., Edwards, M.W., Yeh, H.C.J., Pannell, L. and Daly, J.W. 1992. Epibatidine: A novel (chloropyridyl)azabicycloheptane with potent analgesic activity from an Ecuadoran poison frog. J. Am. Chem. Soc. 114(9): 3475–3478.
[29] Daly, J.W., Garraffo, H.M., Spande, T.F., Decker, M.W., Sullivan, J.P. and Williams, M. 2000. Alkaloids from frog skin: the discovery of epibatidine and the potential for developing novel non-opioid analgesics. Nat. Prod. Rep. 17(2): 131–135.
[30] Bradley, D. 1993. Frog venom cocktail yields a one-handed painkiller. Science 261(5125): 1117–1117.
[31] Badio, B. and Daly, J.W. 1994. Epibatidine, a potent analgetic and nicotinic agonist. Mol. Pharmacol. 45(4): 563–569.
[32] Watt, A.P., Verrier, H.M. and O'Connor, D. 1994. Resolution of synthetic (+) and (-)-epibatidine by chiral high performance liquid chromatography and identification of the natural enantiomer. J. Liq. Chromatogr. 17(6): 1257–1264.
[33] Salehi, B., Sestito, S., Rapposelli, S., Peron, G., Calina, D., Sharifi-Rad, M. et al. 2018. Epibatidine: a promising natural alkaloid in health. Biomolecules 9(1): 6.
[34] Broka, C.A. 1993. Total synthesis of epibatidine. Tetrahedron Lett. 34(20): 3251–3254.
[35] Fletcher, S.R., Baker, R., Chambers, M.S., Hobbs, S.C. and Mitchell, P.J. 1993. The synthesis of (+)-and (–)-epibatidine. J. Chem. Soc. Chem. Comm. (15): 1216–1218.
[36] Clayton, S.C. and Regan, A.C. 1993. A total synthesis of (±)-epibatidine. Tetrahedron Lett. 34(46): 7493–7496.
[37] Huang, D.F. and Shen, T.Y. 1993. A versatile total synthesis of epibatidine and analogs. Tetrahedron Lett. 34(28): 4477–4480.
[38] Olivo, H.F. and Hemenway, M.S. 2002. Recent syntheses of epibatidine. A review. Org. Prep. Proced. Int. 34(1): 1–25.
[39] Carroll, F.I. 2009. Epibatidine analogs synthesized for characterization of nicotinic pharmacophores—A review. Heterocycles 79: 99–120.
[40] Oliveira Filho, R.E. de and Omori, A.T. 2015. Recent syntheses of frog alkaloid epibatidine. J. Braz. Chem. Soc. 26(5): 837–850.
[41] Zhang, C. and Trudell, M.L. 1996. A short and efficient total synthesis of (+/-)-Epibatidine. J. Org. Chem. 61(20): 7189–7191.
[42] Brieaddy, L.E., Liang, F., Abraham, P., Lee, J.R. and Carroll, F.I. 1998. New synthesis of 7-(tert-butoxycarbonyl)-7-azabicyclo [2.2.1] hept-2-ene. A key intermediate in the synthesis of epibatidine and analogs. Tetrahedron Lett. 39(30): 5321–5322.
[43] Liang, F., Navarro, H.A., Abraham, P., Kotian, P., Ding, Y.S., Fowler, J. et al. 1997. Synthesis and nicotinic acetylcholine receptor binding properties of exo-2-(2′-fluoro-5′-pyridinyl)-7-azabicyclo[2.2.1] heptane: A new positron emission tomography ligand for nicotinic receptors. J. Med. Chem. 40(15): 2293–2295.
[44] Carroll, F.I., Liang, F., Navarro, H.A., Brieaddy, L.E., Abraham, P., Damaj, M.I. et al. 2001. Synthesis, nicotinic acetylcholine receptor binding, and antinociceptive properties of 2-exo-2-(2′-substituted 5′-pyridinyl)-7-azabicyclo [2.2.1] heptanes. Epibatidine analogues. J. Med. Chem. 44(13): 2229–2237.

[45] Avalos, M., Parker, M.J., Maddox, F.N., Carroll, F.I. and Luetje, C.W. 2002. Effects of pyridine ring substitutions on affinity, efficacy, and subtype selectivity of neuronal nicotinic receptor agonist epibatidine. J. Pharmacol. Exp. Ther. 302(3): 1246–1252.

[46] Badio, B., Garraffo, H.M., Plummer, C.V., Padgett, W.L. and Daly, J.W. 1997. Synthesis and nicotinic activity of epiboxidine: an isoxazole analogue of epibatidine. Eur. J. Pharmacol. 321(2): 189–194.

[47] Xu, R., Chu, G. and Bai, D. 1996. Total synthesis of (±) epibatidine. Tetrahedron Lett. 37(9): 1463–1466.

[48] Jared, C., Antoniazzi, M.M., Jordão, A.E.C., Silva, J.R.M.C., Greven, H. and Rodrigues, M.T. 2009. Parotoid macroglands in toad (*Rhinella jimi*): their structure and functioning in passive defense. Toxicon. 54(3): 197–207.

[49] Wu, X.Y., Gao, H.M. and Wang, Z.M. 2010. Advances on chemical constituents in crude medicines from toads. Chin. J. Exp. Tradit. Med. Formulae 16: 207–214.

[50] Yang, Q., Zhou, X., Zhang, M., Bi, L., Miao, S., Cao, W. et al. 2015. Angel of human health: current research updates in toad medicine. Am. J. Transl. Res. 7(1): 1–14.

[51] El-Seedi, H.R., Yosri, N., El-Aarag, B., Mahmoud, S.H., Zayed, A., Du, M. et al. 2022. Chemistry and the potential antiviral, anticancer, and anti-inflammatory activities of cardiotonic steroids derived from toads. Molecules 27(19): 6586.

[52] Qi, J., Zulfiker, A.H.M., Li, C., Good, D. and Wei, M.Q. 2018. The development of toad toxins as potential therapeutic agents. Toxins 10(8): 336.

[53] Asrorov, A.M., Kayumov, M., Mukhamedov, N., Yashinov, A., Mirakhmetova, Z., Huang, Y. et al. 2023. Toad venom bufadienolides and bufotoxins: An updated review. Drug Dev. Res. 84(5): 815–838.

[54] Wang, D.L., Qi, F.H., Tang, W. and Wang, F.S. 2011. Chemical constituents and bioactivities of the skin of *Bufo bufo gargarizans* Cantor. Chem. Biodivers 8(4): 559–567.

[55] Qi, J., Tan, C.K., Hashimi, S.M., Zulfiker, A.H.M., Good, D. and Wei, M.Q. 2014. Toad glandular secretions and skin extractions as anti-inflammatory and anticancer agents. Evid Based Complement Alternat. Med.: ID 312684.

[56] Qi, F., Li, A., Zhao, L., Xu, H., Inagaki, Y., Wang, D. et al. 2010. Cinobufacini, an aqueous extract from *Bufo bufo gargarizans* Cantor, induces apoptosis through a mitochondria-mediated pathway in human hepatocellular carcinoma cells. J. Ethnopharmacol. 128(3): 654–661.

[57] Coussens, L.M. and Werb, Z. 2002. Inflammation and cancer. Nature 420(6917): 860–867.

[58] Li, M., Qin, Y., Li, Z., Lan, J., Zhang, T. and Ding, Y. 2022. Comparative pharmacokinetics of cinobufacini capsule and injection by UPLC-MS/MS. Front Pharmacol. 13: 944041.

[59] Liu, R., Zhang, Y., Zhang, R., Zeng, H., Shen, Y., Li, X. et al. 2024. Identification of antitumor constituents from toad skin and toad venom by UPLC-QTOF/MS in-depth chemical profiling combined with bioactivity-based molecular networking. Arab. J. Chem. 17(1): 105427.

[60] Lichtstein, D., Gati, I., Babila, T., Haver, E. and Katz, U. 1991. Effect of salt acclimation on digitalis-like compounds in the toad. Biochim. Biophys. Acta - Gen Subj. 1073(1): 65–68.

[61] Shimada, K., Ro, J., Ohishi, K. and Nambara, T. 1985. Isolation and characterization of cinobufagin 3-glutaroyl-L-arginine ester from *Bufo bufo gargarizans* Cantor. Chem. Pharm. Bull. 33(7): 2767–2771.

[62] Cunha-Filho, G.A., Resck, I.S., Cavalcanti, B.C., Pessoa, C.Ó., Moraes, M.O., Ferreira, J.R. et al. 2010. Cytotoxic profile of natural and some modified bufadienolides from toad Rhinella schneideri parotoid gland secretion. Toxicon. 56(3): 339–348.

[63] Heasley, B. 2012. Chemical synthesis of the cardiotonic steroid glycosides and related natural products. Chem. Eur. J. 18(11): 3092–3120.

[64] Michalak, M., Michalak, K. and Wicha, J. 2017. The synthesis of cardenolide and bufadienolide aglycones, and related Steroids bearing a heterocyclic subunit. Nat. Prod. Rep. 34 (4): 361–410.

[65] Wu, W.Y., Fan, T.Y., Li, N.G., Duan, J.A. and Shi, Z.H. 2020. Total synthesis, chemical modification and structure-activity relationship of bufadienolides. Eur. J. Med. Chem. 189: 112038.

[66] Shimizu, S., Hagiwara, K., Itoh, H. and Inoue, M. 2020. Unified total synthesis of five bufadienolides. Org. Lett. 22(21): 8652–8657.

[67] Tichvon, C., Zviagin, E., Surma, Z. and Nagorny, P. 2024. Synthesis of bufadienolide cinobufagin via late-stage singlet oxygen oxidation/rearrangement approach. Org. Lett. 26(12): 2445–2450.

[68] Hutchinson, D.A., Mori, A., Savitzky, A.H., Burghardt, G.M., Wu, X., Meinwald, J. et al. 2007. Dietary sequestration of defensive steroids in nuchal glands of the Asian snake *Rhabdophis tigrinus*. Proc. Natl. Acad. Sci. USA 104(7): 2265–2270.

[69] Yoshida, T., Ujiie, R., Savitzky, A.H., Jono, T., Inoue, T., Yoshinaga, N. et al. 2020. Dramatic dietary shift maintains sequestered toxins in chemically defended snakes. Proc. Natl. Acad. Sci. USA 117(11): 5964–5969.

[70] Inoue, T., Nakata, R., Savitzky, A.H., Yoshinaga, N., Mori, A. and Mori, N. 2021. New insights into dietary toxin metabolism: diversity in the ability of the natricine snake *Rhabdophis tigrinus* to convert toad-derived bufadienolides. J. Chem. Ecol. 47: 915–925.

[71] Ermakova, A.O., Dunbar, F., Rucker, J. and Johnson, M.W. 2022. A narrative synthesis of research with 5-MeO-DMT. J. Psychopharmacol. 36(3): 273–294.

[71] Vigerelli, H., Sciani, J.M., Eula, M.A.C., Sato, L.A., Antoniazzi, M.M., Jared, C. et al. 2018. Biological effects and biodistribution of bufotenine on mice. Biomed Res Int. 2018: 1032638.

[73] Obreshkova, D., Kandilarov, I., Angelova, V.T., Iliev, Y., Atanasov, P. and Fotev, P.S. 2017. Pharmaco-toxicological aspects and analysis of phenylalkylamine and indolylalkylamine hallucinogens. Pharmacia 64(1): 32–47.

[74] Zhao, C., Chen, M., Sun, S.L., Wang, J.J., Zhong, Y., Chen, H.H. et al. 2021. Bufotenine and its derivatives: synthesis, analgesic effects identification and computational target prediction. Chin. J. Nat. Med. 19(6): 454–463.

[75] Ginnon, W.F., Benigni, J.D., Suzuki, J. and Daly, J.W. 1967. The synthesis of dehydrobufotenine. Tetrahedron Lett. 8(16): 1531–1533.

[76] Peat, A.J. and Buchwald, S.L. 1996. Novel syntheses of tetrahydropyrroloquinolines: Applications to alkaloid synthesis. J. Am. Chem. Soc. 118(5): 1028–1030.

[77] Stoffman, E.J. and Clive, D.L. 2010. Synthesis of 4-haloserotonin derivatives and synthesis of the toad alkaloid dehydrobufotenine. Tetrahedron 66(25): 4452–4461.

[78] Tian, Z., Liao, A., Kang, J., Gao, Y., Lu, A., Wang, Z et al. 2021. Toad alkaloid for pesticide discovery: Dehydrobufotenine derivatives as novel agents against plant virus and fungi. J. Agr. Food Chem. 69(34): 9754–9763.

[79] Venemalm, L., Estévez, C., Venemalm, L., Alvarez, M. and Joule, J.A. 1993. Synthesis of a 1,3,4,5-Tetrahydropyrrolo [4,3,2-*de*] quinoline. Tetrahedron Lett. 34(34): 5495–5496.

[80] Tokuyama, T., Daly, J. and Witkop, B. 1969. Structure of batrachotoxin, a steroidal alkaloid from the Colombian arrow poison frog, *Phyllobates aurotaenia*, and partial synthesis of batrachotoxin and its analogs and homologs. J. Am. Chem. Soc. 91(14): 3931–3938.

[81] Albuquerque, E.X., Daly, J.W. and Witkop, B. 1971. Batrachotoxin: Chemistry and Pharmacology: This novel steroidal alkaloid is a valuable tool for studying ion transport in electrogenic membranes. Science 172(3987): 995–1002.

[82] Tonggu, L., Wisedchaisri, G., Gamal El-Din, T.M., Lenaeus, M.J., Logan, M.M., Toma, T. et al. 2024. Dual receptor-sites reveal the structural basis for hyperactivation of sodium channels by poison-dart toxin batrachotoxin. Nat. Comm. 15(1): 2306.

[83] Dumbacher, J.P., Beehler, B.M., Spande, T.F., Garraffo, H.M. and Daly, J.W. 1992. Homobatrachotoxin in the genus Pitohui: chemical defense in birds? Science 258(5083): 799–801.

[84] Dumbacher, J.P., Spande, T.F. and Daly, J.W. 2000. Batrachotoxin alkaloids from passerine birds: a second toxic bird genus (Ifrita kowaldi) from New Guinea. Proc. Natl. Acad. Sci. USA 97(24): 12970–12975.

[85] Dumbacher, J.P., Wako, A., Derrickson, S.R., Samuelson, A., Spande, T.F. and Daly, J.W. 2004. Melyrid beetles (Choresine): a putative source for the batrachotoxin alkaloids found in poison-dart frogs and toxic passerine birds. Proc. Natl. Acad. Sci. USA 101(45): 15857–15860.

[86] Imhof, R., Gössinger, M.E., Graf, W., Berner, H., Berner-Fenz, M.L. and Wehrli, H. 1972. Steroids and sex hormones. Part 245. Partial synthesis of batrachotoxinin A. Preliminary communication. Helv. Chim. Acta 55(4): 1151–1153.

[87] Kurosu, M., Marcin, L.R., Grinsteiner, T.J. and Kishi, Y. 1998. Total synthesis of (±)-batrachotoxinin A. J. Am. Chemi. Soc. 120(26): 6627–6628.

[88] Logan, M.M., Toma, T., Thomas-Tran, R. and Du Bois, J. 2016. Asymmetric synthesis of batrachotoxin: Enantiomeric toxins show functional divergence against NaV. Science 354(6314): 865–869.

[89] Guo, Y., Guo, Z., Lu, J.T., Fang, R., Chen, S.C. and Luo, T. 2020. Total synthesis of (−)-Batrachotoxinin A: A Local-Desymmetrization Approach. J. Am. Chem. Soc. 142(8): 3675–3679.
[90] Watanabe, Y., Morozumi, H., Mutoh, H., Hagiwara, K. and Inoue, M. 2023. Total synthesis of (−)-batrachotoxin enabled by a Pd/Ag-promoted suzuki–miyaura coupling reaction. Angew. Chem. Int. Ed. 135(40): e202309688.
[91] Ligabue-Braun, R., Verli, H. and Carlini, C.R. 2012. Venomous mammals: a review. Toxicon. 59(7-8): 680–695.
[82] Wood, W.F. 1999. The history of skunk defensive secretion research. Chem. Educ. 4(2): 44–50.
[93] Wood, W.F., Sollers, B.G., Dragoo, G.A. and Dragoo, J.W. 2002. Volatile components in defensive spray of the hooded skunk, *Mephitis macroura*. J. Chem. Ecol. 28: 1865–1870.
[94] Weinstein, S.B., Malanga, K.N., Agwanda, B., Maldonado, J.E. and Dearing, M.D. 2020. The secret social lives of African crested rats, *Lophiomys imhausi*. J. Mammal. 101(6): 1680–1691.
[95] Orlov, S.N., Klimanova, E.A., Tverskoi, A.M., Vladychenskaya, E.A., Smolyaninova, L.V., Lopina, O.D. 2017. Na^+_i, K^+_i-dependent and-independent signaling triggered by cardiotonic steroids: facts and artifacts. Molecules 22(4): 635.
[96] Zhang, H., Sridhar Reddy, M., Phoenix, S. and Deslongchamps, P. 2008. Total synthesis of ouabagenin and ouabain. Angew. Chem. Int. Ed. 47(7): 1272–1275.
[97] Renata, H., Zhou, Q., Dunstl, G., Felding, J., Merchant, R.R., Yeh, C.H. and Baran, P.S. 2015. Development of a concise synthesis of ouabagenin and hydroxylated corticosteroid analogues. J. Am. Chem. Soc. 137(3): 1330–1340.
[98] Mukai, K., Kasuya, S., Nakagawa, Y., Urabe, D. and Inoue, M. 2015. A convergent total synthesis of ouabagenin. Chem. Sci. 6(6): 3383–3387.
[99] Khatri, H.R., Bhattarai, B., Kaplan, W., Li, Z., Curtis Long, M.J., Aye, Y. et al. 2019. Modular total synthesis and cell-based anticancer activity evaluation of ouabagenin and other cardiotonic steroids with varying degrees of oxygenation. J. Am. Chem. Soc. 141(12): 4849–4860.
[100] Sun, J., Chen, Y., Ragab, S.S., Gu, W., Tang, Z., Tang, Y. and Tang, W. 2023. Total syntheses of polyhydroxylated steroids by an unsaturation-functionalization strategy. Angew. Chem. Int. Ed. 135(24): e202303639.

CHAPTER 7

Toxic Marine Metabolites

7. INTRODUCTION: THE UNDERWATER WORLD IS A BATTLEGROUND OF CHEMICAL WARFARE, WITH SEA INHABITANTS, FROM ALGAE TO FISH, ACTIVELY ENGAGING IN THESE "CHEMICAL FIGHTS"

The underwater world, a vast and awe-inspiring battleground of chemical warfare, is a "blue desert" 2.4 times larger than the mainland, yet largely unexplored. Its unique and awe-inspiring features, such as extreme pressure, temperature, and medium (salinity) conditions, have always captivated humanity's attention. Poets, writers, and painters draw inspiration from it, and many people who depend on the sea get their food from it. For scientists, the ocean is a treasure trove of knowledge. It is a massive complex where various unions of life forms occur. Marine organisms have evolved with unique metabolic and physiological capabilities to ensure survival in their extreme habitat and offer the full potential to produce their metabolites. Oceans, home to more than 80% of diverse plant and animal species worldwide, are a testament to marine life's remarkable adaptability and resilience, a source of inspiration and wonder for all of us.

Thus, various sea inhabitants, from microalgae to fish, use different toxic substances, formed into various types of organic compounds, for their defense (and also to obtain their food). These compounds are unique in diversity, structural, and functional features. Proteins, polypeptides, alkaloids, hydrocarbons, steroids, terpenes, lipids, and carbohydrates are marine organisms' arsenal of "weapons". Marine organisms include microorganisms (bacteria, cyanobacteria, and microalgae), sponges, soft corals, tunicates, gorgonians, bryozoans, mollusks, nudibranchs, sea hares, fishes, etc. Right here, it is necessary to mention that some of these inhabitants are served as a marine delicacy in restaurants, and some contain phycotoxins that can be paralyzing, diarrheal, amnesiac, cytotoxic, hepatotoxic, and neurotoxic [1]. In addition, several small molecules have been tested for different bioactivities (e.g., anticancer, anti-inflammatory, and antioxidant, as well as for treating neurodegenerative diseases), and results have shown their potential for possible industrial applications [2–4].

7.1 Marine Protists and Invertebrates: A Rich Source Offering Diverse Complex Phycotoxins

The ocean is rich in biodiversity; marine protists and invertebrates constitute more than 90% of oceanic biomass. They provide a great deal of chemical diversity, dependent on various biosynthetic pathways. Their metabolites habitually disclose astonishingly novel carbon skeletons, different from those in terrestrial biota. Moreover, marine chemical diversity does not overlap to any great extent with terrestrially derived chemical diversity. Many of these bioactive marine natural products serve as chemical defense tools against predators, competitors, and other ecological pressures, and their true origin appears to be the microorganisms living in concert with invertebrates.

Being sessile, soft-bodied, moving slowly, and subject to potential parasite predation and detrimental microbial colonization, most invertebrates use symbiosis with some microorganisms that metabolize toxic substances. These latter facilitate a form of chemical defense. As a result, such invertebrate-microorganism associations provide defensive toxic molecules to unarmed invertebrates and a haven for microorganisms [5]. Now, let's examine these microorganisms, smaller sea inhabitants that can sometimes produce more extensive and more complex molecules.

7.1.1 Microalgae Producing Oxygenated Polycyclic Brevetoxin, Ciguatoxins, and Maitotoxin Molecules

The massive reproduction of some toxic single-celled microalgae (approximately only 60 species of the almost 4,000 marine phytoplankton species) of the order Dinoflagellata (dinoflagellates) can lead to the loss of large numbers of fish and the poisoning of people since these microalgae can "communicate" their toxicity to other marine organisms and humans via the food chain. Sometimes, when many of these organisms accumulate, they give the water a red or brown color ("red tides") due to their chromatophores. The expression "red tide" is a synonym for toxic algae blooms and has been known since ancient times.

Dinoflagellates generally live as isolated cells, have a spheroidal shape, and their greatest diameter varies from 5 μm to 2 mm. Almost all dinoflagellates are surrounded by a shell made up of imbricated plates. They are mobile, having two flagella at one end and the other in a transverse groove. The coordinated beating of these flagella gives them a characteristic spiral swimming movement. The secondary metabolites of harmful dinoflagellates impress scientists with their spectacular structures that almost always belong to a class of characteristic compounds – polycyclic ethers.

The dinoflagellate species *Karenia brevis* (formerly known as *Gymnodinium breve* and *Ptychodiscus brevis*) produce a series of oxygenated molecules with a complex structure, to which they gave the name brevetoxins (PbTxs). They usually have multiple cyclic ether rings, often aligned in a ladder frame, with a long carbon chain backbone bearing many hydroxyl groups. The first two most toxic components isolated from these organisms are brevetoxin A (PbTx-A or BTA-A; $C_{49}H_{70}O_{13}$; MW = 867.09 g/mol) and brevetoxin B (PbTx-B or BTX-B; $C_{50}H_{70}O_{14}$; MW = 895.08 g/mol) (Figure 7.1), whose structures were established in the

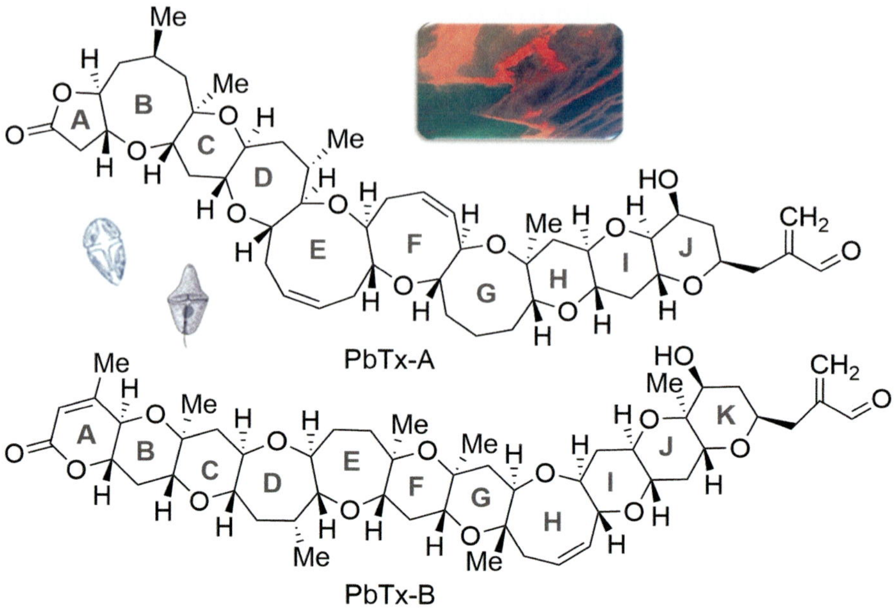

Figure 7.1. Structures of early brevetoxins and images of *Gymnodinium* dinoflagellate and red tide. PbTx-B is a needle-shaped crystalline substance with m.p. 270ºC. PbTx-A displays double melting points with an m.p. 197–199°C and 218–220°C. At least ten different PbTx derivatives have been characterized [8].

1980s [6, 7]. Both are the most common neurotoxic shellfish poison with LD_{50} = 0.20–0.25 mg/kg. PbTx-B is the principal constituent of the venom, while PbTx-A is the most potent of the PbTxs. Both lipophilic toxins possess a lactone ring, an α,β-unsaturated aldehyde side chain, and 9 and 10 consecutive ether rings, respectively. However, the PbTx-A molecule has two oxocene (E ring) and one oxocane (F and G rings) skeletons, and it is a 90° bend at the eight-membered oxygen-containing G ring. In contrast, PbTx-B is essentially planar, in which the G ring holds the boat-chair conformation.

These ladder-like polyether marine biotoxins have been responsible for many fish, mammal, and mollusk kills. Tainted seafood has caused widespread food poisoning in humans, known as neurotoxic shellfish poisoning (NSP), through ingestion, inhalation, or contact. However, they are not toxic to shellfish. That is why shellfish consumption is the leading cause of NSP. Symptoms of NSP include nausea, vomiting, diarrhea, paresthesia, cramps, bronchoconstriction, paralysis, seizures, and coma. PbTxs bind with high affinities to segment 5 (S5) of a voltage-sensitive sodium channel protein, resulting in persistent channel activation favoring the passage of sodium ions at levels higher than those that occur in a normal physiological state.

The difficulty in determining the structures of such substances is that their concentration is small, and the amount of isolated products is a few milligrams. Thus, for example, when extracting (ether, pH 5.5) from 50 L of the cultivation of these algae, only 90 mg of the PbTxs mixture was obtained. When doing chromatographic separations, using different column supports and a 5% mixture of methanol with diisopropyl ether, 0.8 mg of PbTx-A and 5.0 mg of PbTx-B were isolated. The complex and unique structure of these molecules, along with their potent bioactivity, attracted the attention of synthetic chemists. The construction of these polyoxygenated heterocycles represents an almost insurmountable challenge that only a few groups of chemists have accomplished.

After many years of work, Prof. Nicolaou's group successfully carried out the first total synthesis of brevetoxin B [9–11]. Their synthesis is based on a convergent strategy by coupling skeleton ABCDEFG **7.1** as a phosphonium salt and tricyclic IJK ring system **7.2a** with an aldehyde function, which was found through O-C and C=C bond disconnection of the oxocene ring (H ring). Prof. Nakata's group also reported the synthesis of this molecule using the same disconnection approach. Still, a tricyclic IJK ring system was slightly modified **7.2b** [12] (Figure 7.2). To prepare segments **7.1** and **7.2a**, Nicolaou's group selected 2-deoxy-D-ribose and D-mannose as commercially available initial materials. In contrast, tri-O-acetyl-D-glucal and 2-deoxy-D-ribose were used to prepare these segments in the Nakata synthesis of brevetoxin B. These simple precursors were employed in crucial epoxide-opening cascade processes that occurred with *endo*-selectivity.

The "union" of these two fragments to give complete skeleton **7.3a–b** was promoted with n-butyl lithium at low temperature, followed by the formation (first with pyridinium *p*-toluenesulfonate and then with $AgClO_4/SiO_2$) of the oxocene ring (H ring), an essential step in the final stages of the total synthesis [10, 12] (Figure 7.3).

Realized $AgClO_4$-induced ring closure of intermediate adduct, monothioacetals **7.3a–b** were reduced with Ph_3SnH in the presence of AIBN to give **7.4a–b**,

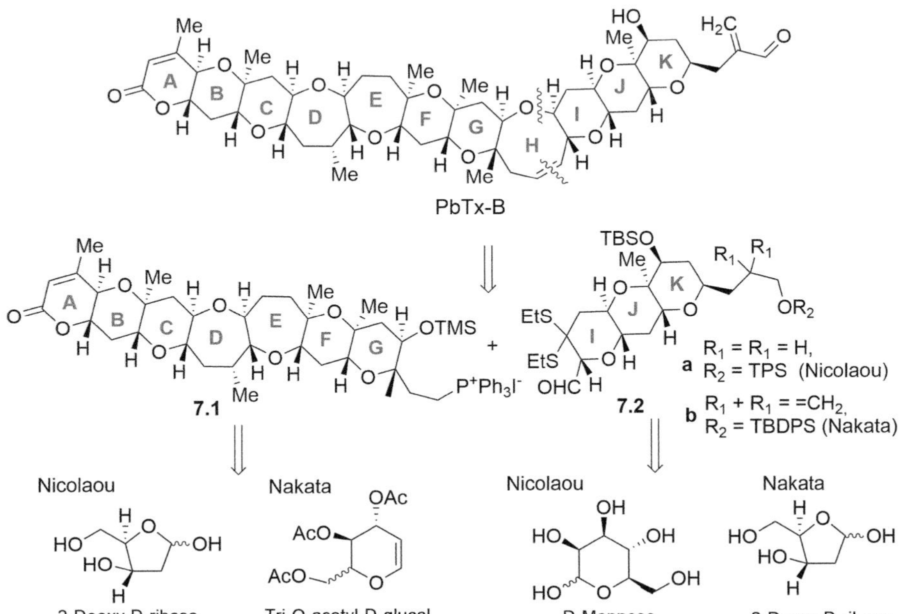

Figure 7.2. The retrosynthetic plan for PbTx-B realized by both groups: Nicolaou et al. (1995) [11] and Nakata et al. (2004) [12]. TBS = *tert*-$BuMe_2Si$; TPS = *tert*-$BuPh_2Si$.

corresponding to the ABCDEFGHIJK-ring. Then, their removal of the TBDPS group and treatment with PCC allowed double oxidation of alcohol and A-ring methylene at the right and left sides to give lactone-aldehyde **7.5**. The final removal of the TBS group of the O-protected PbTx-B **7.5** furnished the desired biotoxin (Figure 7.3). Despite the high yields at all stages (> 90%) and the efficient convergent approach to synthesizing this marine product, the overall yield turned out to be quite small.

Isao Kadota and coworkers realized another retrosynthetic analysis of brevetoxin B [13]. They retrosynthetically break down the polycyclic ether framework of PbTx-B into the ABCDEFG-IJK ester **7.11** and then into the ABCDEFG-ring segment **7.9** and the bicyclic J-K fragment **7.10**. Both heptacycle and bicycle derivatives could be prepared from available materials **7.6**,**7.7**, and **7.8** via the intramolecular allylation of α-acetoxy ethers and subsequent ring-closing metathesis. The missing H ring is formed *in situ* at crucial stages (Figure 7.4).

Thus, convergent construction of the ABCDEFG-ring framework **7.9** was achieved from the α-chlorosulfide **7.6** and the alcohol F-G ring segment **7.7** in 11 linear synthetic steps, where the coupling of the B-C and F-G ring segments and ring-closing metathesis using the first-generation Grubbs catalyst (GC-1) are vital.

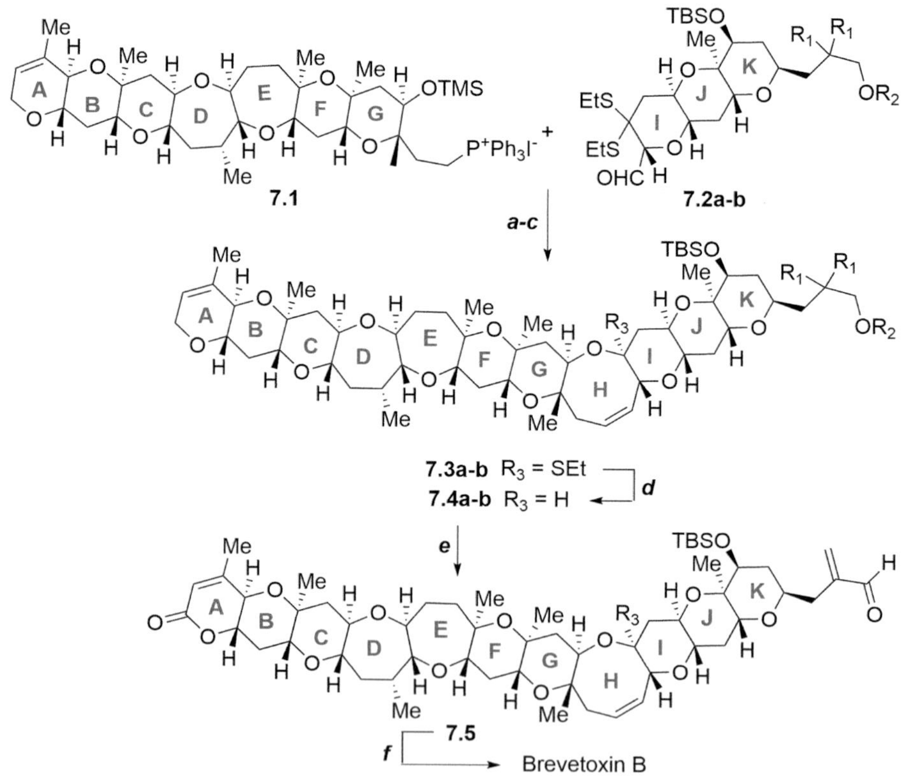

Figure 7.3. The final stages of total synthesis of PbTx-B (Nicolaou 1995; Nakata 2004): (a) *n*-BuLi, HMPA, THF, –78°C to rt; (b) PPTS, CH_2Cl_2-MeOH, rt; (c) $AgClO_4$, $NaHCO_3$, SiO_2, 4 Å MS, $MeNO_2$, rt; (d) Ph_3SnH, AIBN, toluene, 110°C; (e) 1. TBAF, THF, rt, 2. PCC, benzene, 80°C; (f) HF·pyridine, CH_2Cl_2, 0°C. PPTS = pyridinium *p*-toluenesulfonate.

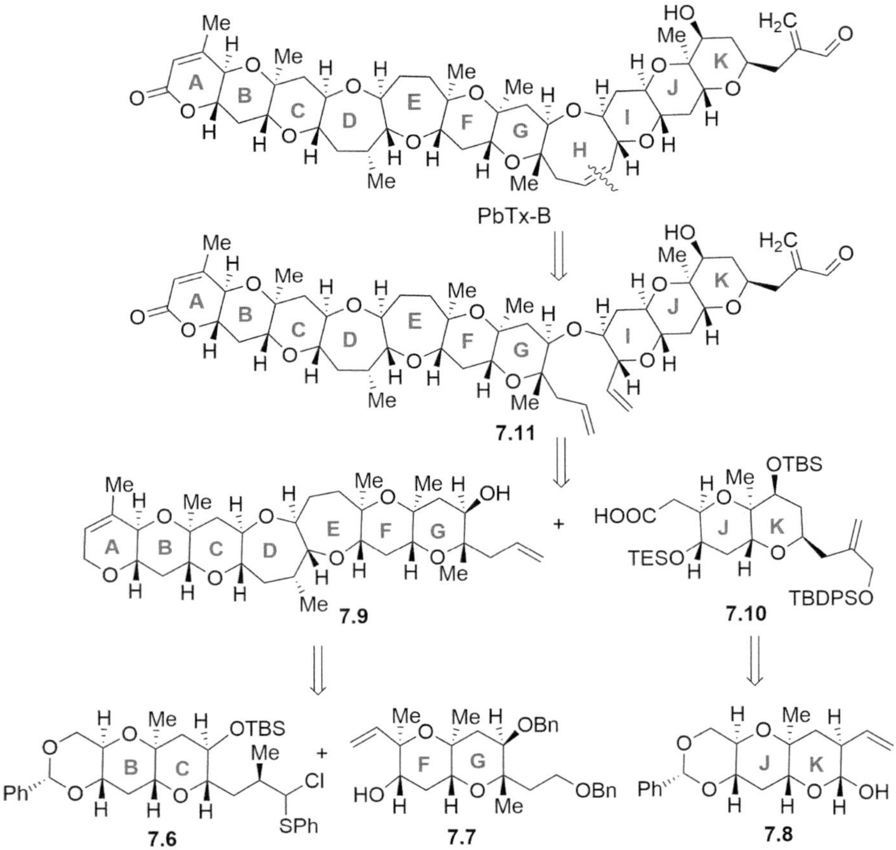

Figure 7.4. Retrosynthetic plan for PbTx-B, realized by Kadota's group (2005).

On the other hand, synthesizing the J-K ring segment **7.10** from precursor **7.8** was performed using eight linear synthetic steps (Figure 7.5). To prepare close precursor **7.11**, a coupling method was developed. It involves an initial esterification reaction of **7.9** and **7.10**, conversion of the obtained respective ester into intermediate allylic stannane ester, and intramolecular allylation [13].

The total synthesis is completed with the ring-closing metathesis of comp. **7.11**, with the Grubbs catalyst (GC-1) to provide the A–K ring skeleton **7.12**. Oxidation of its A-ring moiety with PCC gave the lactone **7.13** in an 81% overall yield. After removing the silyl protective groups with HF·pyridine, selective oxidation of the resulting allylic alcohol with MnO_2 provided brevetoxin B in 84% overall yield.

The present methodology, like of Nicolaou and Nakata, demonstrates the power of synthetic chemists, who have considerably improved the state of the art of total synthesis today. Regarding the landmark total synthesis of brevetoxin A, only two reports by Nicolaou's group [14] and Crimmins's group [15, 16]. Both groups used a similar synthetic strategy for constructing PbTx-A based on the C=C and O-C bond disconnection of the F-ring generating key advanced polycycles: left BCDE and right GHIJ skeletons [17]. Let's leave this retro analysis and convergent synthesis to study on your own and look at other large oxygenated molecules.

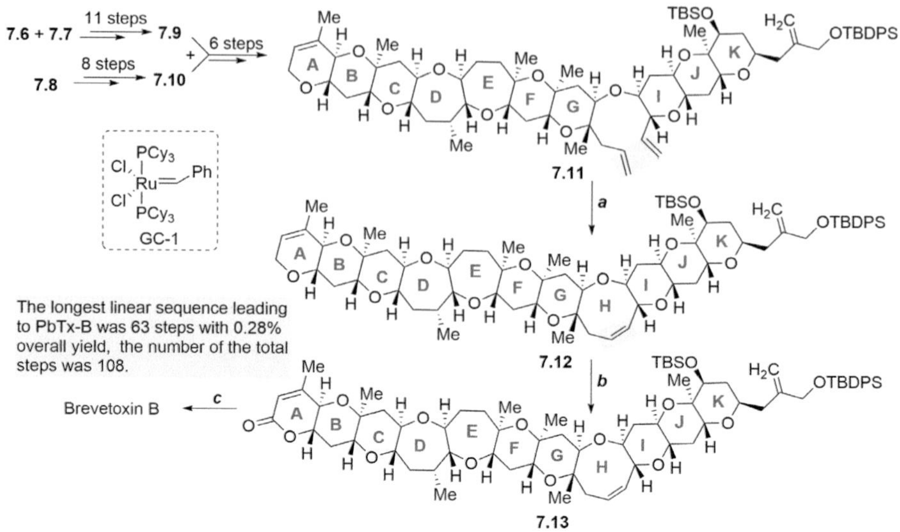

Figure 7.5. Completion of the total synthesis of brevetoxin B (Kadota 2005): (a) GC-1, benzene, 40°C; (b) PCC, benzene, 80°C, 81% (two steps); (c) 1. HF·pyridine, CH_2Cl_2, 0°C, 2. MnO_2, Et_2O, rt, 84%.

Another toxic dinoflagellate that attracts the attention of researchers is *Gambierdiscus toxicus*, which settles on the seaweed that mollusks and some coral fish feed on. This tiny naked dinoflagellate is the leading cause of ciguatera, an illness caused by eating fish that contain toxins several toxins, such as gambierol, gambieric acids (GA), and ciguatoxin-1 (P-CTX-1, $C_{60}H_{86}O_{19}$, MW = 1,110.60 g/mol) (Figure 7.6), present in this marine microalgae. The latter lipophilic cyclic polyether highly oxygenated molecule, isolated in 1980, is a potent neurotoxin with an LD_{50} in mice = 0.25 μg/kg, a hundred times more toxic than PbTxs. It is the leading cause of ciguatera fish poisoning, which produces characteristic gastrointestinal, neurological, and cardiovascular symptoms. Like brevetoxins, ciguatoxins are described to specifically activate voltage-gated sodium channels by interacting with their receptor site 5 [18]. In contrast, gambierol modulates voltage-gated potassium channels; Kv1.2 is the most sensitive voltage-gated potassium channel subtype with an almost entire block (> 97%) and a half maximal inhibitory concentration (IC_{50}) of 34.5 nM [19]. Interestingly, GAs did not show binding capacity to voltage-gated ion channels. Instead, they are known to impart extraordinarily potent antifungal activity against *Aspergillus niger*, which is approximately 2,000 times greater than that of amphotericin B [20].

Maitotoxin-1 (MTX-1), a substantial disulphated polyether molecule, was also found in harmful microalgae *G. toxicus* (Figure 7.7). Its elegant molecular structure represents the most significant and most complex secondary metabolite ever isolated. MTX-1 contains 32 oxygenated rings of different sizes, 98 stereocenters, 28 hydroxy groups, and two sulfonic groups, and its molecular weight is 3,425 Da [21–23]. So far, MTX-1 is the most giant non-polymeric carbon chain molecule. For this reason, the stereochemical studies of Japanese chemists lasted almost 16 years to determine its structure. To establish the chemical structure of this toxin, they used a set of

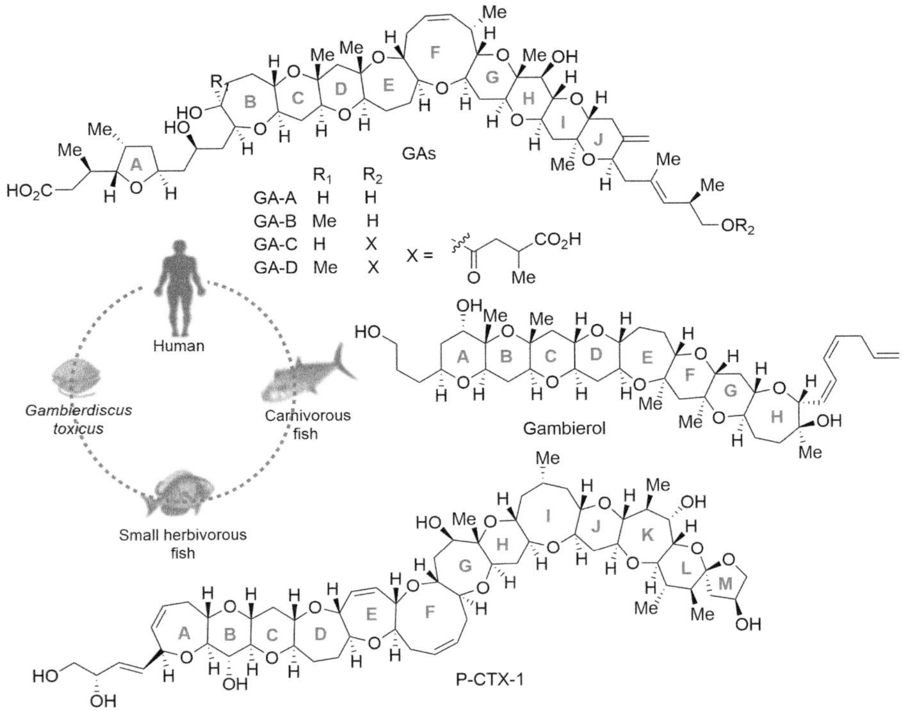

Figure 7.6. Structure of cyclic polyether molecules from *Gambierdiscus toxicus*, responsible for ciguatera fish poisoning, and a diagram illustrating the interrelationships between harmful microalgae and shellfish, carnivorous fish, and humans. Dinoflagellate *G. toxicus* was first linked to the production of ciguatera toxins in the Pacific Ocean and, thus, must be marked as P-CTX-1. Given that there are numerous variants, it has been proposed that a letter code prefix be used to indicate ciguatoxin homologs isolated from the Pacific Ocean (P-CTXs), Indian Ocean (I-CTXs), and Caribbean Sea (C-CTXs).

modern physical and traditional chemical methods with only 8.1 mg of the substance isolated from the culture of *G. toxicus* (5,000 L).

Besides its complex structure, MTX-1 is a highly potent toxin with an LD_{50} = 50 ng/kg. It causes cell death through massive calcium influx, i.e., activates Ca^{2+} influx in various cells, ultimately leading to cellular toxicity. Remember that P-CTX-1 acts mainly through the activation of voltage-gated sodium channels and the blockade of voltage-gated potassium channels. However, both share a part of their chemical structure, with many ether rings forming a ladder shape.

The isolation and biological evaluation of MTX-1 inspired numerous studies directed toward its structural elucidation and synthesis. In particular, the Nicolaou research group has been involved in synthesizing this molecule, reporting the preparation of different skeletons and domains of MTX-1 [24, 25]; however, its total synthesis has yet to be achieved.

CTX and MTX were also detected in the oral glands of the moray eel *Gymnothorax javanicus*, but they are not produced by moray eels themselves; they can be accumulated via the food chain by large specimens of moray eels, such as the giant moray *G. javanicus*.

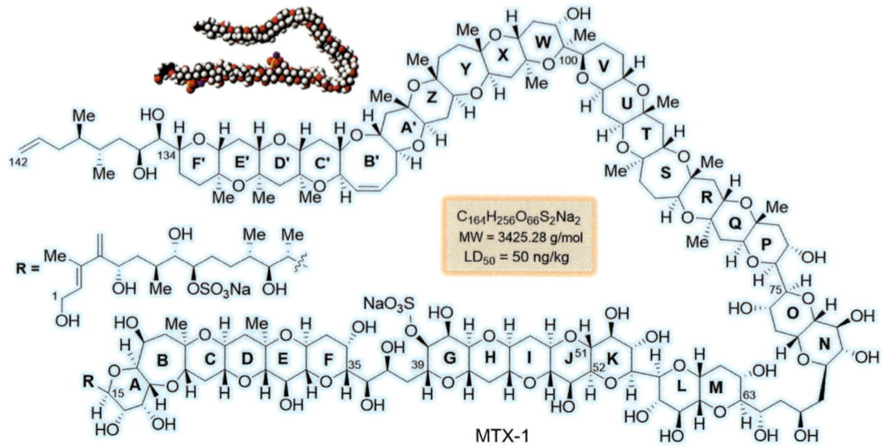

Figure 7.7. Structure of MTX-1. The gross structure of MTX was elucidated in 1993 [21]. The relative stereochemistry of the fused rings and the directly connected ether rings was reported in 1994 [22]. Finally, its complete relative stereochemistry has been established in 1996 [23].

7.1.2 Toxic Nitrogen- and Oxygen-Containing Molecules from the Sponges

Of all the marine invertebrates, sponges are the ones that achieve the most significant number and diversity. Its soft body is supported by numerous tiny crystalline spicules or "organic" fibers forming a skeleton whose constant component is a corneal substance called spongin (hence, the name Spongia). They are multicellular lower animals, incapable of movement, and often look like strangely shaped plants (flat balls, cups, glasses, or fans); besides, they always give off an unpleasant odor. The body of sponges is a porous structure with numerous openings on the surface (hence comes the second name of the phylum, Porifera, from the Latin *porus* – pore and ferre – to carry). Each pore is a canaliculus excavated into a tubular cell called a porocyte.

Due to the vulnerability of their body, sponges have developed chemical defenses to "not be eaten". Usually, other animals rarely attack or eat sponges because of the toxic secretions and unpleasant odors produced by their chemical defense substances of diverse chemical nature. So-called unpalatable metabolites play an anti-predatory role, and they may be more effective in deterring predation than toxic substances. Among these secondary metabolites, alkaloids are good toxicants acting as allomones, protecting them against predators and fouling organisms.Since the chemical defense of diverse *Aplysina* (syn. *Verongia*) sponges against predatory fishes has been well-studied, let's discuss these sponges and their metabolites first [26]. Among their interesting and major metabolite components are brominated isoxazoline alkaloids, including aplysinamisin-1, aerophobin-2, fistularin-3, aerothionin or homoaerothionin, which have been isolated from sponges *Verongia aerophoba*, *Aplysina fistularis* (*syn. Verongia thiona*), *A. aerophoba*, and *A. cavernicola*, known as the yellow tube sponges (Figure 7.8) [27–30]. This striking yellow color is due to the trihydroxy-quinolone pigment

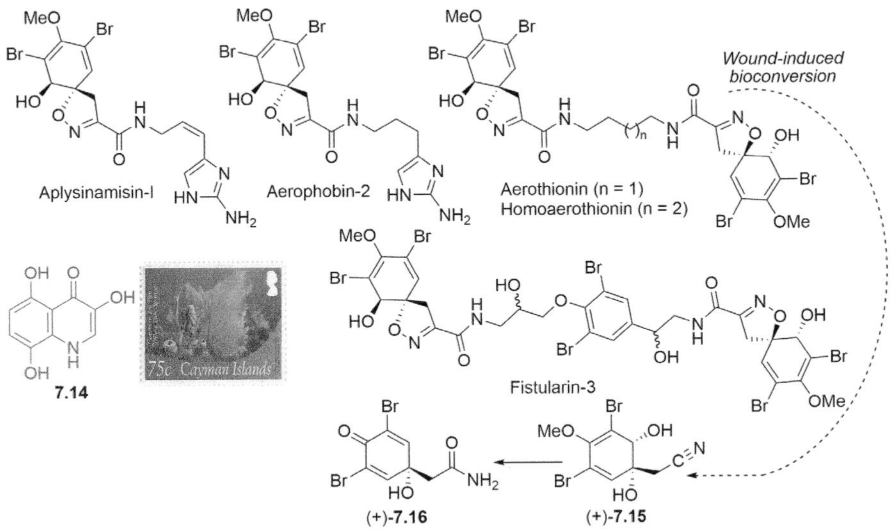

Figure 7.8. Structures of brominated isoxazoline alkaloids isolated from *Aplysina* sponges (stamp image) that accumulate these alkaloids in up to 10% dry weight concentrations. These metabolites are mainly stored in specialized spherulous cells but are not associated with the sponge skeletons. They act as constitutive defense compounds against potential fish predators and bacterial invasion following wounding, i.e., mechanical injury [30].

called uranidine **7.14**. Structurally, these alkaloids are bromotyrosine derivatives of the spirocyclohexadienylisoxazoline type. Aplysinamisin-1 and aerophobin-2 are mono-spirocyclohexadienylisoxazoline derivatives, while fistularin-3 and (homo) aerothionins belong to the bis-spirocyclohexadienylisoxazoline system. Interestingly, all these spirocyclic compounds could be enzymatically converted into the dienone called verongiaquinol (**7.16**) via nitrile aeroplysinin-1 within less than 1 minute upon mechanical sponge damage [28–30].

The formation of 2-((1*S*,6*R*)-3,5-dibromo-1,6-dihydroxy-4-methoxycyclohexa-2,4-dien-1-yl) acetonitrile, (**7.15**), called aeroplysinin-1, and the corresponding dienone amide **7.16** is part of the chemical defense system of *A. cavernicola* [31]. It is noteworthy that this dienone was first obtained as a synthetic product after the treatment of aeroplysinin-1 with trifluoroacetic acid. The latter two compounds displayed potent antibiotic activity. Moreover, verongiaquinol **7.16** showed higher deterrent activity against fish, such as the polyphagous Mediterranean *Blennius sphinx*.As a result, they were also shown to be biologically intensely active against marine bacteria and algae. Thus, pre-formed brominated isoxazoline alkaloids that are constitutively present are transformed into highly toxic molecules like terrestrial plant cyanogenic glycosides and glucosinolates (Sections 3.2 and 3.3). This "activated defense" case is called a wound-induced bioconversion, which is ecologically relevant in protecting the wounded sponge tissue from invading bacterial pathogens. The conversion is fast (e.g., seconds) and is often enzyme-catalyzed.

As marine sponges provide a natural and renewable source of promising antitumor agents, cytotoxicity studies of sponge metabolites form part of today's pharmacological research. Thus, the representative spirocyclohexadienylisoxazole alkaloids, aerothionin, and homoaerothionin were tested as potential anticancer agents. Aerothionin exhibited cytotoxic activity toward the cervical cancer (HeLa) and breast cancer (MCF-7) cells but did not show antibacterial activity. On the other hand, both sponge-derived metabolites can modulate voltage-dependent calcium channels and were active against a chloroquine-resistant strain of *Plasmodium falciparum*, a cause of malaria, presenting promising activity. Interestingly, both alkaloids also showed a significant antitumor activity toward pheochromocytoma cells, which cause a rare tumor that forms in the middle of the adrenal gland [32]. Notably, they have an excellent therapeutic window showing an acceptable toxicity on normal tissue cells.

(+)-Aeroplysinin-1 (**7.15**) has potent antibiotic effects on Gram-positive bacteria (*Bacillus cereus, Bacillus subtilis, Staphylococcus aureus, Staphylococcus albus*, etc.) and several dinoflagellate microalgae that cause toxic blooms. In contrast, it did not affect Gram-negative bacteria, such as *Pseudomonas aeruginosa*, or yeasts, including *Saccharomyces cerevisiae*. In preclinical studies, (+)-aeroplysinin-1 has also shown promising anti-inflammatory, anti-angiogenic, and antitumor effects [33]. On the other hand, the biological properties of the laevorotatory antipode of aeroplysinin-1, isolated from sponge *Ianthella ardis*, are practically unknown.

Based on an earlier study [34], a biosynthetic pathway leading to brominated isoxazoline alkaloids, such as aerothionin and its analogs, can be proposed, as depicted in Figure 7.9 [35].

It is believed that the presence of flavin-dependent halogenase (FDH) genes in the *Aplysina* sponges allows them to produce bromotyrosine **7.17** from L-phenylalanine (Phe) and L-tyrosine (Tyr) via bromination reactions.

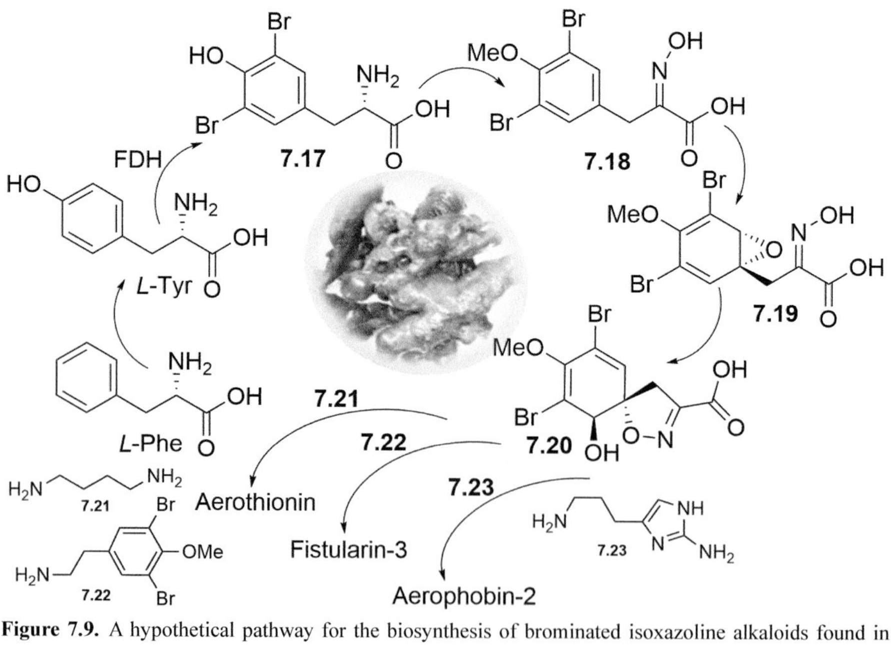

Figure 7.9. A hypothetical pathway for the biosynthesis of brominated isoxazoline alkaloids found in *Aplysina* sponges. An image of *A. cavernicola* is included.

The latter is converted into O-methylated oxime Tyr **7.18**, undergoing an arene oxide reaction to give an oximino epoxide **7.19**. Its epoxide ring opening leads to the construction of the spirocyclohexadienylisoxazoline system **7.20**, which could react with putrescine **7.21** and bromotyramine **7.22** or 2-amino-homohistamine **7.23**, completing the respective bis-spirocyclohexadienylisoxazoline or mono-spirocyclohexadienylisoxazoline systems that aerothionin and fistularin-3 or aerophobin-2 contain (Figure 7.9).

The abovementioned brominated isoxazoline alkaloids are involved in another exciting phenomenon: sponges *A. aerophoba,* living in shallow waters (i.e., visible to solar irradiation), can assist cyanobacteria in developing on their surfaces. The omnivorous gastropod mollusk *Tylodina perversa* mostly consumes the cyanobacteria, eating only a fraction of these sponges. By eating these bacteria, the mollusk captures (sequesters) these bromo derivatives and uses them as chemical defense agents against predators [36]. Interestingly, the mollusk also sequesters the sponge pigment uranidine **7.14** to have the same yellow coloration as its prey, becoming cryptically colored.

Marine pederin analogs, such as mycalamides, onnamides, and theopederins, were isolated from cold-water sponges of the genera *Mycale* and *Theonella*, which are common and widely distributed across the oceans. These sponges produce diverse compounds, including polyketides, cyclic and linear peptides, alkaloids, pigments, lipids, and sterols [37, 38]. Among numerous mostly toxic metabolites, our interest is focused only on the representative polyketide-like pederin molecules (Figure 7.10).

The first mycalamide polyketides, mycalamides A and B, were isolated from a marine sponge of the genus *Mycale* collected in New Zealand in 1989–1990. At the same time, the first member of the onnamide polyketide group, onnamide A, was found in *Theonella* sponges collected on the coast of Japan. Theopederins A and C, the theopederin polyketide group members, were also isolated from *Theonella* sponges in subsequent years (1992–1999) [39, 40].

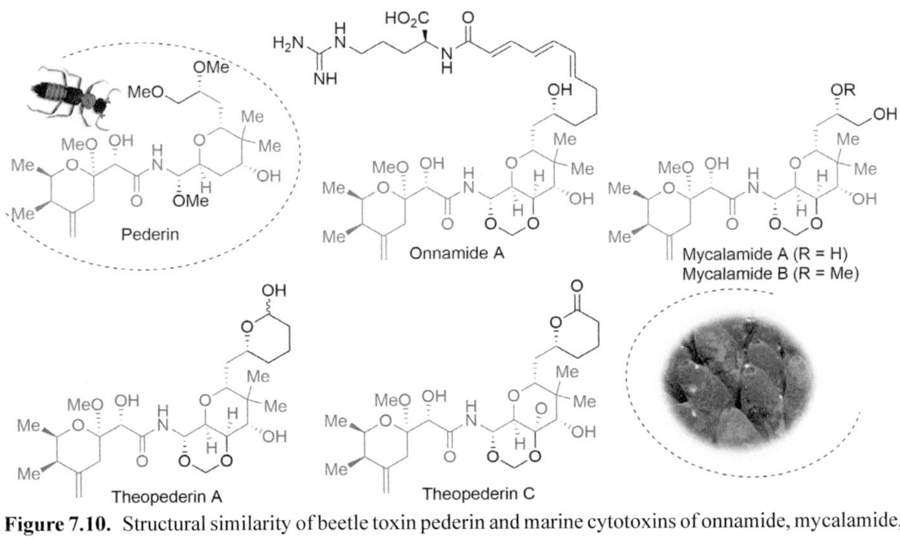

Figure 7.10. Structural similarity of beetle toxin pederin and marine cytotoxins of onnamide, mycalamide, and theopederin groups.

These marine metabolites comprise the left-hand pederic acid fragment and the central tetrahydropyran ring, which is present in the beetle toxin pederin (Section 5.2.3). The main difference is that the N-acyl aminal group of these compounds is constrained in a ring. These polyketide metabolites showed remarkably cytotoxic and antiviral activities. Pharmacologically, such compounds are best classified as protein synthesis inhibitors. Their antitumor effects are at the center of attention of pharmacologists. For example, mycalamide A is a potent cytotoxic agent with an IC_{50} value of 3.0 nM against the P388 cancer cell line. Additionally, it displays cytotoxicity against HL60, HepG2, A2780, and A2780AD cancer cell lines with IC50 values of 4.49 nM, 0.77 nM, 1.06 nM, and 18.90 nM, respectively. Besides these antitumor effects, mycalamide A can prevent the malignant transformation of cells, inhibiting nuclear factors. Thus, it shows promising potential for cancer prevention and cytotoxic therapy [41]. Moreover, mycalamide A shows a potent cytopathic inhibitory activity against Polio type I viruses and herpes simplex virus type-1.

Like mycalamide A, onnamide A strongly induces apoptosis in HeLa cells by modulating transforming growth factor-β, which regulates tumor initiation processes [42]. Theopederin A also displays a remarkable cytotoxic effect on P-388 murine leukemia cancer cells, with IC_{50} values of 0.05 nM.

Are mycalamides, onnamides, and theopederins defensive metabolites of sponges? Because sponges cannot move and lack physical defenses, they are highly susceptible to highly mobile marine predators such as fish, turtles, and invertebrates. Thus, it is possible to assume that such toxic metabolites serve as chemical weapons; however, studies have yet to be carried out due to the high structural complexity of these defense compounds and the complex defense mechanisms and defense dynamics of sponges [43, 44].

The Australian orange sponges of the genus *Axinella* and the spider sponge of the *Trikentrion* contain much less complex metabolites. These are substituted indole derivatives, i.e., polylalkylated cyclopent[*g*]indole alkaloids. They were called herbindoles and trikentrins (Figure 7.11). The structural analogy of herbindoles and trikentrins produced by different sponge genera can be observed. The methyl groups on the cyclopentane fragment can have the *cis* or *trans*-configuration. The absence of substituents in the C-2 and C3-positions of the indolic ring is also worth noting, and the butenyl fragment at C5 or C6 is always *Z*-configuration.

These marine alkaloids, isolated between 1986 and 1990, were shown to exert cytotoxicity against KB cells (nasopharyngeal carcinoma), general antifeedant activity against generalist fishes, including wrasses, triggerfishes, and sergeant majors (herbindoles), and antimicrobial activity against Gram-positive bacterium *Bacillus subtilis* (trikentrins) [45, 46].

The infrequent structure of these alkaloids and the possibility of finding other biological properties have motivated several research groups to obtain them [47]. However, the described syntheses of these falsely simple-looking alkaloids are lengthy, low-yielding, stereo-random, and comprise challenging separations of diastereomeric intermediates. Here is just one example of the enantioselective synthesis of (+)-*cis*-trikentrin A realized by Rajan Babu and coworkers in 2012 [48]. They converted 4-ethyl-7-vinylindole **7.26** into (+)-*cis*-trikentrin A through a linear strategy using

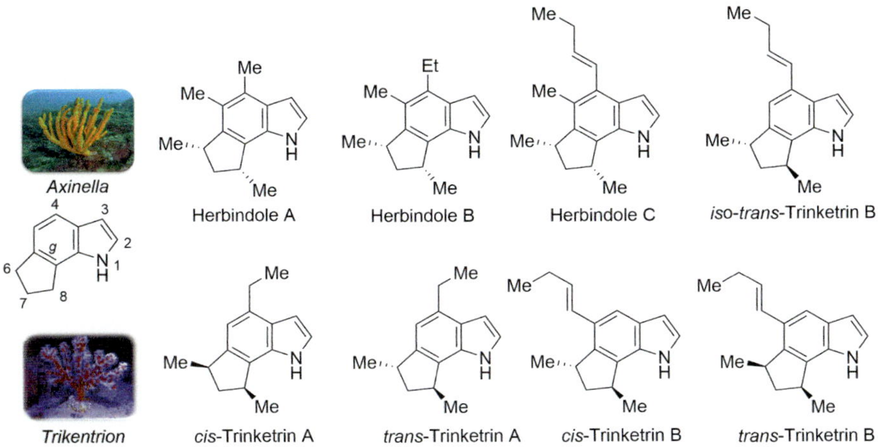

Figure 7.11. *Axinella* and *Trikentrion* sponges as sources of herbindoles and trikentrins based on the dimethylcyclopent[*g*]indole skeleton.

nine steps. This key indole material was prepared from *o*-bromonitrobenzene **7.24** in a two-step procedure involving a Ni(II)-catalyzed asymmetric hydrovinylation reaction of indole ring **7.25** (Figure 7.12). Thus, 4-ethyl-7-vinylindole **7.26** was firstly protected by the Ts-group (comp. **7.27**). The *N*-tosylated indole was exposed to hydroboration (9-BNN, then H_2O_2 to give **7.28**), followed by oxidation to prepare the carboxylic acid **7.29**.

Its intramolecular Friedel-Crafts reaction (formation of **7.30**), followed by a Wittig reaction, gave indole **7.31** with the *exo*-methylene bond. The latter was modified to the needed *cis*-dimethyl configuration (comp. **7.32**) through the selective isomerization and hydrogenation processes in the presence of Pd and Ir catalysts. Removal of the N-Ts indole **7.32** using TBAF in refluxing THF afforded the natural product (+)-*cis*-trikentrin. Let us note that the developed synthesis is efficient and relatively short and uses available commercial bromonitrobenzene. However, the implementation of the developed protocols requires the appropriate laboratory infrastructure.

Various species of sponges produce cytotoxic metabolites of the macro-lactonic system (Section 4.2.2), i.e., macrolides, to protect themselves that can find medical or biochemical applications. The first macrolides isolated from the sponge were latrunculins A and B. Their name comes from the sponge where it was extracted. Thus, from 1975 to 1980, the branching red-colored sponges *Latrunculia magnifica*, which do not allow the fish to be damaged or eaten, were studied by Kashman and coworkers [49, 50]. Interestingly, it was observed that after the sponge was clasped by hand, a reddish fluid appeared, accompanied by a strong odor. This "juice" causes fish to run away instantly in the sea. Squeezing *L. magnifica* into an aquarium with fish produces poisoning and death within 4–6 minutes. The toxic components (LD_{50} = 3.4–10.5 μg/kg, mice) cause excitation of the fish within seconds, followed by hemorrhage, loss of balance, and, finally, after a few minutes, death. After several years of investigating the toxic components of extracts from these sponges, occurring at a depth of 6.0 to 30.0 m in the Red Sea, they were able

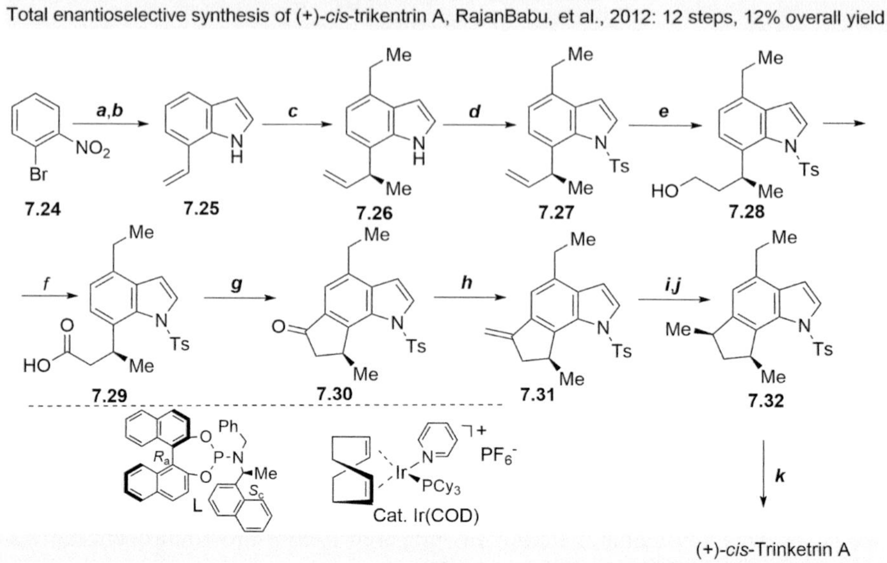

Figure 7.12. Enantioselective synthesis of (+)-*cis*-trikentrin A: (a) CH_2=CHMgBr, THF, –50°C; (b) CH_2=CHSn(*n*-Bu)$_3$, Pd(0), KF, toluene, reflux, 42% in two steps; (c) CH_2=CH_2 (1 atm), L-[(allyl)Ni-Br]$_2$, NaB(Ar)$_4$, CH_2Cl_2, –78°C, Ar = 3,5-$(CF_3)_2$-C_6H_3, 99% (96%, ee); (d) TsCl, KH, DME, 18-crown-6, 53%; (e) 9-BBN, H_2O_2, 92%; (f) 1. TPAP, NMO, 2. $NaClO_2$, $NaHPO_4$, 85% overall; (g), 1. $(COCl)_2$, DMF, 2. $AlCl_3$, CH_2Cl_2, 74% overall; (h) $Ph_3P^+MeI^-$, 94%; (i) CH_2=CH_2 (1 atm), cat. [(allyl)PdCl]$_2$, (*o*-tol)$_3$P/AgOTf, 95%; (j) H_2, cat. Ir(COD), CH_2Cl_2, 95%; (k) *n*-Bu_4NF, THF, reflux, 16 hours, 84%. DME = dimethoxyethane, 9-BBN = 9-borabicyclo[3.3.1]nonane, TPAP = tetrapropylammonium perruthenate, NMO = 4-methylmorpholine 4-oxide, COD = 1,5-cyclooctadiene.

to conclude that the latrunculins A and B are 2-thiazolidinone-bearing marine 16- and 14-membered macrolide skeletons, respectively (Figure 7.13). Latrunculins are the major extractable toxins of *Latrunculia magnifica* (now *Negombata magnifica*), constituting up to 0.35% of the dry weight of the sponge [50, 51].

Besides potent ichthyotoxic activity, latrunculins can form a complex with G-actin and disturb actin filament organization. Both present similar effects *in vitro* [52]. However, due to its structural features, latrunculin A is a cell-permeable macrolide that disrupts microfilament organization in cultured cells by forming a 1:1 complex with monomeric G-actin (K_D = 0.2 μM) without affecting the microtubules. In contrast, latrunculin B is a unique marine toxin that inhibits actin polymerization and disrupts microfilament organization. It is 10 to 100-fold more potent than cytochalasins (Section 4.2.2). Thus, latrunculin A is a promising lead for actin-binding drug development in cancer therapy, showing IC_{50} values of 1.9 μM, 2.7 μM, and 0.9 μM for cell lines SW-620, NCIH-460, and KB3-1, respectively [53]. The disruption of actin filaments of the cytoskeleton by latrunculin A may also result in the modulation of Ca^{2+} channels, specifically intracellular Ca^{2+} channels. Its effects are similar to those induced by cytochalasin D, a well-known actin disruptor.

Interestingly, as a measure against self-toxination, the sponges keep latrunculins in membrane-bound vacuoles that also function as secretory and storage vesicles free of actin. Like brominated isoxazoline alkaloids of *Aplysina* sponges, latrunculins can

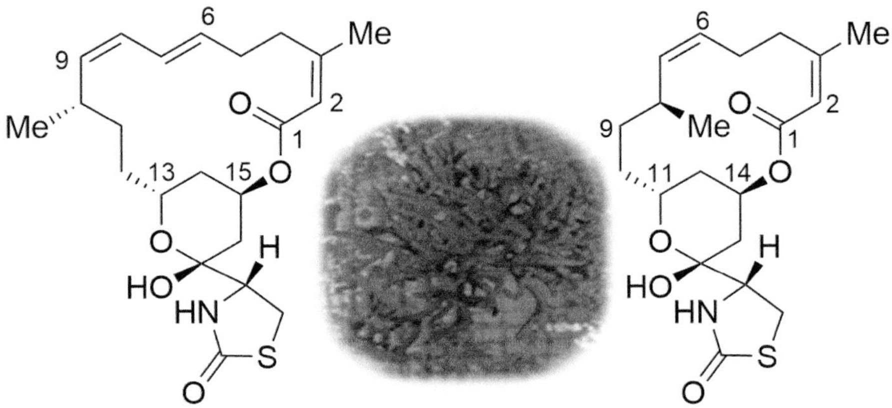

(+)-Latrunculin A (+)-Latrunculin B

Figure 7.13. Molecular structures of latrunculins A and B and their source, sponge *Latrunculia magnifica*. Both latrunculins A and B are colorless oils, showing $[\alpha]_D^{24} = +152°$ (c = 1.2, $CHCl_3$) and $[\alpha]_D^{24} = +112°$ (c = 0.48, $CHCl_3$), respectively.

be sequestered by "sea slugs" (gastropod mollusks) of the genus *Chromodoris*. They selectively transport and accumulate latrunculin A in the sites of the mantle that are most exposed to potential predators [53].

Despite the remarkable properties of the latrunculins and their widespread use as biochemical tools, their availability is relatively limited. Access to them by harvesting and extracting the producing sponge is restricted. This is also due to the complex preparation of these toxic metabolites in a laboratory. Even though the total syntheses developed by the Smith group [54, 55], White and Kawasaki [56, 57], and the Fürstner group [58, 59], both marine sponge products are commercially available but quite expensive; 100 μg of latrunculin A is sold for $260, and 5 μg of latrunculin B is marketed for $545 [60]. These syntheses also provided interesting latrunculin derivatives [61].

All three groups proposed different retroanalyzes of the latrunculin structure oriented toward the convergent synthesis strategy. The retrosynthetic analysis of latrunculin A, conducted by Prof. Amos Smith III and coworkers, was chosen as an example (Figure 7.14). They planned the cleavage of the macrocyclic ester moiety and scission of the *cis* double bond C8=C9.

The southern hemisphere 2-oxo-4-thiazolidin-pyrane aldehyde **7.37** would be accessible from β-hydroxyketone **7.36** via hydrolysis and reorganization processes, and the northern hemisphere diene acid with the phosphonium salt moiety **7.38** would be available from hydroxyacetylene **7.35**. Finally, aldol **7.36** would be derived from acetaldehyde **7.33** and methyl ketone **7.34** (Figure 7.14) [55]. We must first prepare the key precursors to assemble the latrunculin A structure from the found blocks, and the same identified blocks **7.33** and **7.34** were obtained from available reagents (±)-2-allylcyclopentanone and ethyl 2-oxo-4-thiazolidine carboxylate, respectively (Figure 7.15). The union of blocks **7.33** and **3.74** was made with the aldol reaction in the presence of dibutylboron triflate (Bu_2BOTf) and Hunig's base (i-Pr_2NEt) at –78°C. The aldol **3.76** is obtained in 87% yield as an inseparable 4:1 mixture of

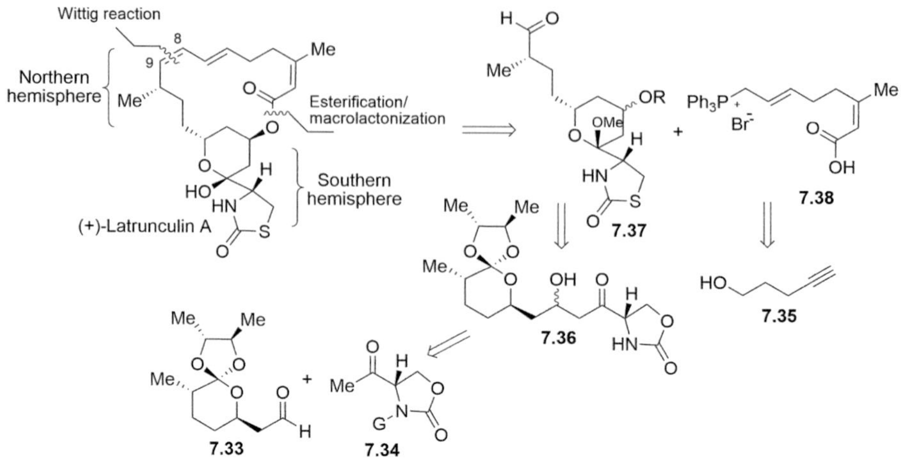

Figure 7.14. Retrosynthetic analysis of latrunculin A via the initial disconnection of the *cis* double bond C8=C9 and the ester bond C-O leading to the key building blocks **7.37** and **7.38** of northern and southern hemispheres. Their second retrosynthetic analysis furnished suitable materials **7.35** and **7.36**, respectively. R = α- or β-TBS (*tert*-butyldimethylsilyl), G = PMB (*p*-methoxybenzyl) or TEOC ([β-(trimethylsilyl) ethoxyl]carbonyl).

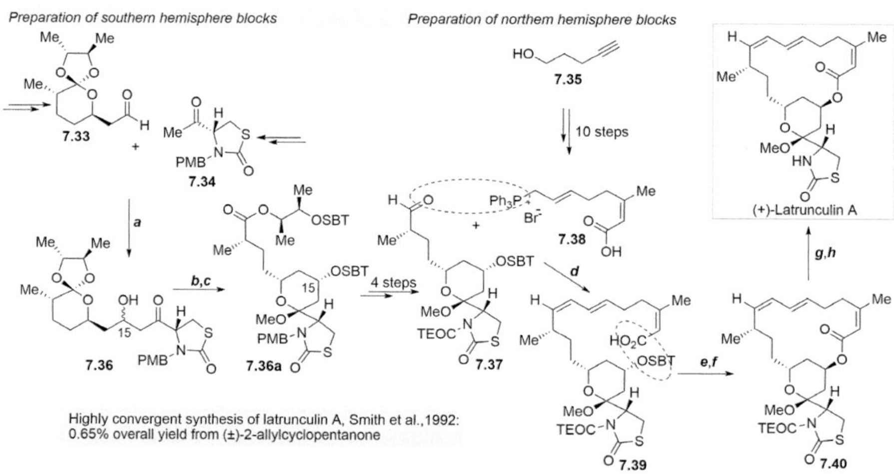

Figure 7.15. Convergent synthesis of (+)-latrunculin A. Simplified scheme: (a) Bu_2BOTf, *i*-Pr_2NEt, 87%; (b) 2N HCl, THF, 78%; (c) 1. CSA, MeOH, 91%, 2. TBSCl, imidazole, 95%; (d) $KN(SiMe_3)_2$, THF, –78°C, 80%; (e) Py·HF, THF, 0°C, 89%; (f) PPh_3, DEAD, 28%; (g) 3N HCl, THF, 0°C to rt, 72%; (h) TBAF, 84%. CSA = (+)-10-camphorsulfonic acid

diastereomers (α- and β-C_{15}-OH). Exposure to aqueous hydrochloric acid in THF promoted a structural reorganization and affected C_{15} equilibration, furnishing a 12:1 mixture of easily separable α- and β-epimers in 78% yield. The hydroxy groups of the major epimer were protected with a TBS group, giving hemiacetal **3.76a**. Then, it was converted into a vital block of the southern hemisphere latrunculin core **3.77** with the changed N-protect group (TEOC) (four steps, 36% overall yield).

The northern hemisphere phosphonium salt **7.38** was prepared using 10 linear steps from commercially available 4-pentyn-1-ol (**7.35**) with a 34% overall yield. The Wittig olefination process between **7.38** and **3.77** smoothly proceeded in the presence of potassium bis(trimethylsily1)amide ($KN(SiMe_3)_2$) to furnish the desired *cis,trans*-diene acid **7.39**. Its desilylation and Mitsunobu macrocyclization (PPh_3, DEAD) then provided **7.40**, which was finally converted into (+)-latrunculin A (Figure 7.15). Thus, the latrunculin A construction was accomplished via a highly convergent route, where the most extended linear sequence was 16 steps. The 0.65% overall yield from commercially available (±)-2-allylcyclopentanone reagent was reached. Notably, similar tactics allowed to prepare (+)-latrunculin B.

The latrunculin B macrolactone ring can also be assembled in another manner, disconnecting it into three key building blocks, **7.34** and **7.41–7.42** (Figure 7.16), that can be efficiently prepared from available reagents. This plan was designed and realized by Alois Fürstner and coworkers [58]. Note 2-oxo-4-thiazolidine ketone **7.34** was also a fundamental block in Smith's latrunculins A and B syntheses. Both other key blocks **7.41–7.42** contain alquine chains to assemble the northern hemisphere of latrunculin B through the novel ring-closing alkyne metathesis (RCAM)/Lindlar reduction process.

The required alquine block **7.41**, containing two chiral centers, was derived from available chiral material, (+)-citronellene **7.43**, using seven steps with 41% overall yield. In contrast, another alquine precursor, **7.42**, was easily obtained from simple ethyl acetoacetate **7.44** in a three-step procedure (54% overall yield). Therefore, the previously described 2-oxo-4-thiazolidine ketone **7.34** was employed in an aldol condensation with alquine aldehyde **7.41** in the presence of $TiCl_4$ to form β-hydroxyketone **7.45** as a diastereomeric mixture (2:1), which was subjected under an acid-catalyzed ring reorganization to deliver hemiacetal **7.46** (Figure 7.17). Its glycosylation with MeOH gave the product **7.47**. This was treated with Tf_2O and reacted with the sodium salt of acid **7.42** that furnished substituted diyene **7.48**.

So, through aldol chemistry and esterification reaction, the substrate **7.48** was made for the formation of the macrocycle **7.49** by the RCAM process in the presence of [Mo{N(*tert*-Bu)(3,5-$Me_2C_6H_3$)$_3$}] precatalyst, which was activated *in situ* with dichloromethane. The obtained macrocycle, a quite strained cycloalkyne, was subjected to a Lindlar reduction to ensure the stereoselective formation of the Z-alkene entity, followed by cleavage of the N-PMB group and the methyl glycoside

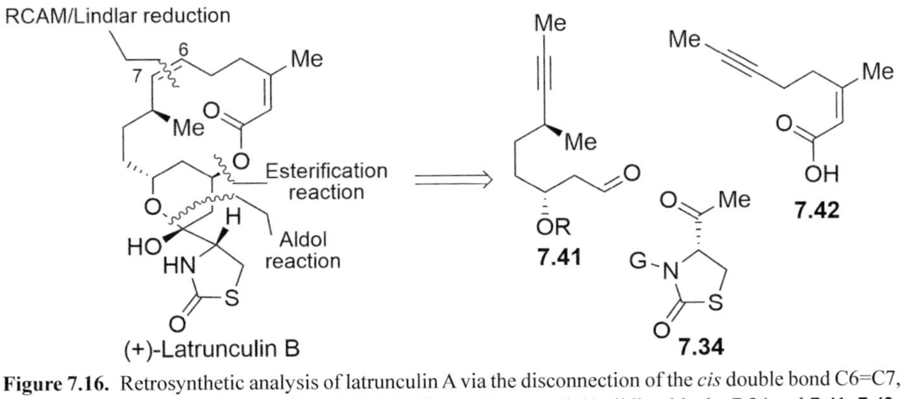

Figure 7.16. Retrosynthetic analysis of latrunculin A via the disconnection of the *cis* double bond C6=C7, the ester bond C-O, and the acetal bond C-O leading to the essential building blocks **7.34** and **7.41–7.42**.

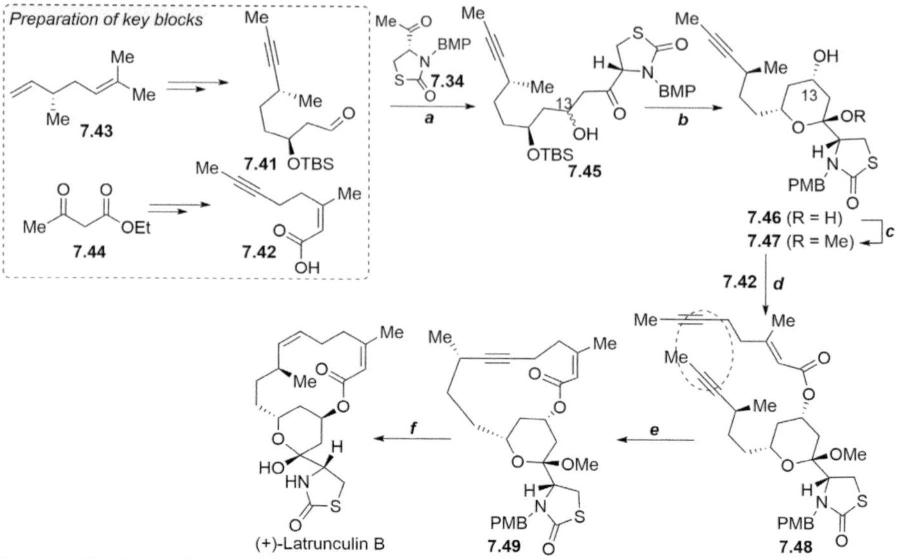

Figure 7.17. Synthesis of (+)-latrunculin B. Simplified scheme: (a) $TiCl_4$, i-Pr_2NEt, CH_2Cl_2, –78°C, 73%; (b) aq. HCl, THF; (c) CSA, MeOH, 65% (over two steps); (d) 1. Tf_2O, Py, CH_2Cl_2, –10°C, 2. Na salt of **7.42**, 15[crown]-5, THF, 58%; (e) 5 mol% [Mo{N(*tert*-Bu)(3,5-$Me_2C_6H_3$)}$_3$], toluene, CH_2Cl_2, 80°C, 70%; (f) 1. H_2, Lindlar catalyst, quant., 2. CAN, MeCN/H_2O, 78%. CSA = (+)-10-camphorsulfonic acid, CAN = ceric ammonium nitrate.

with CAN, providing the desired (+)-latrunculin B (Figure 7.17) [58]. This route is more productive and inherently flexible than other published ones; however, it limits the use of Mo-catalysts, which are still commercially unavailable reagents.

Despite marine sponges contributing nearly 30% of all marine natural products and being the richest source of new marine natural products, we will start other marine invertebrates. Another group of marine sessile invertebrates, which we must mention, are corals belonging to the phylum Cnidaria. Despite morphological similarities, corals and sponges are entirely different from each other. In this section, we have seen some toxic metabolites of sponges, which they themselves produce to defend themselves against serious predators. Will it be the same tactic that corals use?

7.1.3 Defensive Metabolites of Corals

Thus, corals, colonial animals, belong to the phylum Cnidaria and are closely related to jellyfish and anemones. The coral body is a hollow sac-like structure known as a polyp. Corals, known as octocorals, have tentacles around the central opening and use them to protect themselves from predators and capture their food from the water. Corals are primarily grouped into stony corals and soft corals. This section starts with the latter. Soft corals (Alcyonacea), formerly known as gorgonian corals, contrast stony corals because they do not create a calcium carbonate skeleton [62]. However, they produce a vast array of skeletal different classes of secondary metabolites.

The famous soft coral is the *Palythoa* species (*Palythoa toxica*). As with any soft coral, they do not have calcium carbonate skeletons but can contain a highly toxic and potentially lethal metabolite known as palytoxin (PTX), identified in 1971 by Prof. Paul J. Scheuer (1915–2003) [63]. When describing the isolated substance, the authors wrote, "Apart from polypeptide and protein toxins, it is the most highly toxic substance known, with a lethal dose (LD_{50}) in mice of 0.15 microgram per kilogram by intravenous injection." Over the decade, chemical, biological, and toxicological studies on this metabolite have elucidated the peculiar properties of this eye-catching marine toxin that, in our time, is considered one of the most toxic non-proteinaceous natural compounds. Its chemical structure was elucidated in 1981 independently by two groups [64, 65].

PTX is a large, complex molecule with a long polyhydroxylated and partially unsaturated aliphatic backbone containing 64 chiral centers. The chemical structure of palytoxin is $C_{129}H_{223}N_3O_{54}$ (MW = 2,680.14 g/mol), in which 115 of the 129 carbons are in a continuous chain (Figure 7.18). In addition, it possesses eight double bonds that can exhibit *cis*/*trans*-isomerism. All that means that PTX can have more than 10^{21} stereoisomers.

PTX and its analogs have also been identified in marine organisms in tropical, subtropical, and temperate regions. Some analogs, such as ostreocin, mascarenotoxins, and ovatoxin, are all produced by dinoflagellates from the genus *Ostreopsis*.

The toxicity of palytoxin is due to its binding (blockage) to the sodium pump (i.e., Na^+/K^+-ATPase channel, a transmembrane protein), which is essential for maintaining cell homeostasis. Thus, PTX seems to bind to the extracellular part of the sodium pump, thereby inhibiting the active transport of sodium and potassium ions across the cell membrane by transforming the pump into a non-specific, permanently open ion channel. This causes the membrane depolarization. Thus, an

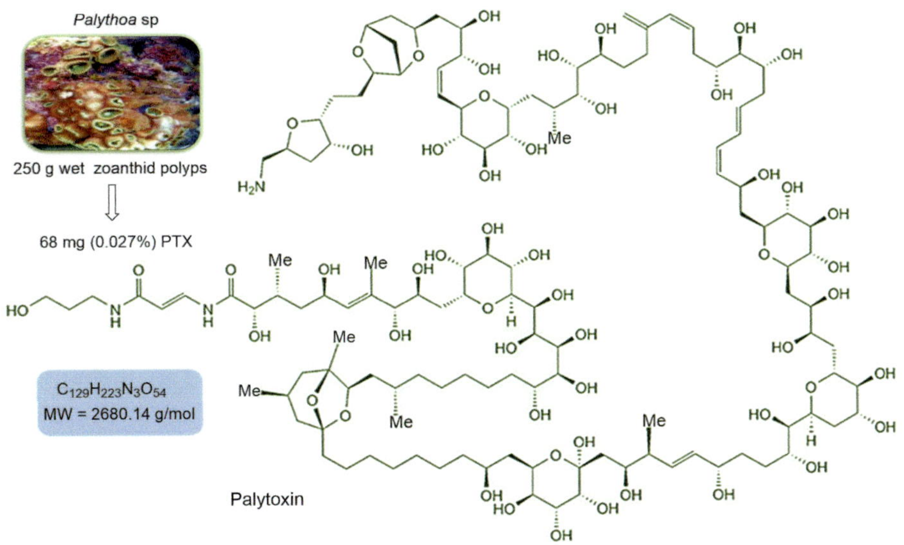

Figure 7.18. Molecular structure of complex polyketide palytoxin exhibiting extreme toxicity in mammals. PTX is one of the strongest known non-proteinaceous toxins occurring in nature, having an LD_{50} of 25–450 ng/kg body weight when injected into mice intravenously. Human symptoms of palytoxin poisoning include abdominal cramps, nausea, diarrhea, paraesthesia, muscle spasms, and respiratory distress. However, at sublethal doses, palytoxin exhibits high anticancer activity.

ion gradient of the cell is destroyed. This depolarization also triggers the activation of voltage-dependent ion channels, including voltage-dependent Ca^{2+} channels [66, 67]. Regarding its preparation, we limit ourselves to commenting only that 13 years after isolating PTX, Kishi's group reported its first synthesis. However, they did it in steps, starting from its acid derivative, obtained in 1989 [68–70], and later synthetic development was not fruitful.

The genus *Sinularia* also includes a group of soft corals. They developed several mechanisms to resist aggressive life conditions involving chemical defense, symbiosis, and allelopathy. As reported, secondary metabolites isolated from *Sinularia* are characterized by their intriguing and diverse structural features from simple cyclopentenone and butenolide derivatives like sinularone A and sinularone I to diterpenoids, such as sinulodurin B or (–)-14-deoxycrassin (Figure 7.19) [71–73].

These metabolites display cytotoxic, antifouling, and antimicrobial properties. For example, sinularone A and sinularone I showed potent antifouling activities against the larvae of the barnacle *Balanus amphitrite* (EC_{50} = 13.86 and 23.50 µg/mL, respectively), also presenting weak toxicity against the barnacle with LC_{50} > 50 µg/mL [71]. Meanwhile, the diterpene sinulodurin B, isolated from the corals *Sinularia levi* and *S. dura*, displayed antiproliferative activity against highly malignant (+SA) mouse mammary epithelial cell growth and viability *in vitro* at a dose of 20 µM [72]. (–)-14-Deoxycrassin, containing a 14-membered carbocyclic nucleus (i.e., cembranolide class of diterpenes), exhibited remarkable antifouling activities against the larvae of the bryozoan *Bugula neritina* and the barnacle *Balanus albicostatus* (EC_{50} for *B. neritina* 3.90 mg/mL and EC_{50} for *B. albicostatus* 21.26 mg/mL), suggesting its potential as an environmentally friendly antifoulant [73]. In addition, its (+)-isomer displayed potent antitumor activity against the human colon (HCT-116), melanoma (SK5-MEL), and kidney carcinoma (A-498) with IC_{50} values of 2.0, 0.5, and 0.2 µg/mL, respectively [74].

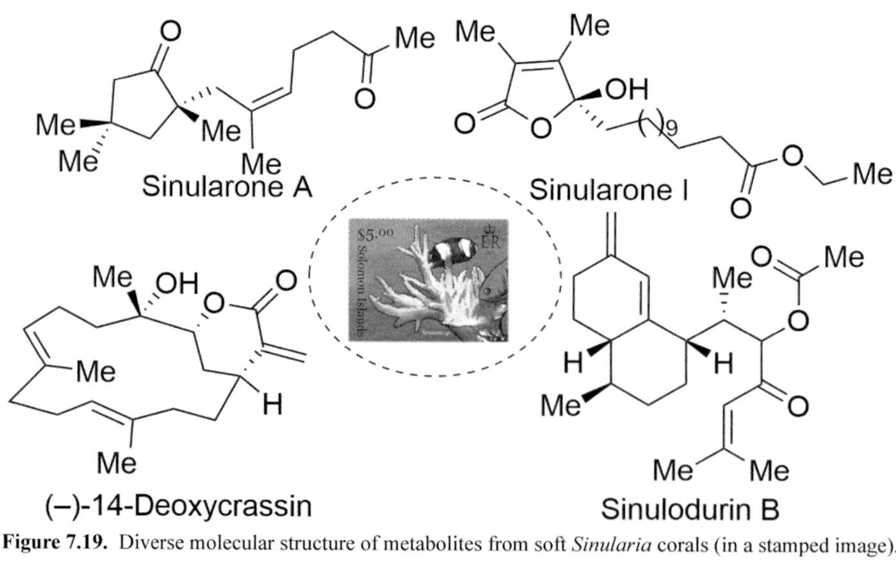

Figure 7.19. Diverse molecular structure of metabolites from soft *Sinularia* corals (in a stamped image).

Given the interesting biological properties and characteristic crassin-like cembrenoid structure of (+)-14-deoxycrassin, it would be worthwhile to examine its unique synthesis realized in 2003 by Li and coworkers (Figure 7.20) [75]. Their overall strategic plan involves an intramolecular lactonization of the pyranone ring and an assembly of the 14-membered cembrane ring with geometrically defined double bonds and substituents resulting from key fragment **7.50** with double bond and fragment **7.51** with the stereogenic center. The latter were readily prepared using chiral reagent (*S*)-limonene in a two-step procedure. Thus, the first reaction, the Horner-Emmons coupling of fragments **7.50** and **7.51**, allowed the construction of an open precursor **7.52**. Those two-step transformations (selective removal of THP and Swern oxidation) afforded the corresponding keto aldehyde ester derivative of intermediate **7.53**, ready for the ring cyclization through the Ti(0)-mediated McMurry coupling reaction.

This cyclization was performed using the $TiCl_4$-Zn system in DME under reflux conditions, furnishing cyclized product **7.54** in a 69% yield. Its desilylation and asymmetric epoxidation in the presence of L-(+)-DET (diethyl tartrate) as a

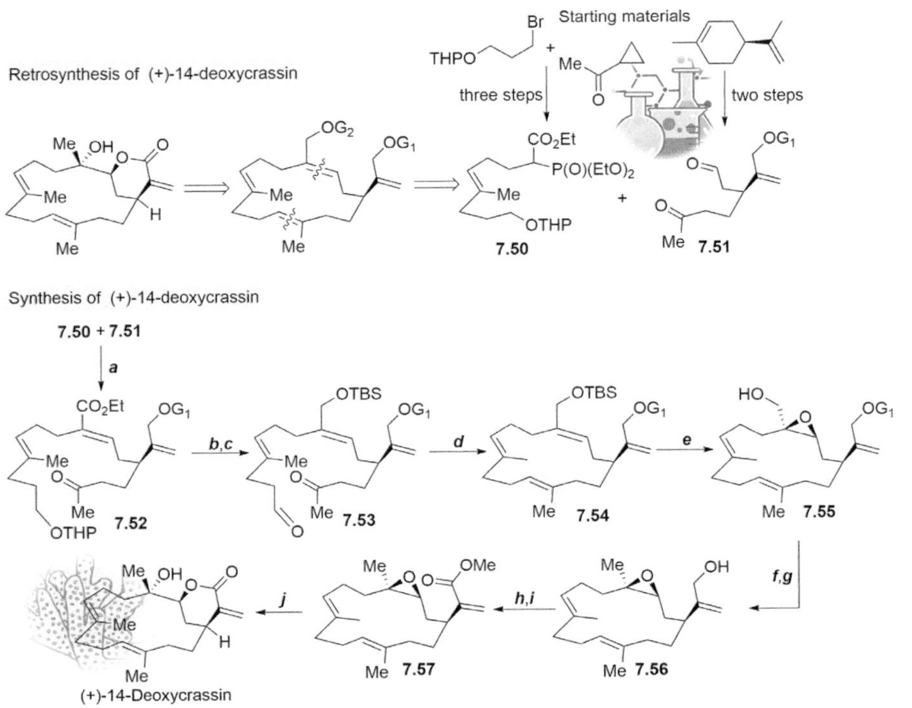

Figure 7.20. Retrosynthesis and asymmetric synthesis of (+)-14-deoxycrassin: (a) *n*-BuLi, THF, –78°C, 77%; (b) 1. DIBALH, CH_2Cl_2, –78°C, 2. TBSCl, imidazole, DMF, 57% overall; (c) 1. $MgBr_2$, Et_2O, rt, 2. Swern oxidation, 58% overall; (d) $TiCl_4$, Zn, pyridine, DME, reflux, 69%; (e) 1. PPTS, EtOH, rt, 2. $Ti(i\text{-PrO})_4$, L-(+)-diethyltartrate, *tert*-BuOOH, CH_2Cl_2, –20°C, 82% overall; (f) 1. I_2, Ph_3P, imidazole, Et_2O–MeCN, 0°C, 2. $NaBH_3(CN)$, THF–HMPA, 60°C, 87% overall; (g) TBAF, THF, 93%; (h) MnO_2, 91%; (i) MnO_2, NaCN, MeOH, AcOH, 58%; (j) $ZnCl_2$, CH_2Cl_2, 0°C to 20°C, 71%. G_1 = TBDPS (*tert*-butyldiphenylsilyl); PPTS = pyridinium *p*-toluenesulfonate, TBS = *tert*-butyldimethylsilyl, THP = tetrahydropyran, HMPA = hexamethylphosphoramide.

chiral mediator afforded the desired epoxy alcohol **7.55**, a close precursor of **7.56**, i.e., pseudoplexaurol, another crassin-like cembrenoid like 14-deoxycrassin, also isolated from the Caribbean gorgonian *Pseudoplexaura porosa*. Thus, the final steps of 14-deoxycrassin synthesis consist of the synthetic conversion of epoxy alcohol **7.56** to the desired lactone derivative, i.e., 14-deoxycrassin, which was obtained via oxidation processes leading to methyl ester **7.57** and $ZnCl_2$ catalyzed lactonization (Figure 7.20) [75]. Resuming facts, the asymmetric synthesis of (+)-14-deoxycrassin from available chiral reagent (*S*)-limonene has resulted in a long and inefficient route (16 steps, ~ 1% overall yield).

Another genus of soft corals, *Cespitularia*, has been well recognized as a rich source of bioactive secondary metabolites, especially diterpenoids [76]. Chemical studies of the EtOH extract of the soft coral *Cespitularia taeniata* collected in Taiwan have resulted in the isolation of nitrogen-containing diterpenoids and sesquiterpenoids, such as cespitulactams A–C (Ce-A–C) and taenialactams A and B (Ta-A–B), respectively (Figure 7.21) [77, 78].

The isolated coral metabolites are cytotoxic molecules and were evaluated against various cancer cell lines. Cespitulactam A, based cyclododeca[*b*]pyrrol-2-one ring with an N-phenethyl substituent, exhibited significant cytotoxicity against human colon adenocarcinoma and medulloblastoma cancer cells with IC_{50} values of 2.72 and 6.34 μg/mL, respectively. At the same time, its analogs B-C were inactive (> 20 μg/mL) [77]. Taenialactams, possessing decahydro-2*H*-benzo[*f*]indol-2-one skeleton, did not also show antitumor properties; however, taenialactones of octahydronaphtho[2,3-*b*]furan-2(4*H*)-one structure **7.58**–**7.59** displayed cytotoxicity. In particular, 8β-methoxyatractylenolide **7.59** exhibited cytotoxicity against human oral epidermoid carcinoma and medullocarcinoma cancer cell lines with ED_{50} values of 10.71 and 7.93 μg/mL [78].

One more rich source for discovering structurally diverse terpenoids and sesquiterpenoids is the genus *Lemnalia* [79]. The first chemical investigation of *Lemnalia* (*L. africana*), conducted in 1974, furnished unique tricyclic sesquiterpenes metabolites called africanol and isoafricanol, containing fused cyclopropane, cyclopentane, and cycloheptane rings, which form the 3/7/5 cyclic system (Figure 7.22A) [80, 81]. Note that a similar tricyclic sesquiterpene known as $\Delta^{9(15)}$-africanene was isolated in 1980 from soft corals *Sinularia polydactyla*, collected at Laing Island (Papua-New Guinea) [82]. These africane-type sesquiterpenoids are anti-inflammatory compounds that suppress the expression of LPS-induced inflammatory mediators independently of NF-κB activation [83].

Besides these africane-type sesquiterpenoids, the most typical metabolites were sesquiterpenes with nardosinane skeletons, based on the 6/6 cyclic system, like lemnacarnol, oxolemnacarnol [84–86], or clavukoellians B–C [87] (Figure 7.22B). Notably, these first are nardosinane molecules with the lactone function, while the clavukoellians possess a maleimide ring, which is rare in marine products. They were tested in antiproliferative assays but did not show considerable effects.

Neolemnane sesquiterpenoids with the 6/8 cyclic system, such as laevinone A and lemneolemnanes A–B (Figure 7.22C), were also found in the *Lemnalia* sp. (*L. laevis*, etc.) [88, 89]. Laevinone A, a diketone molecule, was extracted from the bodies of the soft coral *L. laevis* (1.2 kg, wet weight) with CH_2Cl_2 and followed a

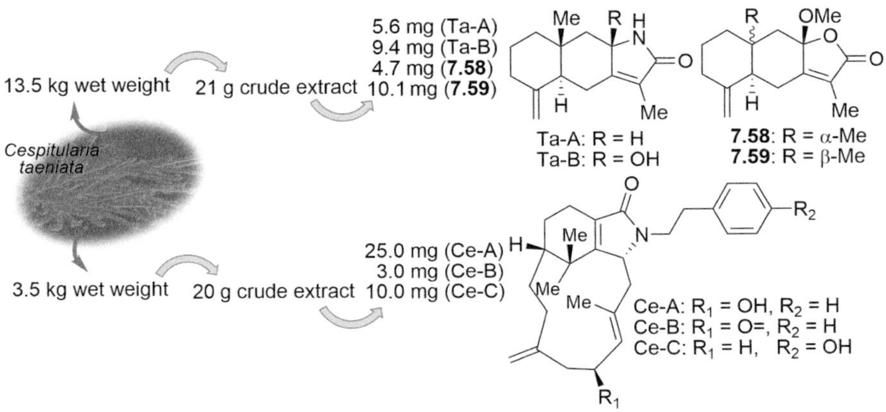

Figure 7.21. Isolation of metabolites from soft *Cespitularia* corals and their molecular structure.

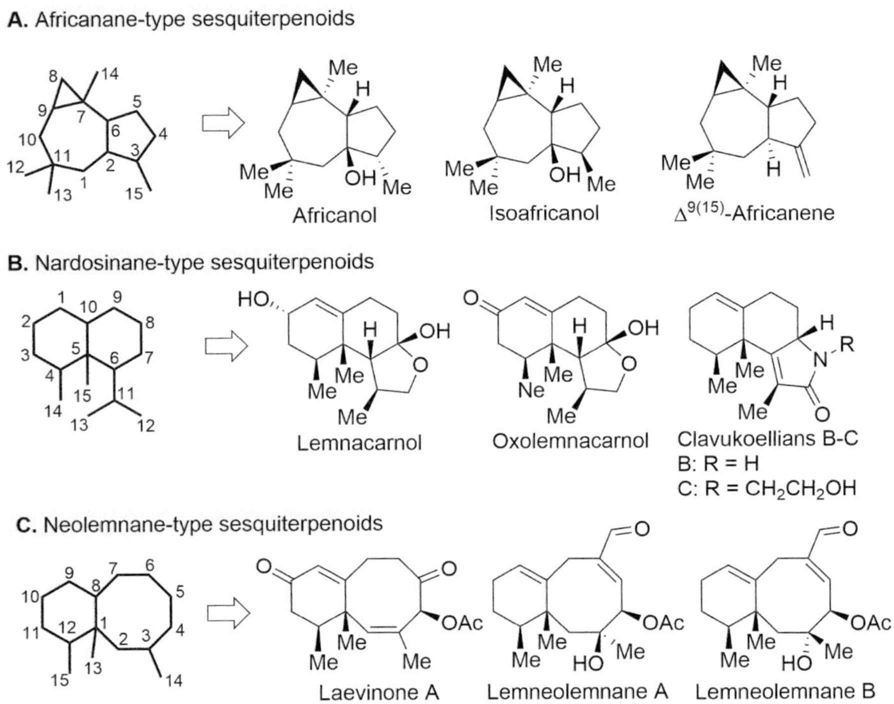

Figure 7.22. Type-carbon skeletons of metabolites from soft coral of the genus *Lemnalia* and their characteristic members.

complex purification process to obtain just 5 mg of the product [88]. Lemneolemnanes A–B are epimers at C-3, featuring an unusual carbon skeleton with a formyl group on C-6. To obtain 7.1–7.3 mg of the substance of these epimers, 7.20 kg (wet weight) of a specimen of *Lemnalia* sp. was treated by common isolation and purification ways [89]. All these were screened for cytotoxic and anti-inflammatory properties; none of the compounds showed significant effects. However, lemneolemnane A

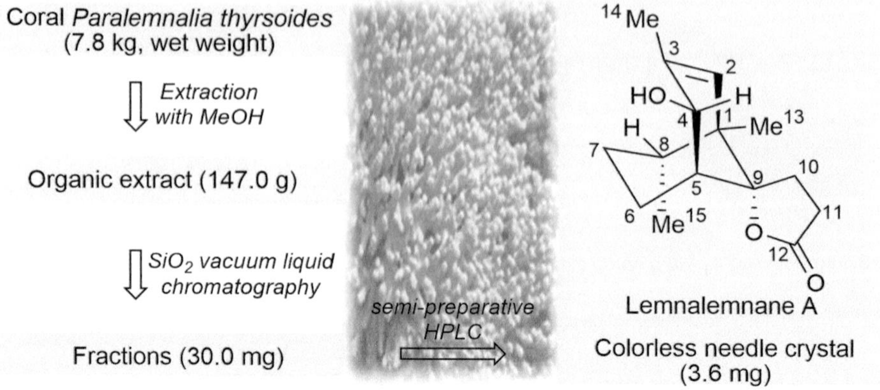

Figure 7.23. Structure of lemnalemnane A and its isolation diagram from the soft corals *Paralemnalia thyrsoides* (in an image). Lemnalemnane A was isolated as a colorless needle crystal with a negative optical rotation in MeOH, $[\alpha]_D^{25} = -59.1$ (c = 2.4, MeOH).

considerably relieved Alzheimer's disease (AD)-like symptoms in the transgenic *Caenorhabditis elegans* model at a concentration of 100 μM/L, indicating that this neolemnane sesquiterpenoid shows potential as a candidate for the development of an anti-AD agent [89]. Its close analog, lemnolin A, isolated from the South China sea soft coral *Lemnalia* sp., displayed antiviral activity against the influenza A (H1N1) virus *in vitro* [90].

A unique sesquiterpenoid, lemnalemnane A, containing an 8/5 cyclic scaffold (Figure 7.23), was identified in 2022 after investigating two similar coral species, *Paralemnalia thyrsoides* and *Lemnalia* sp. This sesquiterpenoid is based on an intriguing basket-like structure with a spiro-bicyclo[3.3.1]nonane-furan core and displayed an interesting anti-inflammatory activity in the transgenic fluorescent zebrafish model [91].

Due to its interesting structure and biological activity, a short and efficient synthesis of lemnalemnane A would be appropriate for acquiring this unusual sesquiterpenoid. There is one approach to this molecule, realized by Ito and coworkers [92]. Their retrosynthetic analysis of lemnalemnane A is outlined in Figure 7.24. The first disconnection suggests the formation of the spironolactone moiety in **7.63**, which must be converted into diketone **7.62**. Its new disconnection would lead to cyclohexanone derivative **7.61**, prepared from (*S*)-carvone (**7.60**).

Ito's synthesis of lemnalemnane A starts with the 1,4-addition reaction of the Gilman reagent (Me_2CuLi) to (*S*)-carvone (**7.60**), followed the treatment with benzyl 3-chloropropiolate (**7.64**) to give (–)-C-alkylated product **7.65** in 99% yield with high diastereoselectivity (Figure 7.25). The next step is the crucial removal of the isopropenyl group of the latter under Kwon's ozonolysis protocol (O_3, $FeSO_4 \cdot 7H_2O$, benzenethiol) to provide dimethyl-alquinylcyclohexanone **7.66**.

Its reduction with the Lindlar catalyst afforded *cis*-alkene **7.61**, and Dieckmann condensation allowed the construction of the bridged bicyclic diketone system (comp. **7.62**). The introduction of the methyl group in this system (comp. **7.67**) was achieved via a two-step procedure (olefin bromination and Pd-catalyzed Heck

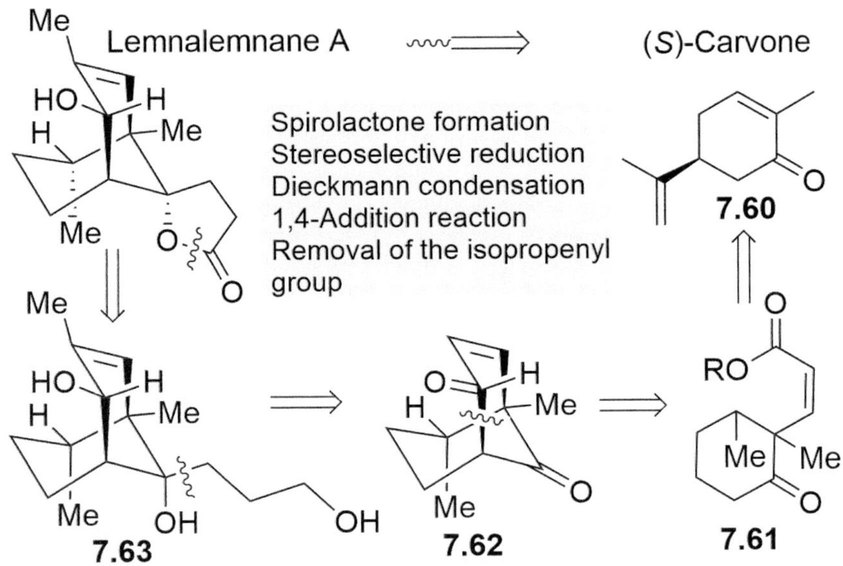

Figure 7.24. Retrosynthetic analysis of lemnalemnane A and found precursors **7.60–7.63** involving the key features of the synthesis.

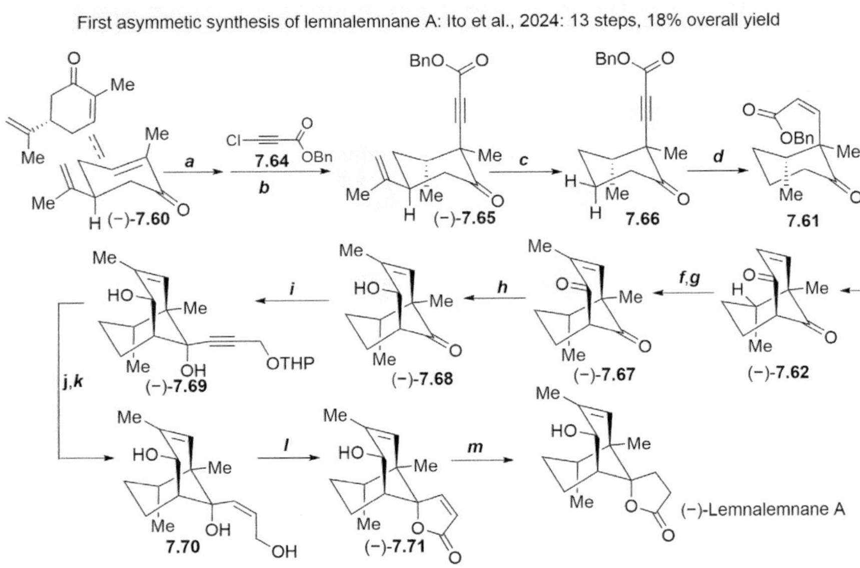

Figure 7.25. Linear synthesis of lemnalemnane A using (*S*)-carvone as an available chiral material [92]: (a) Me_2CuLi, Et_2O, −78°C, (b) Et_2O, DME, −78°C, 2 hours, 99%; (c) O_3, MeOH, then $FeSO_4 \cdot 7H_2O$, benzenethiol, −78°C, 1 hour, 71%; (d) H_2, Lindlar cat., quinoline, EtOH, rt, 1.5 hours, 88%; (e) *tert*-BuOK, THF, 0°C, 30 minutes, 88%; (f) Br_2, Et_3N, CH_2Cl_2, 0°C to rt, 1.5 hours, 94%; (g) $ZnCl_2$, MeMgBr, $Pd(PPh_3)_4$, rt to 50°C, 1 hour, 97%; (h) L-Selectride®, THF, −78°C, 30 minutes, 98%; (i) CH≡CH-CH_2-OTHP, EtMgBr, THF, rt to 50°C, 1 hour, 73%; (j) TsOH, MeOH, rt, 30 minutes, 81%; (k) H_2, Lindlar cat., quinoline, EtOH, rt, 1.5 hours, 84%; (l) TEMPO, $PhI(OAc)_2$, CH_2Cl_2, rt, 2 hours, 79%; (m) $NiCl_2 \cdot 6H_2O$, $NaBH_4$, MeOH, rt, 2 hours, 96%. TEMPO = 2,2,6,6-tetramethylpiperidine 1-oxyl, THP = tetrahydropyran.

coupling reactions). Treatment of (−)-**7.67** with L-Selectride provided the desired alcohol (−)-**7.68** in 98% yield with excellent diastereoselectivity. Completion of the spirolactone ring in the last steps was made by steps with the introduction of the C-3 alquinyl fragment (comp. **7.69**), the Lindlar reduction reaction (comp. **7.70**), and selective oxidation of **7.70** to afford unsaturated spirolactone **7.71**. Finally, the 1,4-reduction of this lactone with sodium borohydride and nickel(II) chloride resulted in the formation of (−)-lemnalemnane A in 96% yield (Figure 7.25). To sum up, this rare sesquiterpenoid was prepared from (*S*)-carvone for the first time in an asymmetric form, achieving 13 steps with an 18% overall yield [92]. The result is remarkable; this methodology can be extended to synthesizing structurally related sesquiterpenoids.

We could provide more beautiful examples of the metabolites of striking corals; what has been seen was selected according to the author's interest. However, our journey through ocean waters must go and does not end with discussing corals' toxic metabolites. In the next section, interesting toxic molecules present in simple organisms called tunicates or ascidians await us.

7.1.4 Tunicates (Ascidians) as Prolific Providers of Toxic Metabolites

Like sponges and corals, tunicates—truly strange creatures—are sessile marine invertebrates. Tunicates are also known as sea squirts, i.e., ascidians. These organisms of the Ascidiacea family are filter feeders of different body colors, such as translucent to blue, green, yellow, red, and brown, with a life span ranging from only two months to one year [93, 94]. Such "sea syringes" forcefully discharge a stream of water through an opening when irritated. Since they are immobile, they must wait for their food, mainly plankton, which sea squirts feed on. However, some predators feed on sea squirts even though they have highly acidic inclusions in their tunics and a high concentration of heavy metals (such as vanadium) and toxic organic substances.

These interesting organisms are often called the "evolutionary connecting link" between invertebrates and chordates. They are considered essential model organisms for several research aspects, such as evolution, development biology, and bioactive compounds. That makes them enormously privileged drug resources in the ocean.

Due to their symbiosis with some microbes, tunicates possess good chemical (deterrent metabolites, acidity) and physical (spicules, tunic toughness) defenses and experience relatively little predation by generalist predators, mainly fish. One of the most studied representatives of tunicates is the tunicates of the genus *Cystodytes*, which produce a series of toxic nitrogen-containing polycyclic molecules. For example, aromatic alkaloids based on fused pyridoacridine skeletons are perhaps the more interesting class of ascidian metabolites [95, 96]. Noteworthy, pyridoacridines are a group of strictly marine-derived alkaloids such as ascididemin [97], cystodytins A–C [98, 97], and kuanoniamines A–D [100] (Figure 7.26), which are the lead models for pharmacological research. Structurally, they all are based on the pyrido[2,3,4-*kl*] acridine ring system.

These secondary metabolites seemed to be the most effective defense mechanism of *Cystodytes* spp. against fish predators but did not deter sea urchins. Ascididemin

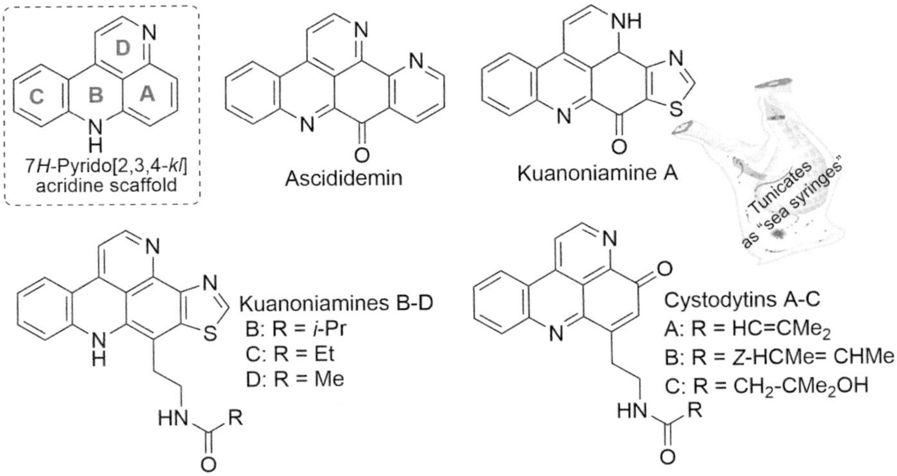

Figure 7.26. Structures of some toxic pyridoacridine alkaloids isolated from tunicates *Cystodytes* and *Didemnum* sp. in a general image. Present marine-derived alkaloids are based on the pyrido[2,3,4-*kl*] acridine ring system, where a 2,7-naphthyridine scaffold stands out.

significantly dissuaded fish predation. This was confirmed in the laboratory feeding assays [101]. It was one of the first pyridoacridine metabolites isolated from ascidians. Kuanonionamines contain the same pentacyclic skeleton as ascididemin, but in kuanonionamines, instead of pyridine, another heterocyclic ring, that of the thiazole, is present. Cystodytins possess a tetracyclic system of 4*H*-pyrido[2,3,4-*kl*] acridin-4-one ring with the ethylamidoalkene side chain. However, all are cytotoxic molecules. Moreover, most pyridoacridines are also known to have antimicrobial, antiviral, antiparasitic, and fungicidal properties.

Ascididemin, yellow solid (mp > 3000ºC), displayed potent cytotoxic activity in A375 (melanoma) and HCT116 or SW480 (colon) cancer cell lines, presenting IC_{50} values of 0.25 μM, 0.10 μM, and 0.38 μM, respectively. This alkaloid was also induced apoptosis strongly in HL-60 and P388 leukemia cells. Its predominant effects are topoisomerase II inhibition, topoisomerase II-mediated DNA cleavage induction, and reactive oxygen species formation. According to the first report on kuanonionamines [100], cytotoxicity (IC_{50}) against KB (human epithelial carcinoma) cells of kuanoniamine A, obtained as yellow needles (mp = 255–258ºC), was found to be one μg/mL. At the same time, kuanoniamine D showed an IC_{50} value of 5 μg/mL, and kuanoniamine B resulted in inactive (IC_{50} > 10 μg/mL). Later, the anticancer activity of kuanoniamines A and C, isolated from the sponge *Oceanapia sagittaria*, was evaluated using a large panel of tumor cell lines, i.e., MCF-7 (breast carcinoma, estrogen-dependent ER+), MDA-MB-231 (breast carcinoma, estrogen-independent ER-), SF-268 (glioma), NCI-H460 (non-small cell lung cancer), UACC-62 (melanoma), and a non-tumor human cell line, MRC-5 (diploid embryonic lung fibroblast) (Table 7.1) [102].

Results in Table 7.1 indicate that kuanoniamine A was more potent than its parent B, inhibiting the growth of all five human tumor cell lines with its low GI_{50} values (≤ 5 μM). Being potent inhibitors of the MCF-7 cells, both alkaloids cause an

Table 7.1. Effect of kuanoniamines A and C on the growth of the human tumor and non-tumor cell lines (cell growth inhibition, GI_{50}, μM).

Compounds	MCF-7	MDA-MB-231	NCI-H460	SF-268	UACC-62	MRC-5
Kuanoniamine A	0.12	0.73	0.91	4.67	1.83	0.58
Kuanoniamine C	0.81	10.23	21.50	33.16	15.78	nd

increase in the apoptotic cells, indicating that their antiproliferative effects on the MCF-7 cancer cell lines could, in part, be related to the phenomenon of apoptosis.

Likewise, cystodytins A–C, isolated from the Okinawan tunicate, *Cystodytes dellechiajei*, showed *in vitro* potent cytotoxicity against mouse lymphocytic leukemia (L1210) cell lines with IC_{50} values of 0.6–2.9 μM [103].

Therefore, synthesizing their analogs has been completed to allow their biological evaluation as potential anticancer and antiparasitic agents. Efficient, green synthetic routes, which offer easy access to significant amounts of ascididemin, kuanoniamines, cystodytines, their analogs, and likely other pyridoacridines, are part of the work of organic and medicinal chemists [104].

The first four-step ascididemin synthesis involves simple starting materials **7.72** and **7.73**. Their oxidative amination gives an amino-derivative **7.74**; then, it is cyclized in acid media, affording tetracyclic quinone **7.75** with the activated methyl group. Reacting it with N,N-dimethylformamide diethyl acetal (DMF-DMA), a hydrozone derivative **7.76** is prepared that cyclocondensation furnished the desired alkaloid ascididemin (Figure 7.27) [105].

This high-yielding four-step approach is still suitable for preparing many related pyridoacridines. In contrast, Echavarren's approach, using α,β-unsaturated hydrazones **7.77** similarly, resulted in a less appropriate procedure (Figure 7.27) [106].

In 2022, Wang and coworkers demonstrated that the biomimetic approach is a suitable strategy for preparing all known cystodytins (A–K). They developed short and efficient synthesis of cystodytins from tryptamine (**7.78**) and 2,5-dihydroxybenzaldehyde (**7.79**) that were found after retrosynthetic analysis

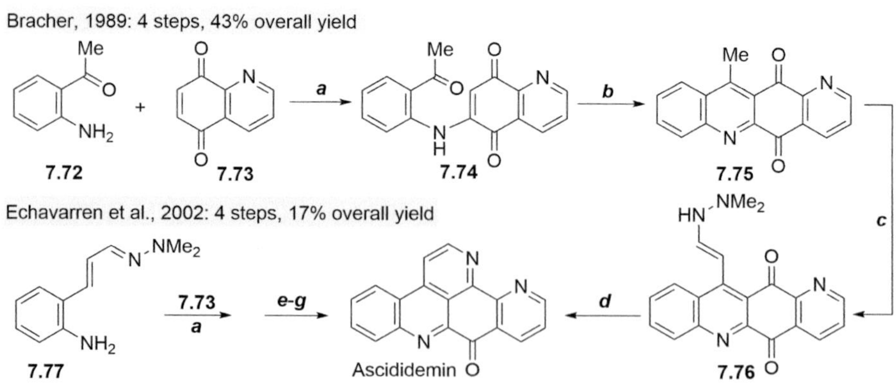

Figure 7.27. Four-step ascididemin syntheses using quinoline-5,8-dione as an available starting material: (a) $CeCl_3 \cdot 7H_2O$, O_2 (air), MeOH, rt; (b) conc. H_2SO_4/AcOH; (c) DMF-DMA, DFM, reflux; (d) NH_4Cl, AcOH; (e) NaH, $(CF_3CO)_2O$, THF, rt; (f) MnO_2, CH_2Cl_2, rt; (g) NH_4Cl, NaOAc, EtOH, reflux.

of cystodytin structure (cystodytins A and B were selected as an example) (Figure 7.28A) [107].

According to the planned strategy, an oxidative amination cyclization of *para*-hydroquinones **7.82**, obtained from **7.79** in a three-step sequence (Henry reaction, reduction, and aminolysis), with tryptamine (**7.78**), to give aminoquinone adducts **7.83**, is a crucial step. This step was made possible due to excellent findings (Figure 7.28B); cooperative catalysis with the copper(I)-TMEDA complex and organic dye Rose Bengal efficiently provided these key intermediates, which were used in the next step without isolation. Their domino cyclization in acetic acid at 60°C under an oxygen atmosphere afforded the respective alkaloids cystodytins A and B. Moreover, chemoselective olefin functionalization (R = –HC=CMe$_2$) in cystodytin A through hydroboration–oxidation provided cystodytin C with a 33% yield [107]. Notably, cytotoxic cystodytins A–B are found to be potent Ca^{2+} release agents and stimulated calcium release from the sarcoplasmic reticulum at 36 and 13 times the potency of caffeine, respectively [98].

Other genera, *Aplidium* and *Eudistoma*, provide many cytotoxic and antiviral alkaloids based on the indole and 9*H*-pyrido[3,4-*b*]indole skeletons, respectively. The green ascidians *Aplidium meridianum* and *A. falklandicum* from cold water (Antarctica) are a rich source of diverse indole molecules [108]. Brominated indoles 3-substituted with a 2-aminopyrimidine ring were first purified and called

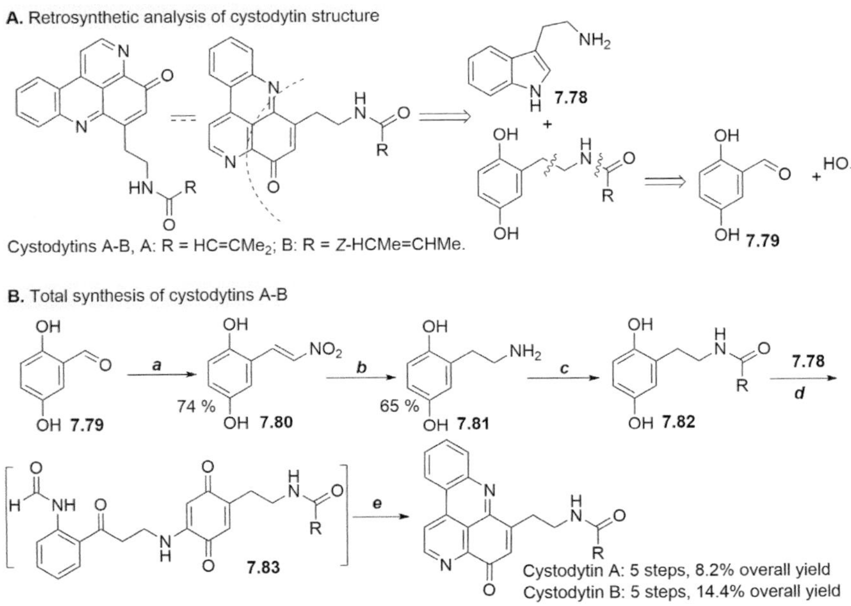

Figure 7.28. Wang's design for synthesizing alkaloids cystodytins. A. Retrosynthetic analysis. B. Five-step synthesis of cystodytins A–B using simple available reagents **7.78**–**7.79** as starting materials: (a) NH_4OAc, CH_3NO_2, HOAc, reflux, 2 hours; (b) H_2, 10% Pd/C, MeOH, 60°C, 48 hours; (c) PyBOP, Et_3N, CH_2Cl_2, N_2, 36 hours, $(Me)_2C$=CHCOOH (for cystodytin A) or *Z*-MeCH=C(Me)COOH (for cystodytin B); (d) CuCl, TMEDA, Rose Bengal, O_2, MeOH, 590 nm LEDs, 30–32°C, 36 hours; (e) HOAc, O_2, 60°C, 1.5 hours. PyBOP = benzotriazol-1-yloxytripyrrolidinophosphonium hexafluorophosphate, TMEDA = tetramethylethylene diamine.

meridianins A–H (Figure 7.29) [109, 110]. In nature, they act as excellent deterrent agents that cause repellency, especially against the Antarctic omnivorous predator, the sea star *Odontaster validus* [111].

In laboratory tests, these brominated 3-(2-aminopyrimidine)-indoles and their N-substituted analogs displayed potent anticancer effects, antimalarial and antitubercular activity [108, 112]. For example, meridianins B and E were active against the human Hep-52 (hepatic adenocarcinoma) cell line, showing IC_{50} values of 1.7 and 1.1 μM, respectively. Additionally, meridianin F exhibited cytotoxicity toward the LMM3 (murine mammary adenocarcinoma cell line) and the Hep 52 cell line, with IC_{50} value of 1.4 and 1.8 μM, respectively. Moreover, meridianins C and G were found as antimalarial agents presenting IC_{50} values in the range of 4.4 to 14.4 μM against both chloroquine-sensitive (D6) as well as -resistant (W2) clones of *Plasmodium falciparum.*

Several synthetic methods of meridianins and their analogs have been reported. Among them, efficient protocols based on the cross-coupling reactions stand out as attractive routes for preparing these molecules. We mention the successful formation of meridianin A, C, D, F, and G through the Masuda borylation-Suzuki coupling sequence, which starts with N-Boc and N-tosyl-protected 3-iodoindoles **7.84** (PG = BOC or Ts), prepared in a one-pot process of selective iodation in the C-3 position of indole ring and subsequent N-protection (Figure 7.30) [113, 114].

The easily accessible, stable-to-storage but reactive in cross-coupling reactions, the 3-iodoindoles **7.84** reacted with pinacolborane (HBpin) in the presence of Pd(0)-catalyst and triethylamine as a base in 1,4-dioxane (Masuda borylation reaction) to provide boronate intermediates **7.85**. Subsequently, 4-chloropyrimidine-2-amine (**7.86**) and cesium carbonate were added to these intermediates, initiating the Suzuki coupling reaction in the same reaction vessel. Notably, the Boc protecting group was cleaved under the Suzuki conditions furnishing the natural product meridianin D in a yield of 25%. In comparison, the tosyl group in **7.87** was cleaved upon the reaction of the tosylate intermediates in the presence of a strong base in an additional step in the one-pot process to give access to meridianin C (48%), meridianin F (66%), and meridianin G (80%). The synthesis of meridianin A was realized similarly, starting from commercially available 4-methoxy-1*H*-indole, and with the demethylation of

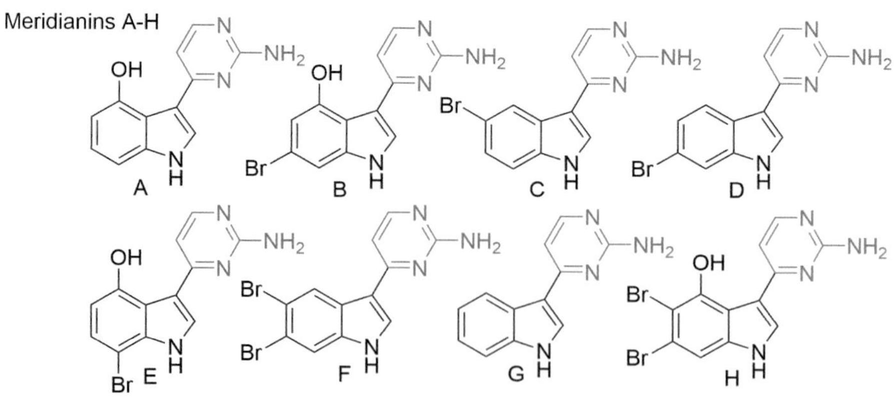

Figure 7.29. Structure of meridianins A–H based on the simple indole ring.

the respective indole **7.87** (R = 4-OMe, G = H) by Py·HCl in the final step (210ºC, 30 minutes), thus producing a yield of 85%. These concise total syntheses of meridianins illustrate the utility of the consecutive Pd-catalyzed Masuda borylation-Suzuki reaction process.

Molecules with 9*H*-pyrido[3,4-*b*]indole skeletons, known as β-carbolines, are widely distributed in nature, including various plants, insects, and marine creatures. Metabolites isolated from the colonial Carobean tunicates *Eudistoma* sp., called eudistomins, can be classified into two groups: tetracyclic alkaloids containing an oxathiazepinotetrahydro-β-carboline ring system (eudistomins C, E, K, and L) and alkaloids basing only on a simple β-carboline ring (eudistomins A, D, G, H, I, J, M, N, 0, P, and Q) [115, 116]. Some characteristic and interesting alkalois of these groups are depicted in Figure 7.31. Both groups are biosynthetically related since they are derivatives of tryptophan and other amino acids, i.e., cysteine (eudistomins C, E) and proline (eudistomins G, H) [117].

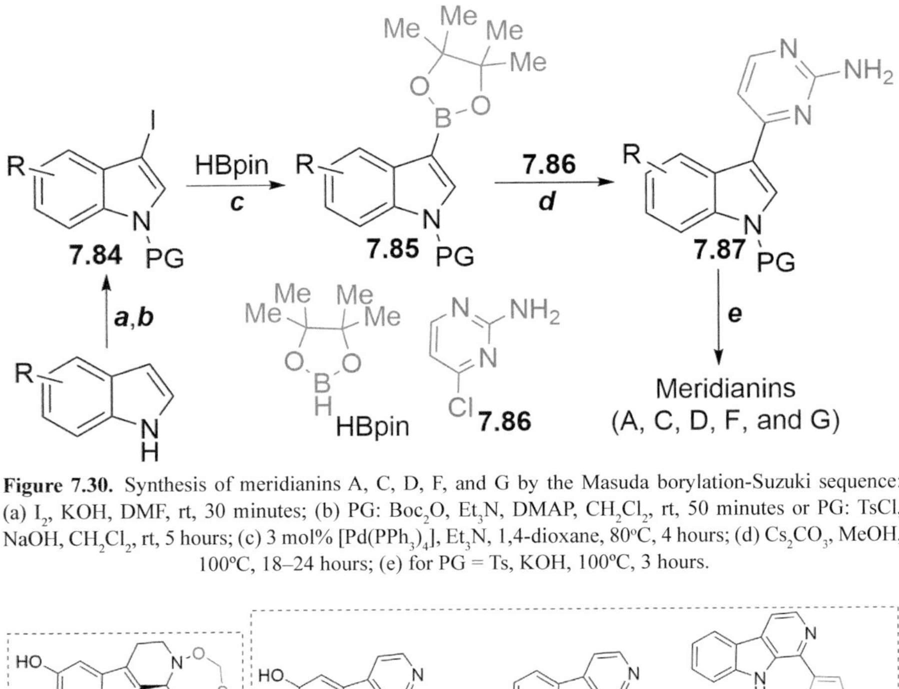

Figure 7.30. Synthesis of meridianins A, C, D, F, and G by the Masuda borylation-Suzuki sequence: (a) I_2, KOH, DMF, rt, 30 minutes; (b) PG: Boc_2O, Et_3N, DMAP, CH_2Cl_2, rt, 50 minutes or PG: TsCl, NaOH, CH_2Cl_2, rt, 5 hours; (c) 3 mol% $[Pd(PPh_3)_4]$, Et_3N, 1,4-dioxane, 80ºC, 4 hours; (d) Cs_2CO_3, MeOH, 100ºC, 18–24 hours; (e) for PG = Ts, KOH, 100ºC, 3 hours.

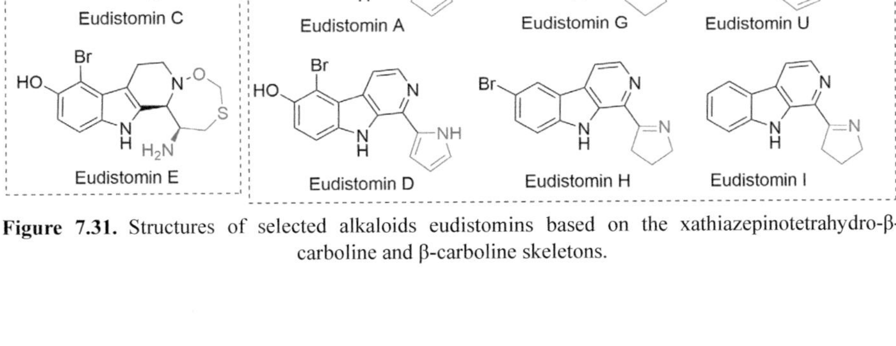

Figure 7.31. Structures of selected alkaloids eudistomins based on the xathiazepinotetrahydro-β-carboline and β-carboline skeletons.

These alkaloids are believed to play an essential role in the chemical defense of the ascidian *E. olivaceum*. In laboratory assays, Davis showed that these alkaloids maintain the epibiont-free surface of ascidian *E. olivaceum* [118]. He extracted eudistomins G and H from colonies of these tunicates and used them in laboratory settlement assays against bryozoan *Bugula neritina* larvae, barnacle *Balanus amphitrite*, and pinfish *Lagadon rhomboids*. The results demonstrated that at concentrations of approximately one-twentieth of those found in the living ascidian, these alkaloids deterred a broad range of invertebrate taxa from settling in laboratory assays. Thus, the pyrroliny1-bearing eudistomins G and H act as antifouling substances. However, these eudistomins and others were ineffective as fish antifeedants.

All isolated eudistomins were assayed against HSV-1 and showed antiviral activity but to a widely varying degree. The most active are those containing the oxathiazepine ring, especially eudistomidins C and E, which are also active *in vitro* against RNA viruses (Coxsackie A-21 virus and equine rhinovirus) and DNA viruses (HSV-1, HSV-2, and Vaccinia virus) in the range of 25–250 ng/disk [119]. The unique structure and the broad-spectrum antiviral activity of eudistomidin C have attracted attention as a lead compound for antiviral medicines. Still, several trials for its clinical development have failed due to the potent cytotoxicity of this marine alkaloid. Its studies on the action mode indicated that it targets 40S ribosome and inhibits protein translation [120]. Eudistomidins D, G, and H also showed antiviral properties. Eudistomidins, possessing a simple 1-substituted β-carboline skeleton, also demonstrated potent cytotoxicity in different model organisms. For example, C-1-(indol-3-yl)-β-carboline (eudistomidin U) is capable of inhibiting the growth of mouse leukemia P388 cells (38% inhibition) at 10 μM; it also showed moderate activity against human C19 (leukemia), CaOV3 (ovarian), and WM266-4 (melanoma) cells presenting IC_{50} values of 15.6 μm/mL, 24.9 μm/mL, and 27.5 μm/mL, respectively, and potent antibacterial activity against Gram-positive bacteria (*S. pyogenes*, *S. aureus*, and *M. smegmatis*) with IC_{50} = 3.4–6.4 μm/mL [121].

C-1-pyrrol-2-yl substituted 5-bromo-6-hydroxy-β-carboline (eudistomidin D) was found to be competitive antagonism to A_1 and A_{2A} adenosine receptors inducing Ca^{2+}-release from the sarcoplasmic reticulum and its synthetic analogs presented better potency as adenosine receptor ligands as compared with caffeine [122]. Interestingly, the biological properties of its close analog, eudistomidin A, have not yet been reported. Eudistomin H and its non-brominated analog (Eudistomin I) were potentially active against a panel of pathogenic bacteria [123].

The beneficial pharmacological efficacy of eudistomins has prompted other researchers to improve their total synthesis further. One of the most simple and short syntheses of eudistomins I and U is Kamal's protocols based on the Pictet-Spengler reaction of methyl L-tryptophanate (**7.88**) and the corresponding aldehydes **7.89–7.90** (Figure 7.32) [124]. They are commercially available starting materials; those condensations under mild reaction conditions (EtOH, reflux) provided the corresponding tetrahydro-β-carboline esters **7.91–7.92** as a diastereomeric mixture. Treating these esters with LiOH allowed to obtain the respective acids **7.93–7.94**, which were then exposed to iodobenzene diacetate, $PhI(OAc)_2$, to give the desired eudistomins I and U.

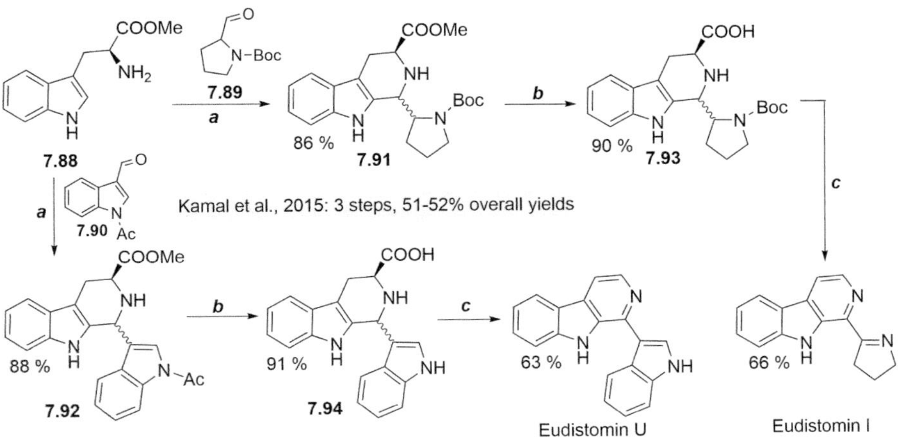

Figure 7.32. Synthesis of eudistomins I and U based on the Pictet-Spengler reaction and $PhI(OAc)_2$-mediated one-pot oxidative decarboxylation and aromatization processes: (a) EtOH, reflux, 12 hours; (b) 2M LiOH, THF, reflux, 12 hours; (c) 1. 2N HCl·dioxane, 0°C to rt, 2. $PhI(OAc)_2$, DMF, rt, 1 hour.

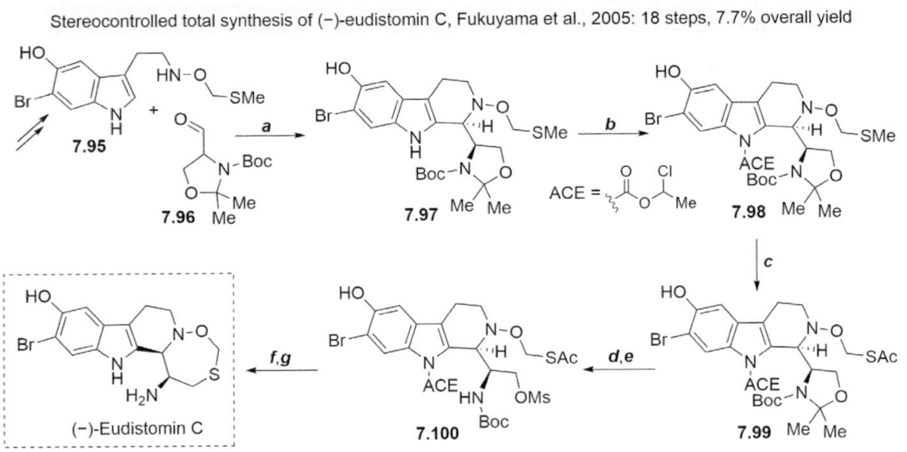

Figure 7.33. Formation of the oxathiazepine ring and the completion of the total synthesis of (–)-eudistomin C in the simplified scheme: (a) cat. Cl_2CHCO_2H, toluene, under an argon atmosphere, 0°C; (b) *n*-BuLi, THF, –78°C; ACE-Cl, 89% (two steps); (c) 1. SO_2Cl_2, CH_2Cl_2, –78°C to rt, 2. AcSH, *i*-Pr_2NEt, THF, 80°C, 95%; (d) aq AcOH, THF, 80°C, 61%; (e) MsCl, *i*-Pr_2NEt, CH_2Cl_2, 0°C, 98%; (f) K_2CO_3, MeOH, reflux, 65%; (g) BBr_3, CH_2Cl_2, –78°C to rt, 78%.

A similar strategy was used to synthesize (–)-eudistomin C. The prepared indole derivative **7.95** and Garner aldehyde **7.96** reacted in the presence of Brønsted acid to give the functionalized tetrahydro-β-carboline skeleton **7.98** with the protecting α-chloroethoxycarbonyl (ACE) group (Figure 7.33) [125]. The latter was transformed into thiol acetate **7.99**. Then, removal of the acetonide and mesylation of the resultant alcohol afforded cyclization precursor **7.100**, and its heating with K_2CO_3 in MeOH allowed methanolysis and the subsequent ring-closing alkylation to provide the desired oxathiazepine ring of eudistomin C.

The developed synthesis allowed the gram-scale preparation of eudistomin C and a variety of its derivatives for their biological studies. However, the selective synthesis of multi-substituted indoles like **7.95** is not necessarily an easy task.

After finishing the discussion on the toxic metabolites of tunicates, we must recognize a couple of metabolites from *Ecteinascidia* and *Trididemnum* tunicates that were converted into new drugs. They are examples of the category with the aspiring title "Drugs From the Sea". Ecteinascidin-743 (ET-743/trabectedin) (Figure 7.34) exemplifies an important landmark in developing marine-derived drugs. Almost 55 years after its discovery and 34 years after the publication of its structure, it became the first marine-derived anticancer drug to reach the market [126]. This abundant active component of toxic extract from the tunicate *Ecteinascidia turbinate* possesses three fused tetrahydroisoquinoline rings. In contrast, plitidepsin (aplidine) (Figure 7.34) from *Aplidium albicans* collected in temperate Mediterranean waters is a cyclodepsipeptide, i.e., cyclic peptide (in which there are one or more ester bonds in place of one or more of a peptide bond) with a pyruvyl amide (2-oxopropyl amide) at the secondary amine of the proline residue. Note that if this fragment is replaced by the lactyl group (2-hydroxypropyl amide), we obtain didemnin B, which was investigated extensively during several phase I and phase II clinical trials against different cancer cells. Still, unfortunately, trials were suspended due to its significant cardiac and neuromuscular toxicity. Interestingly, plitidepsin was found over ten years later than didemnin B discovered in the Caribbean tunicate *Trididemnum solidum*.

Yondelis® (trabectedin, ET-743) is a potent antitumor agent that interferes with DNA minor grooves interacting with guanine at N-2 in GC-rich sequences.

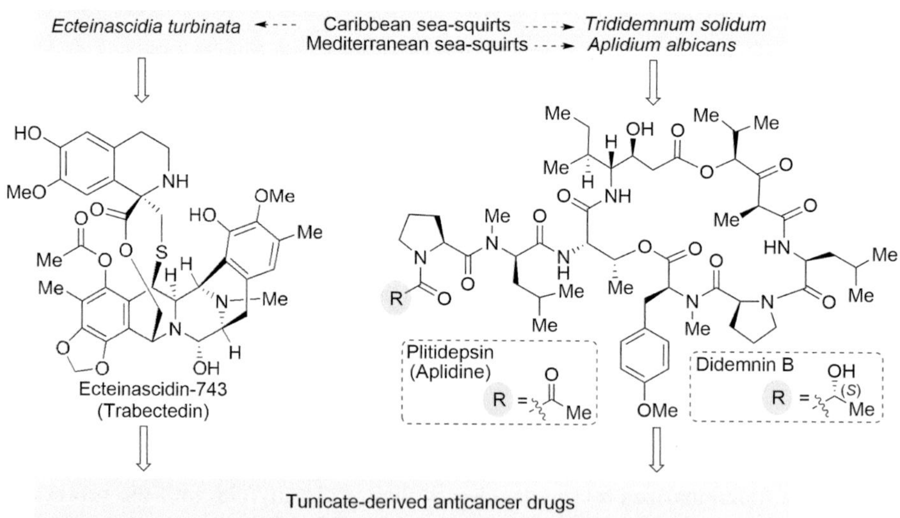

Figure 7.34. Structures of ecteinascidin-743 (ET-743/trabectedin; marketed under the trade name Yondelis® by PharmaMar/Johnson & Johnson/OrthoBiotech), derived from *Ecteinascidia turbinata* and aplidine (also known as dehydrodidemnin B or plitidepsin, marketed by PharmaMar, S.A. under the trade name aplidin®) from *Aplidium albicans*. The time between the natural molecule's isolation and the commercial product's approval was 17 years for trabectedin (Yondelis®) and 22 years for dihydrodidemin B (Aplidin®).

In contrast, conventional alkylating drugs generally bind to guanine at N-7 or O-6. Thus, trabectedin is an unusual DNA-alkylating agent. It can be used in multiple tumor-type treatments, including soft tissue sarcomas, melanomas, and ovarian cancer [127]. It is noteworthy that its antiproliferative activity is 10–1,000-fold higher than that of taxol *in vivo* [128].

Having a cyclic depsipeptide structure, aplidin® strongly inhibits protein synthesis and can be used in treating hematological malignancies, such as leukemia and lymphoma [127, 129]. Aplidin induces dose-dependent cell cycle arrest and an acute apoptotic process. Furthermore, according to the prediction of the *in silico* study, it may have the ability to bind the main protease (Mpro) of coronaviruses, which is an attractive drug target against SARS-CoV-2.

The supply of these two drugs from tunicates is limited by the difficulties of collecting sporadically dispersed source organisms and the need for feasible aquaculture conditions. Therefore, supply depends entirely upon the multistep total synthesis of these substances. Despite the reported five total syntheses and several formal synthesis routes of trabectedin being complex multistep strategies, they are not limited to delivering only a few milligrams of active substance but to providing proper quantities of active pharmaceutical ingredients [130, 131].

The Corey group published the first total synthesis of ET-743 in 1996, which contained more than 17 linear steps [132]. A detailed analysis of this synthesis was examined in the work [130]. Here, we pay attention to using the simple multicomponent reaction approach, such as the Strecker reaction, in the complex, lengthy, waste-generating construction of trabectedin. Thus, as shown in Figure 7.35, the Strecker reaction of polyfuncionalized tetrahydroisoquinoline **7.101** and aldehyde **7.102** in the presence of KCN in acetic acid smoothly afforded α-aminonitrile **7.103** in a 61% yield. This essential precursor was then converted into a complex polycyclic scaffold of trabectedin using 17 linear steps.

Like trabectedin production, the contemporary source of aplidin depends on total synthesis, which can involve more than 20 steps and requires many protecting groups and expensive reagents. This makes it expensive to use aplidin in research and clinical applications. Another approach combines microbial and chemical synthesis to produce aplidine (plitidepsin). A possible solution is producing plitidepsin through bioengineering of aplidine biosynthetic pathway or semisynthesis from didemnin B. Although the origin of widening in tunicates remains unknown, two marine

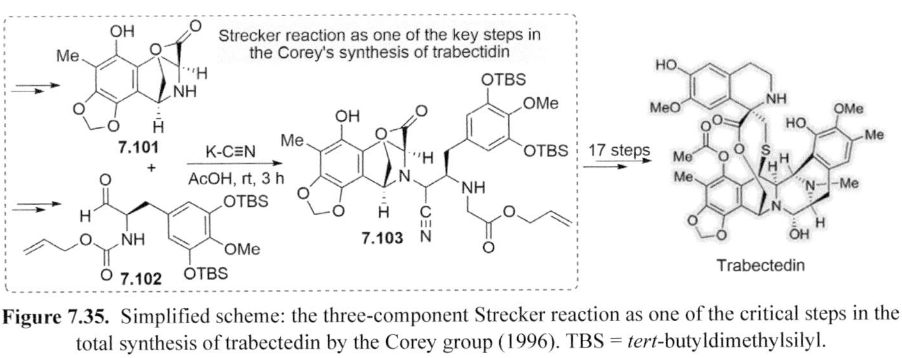

Figure 7.35. Simplified scheme: the three-component Strecker reaction as one of the critical steps in the total synthesis of trabectedin by the Corey group (1996). TBS = *tert*-butyldimethylsilyl.

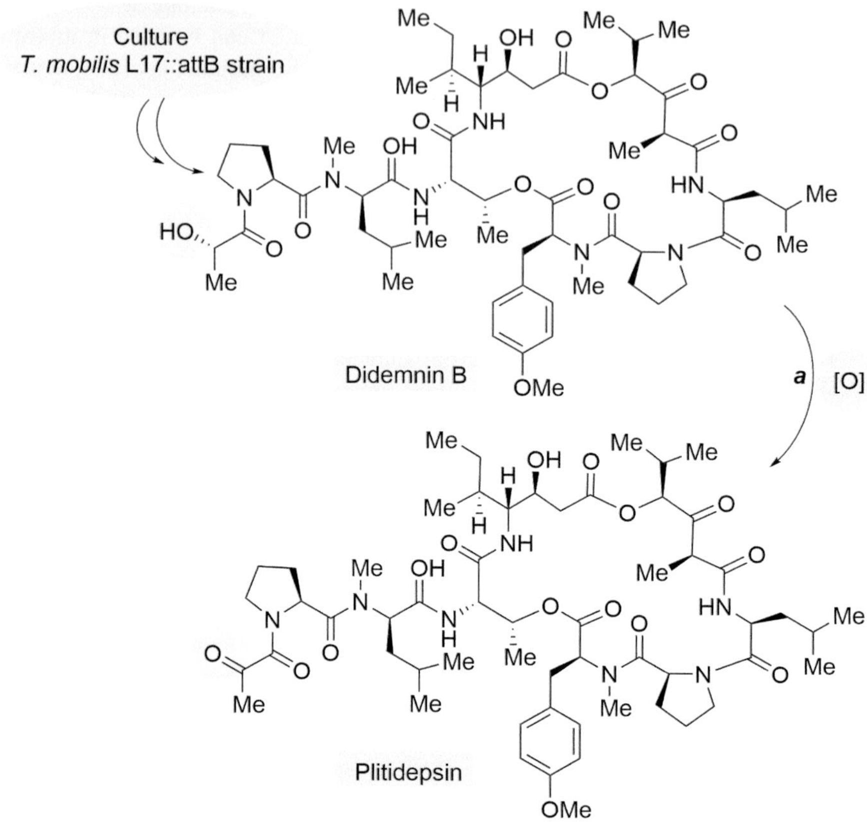

Figure 7.36. Combination of microbial and chemical synthesis of plitidepsin: (a) NaClO, TEMPO, toluene/EtOAc/H_2O, 0°C, 1 hour, 100%.

α-proteobacteria, *Tistrella mobilis*, and *T. bauzanensis*, capable of producing didemnin B, were identified in 2011–2012. Thus, by optimizing the cultivation conditions of the *T. mobilis* L17 strain and engineering it to obtain the *T. mobilis* L17::attB strain, Tang and coworkers managed to get didemnin B in an appreciable yield [133]. With enough didemnin B, they began designing chemical strategies to convert it to plitidepsin. This conversion consists of the change of the lactyl group by the pyruvyl one on the proline residue, i.e., a regioselective oxidation reaction of the hydroxy group at the N-proline ring (Figure 7.36). Screening of regioselective oxidants for this conversion allowed us to find that the known oxidant sodium hypochlorite (NaClO) and radical catalyst 2,2,6,6-tetramethylpiperidinyloxy (TEMPO) provided only the desired product plitidepsin with high yield after one hour at 0°C.

Therefore, this combination of microbial and chemical synthesis offers an efficient and economically feasible approach to producing plitidepsin, providing a promising basis for large-scale industrial production of this tunicate-derived anticancer drug.

7.1.5 Mollusks and Their Self-Defense from Predators and Pathogens

Marine mollusks (phylum Mollusca) have become the focus of many chemical studies aimed at isolating and identifying novel natural products [134]. Two classes, Gastropoda and Bivalvia, are the most studied. In such mollusks, defensive substances may be *de novo* biosynthesized by the organisms, obtained from their diet (biotransformed or not), or perhaps even produced by symbionts. In particular, many nudibranch gastropods derive defensive metabolites from the sponges they eat (Section 7.1.2). These metabolites have incredible structural diversity and display various biological activities [135–137].

In this context, some examples will be examined forward. Marine gastropod predator-prey interactions are interesting phenomena from which we learn about the sequestration, detoxification, and biotransformation of diet-derived secondary metabolites and their role as chemical mediators in mollusks [138]. It is known that herbivorous marine gastropods can devour chemically-defended prey, such as cyanobacteria and macroalgae. Likewise, carnivorous gastropods consume chemically-defended herbivores or chemically-defended filter-feeding sessile invertebrates, such as bryozoans, sponges, tunicates, and even other gastropods. Nudibranchs, sometimes called sea slugs, are an exciting group of soft-bodied carnivorous gastropod mollusks.

The Spanish dancer nudibranch, *Hexabranchus sanguineus*, a brightly colored shell-less sea slug, sequesters secondary metabolites, called halichondramide, dihydro- and ttrahydrohalichondramides (Figure 7.37), from its sponge prey, e.g., a Pacific *Halichondria* sponge. These metabolites act as defensive molecules against common predators of sponges, such as reef fish (*Thalassoma lunare*) and a reef hermit crab (*Dardanus megistos*) [139]. However, the nudibranch does not store halichondramide, the most abundant compound in *Halicondria* sp. The latter can convert halichondramide into tetrahydrohalichondramide, keep it and other components in the mantle and digestive glands, and use them as antifeeding agents. This biotransformation can take place in the digestive system and involves two reactions: double bond hydrogenation and carbonyl reduction, reactions that P450 enzymes may carry out.

Halichondramides, rare trisoxazole macrolides, were discovered as antifungal and toxic agents in 1988. Early studies on their cytotoxicity showed that tetrahydrohalichondramide was more active than halichondramide in the sea urchin model with IC_{50} values of 1.0 and 4.0 μg/mL, respectively. Dihydrohalichondramide showed potent antitumor activity against the L1210 cell line (IC_{50} = 0.03 μg/mL). Later, it was found that halichondramide strongly inhibits A549 (human lung), A549-PA (paclitaxel-resistant human lung), HCT116 (human colon), MDA-MB-231 (human breast), SK-HEP-1 (human liver), and SNU601(human stomach) solid cancer cells, with IC_{50} values of 0.045 μM, 0.036 μM, 0.19 μM, 0.03 μM, 0.12 μM, and 0.15 μM, respectively. However, its antiproliferative effect was inferior to (19*Z*)-halichondramide, a configurational isomer of the double bond (Figure 7.38) [140].

This isomer was isolated from the sponge *Chondrosia corticata*. Both isomers act as actin-depolymerizing agents, but the actin-depolymerizing activity by (19*Z*)-

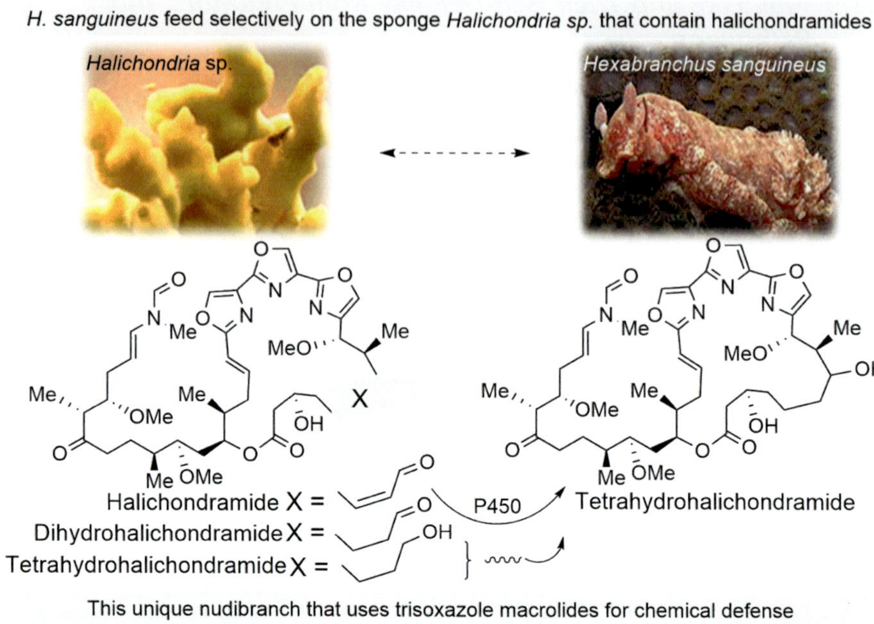

Figure 7.37. Sequestration of sponge secondary metabolites and biotransformation by nudibranchs. *H. sanguineus* seems to be the only nudibranch source of trisoxazole macrolides, but more than 30 similar trisoxazole macrolides have been isolated from sponges of the genera *Halichondria* and *Mycale*. The fascinating structure of such trisoxazole macrolides is apparently of hybrid polyketide-non-ribosomal peptide biosynthesis. The producer is most likely an as yet unidentified microorganism.

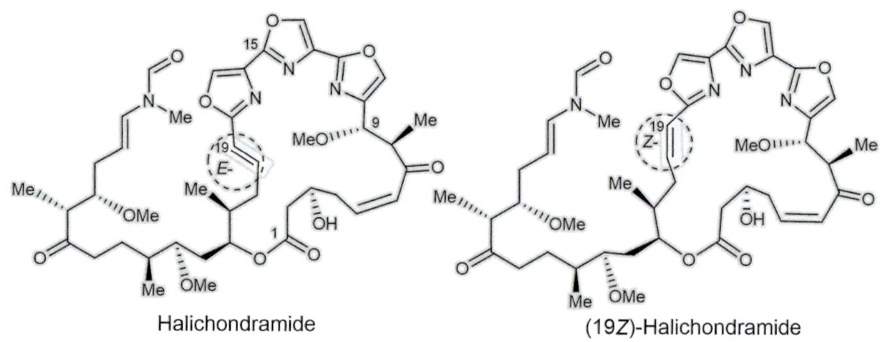

Figure 7.38. Structural characteristics of halichondramide and (19*Z*)-halichondramide as configurational isomers.

halichondramide was four times more potent than that of halichondramide. However, this tendency is the opposite in the antifungal tests, i.e., halichondramide resulted in more activity against fungal strains than (19*Z*)-halichondramide [141].

Their common name, sea snails, is a testament to their slow-moving nature. The toxic metabolites extracted from these creatures are alkaloidal compounds, first identified as lamellarins (Lam) A to D by Faulkner and his team in 1985 [142]. Notably, the specimens of *Lamellaria* sp. were collected by hand during a night dive

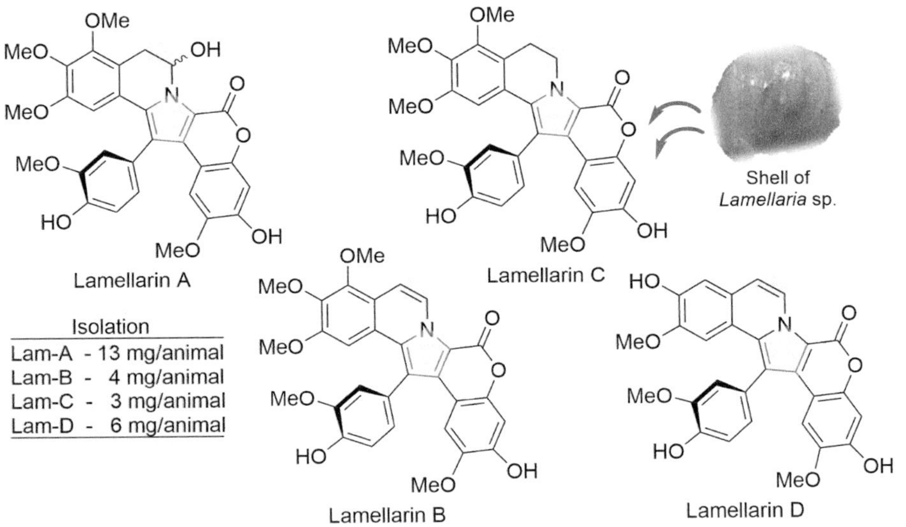

Figure 7.39. Structures of the first isolated lamellarins A-D in 1985 from *Lamellaria* sp. This was just the beginning of a remarkable journey. The lamellarin family, including these and others, was later isolated from an Australian colonial ascidian, *Didemnum* sp., and others. The family of fused and non-fused lamellarins has grown astonishingly, reaching 35 members in 2001, and now, more than 70 have been identified [143–145]. This rapid expansion underscores our research's vast potential and exciting future.

(–5 m) in the waters of Palau. These specimens were then stored in methanol (1 L) at 0–5°C for three years before the isolation and purification of fused lamellarins A-D. Each specimen contained approximately 4–13 mg of these alkaloids, which are pyrrolo[2,1-*a*]isoquinoline-based molecules (Figure 7.39). Lam-A and lam-C belong to the saturated isoquinoline lamellarin group (type I, without double bond), and lam-B and lam-D fit into the isoquinoline lamellarin group (type II, with olefin function). All lamellarins are generally considered DOPA-derived pyrrole alkaloids because the biogenetic origins of lamellarins are believed to be aromatic amino acids such as tyrosine and 3,4-dihydroxyphenylalanine (DOPA).

Note that lam-A exists in solution as a 1:1 mixture of two diastereomers due to restricted rotation around the C_{Ar}–C_{Pyr} single bond and a stereogenic center C_{Isoqu} attached by hydroxy OH group. Molecular mechanics calculations (MM2) suggest that the barrier of rotation was highly high (> 600 kcal/mol). However, all the isolated lamellarins were optically inactive. Thus, its main pentacyclic array is planar, and the aromatic ring attached to the pyrrole carbone is rotated 90° to the central plane [142].

Its isolation from the ascidian *Didemnum* sp. and mollusk *Lamellaria* sp. suggested that lam-D and its analogs are most likely sequestered from the ascidian in the diet of this mollusk.

The first biological test, the fertilized sea urchin egg assay, provided crucial insights into the properties of the lamellarins. At a 19 g/mL concentration, lam-D exhibited a significant 78% inhibition of cell division. In contrast, lam-C only produced a 15% inhibition, and lamellarins A and B were found to be inactive. This test began lam-D's recognition as a marine alkaloid with a broad spectrum of important biological and pharmacological properties, including anticancer activities.

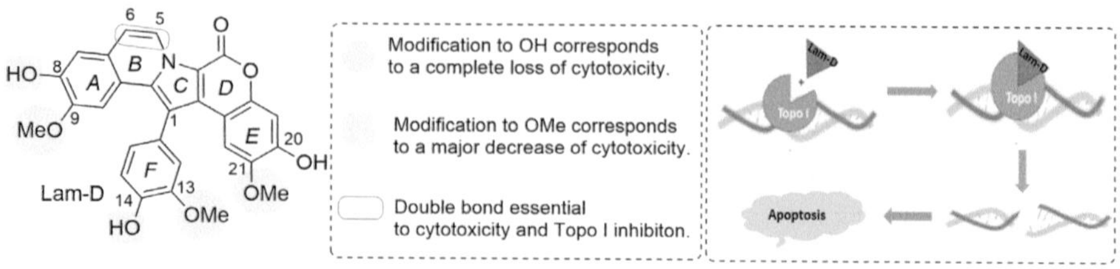

Figure 7.40. Structure of lamellarin D built on the 14-phenyl-6*H*-[1]benzopyrano[4′,3′:4,5] pyrrolo[2,1-*a*] isoquinoline system, its essential pharmacophoric elements for being a lead compound, and a diagram of antitumor mechanism for lam-D as a topoisomerase I (topo I) inhibitor. Lam-D binds to the topo I and DNA complex to form a ternary complex that stabilizes the structure, thereby preventing DNA re-ligation and causing DNA damage, eventually resulting in cell apoptosis.

Lamellarin D provides a robust pharmacological tool to study topoisomerase I inhibition and the perturbation of mitochondrial metabolism, acting directly on cancer cell mitochondria to induce apoptosis.

Attempts to correlate the structures of lamellarin alkaloids with their cytotoxic activities compared with that of the first isolated lam-D generated some following general observations: (a) Full pentacyclic structure of lamellarins is essential for their biological activity; (b) The increase in the number of methylations and/or methoxylations appears to cause a decrease in the antitumor activities; (c) The hydroxy groups at the C-8 and C-20 positions of lamellarin D are vital structural requirements for cytotoxic activity (Figure 7.40).

Owing to their interesting structural features and promising biological activities, many synthetic chemistry groups have developed efficient strategies for synthesizing lamellarin alkaloids and related analogs. In 1997, Ishibashi's group accomplished the first total synthesis of lam-D using an N-substituted benzylisoquinoline derivative **7.107** and benzylisoquinoline **7.106** as the vital ring construction that was possible due to the retrosynthetic analysis of lamellarin D (Figure 7.41A) [146]. Envisaging that the lamellarin ring system could be constructed by the N-ylide-mediated pyrrole ring formation and subsequent lactonization, they chose commercially available ethyl bromoacetate **7.104** and benzoate **7.105** as simple starting materials. Benzylisoquinoline **7.106**, needed for the syntheses of lam-D, was prepared according to a well-known procedure using 3-(benzyloxy)-4-methoxybenzaldehyde (**7.108**). On the other hand, methyl 2,4-dihydroxybenzoate (**7.109**) was used for the benzoate **7.105** preparation (Figure 7.41B). Then, the condensation of the prepared benzylisoquinoline **7.106** with benzoate **7.105** via metalation in the presence of LDA led to the formation of **7.110**, which was reacted with bromoacetate **7.104** to induce the quaternization process, yielding **7.107**. Its treatment with excess triethylamine in MeOH at reflux, followed by hydrogenolysis of its benzyl groups, afforded lamellarin D, formally, in five steps in 18% overall yield.

However, this approach would not be considered a successful attempt because some steps caused severe problems, such as elevated temperature and prolonged reaction. Of course, various approaches exist to synthesize it more efficiently today [147–150]. The designed routes to the lamellarins generally use the functionalized pyrroles as starting materials. For example, Chandrasekhar's design and synthesis of lam-D [149] can be commented on. According to the following retrosynthesis

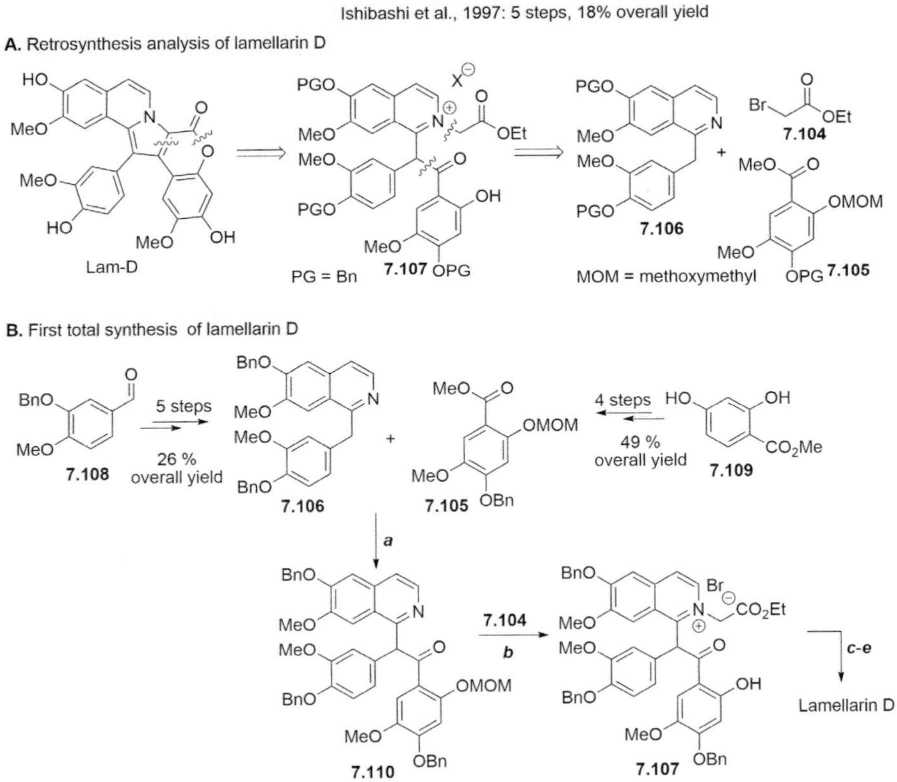

Figure 7.41. Design and synthesis of lamellarin D by Ishibashi's group: (a) LDA, THF, –78°C to –15°C, 1 hour, rt, 3 hours, 63%; (b) 50°C; (c) cat. HCl, MeOH; (d) Et_3N, CH_2Cl_2, 34% (overall, three steps); (e) H_2, $Pd(OH)_2/C$ (Parlman catalyst), EtOAc, 82%.

analysis, bromopyrrole derivative **7.114** and appropriate boronic acid **7.115** can be identified, and then the first could lead to 2,3-diaryl pyrrole-5-carboxylate **7.113**, key intermediate. Original disconnection of the latter offers starting materials, diarylalkyne **7.112** and enamide **7.111** (Figure 7.42A).

The synthesis starts with the [3+2]-annulation reaction between starting reagents in the presence of 5 mol% of $[Ru(p\text{-cymene})Cl_2]_2$ as a catalyst and $Cu(OAc)_2{\cdot}H_2O$ as an oxidant in PEG-400, furnishing 2,3-diaryl pyrrole-5-carboxylate **7.113** in 90% yield (Figure 7.42B). Bromopyrrole **7.114** was prepared with an 82% yield using a bromination reaction with NBS in DMF. Key Suzuki coupling reaction of **7.114** and **7.115** give tetrasubstituted pyrrole-5-carboxylate **7.116**, which was treated with *p*-TsOH in MeOH to construct pyrrolocoumarone skeleton **7.117** via the one-pot two-step lactonization by MOM deprotection process. Then, to complete the pentacyclic system **7.118** (i.e., lamellarin D trimethyl ether) alkaloid, the N-alkylation of **7.117** with bromoacetaldehyde diethyl acetal, followed by intramolecular TfOH-mediated cyclization, was carried out. Finally, the deprotection of the methoxy groups with BCl_3 in CH_2Cl_2 produced lamellarin D in 79% yield [149]. Thus, the desired alkaloid was prepared in seven steps with a 29% overall yield. Such an efficient and concise modular synthetic strategy is suitable for synthesizing various lamellarin analogs.

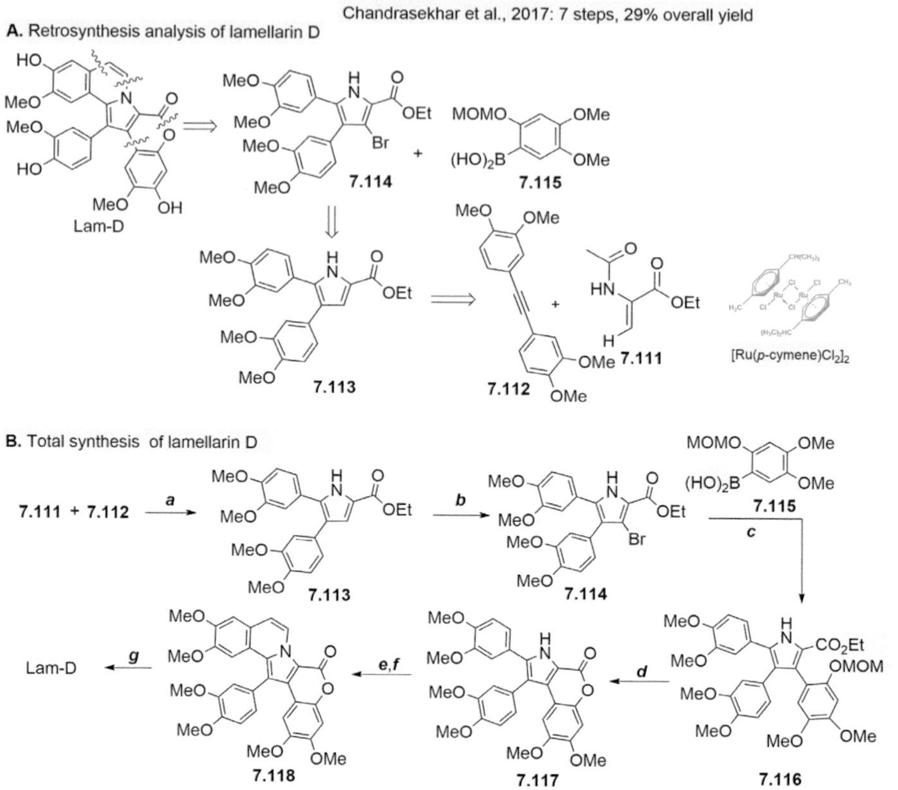

Figure 7.42. Design and synthesis of lamellarin D by Chandrasekhar's group: (a) 5 mol% [Ru(*p*-cymene)$Cl_2]_2$, $Cu(OAc)_2{\cdot}H_2O$, PEG-400, 100°C, 24 hours, 90%; (b) NBS, DMF, 0°C to rt, 12 hours, 82%; (c) $Pd(dba)_2$, dppf, DME·H_2O, reflux, 18 hours, 85%; (d) *p*-TsOH, MeOH, reflux, 16 hours, 95%; (e) Cs_2CO_3, DMF, 80°C, 38 hours, 80%; (f) TfOH, CH_2Cl_2, 0°C, 24 hours, 94%; (g) BCl_3, CH_2Cl_2, –78°C to rt, 4 hours, 79%. dba = dibenzalacetone, dppf = 1,1′-bis(diphenylphosphino) ferrocene, NBS = N-bromosuccinimide, DME = dimethoxyethane.

Regarding the Bivalvia class, the butter clam *Saxidomus giganteus*, a bivalve mollusk, should be mentioned. Clams are well protected by a rigid shell comprising two pieces, and additional protection consists of "burying themselves" in the sand or mud. Their common predators are sting rays, fish, sea otters, and some crabs. However, these mollusks sequester the paralyzing poison from the planktonic source. Among dangerous venom mixtures, saxitoxin (STX) and neo saxitoxin (N1-hydroxy saxitoxin and neoSTX) isolated in 1957 are best studied. The peculiar structure of these toxins, which was established by X-ray only in 1975, comprises a tricyclic heterosystem containing a saturated purine ring condensed with the pyrrolic cycle (Figure 7.43) [151].

Neurotoxins with analogous structures have been isolated from the soft clams *Mya arenaria* collected by "red tide". These toxins are called gonyautoxin II (GTX-II) and gonyautoxin III (GTX-III) and are the 11α- and 11β-hydroxy derivatives of STX. Likewise, 11α- and 11β-hydroxy derivatives of neoSTX are known as gonyautoxin I (GTX-I) and gonyautoxin IV (GTX-IV). These are PSP toxins that target voltage-

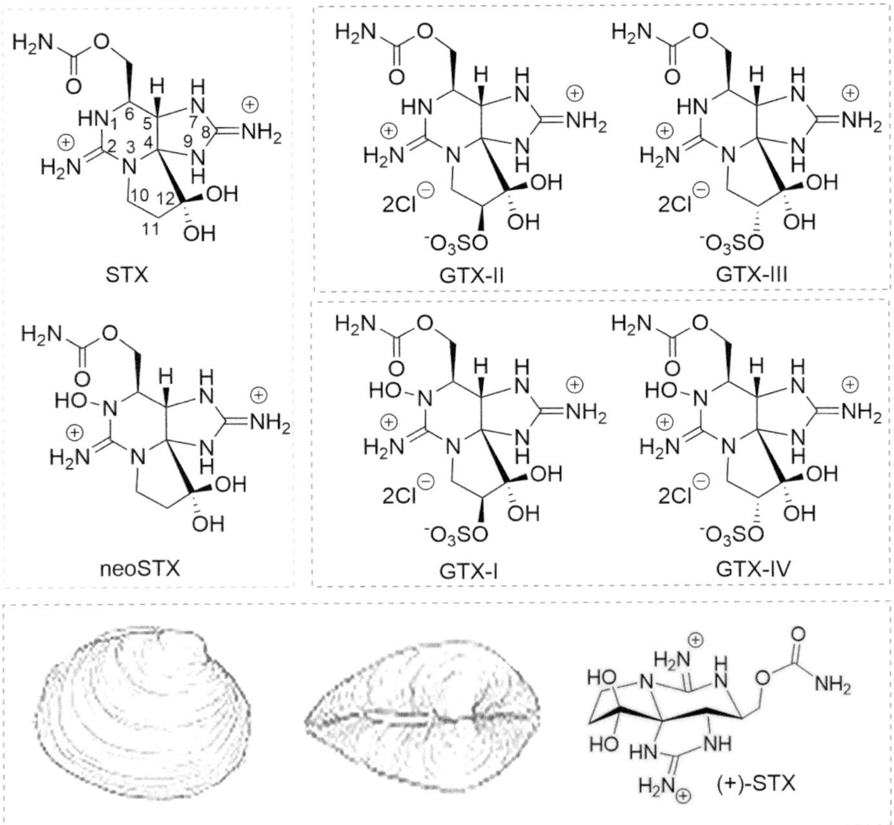

Figure 7.43. Chemical structure of saxitoxin based on a 3,4-propinoperhydropurine skeleton with two guanidinium units and its representative natural N1-hydroxy and/or sulfate congeners. Since STX's discovery in 1957, 57 analogs have been identified in cyanobacteria and dinoflagellate organisms. Saxitoxin is optically active, presenting a specific rotation $[\alpha]_D^{25}$ of + 130.0°. An image of the *Saxidomus giganteus* shells and conformation of (+)-STX.

gated sodium channels (Section 7.1.1). Their toxicity equivalency factors (TEF) are known today as being the most potent neoSTX, STX, and GTX-I. We remember that neurotoxins that we have seen before, such as saxitoxins, interrupt the communication between neurons and muscle cells since they inhibit the production of acetylcholine, acting as outer pore blockers of voltage-gated sodium channels. The neuron must send a nerve impulse to the axon to generate the neurotransmitter. As mentioned above, this activity depends on the flow of sodium and potassium present throughout the network of neurons. When the impulse reaches an axon terminal, acetylcholine is released. These toxins block the internal flow of sodium ions through the membrane channels. In this way, they cut off the impulse and inhibit the secretion of the neurotransmitter. Rodent studies disclosed that saxitoxin and neosaxitoxin LD_{50} were intravenously 3–9 µg/kg and 6.0 µg/kg, respectively.

The study of saxitoxin's structure has not been without its challenges. Saxitoxin dihydrochloride, a white, hygroscopic material with a molecular formula of $C_{10}H_{19}N_7O_4Cl_2$, revealed its diprotic nature through titration studies, with measured aqueous pKa values of 8.24 and 11.60. Its polar and amorphous nature initially impeded crystallization efforts and attempts at structure elucidation. However, through perseverance and innovative techniques, X-ray crystal structures of STX were eventually obtained, providing a clear understanding of its basic structure. This breakthrough was further confirmed after Kishi's successful total synthesis in 1977 [152], marking a significant milestone in studying saxitoxin and its derivatives.

As these clams sequester STX from the planktonic source, biosynthesis of these molecules occurs in cyanobacteria, *Cylindrospermopsis raciborskii* and *Anabaena circinalis*, or dinoflagellate *Alexandrium tamarense,* and the blue-green alga *Aphanizomenon flos-aquae*. It could implicate arginine, acetyl-CoA, S-adenosyl methionine, and carbamoyl phosphate as STX biosynthetic precursors. Thus, STX biosynthesis is somewhat unusual since STX is produced by organisms from two different kingdoms of life, cyanobacteria and dinoflagellate [152, 153].

Regarding the total synthesis of STX and its congers, like their biosynthesis, chemical synthesis remains a fascinating, contemporary challenge. The initial syntheses of racemic STX realized in 1977 (Kishi's group) and 1984 (Jacobi's group) were based on the successive construction of the STX ring skeleton with the introduction of basic polar guanidine functionalities in the late synthesis stages to avoid as much as possible technical difficulties in handling them [154, 155]. We will examine the work of Jacobi and coworkers [155], who were remarkable in design, elegance, and conciseness. Their design was based on the [3+2]- cycloaddition reaction of azomethine ylide **7.121** and reductive N-N cleavage of hydrazide **7.120** found through a disconnection of alcohol **7.119** (Figure 7.44A).

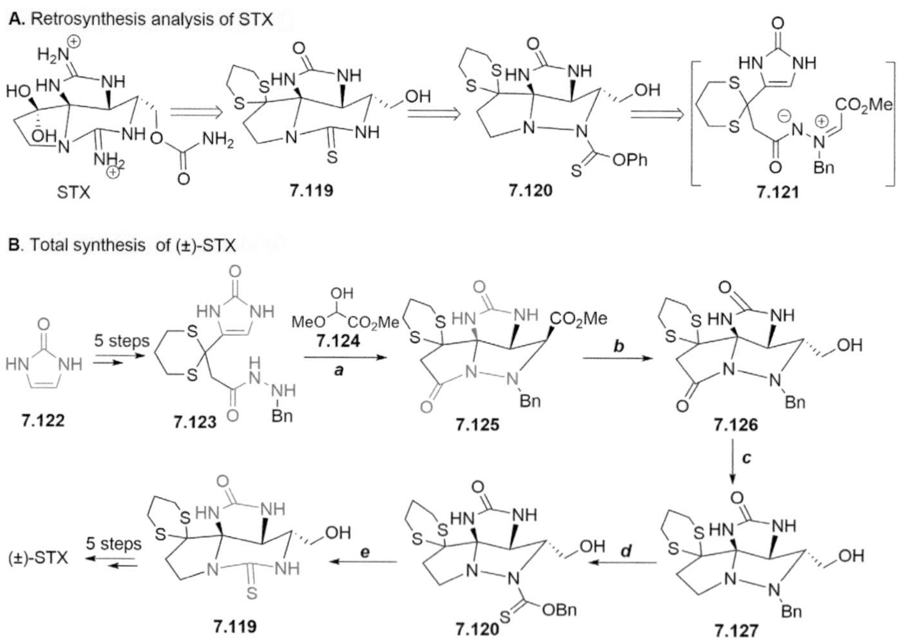

Figure 7.44. Design and simplified synthesis of racemic STX by Jacobi's group: (a) $BF_3{\cdot}OEt_2$, 65–75%; (b) MeONa, $NaBH_4$, 70%; (c) $BH_3{\cdot}SMe_2$, 98%; (d) 1. Pd/C, H_2, HCO_2H, 80%; (e) Na/NH_3, 75%.

The plan involved selecting commercially available imidazolin-2-one (**7.122**), which was then elaborated to the desired hydrazide **7.123** in 5 steps (Figure 7.44B). The obtained hydrazide and methyl-glyoxylate hemiacetal **7.124** were condensed under acid catalysis, generating *in situ* reactive azomethine imine **7.121**. This imine underwent an intramolecular [3+2]-cycloaddition reaction across the tethered imidazolinone to provide the cycloadduct **7.125**.

The preparation of the key precursor **7.119** was achieved in four steps. The crucial reductive N-N cleavage process **7.120** was performed with Na in NH_3. Previously, methyl carboxylate derivative **7.125** was converted into a keto-alcohol **7.126**, and its selective BH_3-mediated hydrazide reduction provided N-benzyl hydrazine **7.127**. The last five steps are similar to the STX synthesis by Kishi; however, in this so facile approach, the desired racemic alkaloid was produced in 0.5–1.0 g quantities without recourse to chromatographic purifications throughout the entire sequence. The first asymmetric synthesis of (+)-STX, performed by Du Bois and coworkers from 2006 to 2008 [152], was a significant milestone, nearly three decades after Kishi described the fundamental synthesis.

As a guanidiun toxin, saxitoxin is always compared with tetrodotoxin (TTX), another selective blocking agent of voltage-gated sodium channels. TTX, contained in pufferfish, has become an extremely popular chemical tool in physiological and pharmacological laboratories since the discovery of its channel-blocking action in the early 1960s. However, as pufferfish are vertebrate animals, we will discuss it in the next section.

7.2 Venomous and Poisonous Fish and Their Toxic Molecules

More than 50 classes of poisonous fish swim in the world's waters; however, their poisons are poorly studied from a chemical point of view. Most studies have explored whole or partially fractioned venom, revealing broad pharmacology, which includes cardiovascular, neuromuscular, cytotoxic, inflammatory, and nociceptive activities. Most venom is large proteinaceous toxins, such as stonustoxin, cardioleputin or verotoxin, enzymes, and peptides [156–159]. Thus, these are the primary metabolites that fish produce. Fish with significant venom include stingrays, lionfish, scorpion fish, stonefish, weever fish, and catfish. Their defensive venoms can kill various organisms. Regarding secondary metabolites, many small bioactive fish venom components remain to be discovered and investigated.

However, fish can accumulate small toxic metabolites of bacteria or dinoflagellates. So-called bioaccumulation and bioconcentration mechanisms explain the presence of paralytic shellfish poisons like saxitoxins or ciguatoxins and maitotoxins in marine invertebrates, especially fish. For example, we have already seen that a toxic compound, ciguatoxin, has been isolated from the venom of the moray eel *Gymnothorax javanius* (Section 7.1.1).

Nevertheless, the most famous case is tetrodotoxin (TTX), isolated from pufferfish, also known as blowfish and globefish (tetraodon fish), of the genera *Arothron* and *Takifugu* (*Arothron tetraodon*, *Takifugu pardalis*, etc.), one of the deadliest natural toxins. To humans, TTX is deadly, up to 1,200 times more poisonous than cyanide. One pufferfish has enough toxin to kill 30 adult humans; no known antidote exists [160]. Many pufferfish species have high concentrations of TTX in the ovaries, followed by the liver and/or skin; the remaining organs contain small amounts of TTX [161]. Curiously, the name "tetrodotoxin" (*tetrodongift* in German) was assigned in 1910 by Y. Tahara, who established the purification of the toxin from pufferfish ovaries (ca. 0.2–4% pure). Only in 1950 did another Japanese researcher isolate the poisonous principle in a crystalline form from the liver and ovaries of puffer fish *Sphoeroides rubripes*. This crystalline form resulted in a tetrodotoxin molecule.

It is interesting to note that, like neotropical frogs, cultured pufferfish do not contain toxins [162]. Still, TTX was detected in numerous wild marine organisms, such as sea snails, slugs, bivalve mollusks, and starfish. Accordingly, the pufferfish's diet consists mainly of invertebrates and algae. Large specimens will crack open and eat clams, mussels, and shellfish with their hard beaks. This deadly toxin is believed to synthesize a bacterial or dinoflagellate species, which puffers eat [163]. Thus, pufferfish are the most notorious TTX source, responsible for numerous human intoxication cases, especially in countries such as Japan, where these fish have been historically consumed as a delicacy known as fugu.

Chemically, TTX contains a polyfunctionalized dioxaadamantane skeleton. Its cyclic guanidine comprises a hemiaminal moiety with nine contiguous stereogenic centers. TTX is a crystalline, weakly basic, colorless substance with a molecular formula of $C_{11}H_{17}O_8N_3$ (MW = 319.27 g/mol) (Figure 7.45). Its functionalized bridge ring structure was elucidated in 1964, and six years later, its absolute stereochemistry was determined. Notably, in acidic conditions, TTX exits as an inseparable mixture of orthoester and lactone; as a result, the purification of the tetrodotoxin in the final steps is a challenging concern. Moreover, its equilibrium mixture is completed with an anhydride form called 4,9-anhydroTTX or 11-norTTX-6-ol [164].

Like STX, tetrodotoxin strongly blocks selectively axonal transmission by lowering the conductance of sodium ions. By blocking sodium ion channels, TTX

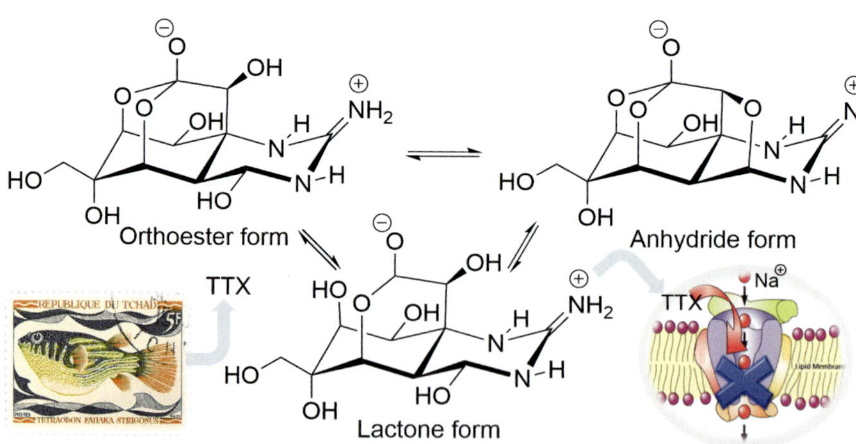

Figure 7.45. Chemical structure of tetrodotoxin (TTX) and its primary source (globefish or fugu in a stamped image).The equilibrium mixture (orthoester, anhydride, and lactone forms) of TTX and its mode action as a potent voltage-gated sodium channel blocker. As a result, TTX displays a strong central analgesic effect.

causes gastrointestinal, neurologic, and cardiac failures and, in numerous cases, death by respiratory paralysis in patients following oral exposure to this neurotoxin at lethal doses of 1–2 mg. The LD_{50} values obtained in mice were 12.5 μg/kg, 10.7 μg/kg, and 332.0 μg/kg for the respective subcutaneous, intraperitoneal, and oral administrations. Its biothreat status is documented by its register as a "Select Agent" by the US Department of Health and Human Services. However, when administered at much lower levels, TTX exhibits therapeutic properties, so far mainly targeted to treating cancer-related, neuropathic, and/or visceral pain [164]. TTX is, in fact, in Phase III trials as an agent (Tectin®) against inadequately controlled pain related to cancer. It is noteworthy that it exhibits potent properties of analgesia without an addiction effect. Tectin® contains TTX at a concentration of 15 μg/mL (47 μM) in 2 mL ampoules for subcutaneous injection [160, 165–167].

Commercial TTX purified from pufferfish is typically sold in 1 mg of pure toxin or 1 mg of the toxin mixed with 5 mg of citrate buffer. The current source of tetrodotoxin is extraction from wild organisms. Moreover, it is not easy to extract tetrodotoxin. For extraction, many wild pufferfish are consumed, which is detrimental to ecological environment protection, and the quality of extracted products is unstable.

On the other hand, the chemical syntheses of TTX still need to have the value of large-scale tetrodotoxin synthesis and preparation as pharmaceutical raw materials. Some relevant works reveal an academic exploration significance that is more remarkable than their practical application value. However, methods for synthesizing TTX should be briefly mentioned. Kihi's team reported the first total synthesis of racemic TTX in 1972. Later, its first asymmetric syntheses by Isobe and Du Bois became known in the early 2000s, and to date, numerous efforts have been made toward its total synthesis, reviewed in detail [168, 169]. Synthetic studies of the TTX skeleton continue to this day because a concise and practical (–)-TTX assembly route is vital to medicinal chemistry research that could identify a congener of the natural product that could be useful in human medicine.

For example, Yokoshima's, Trauner's, and Qi's groups developed new synthetic methods to enable the rapid assembly of natural (–)-tetrodotoxin using naturally occurring materials such as D-glucose (**7.128**), D-isoascorbic acid (**7.129**), *myo*-inositol (**7.130**), dihydro-4*H*-cyclopenta[*d*][1,3]dioxol-4-one derivative **7.131**, available from α-methyl-D-mannoside, or furfuryl alcohol (**7.132**) easily obtainable from furfural. In the first racemic TTX synthesis, aromatic dihydroxy acetophenone compound **7.133** was used as the starting material (Figure 7.46) [170–172]. In 2017, Fukuyama and coworkers also used aromatic starting materials (5-TMS-cyclopentadiene and *p*-benzoquinone) in their total synthesis of (–)-tetrodotoxin and 11-norTTX-6(*R*)-ol that contains 32 lineal steps [173].

Analyzing the scheme presented in Figure 7.46, we can see that the significant synthetic progress over the years is noticeable. From 72 steps with 1.22% overall efficiency (total synthesis by Isobe et al. 2003) [168] to 22 steps and 11% overall efficiency (total synthesis by Trauner et al. 2022) [171] is an enormous distance. Properly using oxygenated biomass materials is a good and vital trend, so D-glucose or furfuryl alcohol are excellent natural raw materials suitable for manufacturing drugs based on the TTX structure. However, one wants to shorten these linear

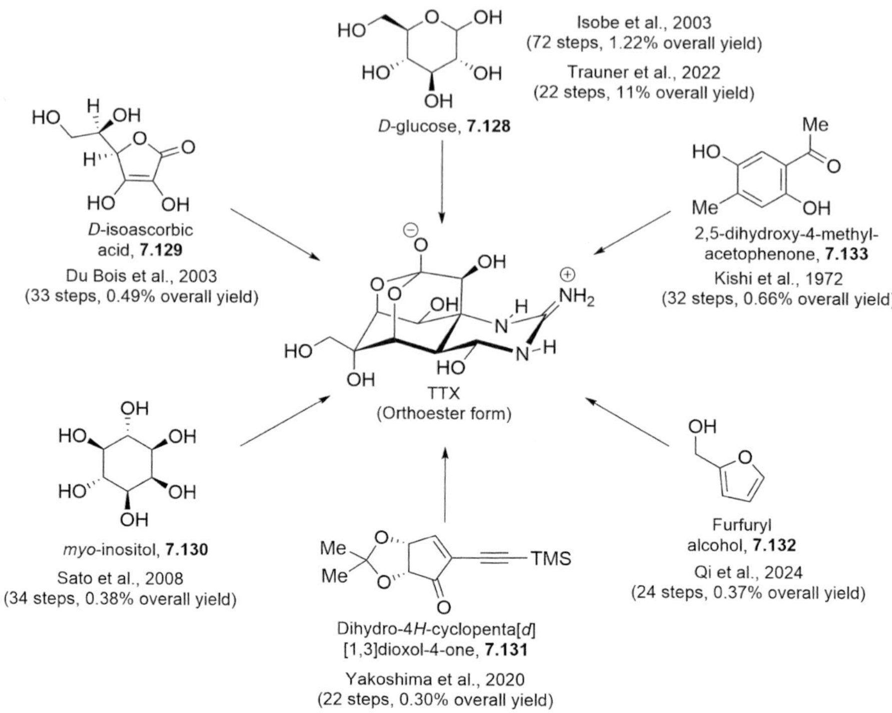

Figure 7.46. Developed enantioselective total syntheses of natural (–)-tetrodotoxin (2003–2024) after the first total synthesis of racemic TTX (Kishi et al. 1972). TMS = trimethylsilyl.

multistep sequence routes or develop them into convergent routes and increase the efficiency of the global process a little more.

After a comprehensive analysis of the complete synthesis of TTX, it is interesting to note that pufferfish inflate like a ball when they sense danger, as indicated by their name, to intimidate an enemy (physical defense). This action presses their secretory glands on the body surface, thus releasing their content, but inflation is unnecessary to relieve the toxin (chemical defense). Although TTX is toxic to other fish when intramuscularly injected, it must directly act on their nervous system and thus is not an ichthyotoxin in the usual sense (which are compounds either toxic to fish or are toxins produced by fish). TTX lacks surfactant and hemolytic properties, while an ichthyotoxin is usually a repellent defense substance with different mechanisms of the repulsive action, i.e., antifeedant activity.

The latter includes pahutoxin (ostracitoxin), produced by boxfishes, e.g., a Hawaiian boxfish, *Ostracion lentiginosus*, of the trunkfish family Ostraciidae. It is a neurotoxin present in the mucous secretions of these fishes, closely related to the pufferfishes [174, 175]. Intraperitoneal injections of this toxin into white mice rapidly initiated ataxia, labored breathing, coma, and death. The LD_{50} of crude toxin is relatively high (0.2 mg/g of mouse). Interestingly, in 1964, Doland Thomson first described the isolation and preliminary biological properties of this toxin [174], naming it ostracitoxin, while Paul Scheuer and his student, Boylan, too isolated and crystallized the same toxin but did not publish their research at that time. They named

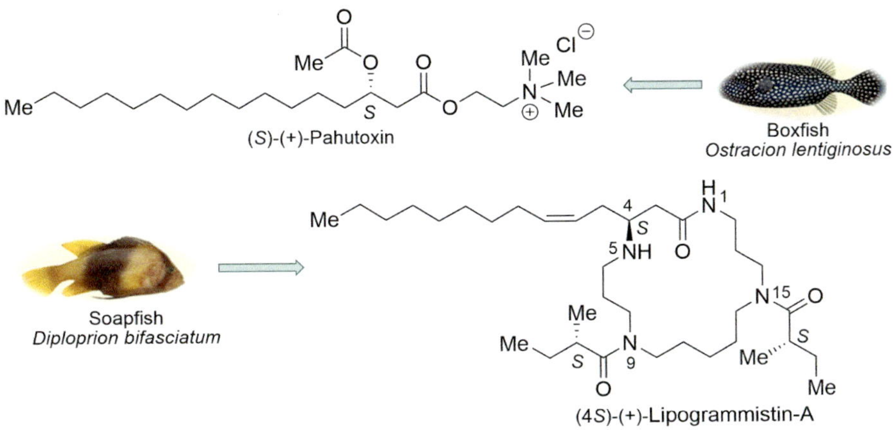

Figure 7.47. Representative lipophilic ichthyocrinotoxins with different structures from skin secretions of the trunkfishes (in an image): pahutoxin, cationic surfactant with the lipophilic alkyl chain and lipogrammistin-A based on an 18-membered lactam ring with the alkenyl chain of *Z*-configuration.

pahutoxin since "pahu" is the Hawaiian word for "box" [175]. Nevertheless, Scheuer and his Boylan reported this toxin's structure, activity, and synthesis three years later, fixing the name proposed as pahutoxin [176]. Note that 60 mg of pahutoxin could be obtained from one adult boxfish.

Structurally, pahutoxin is a simple compound of the cationic surfactant class (Figure 7.47). As can be seen from Figure 7.47, it is a choline ester of 3-acetoxypalmitic acid (i.e., 3-acetoxyhexadecanoic acid). Crystalline pahutoxin melts at 74–75°C, displaying a specific rotation $[\alpha]_D^{25}$ of + 3.05° [176]. However, the absolute configuration of dextrorotatory pahutoxin was determined only in 1989 when Tai and coworkers synthesized optically pure pahutoxin and its antipode from (*S*)- and (*R*)-3-hydroxyhexadecanoic acids, showing that dextrorotatory natural pahutoxin has (*S*)-configuration [177]. The presence of pahutoxin and homopahutoxin in the mucous secretion of the Japanese boxfish *Ostration immaculatus* was also reported [178].

Surfactants are molecules that comprise hydrophobic and hydrophilic domains within the same molecule. Pahutoxin is a surfactant, but there is evidence that its strong ichthyotoxic effect is receptor-mediated due to the acetylcholine fragment. Noteworthy, the chemical structure of pahutoxin was the first to be determined among the repellents of marine animals. This metabolite possesses ichthyotoxic and hemolytic properties, unlike TTX, but is also a nonpeptide toxin. Interestingly, boxfish are tolerant to its pahutoxin, which has high fish lethality (ichthyotoxicity) with an LC_{50} 2.87 μM in laboratory tests using a commercially bred marine bony fish, *Sparus aurata*, as a model for a potential predator [179].

Soapfishes are so named because they discharge a surfactant secretion that makes seawater foam in a small container where they are kept. Mucous secretion obtained from their skin caused ciguateric symptoms in cats when administered orally. The main toxic components are hydrophilic peptide toxins called grammistins, with molecular weights around four kDa. However, lipophilic toxins were also detected in the mucus of these fishes. A lipophilic toxin, lipogrammistin-A, isolated in 1993 from the mucous secretion of the soapfish *Diploprion bifasciatum*, is a good example (Figure 7.47) [180]. It was the first isolated macrocyclic polyamine from a marine animal. Lipogrammistin-A is an 18-membered polyamine lactam decored with the Z-undec-2-en-1-yl fragment, which exists in solution as a conformational mixture of four different states. Notablly, 277 mg of lipogrammistin-A as a pale yellow viscous oil was obtained from the acetone extract of the mucus (52.1 g) collected from nine specimens of *D. bifasciutum*. Lipogrammistin-A is also a dextrorotatory molecule with a specific rotation $[\alpha]_D^{25}$ of + 18.5° [180]. The absolute stereochemistry of this macrocyclic polyamine lactam (4*S*,9*S*,15*S*) was established by chemical degradation and total synthesis five years after its isolation [181]. In this total synthesis, Tachibana's group selected tetradeca-2,5-diyn-1-ol, a little accessible reagent, as a primary material. The preparation of the desired product thus resulted in somewhat slow and ineffective synthesis.

Fukuyama and coworkers developed a more suitable protocol based on the macrocyclization process using alkylation with 2-nitrobenzenesulfonamides as an activating/protecting group (Ns-strategy) [182, 183]. Their synthesis started with N-protected 3-aminopentanedioic acid (**7.134**), which was transformed into aldehyde **7.135** in three steps involving an exchange of the Cbz-protecting group with a Boc group and a reduction reaction (Figure 7.48). The latter, without purification, was subjected to the Wittig reaction with (1-nonyl) triphenylphosphonium iodide ($C_{38}H_{17}CH{=}PPh_3I$, **7.136**) to afford the *cis*-alkene **7.137** (*Z*: *E* = 6:1), followed by the Boc→Ns conversion process.

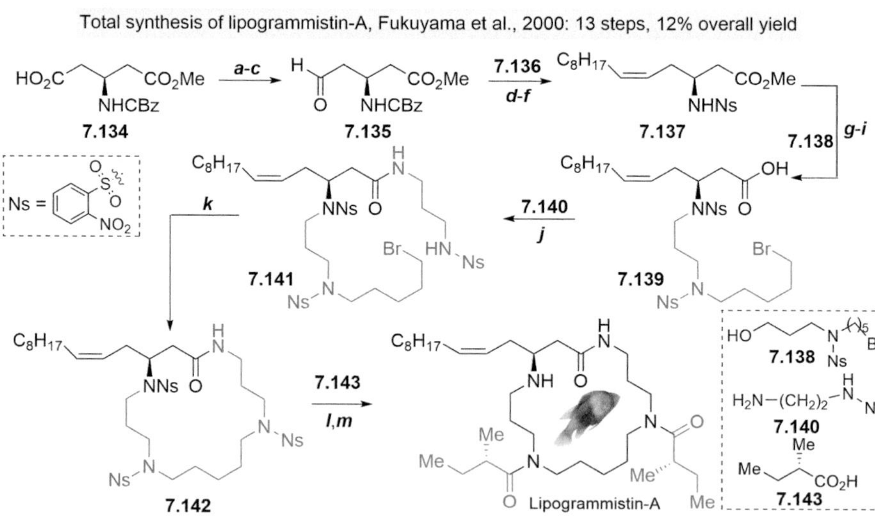

Figure 7.48. Total synthesis of lipogrammistin-A by Fukuyama's group: (a) H_2 (1 atm), Pd/C, Boc_2O, MeOH, rt, 1 hour; (b) DCC, EtSH, DMAP, MeCN, rt, 1 hour; (c) Et_3SiH, Pd/C, acetone, rt, 30 minutes; (d) **7.136**, THF, –78°C to rt, 30 minutes; (e) $SOCl_2$, rt, 30 minutes; (f) Ns-Cl, Et_3N, CH_2Cl_2, 0°C, 10 minutes; (g) **7.138**, DEAD, Et_3N, benzene, rt, 10 minutes; (h) Ti(*i*-PrO)$_4$, allyl alcohol, 95°C, 15 hours; (i) Pd(PPh_3)$_4$, pyrrolidine, CH_2Cl_2, rt, 30 minutes; (j) PivCl, Et_3N, CH_2Cl_2, rt, 30 minutes, **7.140**, Et_3N, rt, 30 minutes; (k) Cs_2CO_3, *n*-BuNI, MeCN, 60°C, 1 hour; (l) HS(CH_2)$_2$OH, DBU, MeCN, rt, 2 hours; (m) **7.143**, BOP-Cl, Et_3N, CH_2Cl_2, rt, 15 hours. DCC = N,N'-dicyclohexylcarbodiimide, DMAP = 4-dimethylaminopyridine, DEAD = diethyl azodicarboxylate, Piv = pivaloyl, DBU = 1,8-diazabicyclo[5.4.0]undec-7-ene, BOP-Cl = bis(2-oxo-3-oxazolidinyl)phosphinic chloride.

Then, a coupling reaction between the derivative **7.137** and N,N-disubstituted 4-nitrobenzenesulfonamide (**7.138**) was achieved under the standard Mitsunobu reaction and some simple transformations on the carboxy function to provide **7.139**. This was converted into a suitable intermediate **7.141** by condensing it with another 4-nitrobenzenesulfonamide derivative **7.140**. Its cyclization, promoted by the presence of Cs_2CO_3 and *n*-BuNI, furnished the desired cyclic product **7.142**. After the removal of the three Ns groups in **7.142** with excess mercaptoethanol and DBU in acetonitrile, the resulting triamine was selectively diacylated with (*S*)-2-methylbutyric acid (**7.143**) and BOP-Cl to give lipogrammistin-A (Figure 7.48). Thus, the authors presented their version of the total synthesis of lipogrammistin-A, initiating with N-Cbz-amino carboxylic diacid **7.134** and achieving in a 13-step sequence in a 12% overall yield.

Numerous species of fishes secrete from their skin a class of toxic metabolites called icthyocrinotoxins, which protect them from being eaten by larger fishes. Pahutoxin and lipogrammistin-A are a small example of these toxins. Other groups are steroid glycosides, including pavoninins and mosesins, which contain a cholestane skeleton with various monosaccharides. The chemistry and biology of these strong antifeedant icthyocrinotoxins isolated from the skin secretion of the peacock sole (*Pardachirus pavoninius*) and the Red Sea Moses sole (*P. rnarmoratus*) could be singly studied. This concludes our story about toxic fish and ocean inhabitants.

7.3 Conclusion

Numerous toxic natural products of marine origin are currently active chemicobiological investigations. The marine environment is home to a greater diversity of animals than the terrestrial environment, and many of these produce interesting organic molecules that are sometimes toxic. Furthermore, the marine toxins that have been studied are among the most powerful naturally occurring poisons, e.g., neurotoxic shellfish poison from dinoflagellate, brevetoxins A and B (LD_{50} = 0.20–0.25 mg/kg), ciguatoxin-1 (LD_{50} = 0.25 μg/kg), and maitotoxin-1 (LD_{50} = 50 ng/kg). They are not only potent neurotoxins but large polycyclic ether-based molecules. Moreover, they can be accumulated via the food chain by large specimens of marine animals, especially fish.

Marine sponges remain one of the most prolific sources of molecules developed to "not be eaten". Unpalatable metabolites from sponges play an anti-predatory role and may be more effective in deterring predation than toxic substances. Among these secondary metabolites, alkaloids are good toxicants acting as allomones, protecting them against predators and fouling organisms. Brominated isoxazoline alkaloids isolated from *Aplysina* sponges function well against predatory fishes, possessing potent antibiotic effects on several dinoflagellate microalgae that cause toxic blooms. Sponge metabolites based on the macro-lactonic system, such as latrunculins A and B, display potent ichthyotoxic activity.

Like brominated isoxazoline alkaloids of *Aplysina* sponges, latrunculins can be sequestered by "sea slugs" (gastropod mollusks) of the genus *Chromodoris*.

Another group of marine sessile invertebrates, corals, provides diverse and strange molecules. The best example is palytoxin, with an LD_{50} of 0.15 μg/kg, one of the most toxic and large non-proteinaceous natural compounds. Compared to it, other metabolites isolated from sponges look like "poor orphans". However, many defensive sponge molecules are attractive from a chemobiological point of view. Sinularones and 14-deoxycrassin, both of terpenic nature, exhibit remarkable antifouling activities. The latter also displays potent antitumor activity. Lemnalemnane A, a sesquiterpenoid, is based on an enjoyable basket-like structure with a spiro-bicyclo[3.3.1]nonane-furan core and shows excellent anti-inflammatory activity.

Tunicates possess good chemical (deterrent metabolites; acidity) and physical (spicules; tunic toughness) defense and deteriorate somewhat little predation by generalist predators, mainly fish. Among deterrent metabolites, pyridoacridine-based alkaloids such as ascididemin, cystodytins, and kuanoniamines stand out as promising lead models for pharmacological research, especially cancer research. A large eudistomins group of ascidian metabolites, based on the simple 9*H*-pyrido[3,4-*b*] indole (β-carboline) skeleton, has also attracted attention as potential compounds for antiviral and antibacterial medicines. However, the best-known toxic metabolites of tunicates are trabectedin (ecteinascidin-743) and plitidepsin (dihydrodidemin B; aplidine). They are structurally different small molecules (polycyclic fused tetrahydroisoquinoline and cyclodepsipeptide systems, respectively) and are examples of the "drugs from the sea" category. Both are marine-derived anticancer drugs with different modes of action. On the other hand, their production depends on total synthesis, which can involve more than 20 steps and requires many protecting groups and expensive reagents. This makes them expensive to use in research and clinical applications.

Mollusk metabolites, especially those of nudibranch gastropods and clams, have incredible structural diversity and display various biological activities. Halichondramides (trisoxazole macrolides), lamellarins (aryl-6*H*-[1] benzopyrano[4′,3′:4,5]pyrrolo[2,1-*a*]isoquinolines), and saxitoxins (guanidinium-containing 3,4-propinoperhydropurines) are significant representents of mollusk metabolites. Halichondramides were discovered as antifungal and toxic agents, while lamellarins, especially lamellarin D, inhibit topoisomerase and perturb mitochondrial metabolism, acting directly on cancer cell mitochondria to induce apoptosis. Saxitoxin and its congers are not proper metabolites of some clams but are sequestered by these bivalve mollusks from the planktonic source. All these are known as paralytic shellfish poisoning toxins that target voltage-gated sodium channels. Among them, saxitoxin and neosaxitoxin are more toxic molecules (LD_{50} = 3–9 μg/kg).

Another potent, selective blocking agent of voltage-gated sodium channels, tetrodotoxin with a guanidinium-based dioxaadamantane skeleton, was first isolated from pufferfish, a marine vertebrate animal. But again, these fish sequester it from bacterial or dinoflagellate species, which puffers consume. It is noteworthy that TTX displays a strong central analgesic effect and is, in fact, in Phase III trials as an agent (Tectin®) against inadequately controlled pain related to cancer. Among other small toxins from poisonous fish, pahutoxin and lipogrammistin-A are distinctive representants of the icthyocrinotoxins group, which are metabolites of the skin secretions of some boxfish and soapfish, and protect them from being eaten by larger fishes. Both are lipophilic toxins despite drastic differences in their molecular architecture. Whereas the first is a choline ester of 3-acetoxypalmitic acid, which acts as a cationic surfactant, the second is an 18-membered polyamine lactam featuring the Z-undec-2-en-1-yl chain, whose biological properties are practically unknown.

Therefore, the marine environment is an exceptional storehouse of novel bioactive natural products, with structural and chemical features generally not found in terrestrial natural products. However, to this day, there are enormous difficulties and expenses involved in collecting and returning large quantities of marine organisms to the laboratory.

Finally, of course, the enormous amount of material accumulated by scientists cannot be fitted into one chapter. We hope the selected material will interest readers and stimulate further research.

7.4 References

[1] Lauritano, C. and Ianora, A. 2020. Chemical defense in marine organisms. Mar. Drugs 18(10): 518.

[2] Molinski, T.F., Dalisay, D.S., Lievens, S.L. and Saludes, J.P. 2009. Drug development from marine natural products. Nature reviews Drug Discovery 8(1): 69–85.

[3] Malve, H. 2016. Exploring the ocean for new drug developments: Marine pharmacology. J. Pharm. Bioallied. Sci. 8(2): 83–91.

[4] Pereira, F. 2019. Have marine natural product drug discovery efforts been productive and how can we improve their efficiency? Expert. Opin. Drug Discov. 14(8): 717–722.

[5] Liu, L., Zheng, Y.Y., Shao, C.L. and Wang, C.Y. 2019. Metabolites from marine invertebrates and their symbiotic microorganisms: Molecular diversity discovery, mining, and application. Mar. Life Sci. Tech. 1: 60–94.

[6] Lin, Y.Y., Risk, M., Ray, S.M., Van Engen, D., Clardy, J., Golik, J. et al. 1981. Isolation and structure of brevetoxin B from the "red tide" dinoflagellate *Ptychodiscus brevis* (*Gymnodinium breve*). J. Am. Chem Soc. 103(22): 6773–6775.

[7] Shimizu, Y., Chou, H.N., Bando, H., Van Duyne, G. and Clardy, J. 1986. Structure of brevetoxin A (GB-1 toxin), the most potent toxin in the Florida red tide organism *Gymnodinium breve* (*Ptychodiscus brevis*). J. Am. Chem. Soc. 108(3): 514–515.

[8] Nicolaou, K.C., Frederick, M.O. and Aversa, R.J. 2008. The continuing saga of the marine polyether biotoxins. Angew. Chem. Int. Ed. 47(38): 7182–7225.

[9] Nicolaou, K.C., Rutjes, F.P.J.T., Theodorakis, E.A., Tiebes, J., Sato, M. and Untersteller, E. 1995. Total synthesis of brevetoxin B. 2. Completion. J. Am. Chem. Soc. 117(3): 1173–1174.

[10] Nicolaou, K.C., Rutjes, F.P.J.T., Theodorakis, E.A., Tiebes, J., Sato, M. and Untersteller, E. 1995. Total synthesis of brevetoxin B. 3. Final strategy and completion. J. Am. Chem. Soc. 117(41): 10252–10263.

[11] Nicolaou, K.C. 1996. The total synthesis of brevetoxin B: a twelve-year odyssey in organic synthesis. Angew. Chem. Int. Ed. Engl. 35(6): 588–607.

[12] Matsuo, G., Kawamura, K., Hori, N., Matsukura, H. and Nakata, T. 2004 Total synthesis of brevetoxin-B. J. Am. Chem. Soc. 126(44): 14374–14376.

[13] Kadota, I., Takamura, H., Nishii, H. and Yamamoto, Y. 2005. Total synthesis of brevetoxin B. J. Am. Chem. Soc. 127(25): 9246–9250.

[14] Nicolaou, K.C., Yang, Z., Shi, G.Q., Gunzner, J.L., Agrios, K.A. and Gärtner, P. 1998. Total synthesis of brevetoxin A. Nature 392(6673): 264–269.

[15] Crimmins, M.T., Zuccarello, J.L., Ellis, J.M., McDougall, P.J., Haile, P.A., Parrish, J.D. et al. 2009. Total synthesis of brevetoxin A. Org. Lett. 11(2): 489–492.

[16] Crimmins, M.T., Zuccarello, J.L., McDougall, P.J. and Ellis, J.M. 2009. Enantioselective total synthesis of brevetoxin A: Convergent coupling strategy and completion. Chem. Eur. J. 15(36): 9235–9244.

[17] Nakata, T. 2005. Total synthesis of marine polycyclic ethers. Chem. Rev. 105(12): 4314–4347.

[18] Nicholson, G.M. and Lewis, R.J. 2006. Ciguatoxins: Cyclic polyether modulators of voltage-gated Iion channel function. Mar Drugs 4(3): 82–118.

[19] Cuypers, E., Abdel-Mottaleb, Y., Kopljar, I., Rainier, J.D., Raes, A.L., Snyders, D.J. et al. 2008. Gambierol, a toxin produced by the dinoflagellate *Gambierdiscus toxicus*, is a potent blocker of voltage-gated potassium channels. Toxicon. 51(6): 974–983.

[20] Nagai, H., Torigoe, K., Satake, M., Murata, M., Yasumoto, T. and Hirota, H. 1992. Gambieric acids: unprecedented potent antifungal substances isolated from cultures of a marine dinoflagellate *Gambierdiscus toxicus*. J. Am. Chem. Soc. 114(3): 1102–1103.

[21] Murata, M., Iwashita, T., Yokoyama, A., Sasaki, M. and Yasumoto, T. 1992. Partial structures of maitotoxin, the most potent marine toxin from the dinoflagellate *Gambierdiscus toxicus*. J. Am. Chem. Soc. 114(16): 6594–6596.

[22] Murata, M., Naoki, H., Iwashita, T., Matsunaga, S., Sasaki, M., Yokoyama, A. and Yasumoto, T. 1993. Structure of maitotoxin. J. Am. Chem. Soc. 115(5): 2060–2062.

[23] Zheng, W., DeMattei, J.A., Wu, J.P., Duan, J.J.W., Cook, L.R., Oinuma, H. et al. 1996. Complete relative stereochemistry of maitotoxin. J. Am. Chem. Soc. 118(34): 7946–7968.

[24] Nicolaou, K.C. and Aversa, R.J. 2011. Maitotoxin: An inspiration for synthesis. Isr. J. Chem. 51(3–4): 359–377.

[25] Nicolaou, K.C., Heretsch, P., Nakamura, T., Rudo, A., Murata, M. and Konoki, K. 2014. Synthesis and biological evaluation of QRSTUVWXYZA′ domains of maitotoxin. J. Am. Chem. Soc. 136(46): 16444–16451.

[26] Puyana, M., Pawlik, J., Blum, J. and Fenical, W. 2015. Metabolite variability in Caribbean sponges of the genus *Aplysina*. Revista Brasileira de Farmacognosia 25: 592–599.

[27] Teeyapant, R., Woerdenbag, H.J., Gross, H.J. and Proksch, P. 1993. Biotransformation of the brominated compounds in the marine sponge *Verongia aerophoba*: evidence for an induced chemical defense? Planta Med. 59(S 1): A641–A642.

[28] Ebel, R., Brenzinger, M., Kunze, A., Gross, H.J. and Proksch, P. 1997. Wound activation of protoxins in marine sponge *Aplysina aerophoba*. J. Chem. Ecol. 23: 1451–1462.

[29] Thoms, C., Wolff, M., Padmakumar, K., Ebel, R. and Proksch, P. 2004. Chemical defense of Mediterranean sponges *Aplysina cavernicola* and *Aplysina aerophoba*. Zeitschrift für Naturforschung C 59(1-2): 113–122.

[30] Thoms, C., Ebel, R. and Proksch, P. 2006. Activated chemical defense in *Aplysina* sponges revisited. J. Chem. Ecol. 32: 97–123.

[31] Lipowicz, B., Hanekop, N., Schmitt, L. and Proksch, P. 2013. An aeroplysinin-1 specific nitrile hydratase isolated from the marine sponge *Aplysina cavernicola*. Mar. Drugs 11(8): 3046–3067.

[32] Drechsel, A., Helm, J., Ehrlich, H., Pantovic, S., Bornstein, S.R. and Bechmann, N. 2020. Anti-tumor activity vs. normal cell toxicity: therapeutic potential of the bromotyrosines aerothionin and homoaerothionin *in vitro*. Mar. Drugs 18(5): 236.

[33] García-Vilas, J.A., Martínez-Poveda, B., Quesada, A.R. and Medina, M.Á. 2015. Aeroplysinin-1, a sponge-derived multi-targeted bioactive marine drug. Mar. Drugs 14(1): 1.

[34] Tymiak, A.A. and Rinehart, L.R., Jr. 1981. Biosynthesis of dibromotyrosine-derived antimicrobial compounds by the marine sponge *Aplysina fistularis* (*Verongia aurea*). J. Am. Chem. Soc. 103(22): 6763–6765.

[35] Kunze, K., Niemann, H., Ueberlein, S., Schulze, R., Ehrlich, H., Brunner, E. et al. 2013. Brominated skeletal components of the marine demosponges, *Aplysina cavernicola* and *Ianthella basta*: analytical and biochemical investigations. Mar. Drugs 11(4): 1271–1287.

[36] Bornancin, L., Bonnard, I., Mills, S.C. and Banaigs, B. 2017. Chemical mediation as a structuring element in marine gastropod predator-prey interactions. Nat. Prod. Rep. 34(6): 644–676.

[37] Habener, L.J., Hooper, J.N. and Carroll, A.R. 2016. Chemical and biological aspects of marine sponges from the family Mycalidae. Planta Medica, 82(09/10): 816–831.

[38] Winder, P.L., Pomponi, S.A. and Wright, A.E. 2011. Natural products from the Lithistida: a review of the literature. Mar. Drugs 9(12): 2643–2682.

[39] Mosey, R.A. and Floreancig, P.E. 2012. Isolation, biological activity, synthesis, and medicinal chemistry of the pederin/mycalamide family of natural products. Nat. Prod. Rep. 29(9): 980–995.

[40] Tammam, M.A. and El-Demerdash, A. 2023. Pederins, mycalamides, onnamides and theopederins: Distinctive polyketide families with intriguing therapeutic potentialities. Curr. Res. Biotechnol. 6: 100145.

[41] Dyshlovoy, S.A., Fedorov, S.N., Kalinovsky, A.I., Shubina, L.K., Bokemeyer, C., Stonik, V.A. et al. 2012. Mycalamide A shows cytotoxic properties and prevents EGF-induced neoplastic transformation through inhibition of nuclear factors. Mar. Drugs 10(6): 1212–1224.

[42] Lee, K.H., Nishimura, S., Matsunaga, S., Fusetani, N., Horinouchi, S. and Yoshida, M. 2005. Inhibition of protein synthesis and activation of stress-activated protein kinases by onnamide A and theopederin B, antitumor marine natural products. Cancer Sci. 96(6): 357–364.

[43] Thoms, C., Schupp, P.J., Custódio, M.R., Lôbo-Hajdu, G., Hajdu, E. and Muricy, G. 2007. Chemical defense strategies in sponges: a review. Porifera research: biodiversity, Innov. Sustain. 28: 627–637.

[44] Rohde, S., Nietzer, S. and Schupp, P.J. 2015. Prevalence and mechanisms of dynamic chemical defenses in tropical sponges. PloS one 10(7): e0132236.

[45] Herb, R., Carroll, A.R., Yoshida, W.Y., Scheuer, P.J. and Paul, V.J. 1990. Polyalkylated cyclopentindoles: cytotoxic fish antifeedants from a sponge, *Axinella* sp. Tetrahedron 46(8): 3089–3092.

[46] Capon, R.J., Macleod, J.K. and Scammells, P.J. 1986. The trikentrins: novel indoles from the sponge *Trikentrion flabelliforme*. Tetrahedron 42(23): 6545–6550.

[47] Silva, Jr. L.F., Craveiro, M.V. and Tebeka, I.R. 2010. Total syntheses of trikentrins and of herbindoles. Tetrahedron 66(22): 3875–3895.

[48] Liu, W., Lim, H.J. and RajanBabu, T.V. 2012. Asymmetric hydrovinylation of vinylindoles. A facile route to cyclopenta[*g*]indole natural products (+)-*cis*-trikentrin A and (+)-*cis*-trikentrin B. J. Am. Chem. Soc. 134(12): 5496–5499.

[49] Kashman, Y., Groweiss, A. and Shmueli, U. 1980. Latrunculin, a new 2-thiazolidinone macrolide from the marine sponge Latrunculia magnifica. Tetrahedron Lett. 21(37): 3629–3632.

[50] Groweiss, A., Shmueli, U. and Kashman, Y. 1983. Marine toxins of *Latrunculia magnifica*. J. Org. Chem. 48(20): 3512–3516.

[51] Khalifa, S., Ahmed, S., Mesbah, M., Youssef, D. and Hamann, M. 2006. Quantitative determination of latrunculins A and B in the Red Sea sponge *Negombata magnifica* by high performance liquid chromatography. J. Chromatogr. B Biomed. Appl. 832(1): 47–51.

[52] Spector, I., Shochet, N.R., Kashman, Y. and Groweiss, A. 1983. Latrunculins: novel marine toxins that disrupt microfilament organization in cultured cells. Science 219(4584): 493–495.

[53] Cheney, K.L., White, A., Mudianta, I.W., Winters, A.E., Quezada, M., Capon, R.J. et al. 2016. Choose your weaponry: selective storage of a single toxic compound, latrunculin A, by closely related nudibranch molluscs. PLoS One 11(1): e0145134.

[54] Smith, III A.B., Noda, I., Remiszewski, S.W., Liverton, N.J. and Zibuck, R. 1990. Total synthesis of (+)-latrunculin A. J. Org. Chem. 55(13): 3977–3979.

[55] Smith, A.B., Leahy, J.W., Noda, I., Remiszewski, S.W., Liverton, N.J. and Zibuck, R. 1992. Total synthesis of the latrunculins. J. Am. Chem. Soc. 114(8): 2995–3007.

[56] White, J.D. and Kawasaki, M. 1990. Total synthesis of (+)-latrunculin A. J. Am. Chem. Soc. 112(12): 4991–4993.

[57] White, J.D. and Kawasaki, M. 1992. Total synthesis of (+)-latrunculin A, an ichthyotoxic metabolite of the sponge *Latrunculia magnifica* and its C-15 epimer. J. Org. Chem. 57(20): 5292–5300.

[58] Fürstner, A., De Souza, D., Parra-Rapado, L. and Jensen, J.T. 2003. Catalysis-based total synthesis of Latrunculin B. Angew. Chem. Int. Ed. Engl. 42(43): 5358–5360.

[59] Fürstner, A. and Turet, L. 2005. Concise and practical synthesis of latrunculin A by ring-closing enyne-yne metathesis. Angew Chem. Int. Ed. Engl. 44(22): 3462–3466.

[60] Sigma-Aldrich. Accessed on 2.05.2024. https://www.sigmaaldrich.com/CO/es/product/mm/428021 and https://www.sigmaaldrich.com/CO/es/product/mm/428020.

[61] Fürstner, A., Kirk, D., Fenster, M.D., Aïssa, C., De Souza, D., Nevado, C. et al. 2007. Latrunculin analogues with improved biological profiles by "diverted total synthesis": preparation, evaluation, and computational analysis. Chem. Eur. J. 13(1): 135–149.

[62] Alarif, W.M., Abdel-Lateff, A., Alorfi, H.S. and Alburae, N.A. 2019. Alcyonacea: a potential source for production of nitrogen-containing metabolites. Molecules 24(2): 286.

[63] Moore, R.E. and Scheuer, P.J. 1971. Palytoxin—new marine toxin from a coelenterate. Science 172(3982): 495–498.

[64] Moore, R.E. and Bartolini, G. 1981. Structure of palytoxin. J. Am. Chem. Soc. 103(9): 2491–2494.

[65] Uemura, D., Ueda, K., Hirata, Y., Naoki, H. and Iwashita, T. 1981. Further-studies on palytoxin. II. Structure of palytoxin. Tetrahedron Lett. 22 (29): 2781–2784.

[66] Ramos, V. and Vasconcelos, V. 2010. Palytoxin and analogs: biological and ecological effects. Mar. Drugs 8(7): 2021–2037.

[67] Pelin, M., Brovedani, V., Sosa, S. and Tubaro, A. 2016. Palytoxin-containing aquarium soft corals as an emerging sanitary problem. Mar. Drugs 14(2): 33.

[68] Armstrong, R.W., Beau, J.M., Cheon, S.H., Christ, W.J., Fujioka, H., Ham, W.H. et al. 1989. Total synthesis of a fully protected palytoxin carboxylic acid. J. Am. Chem. Soc. 111(19): 7525–7530.

[69] Armstrong, R.W., Beau, J.M., Cheon, S.H., Christ, W.J., Fujioka, H., Ham, W.H. et al. 1989. Total synthesis of palytoxin carboxylic acid and palytoxin amide. J. Am. Chem. Soc. 111(19): 7530–7533.

[70] Suh, E.M. and Kishi, Y. 1994. Synthesis of palytoxin from palytoxin carboxylic acid. J. Am. Chem. Soc. 116(24): 11205–11206.

[71] Shi, H., Yu, S., Liu, D., Van Ofwegen, L., Proksch, P. and Lin, W. 2012. Sinularones A–I, new cyclopentenone and butenolide derivatives from a marine soft coral *Sinularia* sp. and their antifouling activity. Mar. Drugs 10(6): 1331–1344.

[72] Radwan, M.M., Manly, S.P., El Sayed, K.A., Wali, V.B., Sylvester, P.W., Awate, B. et al. 2008. Sinulodurins A and B, antiproliferative and anti-invasive diterpenes from the soft coral *Sinularia dura*. J. Nat. Prod. 71(8): 1468–1471.

[73] Wang, J., Su, P., Gu, Q., Li, W.D., Guo, J.L., Qiao, W. et al. 2017 Antifouling activity against bryozoan and barnacle by cembrane diterpenes from the soft coral *Sinularia flexibilis*. Int. Biodeter Biodegr. 120: 97–103.

[74] Rodríguez, A.D. and Martinez, N. 1993. Marine antitumor agents: 14-deoxycrassin and pseudoplexaurol, new cembranoid diterpenes from the Caribbean gorgonian *Pseudoplexaura porosa*. Experientia 49(2): 179–181.

[75] Liu, Z., Peng, L., Li, W.Z. and Li, Y. 2003. Asymmetric total synthesis of pseudoplexaurol and 14-deoxycrassin, two antitumor marine cembrane diterpenoids. Synlett. 2003(13): 1977–1980.

[76] Elshamy, A.I., Nassar, M.I., Mohamed, T.A. and Hegazy, M.E.F. 2016. Chemical and biological profile of *Cespitularia* species: a mini review. J. Adv. Res. 7(2): 209–224.

[77] Shen, Y.C., Lin, Y.S., Kuo, Y.H. and Cheng, Y.B. 2005. Cespitulactams A, B, and C, three new nitrogen-containing diterpenes from *Cespitularia taeniata* May. Tetrahedron. Lett. 46(46): 7893–7897.

[78] Cheng, Y.B., Chen, C.Y., Kuo, Y.H. and Shen, Y.C. 2009. New Nitrogen-Containing Sesquiterpenoids from the Taiwanese Soft Coral *Cespitularia taeniata* May. Chem. Biodiversity 6(8): 1266–1272.

[79] Wu, Q., Sun, J., Chen, J., Zhang, H., Guo, Y.W. and Wang, H. 2018. Terpenoids from marine soft coral of the genus *Lemnalia*: Chemistry and biological activities. Mar. Drugs 16(9): 320.

[80] Tursch, B., Braekman, J.C., Daloze, D., Fritz, P., Kelecom, A., Karlsson, R. et al. 1974. Chemical studies of marine invertebrates. VIII. Africanol, an unusual sesquiterpene from *Lemnalia africana* (Coelenterata, Octocorallia, Alcyonacea). Tetrahedron Lett. 15(9), 747–750.

[81] Karlsson, R. 1976. Measurements of bijvoet differences. Structure and absolute configuration of africanol, a sesquiterpene. Acta Cryst. 32(9): 2609–2614.

[82] Braekman, J.C., Daloze, D., Tursch, B., Hull, S.E., Declercq, J.P., Germain, G. et al. 1980. Chemical studies of marine invertebrates. XXXVIII. $\Delta^{9(15)}$-africanene, a new sesquiterpene hydrocarbon from *Sinularia polydactyla* (Coelenterata, Octocorallia, Alcyonaceae). Experientia; 36: 893.

[83] Kawano, S., Matagawa, T., Matsuda, Y., Koyama, T., Miura, K., Nakata, M. et al. 2022. Biological evaluation for anti-inflammatory effect of africane-type sesquiterpenoids. Bioorg. Med. Chem. 68: 116857.

[84] Tursch, B., Colin, M., Daloze, D., Losman, D. and Karlsson, R. 1975. Chemical studies of marine invertebrates. XII. Lemnacarnol, a novel sesquiterpene from the soft coral *Lemnalia carnosa* (Octocorallia, Alcyonalea). Bull. Soc. Chim. Belg. 84(1-2): 81–82.

[85] Bowden, B.F., Coll, J.C., Mitchell, S.J., Skelton, B.W. and White, A.H. 1980. Studies of Australian soft corals. XXII.* The structures of two novel sesquiterpenes and a nor sesquiterpene from *Lemnalia africana*, confirmed by a single-crystal X-Ray study. Aust. J. Chem. 33: 2737–2747.

[86] Bowden, B.F., Coll, J.C., Mitchell, S.J., Nemorin, J.L.E. and Sternhell, S. 1980. Studies of Australian soft corals. XXIII the co-occurrence of bicyclogermacrene and lemnacarnol derivatives in Parerythropodium fulvum. Tetrahedron Lett. 21(32): 3105–3108.
[87] Wang, Q., Hu, Z., Luo, X., Liu, J., Li, G., Cao, S. and Liu, Q. 2019. Clavukoellians A–F, highly rearranged nardosinane sesquiterpenoids with antiangiogenic activity from *Clavularia koellikeri*. J. Nat. Prod. 82(5): 1331–1337.
[88] El-Gamal, A.A.H., Chiu, E.P., Li, C.H., Cheng, S.Y., Dai, C.F. and Duh, C.Y. 2005. Sesquiterpenoids and norsesquiterpenoids from the Formosan soft coral *Lemnalia laevis*. J. Nat. Prod. 68(12): 1749–1753.
[89] Zong, Y., Jin, T.Y., Yang, J.J., Wang, K.Y., Shi, X., Zhang, Y. et al. 2024. Lemneolemnanes A–D, four uncommon sesquiterpenoids from the soft coral *Lemnalia* sp. Mar. Drugs 22(4): 145.
[90] Liu, M., Li, P., Luo, X., van Ofwegen, L., Tang, X. and Li, G. 2021. Sesquiterpenoids from the soft coral *Lemnalia* sp. Nat. Prod. Res. 35(21): 3752–3756.
[91] Han, X., Wang, Q., Luo, X., Tang, X., Wang, Z., Zhang, D. et al. 2021. Lemnalemnanes A–C, Three Rare Rearranged Sesquiterpenoids from the Soft Corals *Paralemnalia thyrsoides* and *Lemnalia* sp. Org. Lett. 24(1): 11–15.
[92] Kobayashi, T., Sugitate, R., Uchida, K., Kawamoto, Y. and Ito, H. 2024. Asymmetric total synthesis of (−)-Lemnalemnane A. Org. Lett. 26(9): 1803–1806.
[93] Holland, L.Z. 2016. Tunicates. Curr. Biol. 26(4): R146–R152.
[94] Shenkar, N. and Swalla, B.J. 2011. Global diversity of Ascidiacea. Plos One 6(6): e20657.
[95] Plodek, A. and Bracher, F. 2016. New perspectives in the chemistry of marine pyridoacridine alkaloids. Mar. Drugs 14(2): 26.
[96] Ibrahim, S.R. and Mohamed, G.A. 2016. Pyridoacridine alkaloids from deep-water marine organisms: Structural elucidation. Bull Fac. Pharm. Cairo. Univ. 54(2): 107–135.
[97] Kobayashi, J., Cheng, J., Nakamura, H., Ohizumi, Y., Hirata, Y., Sasaki, T. et al. 1988. Ascididemin, a novel pentacyclic aromatic alkaloid with potent antileukemic activity from the Okinawan tunicate *Didemnun* sp. Tetrahedron Lett. 29(10): 1177–1180.
[98] Kobayashi, J., Cheng, J.F., Nakamura, H., Ohizumi, Y., Walchli, M.R., Hirata, Y. and Sasaki, T. 1988. Cystodytins A, B, and C, novel tetracyclic aromatic alkaloids with potent antineoplastic activity from the Okinawan tunicate *Cystodytes dellechiajei*. J. Org. Chem. 53(8): 1800–1804.
[99] Kobayashi, J., Tsuda, M., Tanabe, A., Ishibashi, M., Cheng, J.F., Yamamura, S. and Sasaki, T. 1991. Cystodytins D-I, new cytotoxic tetracyclic aromatic alkaloids from the okinawan marine tunicate *Cystodytes dellechiajei*. J. Nat. Prod. 54(6): 1634–1638.
[100] Carroll, A.R., Scheuer, P.J. 1990. Kuanoniamines A, B, C, and D: Pentacyclic alkaloids from a tunicate and its prosobranch mollusk predator *Chelynotus semperi*. J. Org. Chem. 55(14): 4426–4431.
[101] López-Legentil, S., Turon, X. and Schupp, P. 2006. Chemical and physical defenses against predators in *Cystodytes* (Ascidiacea). J. Exp. Mar. Bio. Ecol. 332(1): 27–36.
[102] Plodek, A. and Bracher, F. 2016. New perspectives in the chemistry of marine pyridoacridine alkaloids. Mar. Drugs 14(2): 26.
[103] Kijjoa, A., Wattanadilok, R., Campos, N., Nascimento, M.S.J., Pinto, M. and Herz, W. 2007. Anticancer activity evaluation of kuanoniamines A and C isolated from the marine sponge *Oceanapia sagittaria*, collected from the Gulf of Thailand. Mar. Drugs 5(2): 6–22.
[104] Fong, H.K. and Copp, B.R. 2013. Synthesis, DNA binding and antitumor evaluation of styelsamine and cystodytin analogues. Mar. Drugs 11(2): 274–299.
[105] Brascher, F. 1989. Total synthesis of the pentacyclic alkaloid ascididemin. Heterocycles 29(11): 2093–2095.
[106] Cuerva, J.M., Cárdenas, D.J. and Echavarren, A.M. 2002. Intramolecular Michael-type addition of azadienes to 1,4-naphthoquinones instead of Aza-Diels–Alder cycloaddition: A synthesis of ascididemin. J. Chem. Soc. Perkin Trans. 1(11): 1360–1365.
[107] Jiang, D., Chen, Y. and Wang, S. 2022. Total Synthesis of Marine Alkaloids Cystodytins A–K. J. Org. Chem. 87(16): 11063–11072.
[108] Xiao, L. 2022. A review: meridianins and meridianins derivatives. Molecules 27(24): 8714.
[109] Franco, L.H., Joffé, E.B.D.K., Puricelli, L., Tatian, M., Seldes, A.M. and Palermo, J.A. 1998. Indole alkaloids from the tunicate aplidium m eridianum. J. Nat. Prod. 61(9): 1130–1132.

[110] Seldes, A.M., Rodriguez Brasco, M.F., Hernandez Franco, L. and Palermo, J.A. 2007. Identification of two meridianins from the crude extract of the tunicate *Aplidium meridianum* by tandem mass spectrometry. Nat. Prod. Res. 21(6): 555–563.

[111] Núñez-Pons, L., Forestieri, R., Nieto, R.M., Varela, M., Nappo, M., Rodriguez, J. et al. 2010. Chemical defenses of tunicates of the genus *Aplidium* from the Weddell Sea (Antarctica). Polar Biol. 33: 1319–1329.

[112] Yadav, R.R., Khan, S.I., Singh, S., Khan, I.A., Vishwakarma, R.A. and Bharate, S.B. 2015. Synthesis, antimalarial and antitubercular activities of meridianin derivatives. Eur. J. Med. Chem. 98: 160–169.

[113] Merkul, E., Schäfer, E. and Müller, T.J. 2011. Rapid synthesis of bis (hetero) aryls by one-pot Masuda borylation–Suzuki coupling sequence and its application to concise total syntheses of meridianins A and G. Org. Biomol. Chem. 9(9): 3139–3141.

[114] Kruppa, M., Sommer, G.A. and Müller, T.J. 2022. Concise syntheses of marine (bis) indole alkaloids meridianin C, D, F, and G and scalaridine A via one-pot masuda borylation-Suzuki coupling sequence. Molecules 27(7): 2233.

[115] Rinehart, K.L., Kobayashi, J., Harbour, G.C., Hughes, Jr. R.G., Mizsak, S.A. and Scahill, T.A. 1984. Eudistomins C, E, K, and L, potent antiviral compounds containing a novel oxathiazepine ring from the Caribbean tunicate *Eudistoma olivaceum*. J. Am. Chem. Soc. 106(5): 1524–1526.

[116] Kobayashi, J., Harbour, G.C., Gilmore, J. and Rinehart, Jr. K.L. 1984. Eudistomins A, D, G, H, I, J, M, N, O, P, and Q, bromo, hydroxy, pyrrolyl and iminoazepino.beta.-carbolines from the antiviral Caribbean tunicate *Eudistoma olivaceum*. J. Am. Chem. Soc. 106(5): 1526–1528.

[117] Shen, G.Q. and Baker, B.J. 1994. Biosynthetic studies of eudistomin H in the tunicate *Eudistoma olivaceum*. Tetrahedron Lett. 35(28): 4923–4926.

[118] Davis, A.R. 1991. Alkaloids and ascidian chemical defense: evidence for the ecological role of natural products from *Eudistoma olivaceum*. Mar. Biol. 111(3): 375–379.

[119] Rinehart, Jr. K.L., Kobayashi, J., Harbour, G.C., Gilmore, J., Mascal, M., Holt, T.G. et al. 1987. Eudistomins AQ, beta.-carbolines from the antiviral Caribbean tunicate *Eudistoma olivaceum*. J. Am. Chem. Soc. 109(11): 3378–3387.

[120] Ota, Y., Chinen, T., Yoshida, K., Kudo, S., Nagumo, Y., Shiwa, Y. et al. 2016. Eudistomin C, an antitumor and antiviral natural product, targets 40S ribosome and inhibits protein translation. Chem. Bio. Chem. 17(17): 1616–1620.

[121] Roggero, C.M., Giulietti, J.M. and Mulcahy, S.P. 2014. Efficient synthesis of eudistomin U and evaluation of its cytotoxicity. Bioorg. Med. Chem. Lett. 24(15): 3549–3551.

[122] Ohshita, K., Ishiyama, H., Oyanagi, K., Nakata, H. and Kobayashi, J. 2007. Synthesis of hybrid molecules of caffeine and eudistomin D and its effects on adenosine receptors. Bioorg. Med. Chem. 15(9): 3235–3240.

[123] Rajesh, R.P. and Murugan, A. 2019. Spectroscopic identification of brominated, non-brominated alkaloids and evaluation of antimicrobial activity of Eudistomin-I, Eudistomin H, from green ascidian *Eudistoma viride*. J. Appl. Pharm. Sc. 9(2): 116–123.

[124] Kamal, A., Tangella, Y., Manasa, K.L., Sathish, M., Srinivasulu, V., Chetna, J. et al. 2015. $PhI(OAc)_2$-mediated one-pot oxidative decarboxylation and aromatization of tetrahydro-β-carbolines: synthesis of norharmane, harmane, eudistomin U and eudistomin I. Org. Biomol. Chem. 13(32): 8652–8662.

[125] Yamashita, T., Kawai, N., Tokuyama, H. and Fukuyama, T. 2005. Stereocontrolled total synthesis of (−)-Eudistomin C. J. Am. Chem. Soc. 127(43): 15038–15039.

[126] Molinski, T.F., Dalisay, D.S., Lievens, S.L. and Saludes, J.P. 2009. Drug development from marine natural products. Nat. Rev. Drug Discov. 8(1): 69–85.

[127] Jimenez, P.C., Wilke, D.V., Branco, P.C., Bauermeister, A., Rezende-Teixeira, P., Gaudêncio, S.P. et al. 2020. Enriching cancer pharmacology with drugs of marine origin. Br. J. Pharmacol. 177(1): 3–27.

[128] Takebayashi, Y., Pourquier, P., Zimonjic, D.B., Nakayama, K., Emmert, S., Ueda, T. et al. 2001. Antiproliferative activity of ecteinascidin 743 is dependent upon transcription-coupled nucleotide-excision repair. Nat. Med. 7(8): 961–966.

[129] Alonso-Álvarez, S., Pardal, E., Sánchez-Nieto, D., Navarro, M., Caballero, M.D., Mateos, M.V. et al. 2017. Plitidepsin: design, development, and potential place in therapy. Drug Des. Devel. Ther. 11: 253–264.

[130] Cuevas, C. and Francesch, A. 2009. Development of Yondelis®(trabectedin, ET-743). A semisynthetic process solves the supply problem. Nat. Prod. Rep. 26(3): 322–337.

[131] Li, D., Meng, J. and Jiang, X. 2023. Total synthesis of trabectedin, lurbinectedin, and renieramycin T. CCS Chem. 5(9): 2152–2158.

[132] Corey, E.J., Gin, D.Y. and Kania, R.S. 1996. Enantioselective total synthesis of ecteinascidin 743. J. Am. Chem. Soc. 118(38): 9202–9203.

[133] Zhang, H., Hui, Z., Cai, M., Huang, S., Shi, W., Liang, M. et al. 2023. Integration of microbial and chemical synthesis for the efficient production of plitidepsin, a promising anticancer and antiviral agent. bioRxiv; 2023-04.

[134] Benkendorff, K. 2010. Molluscan biological and chemical diversity: Secondary metabolites and medicinal resources produced by marine molluscs. Biol. Rev. 85(4): 757–775.

[135] Avila, C. and Angulo-Preckler, C. 2020. Bioactive compounds from marine heterobranchs. Mar. Drugs 18(12): 657.

[136] Chen, Z.H., Guo, Y.W. and Li, X.W. 2023. Recent advances on marine mollusk-derived natural products: chemistry, chemical ecology and therapeutical potential. Nat. Prod. Rep. 40(3): 509–556.

[137] Ngandjui, Y.A.T., Kereeditse, T.T., Kamika, I., Madikizela, L.M. and Msagati, T.A.M. 2024. Nutraceutical and medicinal importance of marine molluscs. Mar. Drugs 22(5): 201.

[138] Bornancin, L., Bonnard, I., Mills, S.C. and Banaigs, B. 2017. Chemical mediation as a structuring element in marine gastropod predator-prey interactions. Nat. Prod. Rep. 34(6): 644–676.

[139] Pawlik, J.R., Kernan, M.R., Molinski, T.F., Harper, M.K. and Faulkner, D.J. 1988. Defensive chemicals of the Spanish dancer nudibranch Hexabranchus sanguineus and its egg ribbons: macrolides derived from a sponge diet. J. Exper. Mar. Biol. Ecol. 119(2): 99–109.

[140] Bae, S.Y., Kim, G.D, Jeon, J.E., Shin, J. and Lee, S.K. 2013. Anti-proliferative effect of (19Z)-halichondramide, a novel marine macrolide isolated from the sponge *Chondrosia corticata*, is associated with G2/M cell cycle arrest and suppression of mTOR signaling in human lung cancer cells. *In vitro* Toxicol. 27(2): 694–699.

[141] Chung, S.C., Lee, S.H., Jang, K.H., Park, W., Jeon, J.E., Oh, H. et al. 2011. Actin depolymerizing effect of trisoxazole-containing macrolides. Bioorg. Med. Chem. Lett. 21(11): 3198–3201.

[142] Andersen, R.J., Faulkner, D.J., He, C.H., Van Duyne, G.D. and Clardy, J. 1985. Metabolites of the marine prosobranch mollusk *Lamellaria* sp. J. Am. Chem. Soc. 107(19): 5492–5495.

[143] Maza, L.J.G., Salgado, A.M., Kouznetsov, V.V. and Meléndez, C.M. 2024. Pyrrolo [2,1-*a*] isoquinoline scaffolds for developing anti-cancer agents. RSC Adv. 14(3): 1710–1728.

[144] Bailly, C. 2004. Lamellarins, from A to Z: a family of anticancer marine pyrrole alkaloids. Curr. Med. Chem. Anticancer. Agents 4(4): 363–378.

[145] Fukuda, T., Ishibashi, F. and Iwao, M. 2020. Lamellarin alkaloids: Isolation, synthesis, and biological activity. The Alkaloids: Chemistry and Biology 83: 1–112.

[146] Ishibashi, F., Miyazaki, Y. and Iwao, M. 1997. Total syntheses of lamellarin D and H. The first synthesis of lamellarin-class marine alkaloids. Tetrahedron 53(17): 5951–5962.

[147] Pla, D., Marchal, A., Olsen, C.A., Albericio, F. and Álvarez, M. 2005. Modular total synthesis of lamellarin D. J. Org. Chem. 70(20): 8231–8234.

[148] Ohta, T., Fukuda, T., Ishibashi, F. and Iwao, M. 2009. Design and synthesis of lamellarin D analogues targeting topoisomerase I. J. Org. Chem. 74(21): 8143–8153.

[149] Lade, D.M., Pawar, A.B., Mainkar, P.S. and Chandrasekhar, S. 2017. Total synthesis of lamellarin D trimethyl ether, lamellarin D, and lamellarin H. J. Org. Chem. 82(9): 4998–5004.

[150] Shirley, H.J., Koyioni, M., Muncan, F. and Donohoe, T.J. 2019. Synthesis of lamellarin alkaloids using orthoester-masked α-keto acids. Chem. Sci. 10(15): 4334–4338.

[151] Wiese, M., D'Agostino, P.M., Mihali, T.K., Moffitt, M.C. and Neilan, B.A. 2010. Neurotoxic alkaloids: saxitoxin and its analogs. Mar. Drugs 8(7): 2185–2211.

[152] Thottumkara, A.P., Parsons, W.H. and Du Bois, J. 2014. Saxitoxin. Angew. Chem. Int. Ed. 53(23): 5760–5784.

[153] Tsuchiya, S., Cho, Y., Konoki, K., Nagasawa, K., Oshima, Y. and Yotsu-Yamashita, M. 2016. Biosynthetic route towards saxitoxin and shunt pathway. Sci. Rep. 6(1): 20340.

[154] Tanino, H., Nakata, T., Kaneko, T. and Kishi, Y. 1977. A stereospecific total synthesis of dl-saxitoxin. J. Am. Chem. Soc. 99(8): 2818–2819.
[155] Jacobi, P.A., Martinelli, M.J. and Polanc, S. 1984. Total synthesis of (.+-.)-saxitoxin. J. Am. Chem. Soc. 106(19): 5594–5598.
[156] Ziegman, R. and Alewood, P. 2015. Bioactive components in fish venoms. Toxins 7(5): 1497–1531.
[157] Cipolari, O.C., de Oliveira Neto, X.A. and Conceição, K. 2020. Fish bioactive peptides: A systematic review focused on sting and skin. Aquaculture 515: 734598.
[158] Harris, R.J. 2024. The piscine arsenal: An updated review of venomous fishes. Rev. Fish Biol. Fish. 34(2): 539–574.
[159] Brighton Ndandala, C., Mustapha, U.F., Wang, Y., Assan, D., Zhao, G., Huang, C. et al. 2023. The perspective of fish venom: An overview of the physiology, evolution, molecular and genetics. Front Mar. Sci. 10: 1085669.
[160] Lago, J., Rodríguez, L.P., Blanco, L., Vieites, J.M. and Cabado, A.G. 2015. Tetrodotoxin, an extremely potent marine neurotoxin: Distribution, toxicity, origin and therapeutical uses. Mar. Drugs 13(10): 6384–6406.
[161] Melnikova, D.I. and Magarlamov, T.Y. 2022. An overview of the anatomical distribution of tetrodotoxin in animals. Toxins 14(8): 576.
[162] Noguchi, T., Arakawa, O. and Takatani, T. 2006. Toxicity of pufferfish Takifugu rubripes cultured in netcages at sea or aquaria on land. Comp. Biochem. Physiol. Part D Genomics Proteomics 1(1): 153–157.
[163] Katikou, P., Gokbulut, C., Kosker, A.R., Campàs, M. and Ozogul, F. 2022. An updated review of tetrodotoxin and its peculiarities. Mar. Drugs 20(1): 47.
[164] Moczydlowski, E.G. 2013. The molecular mystique of tetrodotoxin. Toxicon. 63: 165–183.
[165] Li, Y. and Ou, X. 2023. The Relative Mechanism of Tetrodotoxin: From Poison to Drug. Highl. Sci. Eng Technol. 74: 1723–1729.
[166] Berde, C.B., Athiraman, U., Yahalom, B., Zurakowski, D., Corfas, G. and Bognet, C. 2011. Tetrodotoxin-bupivacaine-epinephrine combinations for prolonged local anesthesia. Mar. Drugs 9(12): 2717–2728.
[167] Bucciarelli, G.M., Lechner, M., Fontes, A., Kats, L.B., Eisthen, H.L. and Shaffer, H.B. 2021. From poison to promise: The evolution of tetrodotoxin and its potential as a therapeutic. Toxins 13(8): 517.
[168] Chau, J. and Ciufolini, M.A. 2011. The chemical synthesis of tetrodoxin: An ongoing quest. Mar. Drugs 9(10): 2046–2074.
[169] Makarova, M., Rycek, L., Hajicek, J., Baidilov, D. and Hudlicky, T. 2019. Tetrodotoxin: history, biology, and synthesis. Angew Chem. Int. Ed. 58(51): 18338–18387.
[170] Murakami, K., Toma, T., Fukuyama, T. and Yokoshima, S. 2020. Total synthesis of Tetrodotoxin. Angew. Chem. Int. Ed. 132(15): 6312–6316.
[171] Konrad, D.B., Rühmann, K.P., Ando, H., Hetzler, B.E., Strassner, N., Houk, K.N. et al. 2022. A concise synthesis of tetrodotoxin. Science 377(6604): 411–415.
[172] Chen, P., Wang, J., Zhang, S., Wang, Y., Sun, Y., Bai, S. et al. 2024. Total syntheses of Tetrodotoxin and 9-epi Tetrodotoxin. Nat. Comm. 15(1): 679.
[173] Maehara, T., Motoyama, K., Toma, T., Yokoshima, S. and Fukuyama, T. 2017. Total synthesis of (−)-Tetrodotoxin and 11-norTTX-6 (R)-ol. Angew. Chem. Int. Ed. 129(6): 1571–1574.
[174] Thomson, D.A. 1964. Ostracitoxin: an ichthyotoxic stress secretion of the boxfish, *Ostracion lentiginosus*. Science 146(3641): 244–245.
[175] Scheuer, P.J. 1964. The chemistry of toxins isolated from some marine organisms. Fortschr. Chem. Org. Naturst. Progress Chem. Org. Nat. Prod. 22: 265–278.
[176] Boylan, D.B. and Scheuer, P.J. 1967. Pahutoxin: A Fish Poison: Structure, activity, and synthesis of a metabolite isolated from the Hawaiian boxfish is described. Science 155(3758): 52–56.
[177] Yoshikawa, M., Sugimura, T. and Tai, A. 1989. Pahutoxin: synthesis and determination of its absolute configuration. Agric. Biol. Chem. 53(1): 37–40.
[178] Fusetani, N. and Hashimoto, K. 1987. Occurrence of pahutoxin and homopahutoxin in the mucus secretion of the Japanese boxfish. Toxicon. 25(4): 459–461.

[179] Kalmanzon, E., Rahamim, Y., Barenholz, Y., Carmeli, S. and Zlotkin, E. 2003. Receptor-mediated toxicity of pahutoxin, a marine trunkfish surfactant. Toxicon. 42(1): 63–71.

[180] Onuki, H., Tachibana, K. and Fusetani, N. 1993. Structure of lipogrammistin-A, a lipophilic ichthyotoxin secreted by the soapfish *Diploprion bifasciatum*. Tetrahedron. Lett. 34(35): 5609–5612.

[181] Onuki, H., Ito, K., Kobayashi, Y., Matsumori, N., Tachibana, K. and Fusetani, N. 1998. Absolute structure and total synthesis of lipogrammistin-A, a lipophilic ichthyotoxin of the soapfish. J. Org. Chem. 63(12): 3925–3932.

[182] Fujiwara, A., Kan, T. and Fukuyama, T. 2000. Total synthesis of lipogrammistin-A: efficient macrocyclization with 2-nitrobenzenesulfonamide. Synlett. 2000(11): 1667–1669.

[183] Kan, T., Fujiwara, A., Kobayashi, H. and Fukuyama, T. 2002. Efficient macrocyclization by means of 2-nitrobenzenesulfonamide and total synthesis of lipogrammistin-A. Tetrahedron 58(32): 6267–6276.

Index

M

N

O

P

Q

R

S

T

For Product Safety Concerns and Information please contact our EU representative GPSR@taylorandfrancis.com Taylor & Francis Verlag GmbH, Kaufingerstraße 24, 80331 München, Germany